Bird Families of the World

A series of authoritative, illustrated handbooks, of which this is the 4th volume to be published

Series editors

C. M. PERRINS Chief editor
W. J. BOCK
J. KIKKAWA

THE AUTHORS Anthony J. Gaston is a research scientist with the Canadian Wildlife Service. He studied co-operative breeding in birds for his D.Phil. at Oxford University. Subsequently, he has studied auks for the Canadian Wildlife Service since 1975, when he began to study Thick-billed Murres in the Canadian Arctic. From 1984, he has also worked on Pacific auks, especially the Ancient Murrelet, on the coast of British Columbia. In addition to studying seabirds, he has conducted research on birds of deserts, jungles, and mountains from North Africa to South Asia and has been involved in wildlife and protected area surveys and environmental impact assessments in India over three decades. Ian L. Jones began his studies of auks with Tony Gaston's Thick-billed Murre project in the Canadian Arctic. After completing his B.Sc. at Carleton University he collaborated with Tony again on his M.Sc. project based at the University of Toronto, in which they studied parent–offspring vocal communication in Ancient Murrelets of the Queen Charlotte Islands. He went on to study the behavioural ecology of Least Auklets on the Pribilof Islands, Alaska for his doctorate at Queen's University, Kingston, followed by postdoctoral research on the evolution of ornaments and social behaviour of *Aethia* auklets based at the University of Cambridge. Recently, he founded a seabird research station at Triangle Island, British Columbia. Ian is presently the Atlantic Co-operative Wildlife Ecology Research Network Chair in seabird ecology at the Memorial University of Newfoundland.

THE ARTIST Ian Lewington has been addicted to birdwatching and drawing birds since the age of five. In 1985 he won British Birds Magazine Bird Illustrator of the Year after which he became a freelance bird illustrator specializing in identification. He illustrated the popular *Rare Birds of Britain and Europe* and the forthcoming *Raptors of Europe*. He has also produced plates for *Birds of the Western Palearctic* (OUP), *Birds of the World* (LYNX), forthcoming field guides covering India, S.E. Asia, Malaysia, Armenia, and Madagascar, and several identification articles in the journals *Birding World*, *Dutch Birding*, *Limicola* and *Var Fagelvarld*. He has travelled widely in search of birds and also has a keen interest in local ornithology having become County Bird Recorder for the Oxford Ornithological Society in 1994.

Bird Families of the World

Bird Families of the World

The Auks
Alcidae

ANTHONY J. GASTON

and

IAN L. JONES

Colour plates by
IAN LEWINGTON

Line drawings by
IAN LEWINGTON and IAN JONES

Oxford   New York   Tokyo
OXFORD UNIVERSITY PRESS
1998

Oxford University Press, Great Clarendon Street, Oxford OX2 6DP

Oxford   New York
Athens   Auckland   Bangkok   Bogota   Bombay
Buenos Aires   Calcutta   Cape Town   Dar es Salaam
Delhi   Florence   Hong Kong   Istanbul   Karachi
Kuala Lumpur   Madras   Madrid   Melbourne
Mexico City   Nairobi   Paris   Singapore
Taipei   Tokyo   Toronto   Warsaw
and associated companies in
Berlin Ibadan

Oxford is a trade mark of Oxford University Press

Published in the United States
by Oxford University Press Inc., New York

© Text: Anthony J. Gaston and Ian L. Jones, 1998;
drawings and colour plates: Oxford University Press

A catalogue record for this book is available from the British Library

Library of Congress Cataloging in Publication Data

Gaston, A. J. (Anthony J.), 1946–
The auks : Alcidae / Anthony J. Gaston and Ian L. Jones ; colour
plates by Ian Lewington ; line drawings by Ian Lewington and Ian Jones.
(Bird families of the world ; 4)
Includes bibliographical references and index.
1. Auks.   I. Jones, Ian L.   II. Title.   III. Series.
QL696.C42G363   1997
598.3'3—dc21   97-22350   CIP

ISBN 0 19 854032 9

Typeset by EXPO Holdings, Malaysia

Printed in Hong Kong, China

Acknowledgements

We are grateful to all of our colleagues who shared our camps and assisted in our studies over the years. Tony Gaston would especially like to acknowledge Thomas Alogut, David Cairns, Don Croll, Garry Donaldson, Leah de Forest, Colin French, John Geale, Grant Gilchrist, Mark Hipfner, Andrea Lawrence, Kara Lefevre, Adami Mangiuk, Josiah Nakoolak, David Noble, David Powell, Stephen Smith, and members of the Laskeek Bay Conservation Society for their help and companionship in the field. Ian Jones would like to thank Christine Adkins, Vernon Byrd, Simon Gawn, Gail Fraser, Anne Harfenist, Mark Hipfner, Fiona Hunter, Gary Kaiser, Hugh Knechtel, Art Sowls, Ian Stevenson, Robert Sundstrom, Jeff Williams, and Steve Zimmerman for their generous assistance with fieldwork, and the City of St Paul, the National Marine Fisheries Service, the Alaska Maritime National Wildlife Refuge of the United States Fish and Wildlife Service, and the United States Coast Guard for providing logistic support for work in the Pribilof and Aleutian Islands. The banding office of the Canadian Wildlife Service kindly supplied us with details of banding recoveries for auks banded in North America, Sam Droege of the U.S. Fish and Wildlife Service provided a copy of the Christmas Bird Count database and Vivian Mendenhall provided us with data from the U.S. Fish and Wildlife Service Alaskan seabird colony catalogue. Ian Jones would especially like to thank the National Geographic Society Committee for Research and Exploration for supporting his work on auklets with several generous grants. Both authors have received support from the National Sciences and Engineering Research Council of Canada and Ian has also benefited from the support of the Vancouver Foundation. Tony Gaston acknowledges release granted by the Canadian Wildlife Service to write this book in his spare time. Many of our colleagues read and commented on earlier drafts of chapters or species accounts: Hugh Boyd, Alan Burger, David Cairns, Harry Carter, Gilles Chapdelaine, John Chardine, George Divoky, Peter G. H. Evans, William Everett, John Fries, Vickie Friesen, Anne Harfenist, Michael Harris, Alasdair Houston, Fiona M. Hunter, Anthony Keith, David Kirk, Sasha Kitaysky, William Montevecchi, Yolanda Morbey, Edward Murphy, S. Kim Nelson, Storrs Olson, Koji Ono, John Piatt, Steven Speich, William Sydeman, Sarah Wanless, Kenneth Warheit, Yutaka Watanuki, and Ron Ydenberg. Special thanks for pictures, tape recordings, or the use of unpublished data, go to Harry Carter, Gilles Chapdelaine, Sharon Dechesne, George Divoky, Scott Hatch, Kara Lefevre, Koji Ono, John Piatt, William Sydeman, and Takaki Terasawa. Timothy Birkhead deserves a special round of applause for reading the entire book. We also thank those, much too numerous to mention, who answered queries by telephone or letter, or allowed themselves to be grilled for information at conferences or on field trips. Few individuals that we approached for assistance or information refused, or even hesitated, to provide assistance. This book is, in a real sense, a co-operative effort in which the authors were merely the self-appointed scribes. We hope that we shall not disappoint the community of alcid biologists.

How do you know but ev'ry bird that cuts the airy way
Is an immense world of delight, clos'd by your senses five?

William Blake, The Marriage of Heaven and Hell

Preface

Wildness and auks are almost synonymous. Unlike their relatives, the gulls, auks have little to do with human affairs and avoid us if they can. They benefit from our activities only in distant and tangential ways and many and grievous are the harms that they have received from us. When we examine their global distribution, with the exception of California and a few colonies in Britain and Newfoundland, the distribution of the auks is like a negative image of the distribution of people. Only the hardiest fishermen venture into those waters of the Aleutians, the Bering Sea, and the Sea of Okhotsk that are the heartland of the kingdom of the auks.

But their wildness is not the only fascination of the auks. More than any other birds, they are the product of both air and sea. The only other birds that have travelled so far in adapting to the marine environment, the penguins, have forsaken flight completely. The auks now, and apparently throughout their history as a group, teeter on the brink between flying and flightlessness, squeezing the last ounce of underwater performance commensurate with retaining the ability to fly. Periodically, they have taken the penguin route and given up the option of flight entirely, but the real defining quality of the auks is the compromise between flight and submarine swimming. Livezey (1988) put it succinctly:

The Alcidae are morphologically committed to a largely aquatic existence. The compromise in wing-shape necessary for wing-propelled diving is substantial and in flighted alcids is associated with their comparatively heavy wing-loading and high wing-beat frequency.

Some variations on this theme recur throughout the book.

For the typical, urban-based birdwatcher, the auks seem an exotic group, occasionally to be glimpsed far off on the horizon. 'Auk at two o'clock going left, might be a razorbill...' is the common experience, or perhaps an exhausted, storm-driven puffin sitting disconsolate on a land-locked reservoir (always, sadly, in winter plumage). Even for those who go to sea in ships, the auks are frustratingly hard to watch, swimming low in the water, diving often at the approach of the bow, or flying rapidly, 'stitching through the waves'; bad weather follows them like a shadow. To appreciate the auk family one must make a determined pilgrimage to one or other of their many remote shrines: the great colony islands that are scattered like a necklace of jewels around the northern borders of the boreal and Arctic seas.

Those who do penetrate the world of auks are rewarded by insights into a group that has developed many strategies for breeding, for socializing, and for rearing their young. In their nesting sites, in their varied social behaviour, and in the strategies adopted by their chicks at departure from the colony, they are a peculiarly diverse group. Such diversity provides fertile material on which to exercise the comparative method and we have followed many others in trying to take advantage of this variation to learn more about the way in which evolution has moulded their adaptation to the marine environment.

Samuel Beckett, not the most optimistic of writers, said 'to be an artist is to fail'. We could say the same of being a scientist, because even the most comprehensive description of nature is inevitably overtaken by newer descriptions that embrace an even greater range of observations. Perhaps we can view truth as a series of Chinese boxes, fitted endlessly within one another. The artist, beginning from the outside, opens each box in turn seeking that innermost essence known as truth. The scientist, on the other hand, works in the opposite direction, opening the boxes from the inside and confronting, with each revelation, a larger and larger universe of knowledge. Look on this book as a box. When you have finished reading it, no, even before we have finished writing it, the universe of knowledge will have expanded to reveal another box that we shall need to open.

Contents

x Contents

Colour plates

Colour plates fall between pages 154 and 155

Abbreviations

♀	female
♂	male
AJG	personal observations by Tony Gaston
Apr	April
asl	above sea level
Aug	August
CBC	Christmas Bird Counts
cm	centimetre(s)
Dec	December
EPC	extra-pair copulation
Feb	February
g	gram(s)
ha	hectare(s)
hr(s)	hour(s)
ILJ	personal observations by Ian Jones
I., Is.	Island(s)
Jan	January
kg	kilogram(s)
kHz	kilohertz (frequency)
km	kilometres
m	metre(s)
Mar	March
min	minute(s)
mm	millimetre(s)
N, S, E, W	north, south, east, west
n	number in sample
Nov	November
Oct	October
ref	reference
s.d.	standard deviation
sec.	second(s)
Sept	September
sp., spp.	species (singular, plural)
ssp.	subspecies

Plan of the book

This book, like others in the series, comprises several chapters dealing with different aspects of the biology of auks, followed by individual accounts of each species. In dealing with the species accounts, we were confronted by the fact that some species have been the subject of many studies, whereas for others even some simple facts of natural history remain unknown. Consequently, the accounts are of very uneven length. For those species about which much information is available we have tried to present an overview based on selected studies, rather than attempting an exhaustive review of all the available information. We have also tried to synthesize the facts to present interesting patterns, or ones that have not previously been identified. Although this is not a primary research study, we hope that, by painting a very broad picture of the family, and by reviewing a range of studies of some species, we can indicate some interesting trends and correlations that may lead to further research and analysis. We have tried to provide, not only an epitaph on past investigations, but a signpost for future ones.

Certain kinds of information are available in abundance in the literature, while others, perhaps no less critical to the survival of the birds, are almost never referred to. For instance, the growth of nestling auks, in terms of their weight increase and the development of their wing feathers, has been described many times for some species. In contrast, the vocalization of the chick, that vital communication link between offspring and parent, is rarely described and almost never quantified. As a result, we found ourselves, especially with the well-known species, passing rather rapidly over large slices of the literature, because they repeated observations already more than adequately recorded for our purposes. By the same token, we include information from certain studies, not necessarily more rigorous or detailed than those passed by, but containing simple observations of things that went unrecorded elsewhere.

Where we have personal experience of a species, we have tended to give precedence to our own observations, rather than those of others, not because we think ours more valid or accurate, but because we can be more certain of the context and ramifications of our own work than that of others. Also, it is possible to give a more lively description of something witnessed at first hand than it is simply to transcribe the work of others. Between us, we have first hand experience of research on nine species of auks, and we have seen in the field all but the Japanese Murrelet. Luckily, our own research experience covers several of the less well known species: Least, Crested, and Whiskered Auklets and Ancient Murrelets. It is no coincidence that these are all Pacific species; all the Atlantic species feature among the better known, although even for those species, large areas of ignorance remain (where do Atlantic Puffins spend the winter and what do they feed on at that time of year, for example?). We have both been working on Pacific auks since 1984, beginning in that year with a joint study on the Ancient Murrelet. Our acquaintance with the Pacific auks provides our main qualification for undertaking this review.

PART I

After an introduction outlining the major characteristics of auks and important features of their history, six chapters give an overview of selected areas of auk biology. We have chosen the topics to cover areas where we have particular expertise or interests, or where the family Alcidae has outstanding potential for synthesis and interpretation of evolutionary biology. The treatment in these chapters is not intended to be exhaustive, but to introduce readers to major problems in auk research and to present a global view of the family. Where information was well established and accepted by the majority of researchers we omitted references, or confined ourselves to one or two primary sources or major reviews. However, we have given references to anything that we thought might not be common knowledge among people studying seabirds. Those who consider references an unsightly waste of space should consider this comment from one of the founders of bird behaviour:

If facts are the stones of which the Palace of Science is constructed, sources are the cement.
                    Armstrong (1942) *Bird Display and behaviour*

Phylogeny and the comparative method

As evolutionary biologists, we should like to know how the physical attributes and behaviour of auks have evolved in response to necessities imposed by their different lifestyles. Hence, we need to know which traits are adaptations and from which selective pressures the adaptations have resulted. This should enable us to answer questions such as: why do puffins have such large ornamented bills? why do *Synthliboramphus* murrelets take their chicks to sea at two days of age? and why are some species black-and-white coloured while others are uniformly dark-plumaged?

A promising approach is to consider some species where a trait, for example coloniality, occurs, and compare those species with others that are not colonial, to try to determine what ecological factor is responsible for the evolu-tion of coloniality. This is known as the comparative method (see, for example Clutton-Brock and Harvey 1979; Harvey and Pagel 1991, and, for a dissenting view, see Bock 1989). However, two factors hamper our ability to apply fully the comparative method to auks. First, the number of auk species is relatively small, so the opportunity to make comparisons is somewhat limited. For example, nesting in trees has evolved only once in the auks, in the Marbled Murrlet, so we have only one case to consider and thus we can only speculate as to which of the many unusual ecological and morphological characteristics of this species led to tree-nesting. The comparative method works best when a trait has evolved independently in several different species. By looking at what ecological or other factors the disparate species have in common we should be able to determine what is likely to be responsible for the evolution of the trait. With only 22 recent auk species, we do not have a lot to work with. The second problem with applying the comparative method to auks is that their phylogeny is poorly understood. We do not know the relationships among the major groups of auks (e.g. we do not know whether the guillemots are more closely related with murres or murrelets (*Brachyramphus*) and we do not understand the species relationships within some groups, such as the auklets.

To understand why we need to know the phylogeny to use the comparative method, consider the following example. Suppose we want to determine why some auk species have plain black-and-white summer plumage. Considering all the living auks, we notice that plain black-and-white auks are predominantly cliff-nesters (e.g. the murres, Razorbill) and we might be tempted to speculate that this plumage type has resulted from selection related to cliff-nesting. However, phylogenetic analyses have shown that these species are more closely related to one another than to the other auks (tribe Alcini). Thus it is as likely that their similar plumage is related to their shared ancestry as it is because of selection related to their cliff-nesting habits. For many comparisons among the auks, the phylogenetic information is equivocal, making it impossible to establish firm explanations for the evolution of different traits. Nevertheless, we believe that much can be learned from the basic associations between ecology and behaviour or morphology, and from the phylogenetic information that we have available. Throughout this book we have tried to work within these constraints, using what comparative evidence is available, and making cautions (but frequent) speculations about evolutionary patterns and ecological associations.

Terminology

A few terms that we use frequently in both parts of the book need to be defined.

AUKS: we use this as the English equivalent to Alcidae, to include all members of the family, not confining it, as is sometimes the case, to the subfamily Alcini.

FORAGING RANGE: it is customary for seabird biologists to distinguish between species that feed close to land and those that feed far out at sea. This division has many fundamental consequences for feeding ecology and reproduction and affects all aspects of a species' biology. Various terms have been used to describe this distinction (coastal, inshore, nearshore, offshore, oceanic, pelagic, etc.). We follow Ashmole (1971) in using *inshore* to denote birds feeding within 8 km of shore, *offshore* for those feeding out to the margin of the continental shelf, and *pelagic* for those beyond the continental shelf.

OCEANOGRAPHY: the distribution of seabirds at sea is determined to a great extent by the nature of ocean waters. Many studies have shown that, despite the apparent lack of visual boundaries at sea, seabirds adhere rather closely to specific water types. On the basis of water temperature and current patterns, the northern hemisphere is divided into four major zones, roughly corresponding to latitude (e.g. Bourne 1963; Ashmole 1971).

The names assigned to these zones vary. We have used *high Arctic* (= polar, Arctic), *low Arctic* (= sub-Arctic, sub-polar), *boreal* (= temperate), and *sub-tropical* (see Chapter 3).

ATLANTIC/PACIFIC: in discussing biogeography, evolution, and conservation, it is convenient to discuss the auks in terms of 'Pacific' and 'Atlantic' species. In actual fact, several species occur throughout the Arctic in summer, but few individuals remain in winter. For the purpose of discussion, we have arbitrarily included those parts of the Arctic Ocean between 100° E and 100° W with the Atlantic sector; the rest with the Pacific.

PART II

This section gives detailed accounts of the biology of each species under a standard set of headings. Longer sections are prefaced with a short summary of the major points. For some species, much literature is available and consequently, given the constraints of space, we had to make choices about what to use. Rather than aim to be comprehensive, we have tried to maximize the geographical spread of information. Where information under a given heading was very sparse, we included it elsewhere, or omitted the topic altogether, to avoid a proliferation of headings with little or no content.

Nomenclature

Assuming that our readership would be cosmopolitan, we were faced with a choice of English names and selected the North American versions mainly for their better correspondence with existing genera of auks (also a much higher proportion of the auks occurs regularly in North America than in Europe). To maintain consistency, we have also used other North American bird and fish names. The British equivalents currently used in Europe are given in the table below. The main potential confusion is over 'guillemot', applied in Britain to both *Cepphus* and *Uria*, while being restricted in North America to *Cepphus*. The use of Tystie, a Scandinavian derivation,

Table: North American and European English names of auks and other species and groups referred to in the text (where different)

North American	European	Scientific
Common Murre	Guillemot	*Uria aalge*
Thick-billed Murre	Brunnich's Guillemot	*U. lomvia*
Dovekie	Little Auk	*Alle alle*
Black Guillemot	same or Tystie	*Cepphus grylle*
Atlantic Puffin	Puffin	*Fratercula arctica*
Loons	Divers	Gaviidae
Jaegers	Skuas	*Stercorarius* spp.
Sandlance	Sand eel	*Ammodytes* spp.

for Black Guillemot would not have solved the problem, because there is no readily available substitute for the Spectacled and Pigeon Guillemots. When we refer to guillemots in this book, we mean members of the genus *Cepphus*. The scientific names of auks have been relatively stable for the past century, so synonymy does not present any significant problem. For taxonomy, we have followed the AOU (1983) and Strauch (1985).

Description

Plumages, including first year, winter, and breeding are described for both sexes together, as there is no significant sexual dimorphism in plumage in any species of auk. Chick down colour is also described, as well as variation that occurs within populations. Some species show conspicuous geographical variation in both plumage and size. These variations are dealt with under a separate heading, as are the timing, duration, and progress of moult.

Although the subspecies is not a concept that finds much favour with biologists at present, and although the analysis of DNA is in the process of providing us with much better information on the limits of populations, we have used those subspecies that are widely accepted as a basis for discussing variation. In fact, especially in the Atlantic, much of the variation on which subspecies have been described is continuous. Hence, our use of subspecies should be looked on as a convenient shorthand for defining segments of the total species population, rather than as an endorsement of any particular taxonomic view.

Measurements and weights

For measurements, we tried to select sources with a good sample size and where a clear attempt had been made to define the nature of the sample (e.g. adult birds taken from burrows while breeding). We have included information on whether measurements were taken on live birds or on study specimens, because this can make a difference, but we have not standardized on one technique. Older sources frequently give means and ranges, whereas more recent authors usually give standard deviations (s.d.) and omit total range. We have given whatever was available (converting standard errors to standard deviations where necessary); numbers in parentheses after the mean are the range unless otherwise indicated.

Range, status, and maps

As the auks are entirely marine in distribution, their breeding areas are confined to a narrow coastal strip and to small offshore islands. Although breeding distributions are well known for most species, the distributions at sea during the non-breeding season are often very poorly known. In fact, a single range map for non-breeding distributions is probably inappropriate for many species, which may have different distributions during breeding, moult, early and late winter, and pre-breeding. We have arbitrarily defined two different non-breeding limits: 'common' = frequently seen, and 'rare' = extreme limits within which the species has been recorded. These can be fairly well defined along coastlines, but their oceanic boundaries are very uncertain in most cases and the lines that we have drawn are 'best guesses', representing the situation typical in mid-winter (Dec–Jan).

For winter distributions, we have consulted the results of Christmas Bird Counts (CBC) in North America, making use of the computerized database held by the U.S. Fish and Wildlife Service. Necessarily, these cover mainly areas where there are observers, and trends over time are hard to interpret because of the more competitive nature of recent counts. We use these numbers only to give broad indications of the species' occurrence in mid-winter.

Field characters
This description highlights characters that enable identification in the field, either at a distance, or from fragmentary carcasses found on beached bird surveys.

Voice
We have described the main vocalizations for each species in words, using our own and published information. Where complex details of calls cannot be easily described, we have presented sonograms, but we have not attempted to include all calls, or to present sonograms for every species.

Habitat, food, and feeding behaviour
This section deals with feeding habitat and with behaviour relating to finding and capturing food. In keeping with other volumes in this series, we have dealt with feeding behaviour alongside diet and feeding habitat, rather than placing it with other behaviour. For a few species populations, the information on diet, especially of chicks, is much too voluminous to be exhaustively catalogued. In such cases, we have selected the most comprehensive data available, in terms of sample sizes, and tempo-

ral spread. Information on diets based on the analysis of stomach contents is presented in a variety of ways: proportion of stomachs containing a given organism (percentage occurrence), the representation of a given organism by number (percentage by number), by volume (percentage by volume), or by wet or dry mass (percentage by mass), even in some cases as a proportion of overall energy equivalent (percentage by energy); composite indices may also be used, such as the index of relative importance (IRI, Sanger 1987a). We have tried to specify the method used when diet statistics are presented. To simplify comparisons among species and areas, we have presented percentage occurrence information wherever possible.

Displays and breeding behaviour
Behaviour is not as easily reduced to simple statistics as many other elements of biology. Words, rather than numbers, are the main currency here. Consequently, the behaviour sections tend to be longer than the amount of study might warrant. In fact, behaviour, especially vocal behaviour, has been much less studied in the auks than many other features of breeding biology.

Breeding and life cycle, and population dynamics
Because auks are most accessible to investigation while breeding, this topic tends to be the one for which most information is available. Sub-headings are defined to make navigation easier in the welter of statistics, but not all types of information are available for all species and some sections are amalgamated in the shorter accounts.

EGGS: these are especially difficult to deal with, because late laid eggs are frequently smaller than those during the peak. Consequently, a collection, or a set of measurements, made early in the season will usually average larger than one made after all laying has occurred. As egg collectors liked to get fresh eggs, museum collections are probably biased towards early-laid eggs, compared with measurements made in the field. Museum collections are further biased by the fact that much egg collecting was carried out for aesthetic reasons. This is most obvious in the case of murres, where the great variety and attractive appearance of the eggs led to huge accumulations of unusual types and markings in private collections that later found their way into museums. Except to establish the limits of variation, such collections are essentially worthless for scientific purposes. We have included some measurements made by us on eggs in the British Museum (Natural History), London (BMNH). We selected samples for which locality information and dates of collection suggested a sample made at a single visit. For species with two-egg clutches, one egg was selected at random for each clutch for measurement.

REPRODUCTIVE SUCCESS: in species that breed in burrows, in crevices, or under boulders, reproductive success is hard to measure without disturbing the breeding birds in the process. For some species, such intrusions are known to depress reproductive success relative to what is achieved in undisturbed sites (Cairns 1980; Gaston et al. 1988a, Rodway 1994). Where observations of reproductive success involved disturbance to the breeders involved, some effect on success must be assumed unless disproven. Some investigators have assessed the effects of their activities, but many published accounts do not include that information. We have inserted caveats where we believed that published information may have been affected by investigator activities, but all reproductive success data are to some extent suspect unless adequate precautions were taken to avoid bias. We have given reproductive success in young reared per pair per year. Hatching success is the proportion of eggs laid that hatch, and fledging success the proportion of hatched eggs that produce a chick that survives to leave the colony voluntarily.

REGIONAL VARIATION: many aspects of biology vary regionally. We have tried to give some idea of this variation, rather than presenting a single pronouncement to characterize the whole species. Contradictory information in the literature does not necessarily imply that one author was wrong and another right; it may simply indicate that different populations behave differently, or even that the sample population may vary its behaviour in response to environmental fluctuations.

INTER-YEAR VARIATION: when people study a species for a few years they frequently detect an extraordinary event, whereas a longer study of the same population may reveal that the 'extraordinary' event on a scale of years is a typical event on a scale of decades. An example occurred to AJG in studies of Thick-billed Murres at Prince Leopold Island, in Arctic Canada. After three years of research in 1975–7 it appears that the peak of egg-laying normally fell between 20 and 30 June. In 1978, laying was 3 weeks later and the season was considered, at the time, to be very unusual. However, in five subsequent years, the timing of breeding was later in every year than in 1975–7, much diminishing the uniqueness of the 1978 event.

MOULT: The formal terminology for moults and plumages (basic, alternate, etc.) is not often used in conversation among bird people. For a group confined to boreal and Arctic areas, with a single discrete breeding season, the more familiar 'breeding' and 'non-breeding', or 'winter' plumage serve very adequately. We have given the more formal terms in those places where clarification may be necessary for specialists.

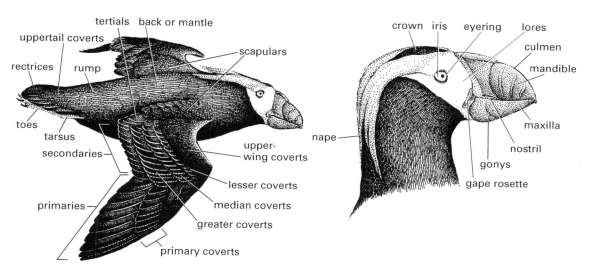

tertials   back or mantle

uppertail coverts   scapulars

rectrices   rump

toes

tarsus

secondaries

upper-
wing coverts

primaries

lesser coverts

median coverts

greater coverts

primary coverts

crown   iris   eyering   lores

culmen

mandible

maxilla

nape

nostril

gonys

gape rosette

Topographical diagram showing the parts of auks referred to in the book.

PART I

General chapters

1

Auks and their world

The auks are, above all marine organisms. Much of the research that is conducted on them, and this certainly includes what we have done ourselves, is carried out on land. Yet in the terrestrial realm, they are usually rather ill at ease. Some are poor at walking and the largest are too heavy to take off easily from the ground. Most of their physical adaptations are moulded by interactions within the marine environment, where they spend the bulk of their lives. In common with many other aquatic birds, their structure is sculpted by water, rather than air. To understand the ecology and behaviour of the auks, it is therefore necessary to understand something of the structure and dynamics of the oceans, specially those surrounding the Arctic.

Physical oceanography

Despite its lack of visual boundaries, the sea is divided into biogeographical regions just as much as the continents. The divisions reflect differences in physical characteristics, such as temperature and salinity, and also the interaction of the water with bottom topography and coasts. The dynamics of the world's oceans are driven by the heat difference between the equator and the poles and the major global wind fields. The latter create the surface currents that continually propel water away from the equator. The warm water carried away from

the equator eventually cools as it approaches the poles. This makes it denser, so that it sinks and, to balance the outward surface flow, currents of cold water return, sometimes at depth (Fig. 1.1). The distribution of these warm and cold currents creates the pattern of surface temperatures that determine, to a large extent, the boundaries of biological communities, including seabirds (Murphy 1936; Reinsch 1976).

Where winds blow onshore at continental margins, water tends to pile up against the coast, eventually sinking; where they blow off-shore, water is carried away from the coast and is replaced by colder water from below. The transport of deep water to the surface means that dissolved minerals (carbonates, sulphates, phosphates) reach the zone of light penetration, where they become available to phytoplankton, which form the basis for most food chains in the sea. Consequently, areas of major oceanic upwelling are generally areas of high biological productivity: examples are found along the coasts of Peru and California, in the Pacific, and off Namibia and Mauretania in the Atlantic.

Upwelling also occurs, on a more modest scale, where any current is deflected by a sharp change in water depth, or by a meeting of water bodies of different density. This type of upwelling may be very local and may alter daily in response to changes in the direction of tidal currents. It may lead to local enhancement of productivity, through water mixing. Where

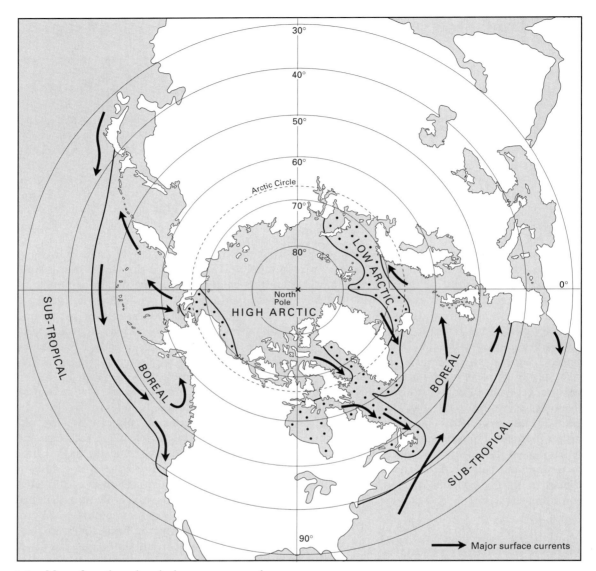

1.1   Map of northern hemisphere oceanography.

tidal currents are very strong, as in areas where a major tidal flow is restricted by a passage between islands, food supplies for birds may be augmented by driving slow-swimming prey, such as zooplankton, towards the surface. Conversely, in areas such as Hudson Bay, where tidal currents are weak and there is no significant upwelling, the water body remains strongly stratified, nutrients are trapped far below the surface, and biological production is poor. These major oceanographic processes are highly influential in determining the distribution of birds, such as auks, that must feed on prey that is highly aggregated (Chapter 4).

Another large-scale feature that is very important in determining auk distributions, is the continental shelf. This is the shallow water area that extends from continental coasts to the edge of the deep ocean basins. Areas of shallow water extending from the coast to the 200-metre isobath are generally regarded as forming the shelf; the steep slopes that descend to the 2000-metre isobath mark the continental slope. Continental shelf waters are highly productive compared with oceanic waters, because they are shallow enough to allow recycling of minerals from the ocean floor into the water column and interchange between benthic and pelagic food-chains. The abysses of the true ocean are far too deep for much of that to occur. Most of the material that reaches the bottom of the ocean deeps will not reappear at the surface until resurrected through geological processes.

Biological oceanography

Primary production, the fixing of atmospheric carbon into biological materials through photosynthesis, is carried out in the oceans mainly by unicellular phytoplankton that lead very brief lives. These tiny plants are preyed upon by tiny animals (primary predators), such as copepods, and these support larger (secondary) predators: jellyfish, arrow-worms, shrimps, and a host of other crustacea. The larger plankton form the principal prey of the more mobile nekton (fishes, squid). As the organisms increase in size, their lives grow longer and their behaviour more complex. The tiers of organisms are known as trophic levels, but the interactions among them are not as simple as the labelling (primary producers, primary, and secondary carnivores, etc.) would suggest. Many organisms change from primary to secondary carnivores as they grow and many organisms feed at several trophic levels. Many fish and squid spend the early part of their lives as part of the plankton. The multitude of interlocking relationships between producers and predators at all levels is known as the food web. The physical characteristics of the ocean, along with its associated food web, comprise the ecosystem. Auks are a part of marine ecosystems, feeding exclusively in that domain.

They take organisms from several trophic levels, but are little preyed upon themselves within the marine environment, hence they can be regarded as top predators, along with whales, seals, and large fishes like tuna and sharks.

We need to note that primary production does not necessarily correspond directly with the amount of biological material (the standing stock) existing at a particular location. They are broadly correlated, but a lot of deviation occurs and this relates to the different rates of turnover of the different trophic levels in the food web. Primary production is affected by temperature, light, and the availability of nutrients. It is high over continental shelves and in areas of large-scale upwellings wherever these occur. Zooplankton standing stocks follow a similar pattern, but tend to be much more dense at high latitudes than in the tropics. The abundance of zooplankton in Arctic waters may be a major reason why auks are found mainly in these waters.

The auks' world

Auks and islands are inseparable. The majority of auks breed on islands, and some species breed exclusively on remote islands well offshore, because islands offer refuge from terrestrial predators and at the same time may be close to foraging areas. Only the two *Brachyramphus* murrelets nest extensively on continental land masses and more than a few kilometres inland. Consequently, the distribution of breeding auks is much influenced by the distribution of suitable breeding islands. Only in the high Arctic, where the presence of connecting sea-ice in late spring makes islands vulnerable to terrestrial predators, does this relationship break down. In our descriptions of distributions and biology we shall mention a number of island groups repeatedly, some of which may be unfamiliar to many readers. The following is a brief description of some major breeding grounds for auks, and some of the lesser known (Fig. 1.2).

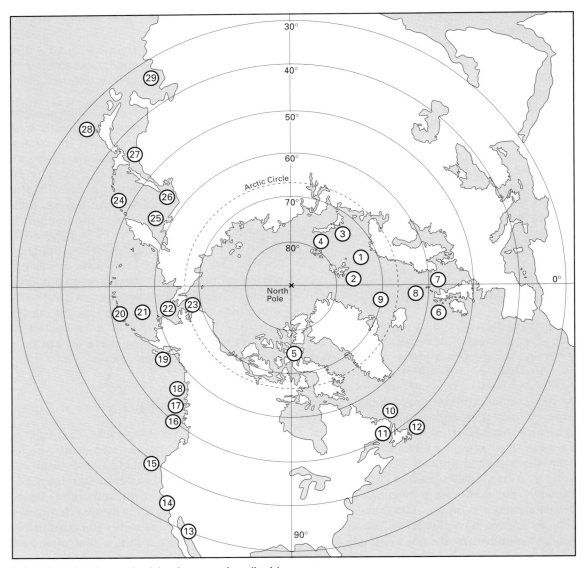

1.2   Map showing major island groups described in gazeteer.

Arctic Ocean

Bear Island (*Bjornøya*): small rocky island (74.5° N, 190.0° E), 200 km S of Spitzbergen, in the centre of Barents Sea (Fig. 1.2, 1).

Spitzbergen (*Svalbard*): an immense archipelago of mountainous, tundra-, and glacier-covered islands (their centre about 78° N, 17° E) on the boundary between low and high Arctic waters (Fig. 1.2, 2).

Novaya Zemlya: large, extensively glaciated, mountainous islands (centred about 74° N, 56° E) forming the eastern boundary of the Barents Sea (Fig. 1.2, 3).

Severnaya Zemlya (Fig. 1.2, 4): barren, low lying islands (about 79° N, 97° E), E of Spitzbergen, with a high Arctic biota.

Canadian Arctic Archipelago (Fig. 1.2, 5), including Prince Leopold Island (74.0° N, 90.1° W, Fig. 1.3), Coburg Island (75.9° N,

1.3 Prince Leopold Island and Barrow Strait, NWT, from the air, July, a major Thick-billed Murre and Black Guillemot colony (photo AJG).

1.4 Coburg Island, NWT, a major Thick-billed Murre colony: note the dark areas of luxuriant vegetation surrounding the whitewashed breeding cliffs (photo John Chardine).

79.4° W, Fig. 1.4), Bylot Island (73.3° N, 78.8° W), and Baffin Island (centred about 70° N, 73° W): network of large and small islands, rugged in the east, but becoming lower and flatter towards the west, mainly with a high Arctic biota and with little vegetation cover above 300 m elevation.

Atlantic Ocean

St Kilda: group of small, craggy, grass-covered islands (57.8° N, 8.6° W), with many huge cliffs, W of the Outer Hebrides, Scotland; former home of hardy islanders who subsisted mainly on seabirds (Fig. 1.2, 6).

Shetlands: cluster of large and small, treeless islands (68° N, 1° W) 180 km N of the Scottish mainland, farmed for sheep, the centre of the North Sea oil industry (Fig. 1.2, 7).

Faeroes: a mountainous island group (62° N, 7° W) 300 km WNW of Shetland and 420 km SE of Iceland, with a small human population engaged in farming and fishing (Fig. 1.2, 8).

Jan Mayen: large (50 km long), uninhabited (by humans) island (71° N, 9° W) 500 km E of Greenland and 650 km N of Iceland, at the edge of the winter pack-ice (Fig. 1.2, 9).

Gannet Islands, Labrador (Fig. 1.2, 10): group of small, low rocky islands, 15 km off the coast of southern Labrador (53.9° N, 56.5° W, Fig. 1.5).

North Shore, Gulf of St Lawrence: coast supports a string of small, barren, uninhabited islands up to 15 km offshore (Fig. 1.2, 11).

E Newfoundland (Witless Bay islands, 47.3° N, 52.8° W (Fig. 1.2, 12); Baccalieu Island, 48.2° N, 52.7° W; Funk Island, 50.1° N, 53.7° W): the most important breeding seabird concentration in eastern North America; islands are mainly small to medium-sized, uninhabited, covered in maritime grasses and stunted spruce and fir trees.

Pacific ocean

Islands of the Sea of Cortez (Fig. 1.2, 13): this nearly enclosed gulf, E of Baja California, Mexico, has sub-tropical waters and many small, arid islands with little or no fresh water and minimal vegetation. Craveri's Murrelet (*Synthliboramphus craveri*) breeds here south to Isla Espiitu Santo (29.5° N, 110.4° W), the southern breeding limit for the family.

Channel Islands (Fig. 1.2, 14): eight islands of varying size offshore from the cities of Los Angeles and Santa Barbara, California, supporting a sparse vegetation of drought-resistant shrubs and annuals. Their ecology has been much altered by introduced mammals. The best known is Santa Barbara Island (33.5° N, 119.0° W).

Farallon Islands (Fig. 1.2, 15): about 43 km off the coast of California (37.6° N, 123° W), a group of small, arid, rocky islets with very little vegetation; the largest seabird colony in California.

Scott Islands (Fig. 1.2, 16): chain of five islands extending 46 km north-west from the northern tip of Vancouver Island; the outermost, Triangle Island (50.9° N, 129.1° W, Fig. 1.6), has steep slopes covered in lush grasses, with some cliffs. It is the largest and

1.5   Gannet Islands, Labrador, home to all the Atlantic Auks except Dovekie (photo ILJ).

1.6 Triangle Island, Scott Group British Columbia: all the auks of British Columbia breed here, including the world's largest Cassin's Auklet colony and British Columbia's largest colony of Tufted Puffins and Common Murres (photo ILJ).

most diverse seabird colony in British Columbia and the largest Cassin's Auklet (*Ptychoramphus aleuticus*) colony in the world.

Queen Charlotte Islands (Haida Gwaii) (Fig. 1.2, 17): a large archipelago stretching 295 km from N to S and located 80–120 km W of the mainland of northern British Columbia; hundreds of large and small islands, the majority covered in tall, dense, evergreen, coastal rainforest (Figs 1.7, 1.8). Many Ancient Murrelets (*Synthliboramphus antiquus*) and Cassin's Auklets breed here, the largest colonies being on Hippa (53.5° N, 132.9° W) and Frederick (54.0° N, 133.2° W) islands.

Alexander Archipelago (Fig. 1.2, 18): SE Alaska: similar in climate and vegetation to the Queen Charlotte Islands, but these islands are more densely packed close to the mainland of the Alaskan panhandle. Forester Island (54.8° N, 133.5° W), at the SW tip of the archipelago, supports the largest seabird colony.

Islands of the Gulf of Alaska (Fig. 1.2, 19): Middleton Island (59.5° N, 146.5° W, Fig. 1.9), Barren Islands (58.9° N, 152.1° W), Semidi Islands (56.1° N, 156.8° W, Fig. 1.10),

Shumagin Islands (55.1° N, 159.7° W): numerous small island groups up to 100 km off the coast of mainland Alaska, rugged, many with high cliffs and stacks, covered in dense tussock grasses, or occasionally with low spruce forest.

Aleutian Islands (Fig. 1.2, 20): a chain of mostly uninhabited, rugged, volcanic islands extending westwards from the Alaska Peninsula through 25° longitude (about 2000 km) and 5° of latitude, covered in dense, waist-deep, tussock grasses and herbs at sea level, but barren and tundra-covered above 300 m elevation. Most of the chain is well south of the maximum extent of sea-ice and thus has no native land mammals, although foxes were introduced to many islands (Bailey 1993). Includes important seabird colonies at Chagulak (Fig. 1.11), Gareloi, Semisopochnoi, Kiska, and Buldir Islands (Fig. 1.12), the latter with 12 species of breeding auk, the most diverse assemblage of auks in the world (Byrd and Day 1986).

Pribilof Islands (Fig. 1.2, 21): an archipelago of four islands (at about 57° N, 170° W),

1.7   The Limestone Islands, Haida Gwaii, British Columbia, covered in dense forest and typical breeding habitat for Ancient Murrelets (photo AJG).

two large (St Paul and St George, more than 15 km long), and two small (Otter and Walrus), with small Aleut communities; mostly rather low-lying although St George Island rises to over 300 m elevation, with tundra-like vegetation. They are regularly beset by sea-ice in winter and have native Arctic Foxes (*Alopex lagopus*), but support large murre and auklet colonies.

St Lawrence (63° N, 170° W) and **St Matthew** islands (60° N 173° W) (Fig. 1.2, 22): two large remote islands in the Bering Sea, the latter uninhabited, with a variety of habitats including Arctic tundra, extensive wetlands, stony barrens, talus slopes, and cliffs, supporting large auklet and murre colonies.

Diomede Islands (Fig. 1.2, 23): two rocky islands (66° N 169° W), Big Diomede (Ratmonov) and Little Diomede, in the Bering Strait between Alaska and Siberia, both supporting huge auklet colonies.

◀ **1.8**   Reef Island shoreline, Haida Gwaii, British Columbia: typical breeding habitat for Cassin's Auklet (close to sea), Ancient Murrelet, and Rhinoceros Auklet (photo ILJ).

1.9   Middleton Island, Gulf of Alaska, showing uplift caused by a recent earthquake that has left murre breeding cliffs far from the shore (photo Scott Hatch).

1.10   Semidi Islands, Gulf of Alaska, showing typical Alaskan tussock-grass, inhabited by Cassin's Auklets and Ancient Murrelets (photo Scott Hatch).

Kuril Islands (Fig. 1.2, 24): a chain of volcanic islands stretching from northern Japan (44° N) to Kamchatka (51° N), similar in topography to the Aleutians but with taller vegetation including shrubs and small trees at lower elevations, particularly at the southern end of the chain. Several islands supported large auk colonies, some now ravaged by introduced predators (Voronov 1982), but this area is one of the least known parts of the auk's world.

Talan Island (Fig. 1.2, 25): a small treeless island in the Sea of Okhotsk (about 59° N,

1.11   Chagulak (foreground) and Amukta (background) islands, Aleutians, showing the rugged volcanic topography typical of the area (photo ILJ).

1.12   Buldir Island, western Aleutians: the main talus with Crested Auklets in flight: this island supports the most diverse colony of auks in the world, with 12 breeding species (photo ILJ).

150° E), offshore from the city of Magadan, Siberia; the best known seabird island in this area, supporting large auklet and puffin colonies (Fig. 1.13).

Iona Island (Fig. 1.2, 26): a small uninhabited rock in the Sea of Okhotsk N of Sakhalin Island (56.4° N, 143.8° E) with a cold war era miniature nuclear powered weather station; the island supports the highest diversity of auk species of any colony in this area.

Teuri Island (Fig. 1.2, 27): a small steep sided island (41.5° N, 139.4° E), 30 km off

1.13   Talan Island, Sea of Okhotsk: home to nine species of auks (photo Scott Hatch).

the coast of Hokkaido, Japan, forested in sheltered places in the interior and with grassy slopes on the periphery; the largest auk colony in Japan, but with a resident population of more than 100 people and currently threatened by commercial fishing, development, and pollution (Fig. 1.14).

Izu Islands (Fig. 1.2, 28): small, mainly uninhabited, islands and reefs SE of Honshu, Japan, supporting a stunted sub-tropical vegetation; Torishima (30.5° N, 140.3° E), the outermost, is 580 km from the mainland.

Yellow Sea Islands (Fig. 1.2, 29): small islands off the coast of China with dense sub-

1.14   Teuri Island, Sea of Japan: the largest Rhinoceros Auklet colony in the world. Tall vegetation conceals extremely fragile soil riddled with burrows; researchers have to tread warily (photo AJG).

tropical vegetation; a few Ancient Murrelets breed from Chenlushan Island, Jaingsu Province in the north (36.4° N, 120.9° E) south to Qingdao Island, Shandong Province (30.2° N, 122.7° E), the south-western limit of the auks in the North Pacific.

Breeding habitat and habits

Compared with many other seabird families, the auks are very inventive, both in the type of breeding habitat that they occupy and in the type of sites that they choose, or construct. None builds a nest, but the puffins, Ancient Murrelet and Cassin's Auklet often dig extensive burrows in soil. Most of the smaller auks make use of rock crevices or cavities among boulders or scree; only the largest, the murres, nest in the open in colonies, where their densely packed ranks form a defence against nest predators. Cliffs are frequently used, especially in the Arctic. The other open nesters, the *Brachyramphus* murrelets, are solitary nesters, either on the horizontal limbs of mature trees (Marbled Murrelet, *B. mormoratus*), or on the ground on remote mountain tops (Kittlitz's Murrelet, *B. brevirostris*).

Like many other small seabirds, some of the auks are nocturnal in their coming and goings to their breeding sites. The *Synthliboramphus* murrelets, Whiskered Auklet (*Aethia pygmaea*) and Cassin's Auklet appear to be invariably nocturnal, but the Rhinoceros Auklet (*Cerohinca monocerata*) and the *Brachyramphus* species are flexible in this character, being largely nocturnal, but diurnal or crepuscular in some parts of their range, or at certain stages of breeding. Like most terrestrial birds, but unlike many seabirds, everything about the breeding strategies of the auks—choice of breeding area and nest-site, and timing of colony visits—suggest the dominant influences of predation and kleptoparasitism.

The auks' year

Being entirely marine in their foraging, auks come to land only in order to reproduce. Their anatomical adaptations to an aquatic existence dictate that they are relatively clumsy on land and some take off from it only with difficulty. This reduces their options for breeding sites and makes them very vulnerable to land-based predators. Consequently, most auks visit their breeding sites only for a limited period prior to and during the breeding season. In colonial species, the first arrivals on land may be preceded by a build-up of birds on the sea adjacent to the colony. The pre-laying period is characterized, in some cases, by a gradual build-up in the numbers of birds attending the colony and the proportion of time that they spend there. In other cases, however, the breeding birds, after a period of gathering at sea, make a more-or-less simultaneous landing at the colony. Numbers frequently fluctuate in a cyclical or irregular manner over periods of 3–7 days and these fluctuations may continue to be evident after incubation has begun. Attendance during pre-laying is often synchronized to a particular time of day, although this pattern becomes less apparent as laying approaches.

Because egg formation is a lengthy process, females tend to spend most of their time away from the colony during the last 10 days before they lay, but male attendance peaks during this period. After completing the clutch, the sexes alternate incubation duty, taking equal shares. Laying, in practically all populations, is confined to a span of about 6 weeks, with most eggs being laid within 2–3 weeks. Incubation periods, ranging from 29 to 45 days, tend to be somewhat longer than for gulls and shorebirds, but substantially shorter than those of petrels (Procellariiformes) laying eggs of the same size (Jouventin and Mougin 1981). Brood patches develop in the last 10 days before egg-laying and begin to refeather as soon as, occasionally before, hatching occurs. The duration of the entire breeding period, from the beginning of incubation to chick departure, varies from a minimum of 35 days in the precocial *Synthliboramphus* species to a maximum of more than 80 days in the Rhinoceros Auklet.

Because most auks do not breed until 3 or more years old, populations contain substantial numbers of non-breeders. Those in their second summer or older often attend the breeding colony to court and select potential breeding sites. The number of pre-breeders at the colony generally rises throughout the breeding season, peaking about the time of hatching and remaining high until most chicks have departed. This is in contrast to petrels, where numbers of non-breeders attending the colony peak during incubation and decline sharply during chick-rearing.

Most breeders leave the colony either with their chicks (where there is parental care after departure), or within 1–2 weeks following chick departure. Juveniles, family parties, and post-breeding adults disperse rapidly from the colony area (except young Whiskered Auklets), but in most cases there is no genuine migration.

A complete moult (except in the puffins, where it involves the body plumage only) follows rapidly on the termination of breeding and involves the shedding of nuptial ornaments and the adoption of a, generally distinctive, non-breeding (winter) plumage. In the auklets, primary moult is sequential and begins during chick-rearing, but in all other species the primaries are dropped simultaneously, resulting in a period of flightlessnes. This period is probably the least known for most species as, except for the guillemots, they tend to remain scattered offshore during the moult.

Following the annual moult, most species shift towards wintering areas, but in some populations of Common Murres (*Uria aalge*) and Black Guillemots (*Cepphus grylle*) breeders return to the area of the breeding colony and commence periodic attendance at the breeding site. This also applies to some Marbled Murrelets. Except where ice formation makes open water inaccessible, some birds usually remain in the vicinity of their breeding sites throughout the winter, although actual visits to the colony may not occur until spring. There is much variation both within and among species in the degree of dispersal and the occurrence of birds at breeding sites outside the breeding season. However, among the most northerly breeders the bulk of the population moves substantially further south. Young birds tend to disperse furthest and there is usually a disproportionately high representation of first year birds in 'wrecks'; the periodic casting ashore of large numbers of weakened birds, often during prolonged storms. Wintering areas and the behaviour of auks on the wintering grounds have generally been much less studied than their activities during the breeding season, because the distribution of auks in winter does not lend itself to investigation. Detailed studies of feeding ecology in winter have mostly been carried out on species and populations that occur in inshore waters. The winter ecology of many species is essentially unknown. Movement towards the breeding colonies begins in February–April, depending on latitude.

Auks and people

Although they choose to breed in the most remote places they can find, auks have not escaped the attention of humankind. Being relatively inept on the ground, and often rather unsuspecting, and providing relatively large, nutritious eggs, they have been the targets of harvesting by people from the earliest times. They became especially vulnerable once seagoing boats had been developed, allowing even their most remote island homes to be plundered periodically. They also suffered from the unintended consequences of human activities: the introduction of predators and competitors, such as rats, cats, and foxes to their nesting islands, and latterly the discharge of oil at sea, whether accidentally or deliberately.

Auk bones are frequent constituents of middens throughout the coasts of the northern hemisphere. The remains of the Great Auk (*Pinguinis impennis*) have been discovered in 40 middens in Norway alone, as well as some in Britain, Iceland, Greenland, and the United

States (Bourne 1993). Excavations in New-foundland dating to 4000 BP contain many Great Auk bones (Montevecchi and Tuck 1987) and they are also found in middens in Florida dating to 1000 BC (Brodkorb 1960). All major colonies of Thick-billed Murres (*Uria lomvia*) in the eastern Canadian Arctic show traces of Eskimo occupation nearby, usually with associated piles of Thick-billed Murre bones. In areas surrounding the Straits of Georgia, British Columbia, Common Murre remains are widespread in Indian middens dating from the pre-European period (Hobson and Driver 1989), while middens in the Aleutians, some dating back as much as 4000 years, contain the remains of auklets and puffins, as well as murres (Friedman 1935).

A variety of techniques were developed for catching auks. One of the most widespread, used by the Bering Sea Inuit (eskimos), the Icelanders and the Faeroese to catch puffins, and by the Inuit of North-west Greenland to catch Dovekies (*Alle alle*) was a 'fleyg'; a net at the end of the long pole (Nelson 1887; Freuchen and Salomonsen 1958; Vaughan 1992). The hunter sheltered behind a stone wall, or depression the ground, and suddenly raised the net in the path of low-flying birds which were unable to turn in time to avoid it. At a Thick-billed Murre colony on eastern Baffin Island, a wall of boulders constructed in a small indentation at the top of the colony cliffs showed that a similar technique had been used by the Inuit of that area. A more primitive version was apparently used to kill Thick-billed Murres at Digges Sound, in north-eastern Hudson Bay, where a small promontory, often overflown by murres commuting to their feeding area, has an Inuit name translating as 'the place of beating with sticks'; according to local people, long poles were used to strike the passing birds and knock them to the ground; this technique was also used at one time in the Faeroes to kill puffins. In Iceland, snares placed on rafts floating offshore were also used to good effect, trapping murres, puffins, and Razorbills (*Alca torda*) (Gardarsson 1982).

The technique of netting birds flying over the colony is a very efficient way to harvest auks, as the pre-breeding component of the population often circles constantly, making them much more vulnerable than the breeders. Similarly, snares placed on boulders used by displaying birds, a method used on St Lawrence Island to capture Least and Crested Auklets (*Aethia pusilla* and *A. cristatella*) and in Thule, Greenland, to capture Dovekies, or on floating rafts, probably selected mainly pre-breeders. The same is probably true of fires set in colonies at night by Haida Indians in British Columbia to attract Ancient Murrelets. The removal of these birds has much less effect on the population than the killing of breeders and may partly explain how early societies managed to coexist successfully with their prey. Gardarsson (1982) estimated that 150 000–200 000 Atlantic Puffins (*Fratercula arctica*) were taken annually in Iceland, without any apparent effect on population levels. In contrast, the shooting of breeding birds at colonies that became widespread in Greenland in this century has been the main cause of the drastic decline suffered by Thick-billed Murre populations there over the past 50 years (Evans and Kampp 1991).

Eggs form an especially useful form of food, coming in their own packaging and remaining good for much longer than meat. Auk eggs are very high in fat, having larger yolks than those of most terrestrial birds. Consequently, the harvesting of auk eggs has been very common throughout their range and may well have been a factor controlling their distribution on inshore islands accessible to permanent human settlements. On St Kilda, harvesting of puffin and murre eggs was a regular activity (Connell 1887), while the Queen Charlotte Islands, and throughout the Alaskan islands the excavation of Ancient Murrelet and auklet burrows for eggs was a routine spring harvest. At Cape Graham Moore, on Bylot Island, ropes are permanently fixed above the colony to assist in the annual harvest of Thick-billed Murre eggs, while Black Guillemot eggs are harvested in western Hudson Bay, where they are traditionally eaten mainly by women.

Where pre-industrial societies exploited seabirds for subsistence purposes, there is little evidence that their activities did substantial damage to populations (although there is little evidence either way). However, commercial exploitation of auk colonies seems to have resulted everywhere in substantial declines. In the Gulf of St Lawrence, the huge auk colonies described by Audubon (1835) were reduced to a mere remnant by the late nineteenth century, while at Funk Island, off Newfoundland, the Great Auk was exterminated largely for its feathers (see species account). The Common Murre population at the Farallon Islands, California, and several large Thick-billed Murre populations in the Russian Far East and Novaya Zemlya, were decimated by egg harvests, perhaps partly because of a misunderstanding about how many times the birds could be expected to lay replacement eggs (Ainley and Lewis 1974). At the Farallon Islands, murre eggs were such a valuable commodity in the booming gold-rush city of San Francisco that competition among eggers led to pitched gun-battles in 1863 (White 1995).

Seabirds are a very ephemeral food source. While they are at their colonies they provide a bonanza of meat and eggs, but for the nine months of the year when they are at sea, they become inaccessible. This led to the development of long-term storage techniques by those groups that were especially dependent on them. At Digges Island, where the enormous Thick-billed Murre colonies probably attracted Inuit for millenia, birds were stored in beehive-shaped huts, loosely built out of boulders. Samuel Prickett, one of the survivors of Henry Hudson's original voyage to the bay named after him, described seeing the birds within 'hanged by the neck' (Asher 1860). The remains of these stone buildings are still visible on the island, although the technique has now been superseded for local people by the ubiquitous modern freezers. The stone huts of Digges Island appear almost identical in structure and function to the 'cleits' formerly used on St Kilda, in the Outer Hebrides, to store Northern Fulmars (*Fulmarus glacialis*) and other seabirds (Fisher 1952).

In the Thule district of North-west Greenland the preservation of Dovekies was even more important, as these birds formed practically the only food supply in autumn until the sea-ice was sufficiently strong for seal hunting. There, the Inuit stuffed them inside sealskins that had been skinned out through the mouth. As many as 7–800 could be crammed into a single skin, which was then sewn up and buried, perhaps to prevent raids by foxes (Freuchen and Salomonsen 1958). When dug up, several months later, the corpses had 'ripened' considerably. According to Fred Bruemmer, the Canadian explorer who sampled them in the 1950s, they were eaten whole (except for the gall gladder; the bones had to be well chewed) and uncooked and tasted 'like strong cheese'.

In addition to their use as food, the skins and ornaments of certain auks were valued for clothing and decoration. Inuit on St Lawrence Island and Aleuts in the Aleutian chain sewed parkas out of auk skins, especially those of Crested auklets and Horned Puffins (*Fratercula corniculata*). The bird skins were particularly valuable to these people, who had no access to Caribou (*Rangifer tarandus*) (Fitzhugh and Crowell 1988). Elsewhere, puffin and Dovekie skins were sewn into inner garments, to be worn under furs. In North-west Greenland, Dovekie skins were made into undershirts, while further south, in Upernavik District, murre skins were made into capes. Dovekie skins had to be softened by chewing; only elderly women did this, as their teeth were worn smooth enough not to damage the delicate skins (Bruemmer 1972). The spectacular beaks of puffins and auklets were also used as ornaments by the Aleuts and Inuit of the Bering Sea region, hundreds sometimes being sewn on the outside of a garment, along with the golden crests of Tufted Puffins (*Fratercula cirrhata*) (Nelson 1887; Vaughan 1992). Such garments were normally used on ceremonial occasions.

Large concentrations of auks provided benefits to people in other ways. The huge colonies of Dovekies in the Thule district of Greenland created a very high population density of Arctic Foxes in the area. This allowed the small number of Inuit living nearby to supply a high proportion of the fox furs exported from Greenland in the late 1940s (Malaurie 1982). The same principle was made use of in the Aleutian Islands, and other islands off Alaska that previously had no native mammalian predators, where foxes were introduced for 'fur farming'. The foxes lived on the breeding seabirds mainly puffins, auklets and murrelets. Unfortunately, these birds, unlike Dovekies, were not adapted to the presence of terrestrial carnivores on their breeding islands. When the birds were wiped out, as they were in some cases, rodents or Rabbits (*Oryctolagus cuniculus*) were sometimes introduced to replace them as a source of food, further upsetting many island eco-systems (Bailey and Kaiser 1993).

Substantial harvesting of auks still occurs in several areas. Traditional harvests of murres and puffins continue in Iceland and the Faeroes, although reduced from former levels (Harris 1984; Norrevang 1986; A. Petersen, personal communication). Relatively small numbers of Thick-billed Murres and their eggs (less than 1 per cent of those laid) are taken by Inuit in the Canadian Arctic, although they form important components of the summer diet at a few settlements. Much larger numbers are taken in West Greenland, where regulations prohibiting the shooting of birds at their colonies were introduced only in 1978 and were still more or less unenforced in 1987. Shooting away from the colonies is still permitted throughout the year in some dis-tricts, although subject to season limits in the more populated areas (Evans and Kampp 1991). The same populations are hunted on a large scale in their wintering areas off Newfoundland and Labrador, with the annual kill estimated at between 300 000 and 800 000 birds during 1977–87 (Elliot *et al.* 1991). Although it is only legal to kill murres, some

Razorbills and Dovekies, and a few puffins, are also shot. Until 1993, there was no bag limit, but season limits and bag restrictions imposed since then are believed to have more or less halved the kill. A similar non-breeding harvest of murres and Razorbills occurred in Norway until 1979, where many young Common Murres from British and Swedish colonies were shot.

Although direct harvest have affected several auk species and were responsible for the exter-mination of the Great Auk, the effects of mammalian predators, introduced either delib-erately, or accidentally, have probably had a much greater impact on auk populations world-wide. This issue is dealt with in detail in Chapter 7. The main agents of destruction have been Arctic Foxes, introduced through-out the Alaskan islands for fur farming, and rats (*Rattus* spp.), the effects of which are less obvious because the introductions were acci-dental and may have taken place long ago in some cases. Raccoons (*Procyon lotor*) and Mink (*Mustela vison*) have also had an import-ant impact in some areas, and rabbits, through their effects on vegetation and soil, may also have caused problems for some burrowing species (Bailey and Kaiser 1993).

The general reorganization of the biosphere to meet human needs and aspirations that has been proceeding with accelerating speed since the termination of the Pleistocene glaciations tends to affect most adversely those species that are specialized competitors with us for food or habitat. To the extent that the seas they occupy were some of the last to come under the influence of industrial society, auks in the Pacific fared relatively well. Those that have been most affected by our activities are the southern murrelets: Japanese (*Synthliboramphus wumizusume*), Craveri's, Xantus' (*S. hypo-leucus*) and Marbled, all at present endangered or threatened in one way or another. It is hard to detect any benefits that have accrued to auks through our hegemony over the bios-phere, apart from the possible relaxation of competition from baleen whales and large, commercially desirable fishes, such as cod

(*Gadus*) and walleye pollock (*Theragra chalcogramma*), for their common prey: zooplankton and small schooling fishes. It is certain that the majority of auk populations are smaller, in many cases much smaller, than they would have been a few centuries ago. It seems unlikely that we shall see much change in that situation, although programmes to eliminate introduced predators from certain important Pacific islands may improve the situation for some species.

The scientific discovery of the auks

When Carl von Linné published the tenth edition of his *Systema Naturae* in 1758 he had six auks available to him to describe: a murre (a Thick-billed Murre, as it subsequently transpired), Atlantic Puffin, Razorbill, Great Auk, Black Guillemot, and Dovekie. All these species were already well known to European scientists. It was believed for a long time that he also described the Common Murre, but the specimen involved came from Spitzbergen and was actually a Thick-billed Murre, so the original description of the Common Murre fell to Pontoppidan (1763). The separation of the two *Uria* species was made by Brunnich (1764) in the same year; hence the British name for the Thick-billed Murre, Brunnich's Guillemot.

Pallas, in his *Spicilegia Zoologica* (1769), described the Crested Auklet and Parakeet Auklet (*Cyclorhynchus psittacula*) and the Tufted Puffin. By 1785, Pennant had described the Ancient and Marbled Murrelets, which were formally introduced into the binomial system of nomenclature by Gmelin (1789), who also described the Whiskered Auklet. Hence, by the end of the eighteenth century, 13 of the 23 currently recognized species had been described, including all those from the Atlantic, plus a number of Pacific species discovered by Steller in the Bering Sea, or brought back by Captain Cook's voyages.

A further five species were added by Pallas in his *Zoographia Rosso-Asiatica*, published posthumously in 1811 (the date of this publication is very important in establishing priority and has been disputed because the plates were delayed and a full edition did not appear until 1831; however, 1811 is the generally accepted date). Vigors named Kittlitz's Murrelet in 1828. It was figured in 1827 by Audubon on the same plate as the Ancient Murrlet and mistakenly captioned as a juvenile of that species. The last auks to be discovered were also the rarest: Japanese (Temminck's or Crested) Murrelet described by Temminck in 1835, Xantus' Murrelet, discovered by Xantus in 1859, and Craveri's Murrelet, described by Salvadori in 1865. Recent genetic analysis shows that at least one other population deserves specific status, the 'Long-billed Murrelet', currently recognized as a race (*B. marmoratus perdix*) of the Marbled Murrelet (Friesen *et al.* 1996*b*).

Naturally, when the definition of the species, and even its biological basis, was poorly understood, many forms were described that were later incorporated within existing species. The Black Guillemot and Common Murre, in particular, received a bevy of species names, later demoted. Genera, on the other hand, waxed and waned. Of the 12 genera that we are using here, only *Alca* survives from Linnaeus (1758). This name was originally applied to all the auks and only gradually became restricted to the Razorbill (and in many classifications, the Great Auk), as fashions in generic definitions changed. In 1760, Brisson invented *Fratercula*, Brunnich created *Uria* in 1764, Pallas *Cepphus* in 1769, Merrem *Aethia* in 1788, and Bonnaterre *Pinguinus* in 1791. In the nineteenth century, Link added *Alle* (1806), Bonaparte *Cerorhinca* (1828), and Kaup *Cyclorhynchus* (1829), before Brandt (1837), in his first study of the family, rounded things off with *Brachyramphus*, *Synthliboramphus*, and *Ptychoramphus*. Many other genera were named, those that had greatest currency being *Plautus* (Dovekie, Great Auk), *Chenalopex* (Razorbill and Great Auk), *Mergulus* (Dovekie and several small Pacific auks), *Lomvia* (=*Uria*), *Phaleris* (auklets), *Mormon* and *Lunda* (puffins and auklets), and

Simorhynchus (auklets). Confusingly, at one time the guillemots were placed in *Uria* while the murres were in *Lomvia*.

The English names present, if anything, even more of a puzzle than the Latin ones. The Atlantic species had garnered any number of epithets before scientific naming began. Nautical accounts from earlier centuries abound with names like 'willocks', 'looms', 'sea parrots', and 'sea pigeons'. Some of these survive in Newfoundland, where Black Guillemots are still known as 'sea pigeons' and puffins as 'sea parrots'. As with other birds, names used for one bird in the Old World became attached to another, not always closely related, in the New World. Loom is a case of point. The English use of the world probably comes from the Scandinavian 'lomvie' (murre) and colonies were known as 'loomeries'. Although applied by English sailors in American waters to murres, it somehow became applied to those other large, fish-eating birds, the British 'divers', which became the North American 'loons' *Gavia* spp., a name in keeping with their maniacal laughing call, often given after dark.

One surprising development was the demise of 'auk' as an English name. In the eighteenth century it was applied to most of the family (Razorbilled Auk, Little Auk, Black-throated Auk, Whiskered Auk, etc.) Its application was gradually eroded to the point where today, with the Great Auk extinct, we have only the British name for the Dovekie (Little Auk) left. No other seabird family has so many types of common name attached and many as large or larger (Pelicans, Cormorants, Albatrosses, Storm-Petrels, Terns, Gulls) have only one or two. Why there should have been such a proliferation of English names is not clear, but they tend to emphasize how distinctive in plumage and morphology many of the auks are, and how familiar to people because of their importance as food.

Biological studies on the auks

The history of ornithology can be divided into several phases. In the beginning came collect-ing, describing, and naming. As we have seen, this phase was essentially complete for the Alcidae by the middle of the nineteenth century and the outstanding exponents for the auks were Steller and Pallas, the former the intrepid collector and the latter the careful describer. Pallas gets most of the credit, because the task of assigning names fell to him, along with its accompanying immortality, but Steller had the satisfaction of actually obtaining the birds and recognizing their uniqueness. He may have enjoyed the greater excitement from his role.

Following the description of species, came their arrangement in systematic hierarchies. This process received an enormous impetus from the idea of evolution through natural selection, so cogently presented in Darwin's *Origin of Species* (1859), but the seeds of such thinking were implicit in pre-existing arrangements that sought to place species into families, tribes, and genera that reflected their relative similarities. The first to do this for auks, and the true fathers of alcid studies, were Coues (1868) and Brandt (1869), the latter developing an arrangement more in keeping with modern ideas.

As systematics and evolution developed, they drew on an increasing variety of evidence. Studies of plumage and external morphology were succeeded by those of internal anatomy, especially skeletal characters. It was at this stage that the superficial resemblances between auks and penguins became overwhelmed by the contradictions in their structure and the similarities between auks, gulls, and shorebirds (Chapter 2). By the end of the nineteenth century the general affinities of the auks were clear, leaving naturalists free to turn their attention to the fields that have now become behaviour and ecology.

Because of the remoteness of their breeding islands, auks do not lend themselves easily to natural history studies. Broad outlines of their breeding habits could be gleaned from cursory visits to nesting colonies, but many details required weeks of painstaking and persistent observations. The fruits of the first wave of natural history endeavours are summarized for the Atlantic auks by the *Handbook of British Birds* (Witherby *et al.* 1941) and for the Pacific species by Bent's (1919) *Life Histories of North American Diving Birds*. Although the *Handbook* was the more scientific (i.e. numerical) in its approach, Bent's work is more interesting now, because he quotes his own notes and those of his correspondents verbatim, making it primary source material.

The difference between Bent's approach and the British *Handbook* reflects the differences in maturity at that time between studies in the two oceans. It is worth remembering that, before the California gold rush of 1849, the Pacific coast of North America was virtually uninhabited by Europeans, let alone naturalists. Moreover, after the initial exploration of the Bering Sea area, at the direction of the Russian imperial court, and especially after the severe reduction in Sea Otter (*Enhydra lutris*) populations that took place by the early 1800s, interest in the biology of the area faded. It did not blossom again until the transfer of Alaska to the United States in 1867. Little was added to knowledge of the biology of the North Pacific auks between the writings of Pallas and those of American naturalists, such as Nelson (1883, 1887) and Stejneger (1885, 1887) nearly a century later.

A new approach altogether was adopted in the 1930s by Lockley (1953) and Perry (1940), both of whom chose to live for extended periods on offshore islands, and carried their notebooks daily to record the comings and goings of the Atlantic Puffins, Common Murres, and Razorbills that populated their island homes. In the same decade Belopol'skii (1957), Kaftanovskii (1951), and colleagues at the Seven Island Sanctuary, on the Russian White Sea coast, set out for the first time to conduct prolonged biological field studies of auks, while a series of British expeditions to Arctic and sub-Arctic islands in the Atlantic fleshed out knowledge of the Arctic-breeding species. The Russian work also continued after World War II.

In Greenland, immediately after World War II, the remarkable Dane, Finn Salomonsen, began banding auks in large numbers, a programme that was to continue almost unbroken until the 1980s. This yielded much new information on the movements of auks. In 1944, a refugee in Sweden, Salomonsen wrote *The Atlantic Alcidae*, a comprehensive account of the Atlantic species, much of it based on personal research. From this point forward, studies of auks have, like the rest of science, swelled from a trickle to a flood.

Following World War II, Russian biologists undertook further studies on auks in Novaya Zemlya (Uspenski 1956). Some of the fruits of their research found a western outlet in Kartaschew's (1960) semi-popular account of the North Atlantic auks. At the same period, a series of British expeditions to Spitzbergen, Bear Island, and elsewhere extended the information available on Arctic species.

In 1952, Robert Storer published a very detailed study of the genera *Uria* and *Cepphus*, extending considerably the information presented by Salomonsen. Also in the 1950s Leslie Tuck, working for the Canadian Wildlife Service, began a wide-ranging study of the murres in eastern North America, including the first visits by a biologist to several of the large Arctic colonies, and the first large-scale banding of Arctic Thick-billed Murres in North America. He summarized his results in a monograph of the genus *Uria* (Tuck 1961). In Europe, the Swiss ornithologist Beat Tschanz (1959, 1960) began a series of remarkable studies on murres which provided exceptionally detailed information on their behavioural adaptations.

The 1960s saw the initiation of scientific studies of auks in Alaska, the first being that of Swartz (1966) on murres and puffins at Cape Thompson. This was followed by studies of auklets and puffins on St Lawrence Island in the Bering Sea by two Canadians, Spencer Sealy and Jean Bédard, which provided much new information on diets and breeding biology of these species (Bédard, 1969*c,d*; Sealy 1968, 1973*c*). Sealy went on to work in British Columbia and provide the first scientific studies of Marbled and Ancient Murrelets

(Sealy 1975*a,b*, 1976), and in the same area Rudi Drent (1965) performed a detailed study of the breeding ecology of the Pigeon Guillemot (*Cepphus columba*). In eastern North America, David Nettleship (1972) conducted the first detailed analysis of the effects of breeding habitat on reproductive success, in a two-year study of the Atlantic Puffin. At the end of the decade, the Seabird Group in Britain carried out a comprehensive inventory of seabird populations, including auks, that provided an important baseline against which to judge subsequent population trends.

By the 1970s, even detailed studies of auks become too numerous to specify. In North America, the need for environmental assessments related to the leasing of offshore areas for oil exploration spawned numerous studies, many in areas hardly visited before, especially Alaska, the centre of alcid diversity and abundance. An atlas of Alaskan seabird colonies, completed in 1978 (Sowls *et al.* 1978) marked a milestone in these surveys. The Pribilof Islands and the Chukchi Sea colonies at Bluff and Cape Lisburne became the focus of many studies, especially of the two murre species (Murphy *et al.* 1986; Hunt *et al.* 1986). Oil exploration was also the impetus for many surveys of offshore seabird distributions around the coasts of North America and Europe (e.g. Johnson *et al.* 1976; Briggs *et al.* 1988; Tasker *et al.* 1988). At the same time, concerns about the potential effects of toxic chemicals, especially organochlorines, on marine birds led to studies of tissue contamination in various auks and the establishment of programmes to monitor their levels (Parslow and Jefferies 1973; Ohlendorf *et al.* 1982; Risebrough *et al.* 1995).

Also in the 1970s, the Point Reyes Bird Observatory began a comprehensive study of the seabird community on the Farallon Islands, off California, including Common Murres, Cassin's Auklets, Pigeon Guillemots, and Tufted Puffins. This was summarized by Ainley and Boekelheide (1990). In Canada, the Canadian Wildlife Service initiated a major study of high Arctic Thick-billed Murre

colonies (Birkhead and Nettleship 1981; Gaston and Nettleship 1981). In Britain, the Edward Grey Institute followed up regular banding of young Common Murres, Razorbills, and Atlantic Puffins by setting a group of students to the task if investigating their population dynamics (Birkhead and Hudson 1977; Lloyd and Perrins 1977; Ashcroft 1979). Tim Birkhead went on to work extensively with David Nettleship on Canadian Arctic and sub-Arctic auks and they subsequently collaborated to edit the second *Atlantic Alcidae* (Nettleship and Birkhead 1985), summarizing the biology of the Atlantic auks to that date. In addition, Michael Harris, working for the Institute of Terrestrial Ecology in Scotland, initiated studies of the Atlantic Puffin that were to culminate in a comprehensive biology of the species in Britain (Harris 1984).

In the 1980s, research on auks continued at a furious pace, the emphasis switching from studies of breeding biology to those related to ecosystems, in response to conflicts between commercial fisheries and auks. This led to work on energetics, using isotope-labelled water to find out how much energy auks used in their normal activities, and hence how much food they needed (e.g. Gabrielsen 1994), to studies of time budgets and diving behaviour using instruments attached to the birds (e.g. Cairns *et al.* 1990), and to the integration of such information in large-scale energy models used to estimate the total intake of food by auks and other seabirds (Chapter 4).

On the pacific coast of North America, still relatively little known, baseline studies of populations, distribution, and breeding biology of auks continued, reaching their apogee in the detailed series of colony inventories carried out in British Columbia by the Canadian Wildlife Service (Rodway 1991). In the United States, especially, the agenda for auks in the 1980s was dominated by studies of Marbled Murrelets, an outgrowth of the ongoing battle to save the old-growth forests in which they nest. This research spawned a battery of new techniques, including the use of video cameras, portable radar, and floating mistnet arrays. The future undoubtedly holds the deployment of satellite transmitters (first attempted in Alaska, 1995) and further high technology instruments, as we delve ever deeper into the lives of the auks.

Apart from the studies alluded to above, there have been several general works relating to seabirds, or specifically to auks, that have been important sources of information for us. For information on populations we have drawn heavily on the two ICBP Technical Reports (Croxall *et al.* 1984; Croxall 1991) and on the recent Pacific Seabird Group symposium volume (Vermeer *et al.* 1993*c*). For the Atlantic species, the information provided in Nettleship and Birkhead's (1985) *The Atlantic Alcidae* and in volume IV of *The Birds of the Western Palaearctic* (Cramp 1985) provide an excellent summary. However, it is a measure of how much research has been going on, that the passage of a single decade has added substantial new information even for the relatively well-known Atlantic species. We are very conscious of the fact that this work, in turn, will be soon overtaken by new results, but we hope that the book itself, by identifying areas of ignorance and opportunities for new research, will advance its own obsolescence.

2

Systematics and evolution

The study of systematics and evolution are very closely allied. The science of systematics involves arranging organisms in their most appropriate groupings, according to specified criteria. Since the Darwinian revolution, this has meant classifying them according to our best assessment of their descent. Organisms grouped together are believed to have had common ancestors more recently than those placed far apart. Today this can be judged on the basis of similarities in their DNA. However, it is worth remembering that systematics did not begin with Darwin. Organisms were classified into hierarchies on the basis of common characteristics long before evolution was a generally accepted fact. It is a tribute to early systematists that many of their arrangements have passed the scrutiny of evolutionists and can be easily discerned in modern classifications. This is true of the auks.

Characteristics of the Alcidae

The following description is adapted and expanded from Verheyen (1958). Auks are confined to the northern hemisphere and with very minor exception all species are entirely marine throughout the year. Physically, auks share a compact, streamlined body, short wings, and a very short tail. The feet are placed far back on the body, webbed, with no hind toe. The claws are narrow, needlelike in some species, and the tarsus is laterally compressed, although not as much as in grebes

and loons. The bill is very variable in shape and may be highly ornamented in the breeding season.

There are 11 primary wing feathers, the outermost being small and non-functional and the longest being, usually, the tenth. Secondaries vary in number from 16 to 21. The feather tracts of the back and belly are continuous (Fig. 2.1) and, in addition to the regular contour feathers, down feathers are present and dense all over the body. There are six to eight, occasionally nine, pairs of mostly very short tail feathers. The preen-gland is feathered.

All species of auks fly with rapid wing-beats and without gliding or soaring, using the spread feet for steering and braking. They generally take off from land with difficulty and use the feet to taxi when taking off from water. They obtain food by wing-propelled underwater swimming. To submerge, they kick with the feet to upend, then give a flick of the wings, so that the wing tips surface briefly as they disappear (in contrast to the initial upwards jump of some cormorants and grebes, and the smooth forward submergence of loons). Air is released from the plumage before diving by forcing it out of the breast feathers using subcutaneous muscles.

On land, auks frequently rest forward on the belly, or on the tarsi, rising onto the feet only in walking. However, the auklets and puffins are more agile and normally stand with the tarsus erect, rather than horizontal. All species are more mobile on land than foot-propelled divers, such as loons and grebes.

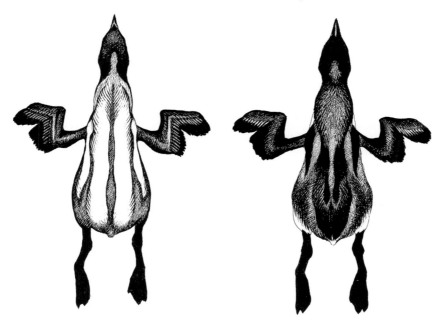

2.1   Initial development of feather tracts: *c.* 14 day nestling Thick-billed Murre.

The anatomy of the auks is generally similar to that of other members of their order (Charadriiformes), except in so far as they are specialized for underwater swimming. For the technically minded, they have the following skeletal characteristics: they are schizorhinal (nasal cavity extends backwards as a slit beyond the front of the premaxilla); the articulation between the cranium and the maxilla is supple; the premaxilla–nasal suture is persistent and there is no naso-frontal fossa; there is no nasal septum; the vomer is not connected with the palatines; the palatines and the maxillo-palatines are independent; and there is only one pair of mandibular foramina.

Nasal glands, which are used to excrete salt and maintain ion balance, are very well developed, as befits marine birds, forming pronounced supra-orbital depressions in the skull, separated by a median crest. Supra-orbital ridges are well developed in the Atlantic genera (*Uria*, *Alca*, and *Alle*), creating a trough on either side of the median crest in which the nasal glands are sunk. This feature is absent in Pacific auks (Fig. 2.2). The temporal fossae are distinct and the plane of the foramen magnum inclined at 45 degrees. The bony septum between the orbits is pierced by a large opening.

The long bones and breast bone are not pneumatized, the sternum and costosternum being elongate and relatively narrow, although this elongation is more pronounced in the murres, guillemots, and *Synthliboramphus* murrelets (Fig 2.3), than in the puffins and auklets. The furculum is simple and U-shaped. The humerus is longer than the ulna or femur, and the latter is generally longer than the tarso-metatarsus. There are 2 + 13 cervical, 6–8 dorsal, 4–6 dorso-sacral, and 8–11 caudal vertebrae, while the synsacrum comprises 11–15 vertebrae. The trachea and oesophagus are on the right of the vertebral column, the trachea being round in section. The oblique septum of the diaphragm is partially muscular.

The digestive system includes a very well-developed proventriculus, a muscular stomach, and a relatively short intestine. A functional crop is present only in auklets,

Atlantic Puffin

Common Murre

2.2　Examples of auk skulls.

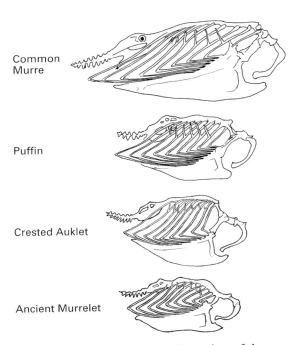

Common Murre

Puffin

Crested Auklet

Ancient Murrelet

2.3　Auk sternums, showing the variety of shapes.

although there is some crop-like development of the lower proventriculus in the puffins. Auklets and Dovekies develop diverticulae in the throat while breeding, in which they transport food for their young.

Most species are highly social while breeding, with a prolonged pre-breeding period and extensive courtship activity in which both vocal and visual displays are prominent. Poor flight ability leads to a paucity of flight displays. Feeding may be solitary or in groups, but is rarely co-operative. Breeding sites are coastal, except in the Marbled and Kittlitz's murrelets, which are also the only solitary nesters, and in the Dovekie, which may breed on mountainsides up to 30 kilometres from the sea.

All species are socially monogamous and the sexes play equal roles in incubation and in rearing chicks to the age when they leave the colony. Nest building is poorly developed or absent, but some species dig burrows and the puffins even line their nests. Most show a strong tendency to return to the colony where they were reared, although this tendency may not be as strong as it seems, because it is much easier to find banded birds on the colony where they were banded than elsewhere.

Clutches consist of one or two eggs and, except in very unusual circumstances, only one brood is reared annually. Eggs are thick-shelled, either elliptical-ovoid or elongate pyramidal and either white, buff, tan, or bluish in ground colour, sometimes with prominent black or brown markings. The membrane inside the shell is very tough at hatching and is often very persistent. Most species, including the puffins, auklets, and Razorbill that lay only a single egg, have two lateral brood patches. The murres and the Dovekie have a single, central patch, set far back on the belly, and the Great Auk probably had only one as well (Fig. 2.4).

Chicks are active within 1–2 days of hatching and capable of thermoregulation either immediately (*Synthliboramphus*), or within a few days (the rest). Hatchlings are covered with a

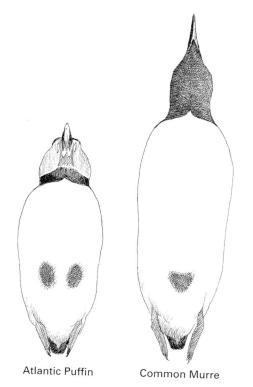

Atlantic Puffin   Common Murre

2.4   Examples of auk brood patches.

dense, woolly, down plumage. All but the *Synthliboramphus* murrelets are semi-precocial and reared at the nest-site for a minimum of 2 weeks. There is no post-fledging parental care, although in the *Synthliboramphus* murrelets, the murres, Razorbill, and Dovekie, the young are cared for until some time after they leave the colony, which they do when only partly grown and incapable of sustained flight.

Temporary nuptial ornaments, including ornamental plumes on the head and deciduous, brightly coloured, horny plates on the bill, are present in several genera. Sexual dimorphism is very slight, with males of most species being slightly larger in some dimensions, especially bill depth. There is no plumage dimorphism and no distinctive immature plumage. First winter birds generally appear similar to winter-plumage adults, although young guillemots are readily distinguishable by their greater mottling.

The systematic position of the Alcidae

Because their anatomy is specialized for underwater swimming, and especially for wing-propulsion, the auks inevitably show striking convergencies with other groups that use the same form of locomotion. This is probably the main reason why, although the limits of the group are very well defined and have been agreed since the earliest classifications, there has been some debate about where to place them in relation to other families of birds.

The family Alcidae was named by Vigors in 1825. Early writers mainly placed the auks alongside other streamlined, dagger-billed, diving fish-eaters, especially the loons, grebes, and penguins. Although the penguins were recognized as belonging to a separate stem fairly early (Huxley 1867), the grouping of the auks with the grebes and loons ('pygopodes') persisted as late as Sclater (1880).

Garrod (1873) recognized that the auks belonged with the gulls and the plover/sandpiper assemblage in the order Charadriiformes, in which he also included pigeons and cranes. In 1892 Gadow, in a general classification of birds, placed the auks alongside the gulls in a suborder, Gaviae, of Charadriiformes. The other suborder, Limicolae, contained plovers, seedsnipes, sandpipers, coursers, thick-knees, sheathbills, and jacanas. Most later authors maintained the arrangement of a gull–auk group within the Charadriiformes (Beddard 1898; Mayr and Amadon 1951). This general grouping is very similar to the recent classifications of Cracraft (1981) and Sibley and Ahlquist (1990). Cracraft gave the alcids the status of a suborder, following the comparative anatomical studies of Hudson *et al.* (1969), whereas Sibley and Ahlquist downgraded them to a subfamily (Alcinae) of the Laridae (Fig. 2.5). Storer (1971) twinned the Alcidae with the extinct Mancallidae in a suborder (Alcae), but most other authors include the flightless mancallines within the same family as the rest of the auks (Olson 1985).

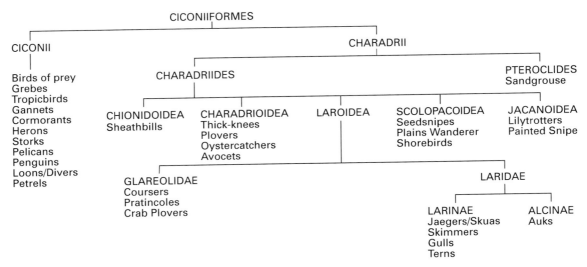

2.5 The auks and related families, according to Sibley and Ahlquist (1990).

Despite the general consensus that the auks belong within the Charadriiformes, there have been dissenting opinions. In particular, Verheyen (1958, 1961) resurrected the grouping of the auks with the penguins, linking them with the petrels (Procellariiformes) in a superorder (Hygronithes). He received some support from work by Gysels and Rabaey (1964) on eye-lens proteins that suggested a closer link between auks and penguins than between auks and other Charadriiformes. However, this grouping has found little support otherwise (Strauch 1978).

Another peculiarity of Verheyen's classification was the incorporation of the diving-petrels (*Pelecanoides*) into the 'Alciformes'. Many authors have recognized the remarkable similarity between the diving petrels of the southern hemisphere and the some of the smaller members of the auk family (Kuroda 1967). The Common Diving Petrel (*P. urinatrix*) is most similar in appearance to Cassin's Auklet, the resemblance extending to the colour of the feet (Fig. 2.6). If they occurred in the same area, one might suspect mimicry. However, the close phylogenetic affinities between the diving petrels and the shearwaters are recognized by practically all other authors (Warham 1990). Their similarities to the small auks are regarded as being the result of convergence and provide a very striking example of how similarity in ecology and behaviour can lead to similarities in external appearance.

Classifications of the Alcidae

The earliest author to attempt a detailed arrangement of the family was Brandt (1837). He recognized two major divisions: (a) the puffins and auklets, and (b) the murres, murrelets, and guillemots, plus the Dovekie, Razorbill, and Great Auk. This division corresponds roughly to one between those species with deep, laterally compressed bills and colourful ornamentation (a) and those with sharp pointed bills and without colourful ornaments (b). Within this scheme, Brandt recognized 8 genera, 10 subgenera, and 20 species (21 in Brandt 1869), with another two possible species (Table 2.1). Coues (1868), who also monographed the family more or less simultaneously in a different continent, tallied 12 genera in three major groups, the divisions being identical with those of Brandt except

Cassin's Auklet

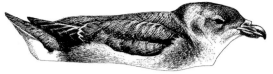
Diving Petrel

2.6   Cassin's Auklet and Common Diving Petrel, showing convergence in external appearance.

Table 2.1   Classification of the auks according to Brandt (1837) (current specific names in parentheses).

Subfamily	Genus	Subgenus	Species
Alcinae	Alca	Plautus	impennis
		Utamania	torda
	Uria	Lomvia	arra (lomvia)
			troile (aalge)
		Grylle	carbo
			grylle
			columba
	Brachyramphus		marmoratus
			kittlitzii (brevirostris)
	Synthliboramphus		antiquus
			temminkii (wumizusume)
			hypoleucus?
	Mergulus		alle
Phalerinae	Ptychoramphus		aleuticus
	Simorhynchus	Tyloramphus	cristatellus
		Phaleris	camatschaticus (pygmaeus)
			pusilla?
		Ombria	psittacula
		Ceratorhina	monocerata
	Lunda	Ceratoblepharum	arctica
			corniculata
		Gymnoblepharum	cirrhata

that the Razorbill and Great Auk were placed in their own subfamily.

The division between the two major groups recognized by Brandt and Coues was not sustained by later authors. The American Ornithologists' Union check-list of 1910 (AOU 1910) listed four subfamilies with 14 genera, while Storer (1960) divided the family into seven tribes with the same 14 genera (Table 2.2). Verheyen (1958), despite his very unorthodox views on the affinities of the Alcidae, arranged them in very much the same way as Storer, except that he sank the *Brachyramphus*

murrelets in the Aethiini and arranged the tribes in five subfamilies. The genera listed by Storer and Verheyen were identical and included all Coues' 12 (albeit with different names in some cases). The most detailed anatomical classification to appear (Strauch 1985) reduced the family to five tribes and 10 genera. Strauch re-established both of the genera split by Storer to the form given by Coues, lumped all the puffins together in a single genus, sank *Cyclorhynchus* into *Aethia*, and recombined the Great Auk with the Razorbill in *Alca* (Table 2.2).

Table 2.2   Some twentieth century classifications of the Alcidae.

AOU 1910		Storer 1960		Strauch 1985	
Subfamily	Genus	Tribe	Genus	Tribe	Genus
Fraterculinae	Fratercula	Fraterculini	Fratercula	Fraterculini	Fratercula
	Lunda		Lunda		Cerorhinca
Aethiinae	Cerorhinca		Cerorhinca		
	Phaleris	Aethiini	Cyclorhynchus	Aethiini	Aethia
	Aethia		Aethia		Ptychoramphus
	Ptychoramphus		Ptychoramphus		
	Synthliboramphus	Synthliboramphini	Synthliboramphus	Brachyramphini	Brachyramphus
	Endomychura		Endomychura	Cepphini	Cepphus
	Brachyramphus	Brachyramphini	Brachyramphus		Synthliboramphus
	Cepphus	Cepphini	Cepphus		
Allinae	Alle	Plautini	Plautus (= Alle)	Alcini	Alle
Alcinae	Plautus	Alcini	Pinguinus		Alca
	(= Pinguinus)				
	Alca		Alca		Uria
	Uria		Uria		

Studies of protein polymorphism by Watada *et al.* (1987) once again suggested that the auks divide into two groups, one consisting of the auklets and puffins, and the other embracing the rest. In their analysis, the Rhinoceros Auklet fell very close to the true puffins. Unfortunately, they considered material from only 12 species. Recent work on mitochondrial DNA and allozymes, including all extant species, mainly supports Strauch's arrangement, but places *Alle* in the same group as *Uria* and *Alca* and maintains separate tribes for *Cepphus* and *Synthliboramphus* (Moum *et al.* 1994; Friesen *et al.* 1996a). Surprisingly, considering their plumage, DNA analysis suggests that the two Pacific Guillemots are more closely related to one another than either is to the Black Guillemot, and that the sympatric Marbled and Kittlitz's murrelets are more closely related than either is to the Long-billed Murrelet, currently considered a sub-species of Marbled Murrelet (Fig. 2.7). Watada *et al.* (1987) and Friesen *et al.* (1996a) considered the puffins (Fraterculini) and auklets (Aethiini) to be sister tribes, more closely related to one another than to the other auks. Otherwise, the molecular data do not indicate any particular associations among the tribes and

suggest rather that they originated from an initial rapid divergence among early members of the family (see below). It is surely of satisfaction to Brandt's ghost to see his original arrangement of the alcids, with the auklets and puffins grouped together, re-established by the latest hi-tech weapons in the armoury of science.

The format of this book dictates that we make some decisions on names and species limits. Current DNA work may, in due couse, allow us to make definitive statements about the course of alcid evolution. However, results obtained so far by different groups of researchers have not always been consistent. No study has clarified the relationships among the major clades (subfamilies). The limitations of DNA sequencing are illustrated by the case of the Aethiini, within which relationships have been rendered less and less clear with incrementally greater amounts of DNA sequenced. It may be some time before we have a classification that is stable and widely accepted. Rather than take sides on matters where we hold no firm opinion, we have chosen to use the classification given in the latest listing of the American Ornithologist's Union (1983). Subfamily names follow Strauch (1985).

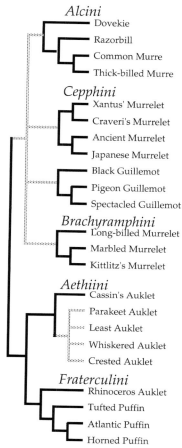

Alcini
Dovekie
Razorbill
Common Murre
Thick-billed Murre

Cepphini
Xantus' Murrelet
Craveri's Murrelet
Ancient Murrelet
Japanese Murrelet
Black Guillemot
Pigeon Guillemot
Spectacled Guillemot

Brachyramphini
Long-billed Murrelet
Marbled Murrelet
Kittlitz's Murrelet

Aethiini
Cassin's Auklet
Parakeet Auklet
Least Auklet
Whiskered Auklet
Crested Auklet

Fraterculini
Rhinoceros Auklet
Tufted Puffin
Atlantic Puffin
Horned Puffin

2.7 Phylogeny of the auks, according to evidence from mitochondrial DNA (redrawn from Friesen *et al.* 1996*a*).

Comparing the various classifications proposed for the auk family, the following points stand out:

(1) the puffins *Fratercula* (including *Lunda*) and the Rhinoceros Auklet *Cerorhinca* form a distinct group (Fraterculini of Storer and Strauch);

(2) the large Atlantic auks, *Alca*, *Pinguinus*, and *Uria*, group together and are frequently linked with *Alle*;

(3) the planktivorous Pacific auklets, *Aethia*, *Cyclorhynchus*, and *Ptychoramphus*, are closely related;

(4) there is general uncertainty about the position of *Cepphus*, *Brachyramphus*, and *Synthliboramphus*, both relative to one another and to other genera. Their comparatively simple, unspecialized bills make them look somewhat alike. However, the tendency for them to be placed together in classifications may be just a convenient lumping of 'all the rest'. The molecular data tend to suggest that they are not closely related.

The fossil record

Because of the general paucity of birds as fossils (compared with mammals, for example) most avian classifications take little account of the fossil record. The auks, being marine and moderately large, have yielded a better fossil record than most groups (fossils preserve well in marine sediments). Although their fossils do not tell us a lot about their phylogeny, they do provide some crucial pieces of evidence. In addition, some fossil assemblages are sufficient to give interesting insights into the evolutionary potential of the family and the type of community structure that occurred in the past. However, it would be unwise to assume that the absence of a particular fossil necessarily implies a gap in distribution at the time.

Possible fossil auks have been identified from Eocene and Oligocene deposits (more than 25 million years ago; Brodkorb 1967; Mlikovsky and Kovar 1987), although there is some uncertainty concerning these assignments (Olson 1985). The first unequivocal records are from the middle Miocene, about 15 million years ago. These earliest fossils come from marine deposits in Maryland and California. On the Atlantic side, the fossils concerned have been assigned to the extinct genus *Miocepphus* (Olson 1985), while in the Pacific they belong to the poorly known genus *Alcodes* (Howard 1968).

By the second half of the Miocene (15–10 million years ago), there are numerous fossils available from sediments associated with both oceans. In the Pacific, these include bones possibly referable to the modern genera *Cepphus*, *Cerorhinca*, *Aethia*, and *Uria* (Howard 1982; Warheit 1992).

These sediments have also yielded remains of *Praemancalla*, a member of the now extinct subfamily Mancallinae, all the species of which were flightless (Howard 1978), and of *Alcodes*, possibly also related to that group, and also incapable of flight (Livezey 1988). Late Miocene (7–4 million years ago) deposits from California have yielded at least five species of the flightless *Mancalla*, probably ranging in size from a little larger than a murre up to about 4 kilograms (Livezey 1988), with at least three and possibly as many as four species coexisting (Howard 1982; Olson 1985). Judging from their skeletons, these birds must have been rather similar in form to the Great Auk, with tiny, penguin-like wings and robust bodies and legs. Livezey (1988) speculated that they must have bred on islands known to have been situated at that period off the coast of southern California. *Mancalla* is the most commonly represented bird in Pliocene marine formations of California (Howard 1982), suggesting that members of the genus were abundant at that time (although they may also have preserved more easily, because of their very large, robust bones). Extinct species of *Cerorhinca*, *Ptychoramphus*, and *Brachyramphus* have also been identified from these deposits.

By the late Pliocene (2–3 million years ago) the California seabird community included two species of *Brachyramphus*, and one each of *Synthliboramphus*, *Ptychoramphus*, and *Cerorhinca*, along with three species of *Mancalla* (Chandler 1990). If we subtract the flightless mancallines, then this community is very similar to today's resident auks, except that *Uria* and *Fratercula* are missing. Their absence is not surprising, as Pliocene climates were, if anything, warmer than those of today and both *Uria* and *Fratercula* reach their southern limit in central California at present.

In the Atlantic, numerous auk bones come from late Miocene and Pliocene sediments in Florida, Maryland, and Virginia. These include the recent genera *Alca*, *Alle*, *Pinguinus*, and *Fratercula*, and several representatives of the extinct *Australca*. Olson (1985) regarded both *Miocepphus* and *Australca* as closest to the recent *Alca–Pinguinus* group. Interestingly, there are no Atlantic fossils of *Cepphus* and the earliest of *Uria* are not found until the late Pleistocene, about 12 000 years ago. However, as in the Pacific, the available sediments come from the southern extremity of the family's present range and both *Uria* and *Cepphus* currently have mainly Arctic distributions. Analysis of mitochondrial DNA suggests that Atlantic and Pacific populations of Thick-billed Murres separated more than 100 000 years ago (Birt *et al.* 1992). Hence, *Uria* must have been present in the Atlantic since at least the mid-Pleistocene. Moreover, Friesen *et al.* (1996*b*) have argued that the two modern murre species diverged about 5 million years ago as a result of separation into Atlantic/Arctic (Thick-billed) and Pacific (Common) populations. In that case, murres must have been present in the Atlantic for much longer than is indicated by the fossil record.

According to Storrs Olson (personal communication) the Lea Creek formation of Maryland, dating from the mid-Pliocene, contains an enormous number of auk bones, along with those of other seabirds, including several species of shearwaters and albatrosses. The remains were apparently accumulated as a result of predation, perhaps by sharks. The seabird and marine mammal community of the time was extremely diverse, including many modern species, and seems to have been boreal in affinities. It contained two puffins, several species of *Alca*, and several of *Australca*, including one species that was almost as large as a Great Auk. It also included species of *Alle* and *Pinguinus*, but no guillemots or murres. This one fauna contained many more species than are currently present in the entire Atlantic.

The specimen of *Australca* that approached the Great Auk in size is of particular interest, because it was apparently capable of flight.

The genus also included at least one other species larger than the modern murres and apparently capable of flight (Olson 1985). In addition, at least one species of *Alca* also approached the Great Auk in size and apparently retained the power of flight (S. Olson, personal communication).

The existence of extinct auks larger than murres but capable of flying contradicts the assumption, frequently made, that the murres are at the maximum size limit for flying wing-propelled divers. In fact, Livezey (1988) pointed out that extrapolating the relationship between wing-loading and body mass among auks suggests that they could reach 2–3 kilograms without exceeding the threshold limit for wing-loading (weight/wing area) among flying birds (2.5 g/cm^2). The very high wing-loading of murres may be related to their enhanced performance in underwater swimming, as demonstrated by their very deep diving capabilities (Chapter 4). The Tufted Puffin, which is as heavy as the smallest populations of Thick-billed Murre, has a wing-loading only half as great, and could presumably be scaled up to create a much larger flying auk than the recent murres, if evolution proceeded in that direction.

The lack of any characteristic Pacific genera from earlier Atlantic faunas shows that the auks of the two oceans evolved largely in isolation. This is in accordance with the fact that the auks do not occur in tropical waters, which might have provided a periodic connection via Central America, or eastwards, via the Tethys Sea. However, the diversity of auks represented in the Pliocene sediments of eastern North America suggests that the current, relatively small, community of auks in the Atlantic is not a historical consequence of the lack of connections with the Pacific. Instead, it is a consequence of extinctions during the Pleistocene, presumably as a result, directly or indirectly, of the ice ages.

Olson (1985), while presenting much new evidence on the history of auks in the Atlantic, concurred with Storer (1960) and Udvardy (1963) that the family probably originated in

the Pacific, a hypothesis also accepted by Bédard (1985). The Atlantic achieved its current size only mid-way through the Cenozoic (about 30 million years ago) and hence the Pacific would have provided a much greater expanse for the evolution of marine birds at the period between the Eocene and Miocene when the auks were differentiating from their charadriiform ancestors. However, by the time that suitable sediments become available, in the mid-Miocene, the fauna of the two oceans had diverged almost completely. Olson suggested that *Fratercula* probably entered the Atlantic via the northern route during a period in the late Miocene when the Bering Sea land bridge was submerged (although there is some uncertainty about this). If so, the similarity in plumage and bill ornamentation between the Atlantic and Horned Puffins suggests either striking convergence, or a very stable phenotype. Recent DNA analysis by V. Friesen (personal communication) and co-workers suggests a more recent differentiation of the Horned and Atlantic Puffins (about 1.5 million years ago) and implies a much more recent colonization of the Atlantic, apparently during the Pleistocene. If that is the case, the Pliocene specimens of puffin found in Atlantic sediments may not be ancestral to the modern Atlantic Puffin. Alternatively, the Horned Puffin may represent a return migration of the Atlantic puffin stock into the Pacific. These movements presumably took place via the Arctic, as there would have been no southern connection between the oceans at the time. The ancestor of the Black Guillemot must have followed a similar trajectory from the Pacific. Evidence from DNA shows that, despite phenotypic similarity, the Black and Pigeon Guillemots are not closely related and that the Pigeon Guillemot is more closely related to its Pacific congener, the Spectacled Guillemot. The divergence of ancestral Black Guillemots' probably took place during the early Pleistocene (Kidd and Friesen 1995).

Bédard (1985), considering both the fossil evidence and the information available on

Tertiary climates, suggested that the auks probably radiated fairly abruptly in the mid–late Oligocene (roughly 25 million years ago). He based this hypothesis on the fact that the North Atlantic grew to significant size only in the Eocene and the observation that the world's oceans underwent a cooling by roughly 7 °C during the Oligocene, presumably enhancing marine productivity. He suggested that the modern genera already present by the late Miocene may have remained unchanged from the original Oligocene radiation.

Bédard's views were speculative, but correspond reasonably well with the observation that many modern genera of marine birds had evolved by the Miocene (Brodkorb 1967). However, the DNA evidence suggests that the original radiation of subfamilies was more recent, perhaps during the Miocene (10–12 million years ago), a position that accords better with the fossil evidence. The rapid differentiation into distinct tribes, followed by a subsequent period of relative stability, is a pattern that appears commonly in the fossil record.

If Bédard's ideas on biogeography are correct, they presuppose an ancestral alcid stock originating in the Pacific, with some elements finding their way into the Atlantic at an early stage, presumably via a southern sea connection, as there was no northern outlet to the Atlantic then. The fact that all the Atlantic, fossil auks (*Miocepphus*, *Australca*, *Alca*, *Pinguinus*, *Alle*) appear to belong to a single lineage, and that none of the Pacific tribes except the Fraterculini are represented as fossils in the Atlantic, suggests that the colonization of the Atlantic took place before some of the present tribes had differentiated. However, if that is the case, it raises the question of why the family radiated into five tribes (auklets, two tribes of murrelets, guillemots, and puffins) plus the mancallines, in the Pacific, but developed no such radiation in the Atlantic. It is worth noting that the cormorants, also, are represented by a much smaller number of species in the North Atlantic than in the North Pacific.

The above scenario corresponds pretty well to evidence of affinities deduced from current phenotypes. The only difference is that, if we accept the current view that murres originated in the Pacific and arrived in the Atlantic no more than a few million years ago, then the separation of the ancestral murre and razorbill must have been at least as far back as the mid-Miocene. In contrast, all recent arrangements of the auks, including the DNA evidence, place *Uria* and *Alca* closer together than they place *Uria* in relation to any Pacific genus. The similarity in their unique chick departure behaviour (Chapter 6) seems too great to be accounted for by convergent evolution and apparently demands a common ancestry.

This contradiction can be resolved if we suppose an Atlantic origin for the murres. The *Uria* specimen from the late Miocene of California suggests that they must have reached the Pacific at about the same time that, according to Olson (1985), the puffins reached the Atlantic. However, the very extensive collection of fossils from the Lea Creek formation does not contain any murre material, suggesting that the genus was absent from the Atlantic in the middle Pliocene. This leaves us with several alternative hypotheses for the evolution of the murres:

(1) murres evolved in the Pacific from stock unrelated to *Alca* and their similarities are the result of convergence;

(2) *Alca*, or an ancestor of *Alca* and *Uria*, originated in the Pacific and the peculiarities of *Alca*/*Uria* biology were already established before *Alca* reached the Atlantic;

(3) *Uria* originated in the Atlantic, reached the Pacific in the Miocene, but then died out in the Atlantic before the period of the Lea Creek fauna. Finally, the genus recolonized from the Pacific during the Pleistocene. It seems likely, on the basis of DNA evidence, that populations of both recent species of *Uria* were in contact via the Arctic Ocean during the Pleistocene (Friesen *et al.* 1996*b*).

The first hypothesis seems unlikely on the basis of morphological, developmental, molecular, and behavioural evidence. Both of the other options seem possible, although the first is simpler; only additional fossil evidence will provide the solution to the problem.

The argument over the ocean of origin of the murres has implications for the location of the ancestral auks. If we consider that *Uria* originated in the Pacific, then all tribes are represented there from their origin and the auks could only have arisen in the Pacific. However, if *Uria* originated in the Atlantic, then all members of the *Miocepphus–Australca–Alca–Pinguinus–Alle* group (Alcini of Strauch) are Atlantic in origin and the tribe could have diverged from an ancestral alcid stock in either ocean. Such a scenario would make it likely that the Alcini diverged from the ancestral stock earlier than other recent tribes, but the DNA evidence gives no indication of this.

A scenario in which the Atlantic auks originated from an undifferentiated ancestral stock, prior to the evolution of other tribes, suggests that the subfamily status accorded to *Mancalla* cannot correspond to an earlier date of differentiation. Hence, *Mancalla* must necessarily be more closely related to the other Pacific auks than to those of the Atlantic (Olson 1985). In fact, if we accept the idea that the Alcini originated before the other tribes, and if we use timing of divergence as our criterion, it would appear that the Alcini has the best claim to subfamily status.

As all the known fossils of the *Alcini* were larger than modern Dovekies, it seems likely that the Dovekie evolved from an ancestor substantially larger than itself. The Dovekie has a typical semi-precocial chick, similar to those of the auklets and puffins. The species could have evolved from the ancestor of the Alcini before the development of the departure behaviour characteristic of modern murres and Razorbill, or it could have reverted to a more typical, semi-precocial development after diverging from the rest of the tribe. Dovekies carry food to their young in a throat pouch, like auklets, and this feature presumably evolved in

parallel in the two groups. However, unlike the auklets, male Dovekie parents accompany their chicks to sea and supply post-fledging parental care (Bradstreet 1982*b*; J.R.E. Taylor, *in litt.*). The reason why such post-fledging care is necessary is not obvious, but its occurrence suggests that such behaviour may have been characteristic of the Alcini from early in their evolution. If so, it was probably also found in the Great Auk and would have been an enabling factor in the evolution of very premature colony departure (see species account).

The origins of the auks needs to be viewed against a background of other potential vertebrate competitors. In the Pacific during the late Oligocene and early Miocene, penguin-like, flightless diving birds of the Pelecaniform family Plotopteridae occurred (Olson and Hasegawa 1979). The largest of these was bigger than any living penguin. It may be no coincidence that both they, and many large fossil penguins of the southern hemisphere, died out at about the time that seals appeared, in the mid-Miocene (Simpson 1976). The Plotopterids must have coexisted with early auks, although their fossils have not been found in association. Their niche may later have been partially filled by the smaller *Mancalla* species.

Warheit and Lindberg (1988) have pointed to the frequent sea-level fluctuations of the Pleistocene, allied to competition with pinnipeds for potential breeding sites, as likely causes for the extinction of *Mancalla*, which did not persist long enough to have encountered early human immigrants into North America. We know that seals and sealions can have dramatic effects on seabirds breeding on low-lying islands. In the Pribilof Islands of the Bering Sea, a colony of hundreds of thousands of murres was eliminated by the encroachment of Northern Sealions (*Eumetopias jubatus*) within about 20 years (Hatch 1993).

The genus *Mancalla* has been reported only from California. Its diversity, and that of other auks in California, at a time when global temperatures were probably higher than those currently typical of the area, suggests that the seasonal cold water upwellings of the California Current system probably occurred then, as they do today. If not, then early auk communities were apparently capable of exploiting much warmer oceans than those that they occupy now (Warheit 1992).

Functional adaptations

Wing-propelled diving

Stettenheim (1959), who investigated the anatomical adaptations of the auks for diving, pointed to a number of distinctions from related Charadriiformes. In the auks the covert feathers are stiffer and more extensive, both above and below the wing, than in gulls or shorebirds, a character that is shared with the diving petrels *Pelecanoides*. This arrangement reinforces the trailing edge of the wing and closes gaps between adjacent flight feathers, making the wing more effective as an underwater paddle. Moreover, in the auks, 'primaries 6–10 form a unit in which little independent movement is possible'. This rigidity is caused by a greater development of connective tissue than in other Charadriiformes.

The above features are specific to wing-propelled swimming. On the other hand, features such as the fusiform body, with the feet set far back, are common to all swimming birds, including those such as cormorants, grebes, and ducks, that are foot propelled. Other characters that are common to other underwater swimmers, but absent in Charadriiformes that do not dive, are the presence of strongly developed vertebral hypophyses on the last cervical vertebra, and the enlarged number of thoracic vertebrae (8–10 compared with 5–7 in other Charadriiformes). An increase in the number of vertebrae allows for a longer body, while retaining flexibility of movement, and is also found in loons, grebes, and cormorants.

The humerus in the auks is an especially robust bone. In the flightless Great Auk and in the extinct mancallines, it became even more developed, with a distinctive oval cross section, which is developed to a lesser extent

in the surviving members of the family (Fig. 2.8). The very robust development of the humerus was presumably a response to the considerable force applied to it in swimming. Even in murres, the largest flying auks, the humerus can apply surprising force, as anyone who has had their fingers battered by the bend of a murre's wing can attest.

The articulating surface of the humerus is much larger in auks than in other Charadriiformes, but similar in relative size to that of diving petrels. This enlargement allows greater force to be transmitted in swimming. However, like other flying birds, but unlike penguins, the auks retain considerable flexibility of movement at the articulation of the humerus (Hudson *et al.* 1969), apparently an inevitable constraint imposed by the needs of flight. There are no sesamoid bones in the wing articulations.

In comparison with gulls and shorebirds, auks have short wings and legs. The relative length of the humerus is also reduced: 'The forearm resembles the handle of an oar, the blade of which is the carpometacarpus' (Kaftanovskii in Kozlova 1957).

The main difference in flight muscles is the relative size of the supracoracoideus (2–3 per

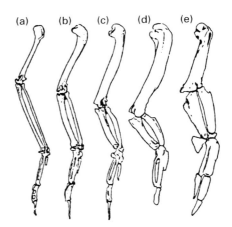

2.8   Comparison of humerus and associated bones in different genera of flying and non-flying seabirds: (a) gull (*Larus*), (b) Razorbill (*Alca*), (c) Great Auk (*Pinguinus*), (d) Lucas Auk (*Mancalla*), and (e) penguin (*Spheniscus*).

cent of body mass versus a maximum of 1.5 per cent in shorebirds and gulls). This muscle raises the wing during the upstroke. The pectoralis muscles, which are the main source of power on the downstroke, are similar in relative size to those of gulls and shorebirds, but more elongated, with the supracoracoideus lying directly below them. This arrangements appears to be dictated by the need for streamlining. The need to keep the cross-sectional area of the body as small as possible may also account for the elongated shape of eggs laid by many species, especially those such as *Synthliboramphus* species, which lay especially large eggs.

The enlargement of the supracoracoideus points to the effort imposed by the upstroke in water, compared with air, where the reduced air pressure above the wing created by the aerofoil in level flight reduces the power needed to raise it. The wing section of auks is less cambered than that of other birds, presumably to reduce the lift engendered by the properties of a hydrofoil (Rayner 1985). The same effect may put a premium on vertical, rather than oblique, diving. In a vertical descent, any lift developed by the hydrofoil would be neutral in relation to overcoming buoyancy. However, buoyancy may not be a problem with deep-diving auks, such as murres. The compression of their contour feathers with depth is likely to make their overall body density greater than that of water below 50 metres (Lovvorn and Jones 1991), which means that during their ascent they cannot rely on buoyancy to rise and must actively swim upwards.

According to Rayner (1985), we still do not know whether the upstroke of an auk delivers any forward propulsion in water, or whether all the propulsion comes on the downstroke. The very jerky forward movement of murres underwater, with a rapid acceleration at each downstroke, suggests that forward propulsion on the upstroke, if any, is relatively small. The leading edge of the primaries is very curved on the downstroke, presumably because of the resistance of the water. The primary feathers are actually very stiff, compared with those of most birds of their size. Like other wing-propelled diving birds, auks do not initiate submergence by jumping upwards before diving head first (Wilson *et al.* 1992).

Swimming by means of paddles formed from modified fore-limbs is not common among aquatic animals. Apart from wing-propelled diving birds, the main exponents are the turtles. Sealions use their fore-limbs as paddles, but the main propulsion comes, as in seals, from the hind limbs, acting essentially as fins. Anatomy suggests that the extinct plesiosaur dinosaurs used flippers for propulsion (Rayner 1985). When extended, the wings of those auks that can fly are relatively much larger than the paddles of other paddle-propelled swimmers. To convert them to a more suitable size for swimming, auks hold them sharply bent at the wrist, reducing the functional surface area (Fig. 2.9). The effective area of the wing in underwater swimming is slightly less than half that when spread in flight (Jouventin and Mougin 1981). The fact that the normal wing area is unnecessary for effective underwater propulsion is emphasized by the fact that most extant auks (all except the auklets) lose all their flight feathers simultaneously during the annual post-breeding moult, becoming flightless for a month or more. Clearly, they are able to feed themselves effectively while swimming with the much smaller area presented by the covert feathers.

Flightlessness

Livezey (1988) examined the circumstances associated with flightlessness in auks and other groups. It is a truism to say that flightlessness evolves when the advantages of flying become outweighed by the improvements to swimming or diving performance that can be achieved by forgoing flight. As we noted above, there is no evidence that a greater body weight than that of recent murres will inevitably lead to flightlessness, but it is striking that the smallest flightless seabird, the Little Penguin (*Eudyptula minor*), is about the size of the largest recent

2.9   Common Murre wing during the downstroke (a) in flight, (b) underwater.

flying auk and that all the flightless auks that are known were larger than murres. It appears that there is a sharply defined size threshold, above which flightlessness becomes an attractive adaptive option.

Comparing the wing skeleton of the Great Auk with those of the mancallines (Fig. 2.8) suggests that the latter had travelled much further down the road of specialization to flightlessness. A major feature of this process, aside from overall shortening of the wing elements, is a reduction in the size of the distal parts (ulna, carpo-metacarpus) relative to the humerus. In flying auks, the ratio of ulna to humerus length is about 0.8, whereas in three species of *Mancalla* it was 0.40–0.42. The ratio for the Great Auk, at 0.54, was intermediate between these extremes and close to that of *Praemancalla* (0.52), a possible ancestor of *Mancalla*. Likewise, the ratio of carpo-metacarpus to ulna length, which is about 0.67 in flying auks, and ranged from 1.19 to 1.22 in *Mancalla*, in the Great Auk was 0.75, not far from that of flying auks (data from Livezey 1988). These features suggest that the Great Auk had not progressed far in adapting to flightlessness, which fits well with the fact that in many respects in appeared to have diverged little from the Razorbill.

Insulation and thermoregulation

The need for streamlining has had an effect on the plumage, as well as the muscular anatomy, of alcids. The contour feathers, heavily interspersed with down, are short, rigid and very densely packed, compared with those of gulls, especially those on the belly. The effect produced is that of a dense pile carpet. Streamlining is further aided, and insulation may also be enhanced, by the fat stored in a thin subcutaneous layer over most of the body, rather than in discrete pockets, as occurs in many birds.

Brood patches are a potential site of heat loss, especially during deep diving, when birds may be subject to several times normal atmospheric pressure. The patches defeather rapidly at the start of incubation and refeather as soon as hatching occurs, sometimes starting before the end of incubation. The patches are small relative to the size of the eggs, so that during incubation only part of the egg's surface is in contact with the patch. Non-breeders either do not form brood patches, or develop only partial patches. The brood patches are often difficult to locate, being completely covered by the stiff anterior feathers which seal them very effectively: presumably a necessary precaution to prevent the ingress of water. In these characters, the auks appear to be much more conservative in the amount of insulation that they shed than most other marine birds, even those such as petrels that lay only a single egg.

The legs, dangling in cold water for most of the birds' life, are particularly vulnerable to heat loss, which is four times as rapid in water as it is in air. Like those of most aquatic and wading birds, they are equipped with heat exchange mechanisms (rete mirabile) in the blood vessels, which reduce heat loss (Calder and King 1974). This arrangement leads to the thermodynamic description of a young Ancient Murrelet as 'a warm cup-cake with two icicles attached' (Gaston 1992a).

Notwithstanding the adaptations described above, the insulation of the auks appears to be relatively poor, considering the climates in which they live. To maintain the typical, high, avian body temperature, they rely principally on a very high rate of metabolism (Johnson and West 1975). Hence, it is not surprising that Basal Metabolic Rates (the metabolic rate of resting, non-digesting birds) among auks are exceptionally high, compared with other birds of the same size (Birt-Friensen et al. 1989; Konarzewski et al. 1993). The auks appears to have made a compromise between the need for insulation (promoted by long feathers with air-spaces), and the need for streamlining and low buoyancy, selecting for shorter feathers with less opportunity for trapping air.

Diving

The auks have developed many physiological adaptations for prolonged diving. Most are common to a range of diving birds. As the auks are the only members of the Charadriiformes to undertake prolonged activity underwater, they probably evolved these adaptations independently from other groups such as ducks, cormorants, and penguins. The modifications include a high blood volume (12 per cent of body mass in the Thick-billed Murre), and high levels of myoglobin (more than 1 per cent body mass), both of which enhance the bird's ability to store oxygen (Croll et al. 1992). Despite these enhancements, many long dives undertaken by Thick-billed Murres exceed the limit for aerobic respiration (about 48 sec), requiring the birds to respire anaerobically for a portion of the time underwater, possibly restricting heart rate and peripheral blood flow at the same time. At present, we cannot account for how murres, and presumably other auks spending long periods underwater, manage to do so persistently without building up an excessive amount of lactic acid. In addition, we do not know how auks cope with the potential collapse of their lungs under the pressure of deep dives and the possibility of nitrogen bubble formation in the bloodstream ('the bends'). There is plenty of room for further research on the physiology of auks.

Adaptations for prey capture

The generic diversity of the auks is based on considerable variation in anatomy. Differences in bill morphology are most striking, at first glance, but there is also considerable variation in skeletal and muscular anatomy and this appears to be mainly related to variation in the demands of locomotion on land, in the air, and underwater.

Because of the demands of wing-propelled swimming, the auks have very small wings relative to their weight (Fig. 2.10). The relationship of wing length to body weight is similar to that of cormorants. However, auks have a much higher weight to wing area ratio, because their wings are narrower (Jouventin and Mougin 1981). Consequently, auks in flight are entirely dependent on flapping to remain aloft and this puts heavy demands on the pectoralis muscles, which are supported by an elongated attachment to the sternum.

Within the *Uria–Alca–Pinguinus–Alle* group (Alcini), differentiation involves mainly body size and bill morphology, not plumage. This suggests an adaptive radiation in response to

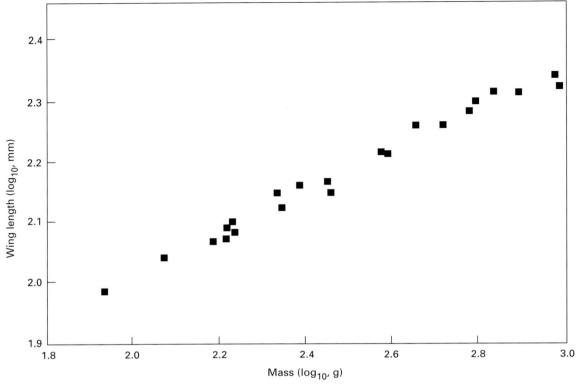

2.10   Wing length of auks in relation to body weight.

the exploitation of a range of different prey. Such a radiation would most likely have occurred soon after the early auks reached the Atlantic because there is no fossil evidence in that ocean for competition from other wing-propelled marine divers (Bédard 1985).

Bédard (1969a) drew attention to parallels between bill morphology and the structure of the mouth lining and tongue and the diet of auks. Those that feed mainly on fish tend to have narrower, more pointed bills (low width/gape ratio), and narrower, more cornified, tongues that those that feed on plankton. The plankton eaters, apart from having broader, shorter bills, have large numbers of fleshy denticles on the palate and the upper surface of the tongue. The convergence of these characters for auks feeding exclusively on plankton is dramatically demonstrated by the parallels between the auklets and the Dovekie. These

adaptations are also shown within the genus *Uria*, with the predominantly fish-eating Common Murre having a narrower tongue and fewer palatal denticles than the Thick-billed Murre which feeds on a greater variety of prey (Spring 1971). Bill depth seems to be unrelated to diet, suggesting that the very deep bills of puffins and auklets, highly decorated during breeding, may have evolved for secondary sexual purposes, rather than as feeding adaptations.

At the time of Bédard's work, there was relatively little information available on diets for many auks, so his conclusions were necessarily speculative. Hobson *et al.* (1995) have examined the diet of auks by stable isotope analysis (Chapter 4). A comparison of their findings with Bédard's gives some support to his generalization (Fig. 2.11): birds with high width/gape ratios feed at low trophic levels (i.e. take plankton rather than fish), and those with

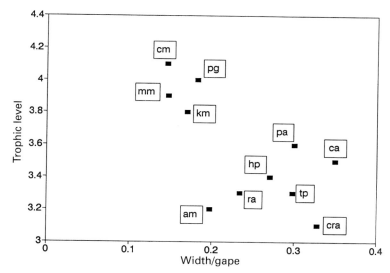

2.11   Trophic feeding level (from Hobson *et al.* 1994) in relation to bill width/gape ratio (data from Bédard 1969*a*). Species indicated: cm = Common Murre, pg = Pigeon Guillemot, mm = Marbled Murrelet, km = Kittlitz's Murrelet, am = Ancient Murrelet, ca = Cassin's Auklet, pa = Parakeet Auklet, cra = Crested Auklet, ra = Rhinoceros Auklet, hp = Horned Puffin, tp = Tufted Puffin.

lower ratios are mainly fish feeders. Only the Ancient Murrelet, feeding largely on plankton according to Hobson *et al.* (1994), falls well away from the general trend, having a narrower bill than would be expected with its diet.

As a rule, we should expect an increase in the range of prey size taken as birds get larger (Cohen *et al.* 1993). Because they swallow prey whole, small auks face an upper limit to the prey they eat: all auks take prey up to the size of euphausiids (shrimp-like crustacea up to about 25 mm long and weighing less than 1 g), but this is as much as Least and Whiskered Auklets can manage, whereas murres may take prey 10 times as long and 100 times as heavy. Consequently, size alone appears to dictate a division between planktivorous and potentially piscivorous auks. If we want to look for specialization in feeding apparatus, we need to compare auks of a similar size.

There is little to suggest any specialization within the family in gape length. When plotted against body mass ($\log_{10}$ transformed to create a straight line relationship) we find only slight variation about the general trend (Fig. 2.12). When we examine bill width, however, two features are apparent. First, the slope of the relationship is not as steep as that for gape length, showing that longer bills tend to be narrower. This accords with Bédard's idea that narrower bills are adapted to taking fish, because the larger auks, by virtue of their size alone, are in a position to take more fish than smaller auks. Second, auks with body weights in the range 150–250 grams show considerable variation in their bill width, with murrelets of the genera *Synthliboramphus* and *Brachyramphus* having much narrower bills than their body weights would imply. The Dovekie and Cassin's Auklet, both planktivores and similar in body size, have much broader bills. If we treat any auks as specialized in their feeding apparatus, it should apparently be the murrelets, which seem to be more adapted to a fish diet than we might expect from their body size.

Although we can discern patterns in the bill morphology of the auks that seem to relate to differences in diet, it is not obvious why a long,

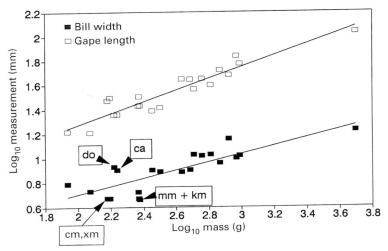

2.12   Gape length and bill width in relation to adult body mass. The straight lines are least-squares regressions for all points. Species indicated: ca = Cassin's Auklet, cm = Craveri's Murrelet, do = Dovekie, km = Kittlitz's Murrelet, mm = Marbled Murrelet, xm = Xantus' Murrelet.

pointed bill should be more efficient for handling fish and a short, stout one for plankton. However, the dagger-like bill of the Common Murre is convergent with those of many other piscivores: loons, grebes, herons, and darters. Likewise, the broad bills of auklets seem to parallel those of certain petrels, such as prions (*Pachyptila*), and diving-petrels, that also specialize in crustacea. We infer that such similarities are adaptive.

Evolution and adaptation

The reality of evolution through natural selection is irrefutable, but assuming that everything is the product of natural selection is not the same as claiming that everything must be perfectly adapted to the task it is carrying out. The auks' wings are an excellent example of this. They are a compromise between the optimal design for underwater swimming and the minimum required for the sort of flight that each species requires in its particular lifestyle. Their small size imposes a cost in terms of flight efficiency. Pennycuick (1987) estimated that a Razorbill requires 2.1 times as much power for flight as a petrel of similar

size, while flying only 28 per cent faster. Hence the Razorbill requires 64 per cent more energy than the petrel to travel the same distance. Moreover, small wings dictate a high gliding speed, making landing rather awkward and requiring the murres to use a 'ballistic' technique that involves a high speed approach, terminating in a steep climb that reaches its apogee at the breeding ledge, allowing the bird, if it has judged things correctly, to touch down lightly. Young birds visiting the colony for the first time spend hours repeating such approach flights, shearing off at the last moment, or touching briefly before falling off. Their constant repetitions suggest that getting the manoeuvre right requires a lot of practice. Thick-billed Murres could never breed in boulder screes away from the coast, because they would never be capable of taking off if they slipped off the top of a boulder. Puffins retain the ability to run about easily on the grassy slopes where they nest, but the price they pay is poorer underwater performance, preventing them from diving to the depths achieved by murres.

Everyone knows that ecology is the science of finding out how everything is connected to everything else. This aspect sometimes gets forgotten when we are dealing with adaptation. Ultimately, if we know exactly how selection operates, we may find that every species attains the ideal compromise for its niche, but we are a long way from such omniscience. Instead, we compose crude stories, invoking one, or at most two, interacting causes to account for patterns of differentiation. In reality, there must be many causes and many effects, the impermanence of the environment creating a fluctuating field of selection pressures, to the tune of which the species' population performs a constant dance. Consider the beak. It must capture prey for the adult organism, but must also be adequate for handling small prey for the chick, it must be adapted for the summer as well as the winter diet, robust enough for fighting, delicate enough for preening feathers, and in some cases it doubles as a signalling device in courtship display. It is hardly surprising that no single generalization, based on the very fragmentary natural history information that we have, is likely to yield an explanation of more than a small proportion of the variation that occurs. It is worth bearing this in mind as we explore feeding adaptations (Chapter 4) and social behaviour (Chapter 5).

3

Distribution and biogeography

Biogeography

The dispersal and evolution of the auks appears to have been constrained by two major barriers: the tropics and the frozen Polar Ocean. Of the two, the tropics seem to have presented the more important obstacle, as they have for the penguins: auks and penguins are the only two seabird families that have failed to colonize both hemispheres.

At present the auks are mainly concentrated in low and high Arctic waters (Fig. 3.1). Only the murrelets of the genus *Synthliboramphus* are found in sub-tropical waters, and a few Cassin's and Rhinoceros Auklets edge into the sub-tropics at the southern extremity of their range. Even in winter, few auks occur outside boreal and low arctic waters. Two of the most southerly breeding auks (Japanese and Xantus' Murrelets) actually move northwards outside the breeding season, so that much of their winter range is to the north of their breeding range.

We should not pass over this avoidance of sub-tropical waters too lightly, because not all seabirds that breed in sub-Arctic and Arctic waters behave in this way. Northern Gannet (*Morus bassanus*), terns (*Sterna* spp.), jaegers (*Stercorarius* spp.), and some gulls, move to tropical and sub-tropical waters in winter, or travel further south to spend the non-breeding period in the austral summer (Manx Shearwaters, *Puffinus puffinus*, Arctic Tern, *Sterna paradisea*). Among southern hemisphere breeders, many travel the other way, to spend the austral winter in boreal latitudes (several shearwaters, *Puffinus* spp., Wilson's Storm-Petrel, *Oceanites oceanicus*, South Polar Skua, *Catharacta maccormicki*).

. Presumably, auks do not winter in tropical waters because their feeding adaptations are specific to cold marine ecosystems. Cairns *et al.* (1991) pointed out that burst swimming speeds of most fishes double between water temperatures of 5° and 15 °C. If the swimming speeds of the warm-blooded auks remain unchanged, underwater pursuit of fish must become more difficult as water temperature increases. At the same time, the relative success of large predatory fish is likely to increase, intensifying competition for the auks, and perhaps increasing the risk of predation from them.

The reason that auks never developed a trans-tropical migration may lie in their general migration strategy. Unlike many species of land birds and waders, they do not exhibit any 'leapfrog' migration, in which northern populations pass over more sedentary southern populations to winter further south. In general, those species and populations that breed furthest north also winter furthest north. Consequently, few auks engage in any systematic long-distance movements, most migrations consisting of a gradual southerly drift. The main exceptions to this generalization are Thick-billed Murres breeding in the Barents Sea and Baffin Bay, many of which move several thousand kilometres to winter off south-west Greenland and Newfoundland, respectively, and those species breeding in

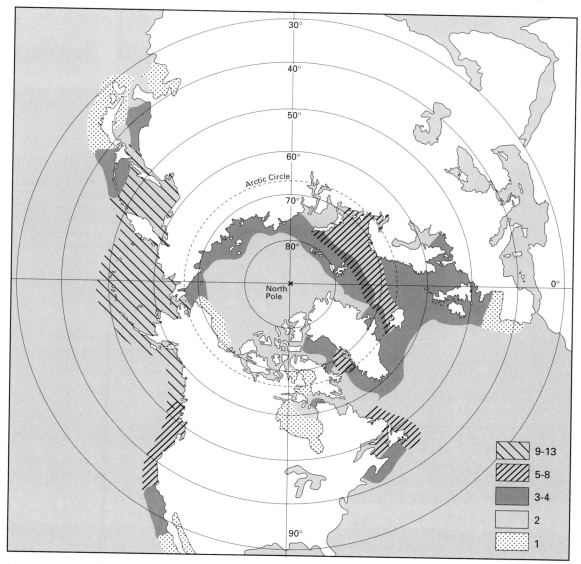

⧄⧄	9-13
▨▨	5-8
▨	3-4
☐	2
⦂⦂	1

3.1   Numbers of auk species in relation to latitude in different parts of the northern hemisphere.

Alaska that winter in cold-current waters off California.

Udvardy (1963) classified the distributions of the Pacific auks according to water type, and whether or not they occurred on both sides of the ocean. He pointed out the remarkable symmetry of some species' distributions, with the Marbled Murrelet and Rhinoceros Auklet occurring in boreal waters on both

sides of the Pacific, but with a major gap in the Bering Sea. We have extended and modified Udvardy's scheme to classify the entire family (Table 3.1).

Only three genera, *Alca*, *Pinguinus*, and *Alle*, all represented by a single recent species, are unique to the Atlantic. A few Dovekies are found in the northern Bering Sea, but breeding has yet to be proven and this seems almost

Table 3.1   Zoogeographical classification of the auks (modified from Udvardy 1963).

Range	Climate zone	Species
Pan-Atlantic	High Arctic Low Arctic[1]	*Alle alle* *Alca torda* *Pinguinus impennis*
	High/Low Arctic	*Fratercula arctica* *Cepphus grylle*
Holarctic	High/Low Arctic Low Arctic	*Uria lomvia* *Uria aalge*
Pan-Pacific	Low Arctic	*Cepphus columba* *Fratercula corniculata* *Fratercula cirrhata* *Brachyramphus brevirostris*
	Boreal[2]/Low Arctic	*Brachyramphus marmoratus* *Synthliboramphus antiquus* *Cerorhinca monocerata*
West Pacific	Low Arctic	*Aethia pygmaea* *Cepphus carbo*
	Boreal/Sub-tropical	*Synthliboramphus wumizusume*
W. Pacific/Beringian	Low Arctic	*Cyclorhynchus psittacula* *Aethia cristatella* *Aethia pusilla*
East Pacific	Boreal/Low Arctic Boreal/Sub-tropical Sub-tropical	*Ptychoramphus aleuticus* *Synthliboramphus hypoleucus* *Synthliboramphus craveri*

[1]   We use low Arctic, following Ashmole (1971), rather than sub-Arctic (Salomonsen 1944), or sub-boreal (Udvardy 1963).
[2]   Boreal refers to climates between low Arctic and sub-tropical.

certain to be a recent range expansion (Bédard 1966). Some fossil material suggests that *Alca* may also have occurred in the Pacific at one time.

There are six endemic genera in the Pacific, of which three (*Ptychoramphus*, *Cyclorhynchus*, and *Cerorhinca*) consist of only a single species (*Cyclorhynchus* is lumped with *Aethia* by some authors). The remaining genera, *Uria*, *Cepphus*, and *Fratercula*, are found in both oceans, but only the two species of murre are circumpolar, the other two genera being represented by different species in the Pacific and Atlantic (Black Guillemots just reach the Chukchi Sea). If we exclude the Chukchi and Barents Seas, then only the Thick-billed Murre, Black Guillemot, Dovekie, and Atlantic Puffin occur in significant numbers in the Arctic Ocean. These four species include the three genera common to both Pacific and Atlantic Oceans and hence presumably best adapted to dispersal via the Arctic Ocean.

Diversity within genera

The auks exhibit much greater variation in morphology, behaviour, and development than other seabird families. This is reflected in their high generic diversity (Fig. 3.2). Half of the extant genera of auk include only one species, according to the classification that we have adopted. This is fairly typical for seabird families. Less typical is the absence of any of those large genera, such as *Larus*, *Sterna*, *Pterodroma*, *Diomedea*, *Oceanodroma*, *Pelecanus*, *Sula*, or *Phalacrocorax*, that push up the average numbers of species per genus for other seabird families.

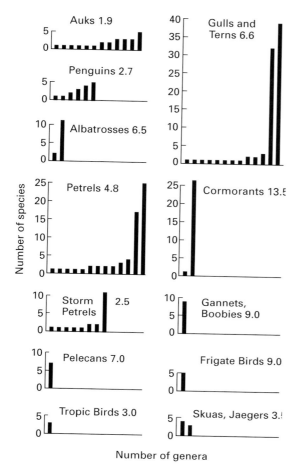

3.2 Numbers of species per genus among seabird families.

populations to develop. The enormous contractions and expansions of range that must have occurred in response to the temperature and sea-level changes of the Pleistocene would have kept populations on the move, and probably reduced the opportunity for speciation, except in so far as it restricted movement between Atlantic and Pacific oceans. In the Atlantic these changes may have brought about the extinction of many species, causing a severe reduction in the diversity of the genus *Alca*, which contained several coexisting species in the Pliocene (Chapter 2).

Pacific distributions

The distributions of most auks in the Pacific, and the size of their populations, must have been much affected by Pleistocene events. The periodic drying out of parts of Beringia would have denied to Kittlitz's Murrelet and the *Aethia* auklets a major part of their present range. In addition, the cutting off of cold Arctic waters would have greatly altered the oceanography of the North Pacific, probably with a resulting loss of productivity. During such periods, populations of auks in the North Pacific probably would have been much reduced. However, Pacific auks must have benefited from the presence of two lengthy island chains (Aleutians, Kurils) in deep water, that would have provided breeding habitat throughout the Pleistocene. The generally westward centre of gravity of the *Aethia* auklets suggests that the Kurils may have been a particularly important refuge. The lack of morphological differentiation between eastern and western populations of Rhinoceros Auklet and Ancient Murrelet suggests that Pleistocene events did not result in a prolonged fragmentation of populations, although this needs to be investigated by genetic analysis. On the other hand, the existence of Asian and North American species of *Cepphus* and *Brachyramphus* (assuming *longirostris* to be a good species), suggests that these genera were split into isolated populations at some time during the Pleistocene.

There may be several reasons for the lack of large genera among auks. Their world oceanic range is smaller than that of any other seabird family, partly because few occur in waters warmer than boreal and partly because land and permanently ice-covered sea cover a much greater proportion of the northern hemisphere than they do in the south. Moreover, their restriction, for the most part, to continental shelf waters, and the lack of oceanic islands in the northern oceans compared with those of the southern hemisphere, both reduce the opportunity for isolated

Atlantic distributions

In the Atlantic, ice sheets and persistent sea-ice cover at the height of the Pleistocene glaciations would have denied to Dovekies and Thick-billed Murres most of their current range. The almost complete absence of cliffs and offshore islands on the east coast of North America south of the Gulf of Maine meant that Razorbills, murres, and Atlantic Puffins had only limited possibilities for retreat when their breeding areas from Newfoundland and the Gulf of St Lawrence northwards were made unavailable. However, on the European side, the cliffs of the English Channel coast, Brittany, northern Spain, and the Channel Islands would have provided suitable retreats. Rockall Bank, which is situated 7° to the west of Scotland and presently emerges only as rock pinnacle barely emerging above the spray zone, would have been exposed at the lowest periods of sea level as a substantial island. It could have been an important refuge area for some Atlantic auks (de Wijs 1978).

Variation within the Atlantic species shows some interesting geographical congruences. Puffins and Common Murres are larger in the Arctic than at lower latitudes (Fig. 3.3), as might be anticipated on the basis of the thermoregulatory advantage conferred by larger body size. Black Guillemots show a similar, but less striking, trend. More surprising is the fact that differentiation within the eastern Atlantic is generally greater than between east and west. In the Puffin, Common Murre, Razorbill, and Black Guillemot the populations in the Canadian maritime provinces, the Gulf of St Lawrence, and the eastern Arctic belong to a single subspecies that is also represented in the eastern Atlantic. By contrast, all these species are represented in the eastern Atlantic by more than one subspecies.

If we assume that recognized subspecies represent populations with a common ancestry, then the present distribution of subspecies suggests that the sub-Arctic of the western Atlantic was colonized relatively recently from the east. This hypothesis fits the geological evi-

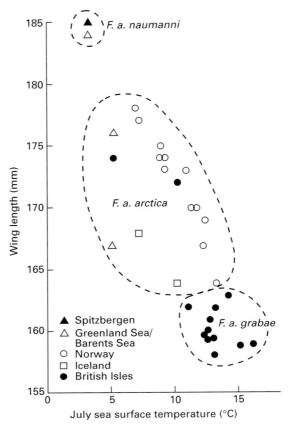

3.3 Wing length of Atlantic Puffin populations in the eastern Atlantic, in relation to sea surface temperatures at their breeding sites.

dence that suggests there were few suitable breeding sites for auks in the western Atlantic at the height of the Pleistocene glaciation. Reconstructions of Pleistocene climates and oceanography also suggest that there was a very steep gradient in ocean temperatures in the western Atlantic between Arctic and tropical waters, leaving little marine habitat for seabirds adapted to cold waters (Dawson 1992).

Another plausible hypothesis is that, in the eastern Atlantic, where temperate and sub-Arctic waters extend over a much greater area than in the west through the warming effects of the Gulf Stream Drift, evolution from an originally pan-Atlantic stock has proceeded more rapidly than in North America. In the eastern Atlantic, Common Murre, Razorbill, and Atlantic Puffin breed over a much broader range of latitude than in the west (Fig. 3.4) and large numbers of auks winter from the Bay of Biscay, at 45° N, to as far north as the Barents Sea, at 75° N. In the western Atlantic, only a few auks winter north of 53° N, so that most populations are to some extent migratory. That populations of murres, Razorbills, Black Guillemots, and Atlantic Puffins breeding in Iceland, Norway, Britain, and the Faeroes can remain essentially sedentary may have allowed the evolution of local genotypes.

The hypothesis that differentiation has occurred mainly since the Pleistocene, facilitated by the existence of sedentary populations, is supported by the fact that the Black Guillemot, the only species to remain partially sedentary in the eastern Canadian Arctic, is the only one to be represented by a distinct subspecies in that area (*Cepphus grylle ultimus*). Whether the pattern of subspeciation exhibited by the Atlantic auks can be attributed to recent colonization of the western Atlantic from the east, or to the larger range of sedentary populations in the east, can only be resolved through detailed genetic studies to establish

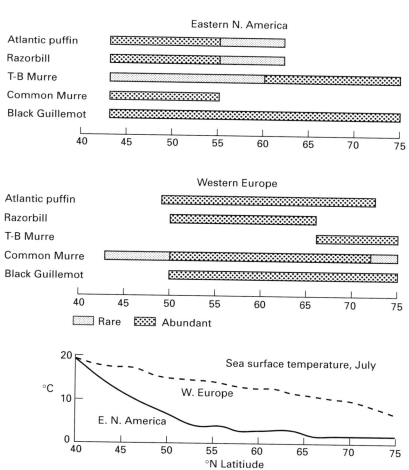

3.4   Latitude range of auks and sea-surface temperatures in July in the eastern and western Atlantic.

whether western Atlantic populations are most closely related to members of the same subspecies in the east, or equally related to all the eastern races.

Much of the variation in the eastern Atlantic is clinal and only weakly differentiated, suggesting a graduated response to broad environmental selection. In contrast, both Black Guillemots and Razorbills have distinctive races in the Baltic Sea, with a sharp change in plumage and measurements between them and their conspecifics in the North Sea. This break corresponds to a rather sharp change in the marine environment between the cold, land-locked Baltic Sea, with its low salinity and heavy winter ice cover, and the much more oceanic and temperate North Sea.

An example of clinal change in gene frequencies that can be observed without recourse to the paraphernalia of DNA analysis is the 'bridled' form of the Common Murre (see species account). This plumage variant occurs almost throughout the Atlantic, but is not found in the Pacific. In the eastern Atlantic it is absent in the small Iberian population and increases from 1–4 per cent in southern Britain to 8–12 per cent in northern Scotland and 19–27 per cent in Shetland. Few bridled birds occur at Heligoland, or in the Baltic, but in Norway the proportion rises from 12 per cent in the south to 25 per cent in Finmark. In the Faeroes 28 per cent are bridled, and in Iceland up to 50 per cent, with a higher proportion on the south-west (20–50 per cent) than on the north coast (4–10 per cent). About half the population is bridled in Spitzbergen and Novaya Zemlya. In eastern North America there appears to be little variation, with 17–25 per cent in Newfoundland and Labrador and 8–16 per cent in the Gulf of St Lawrence (Birkhead 1984; Fig. 3.5).

The continuous nature of the cline in bridling indicates that the polymorphism is maintained by selection, presumably related to climate or water temperature. Birkhead (1984) showed that the proportion of bridling is better correlated with temperatures at the breeding site than those occurring in the wintering area, hence selection apparently operates in summer. Furness (1987) comments on the similarity between the cline in bridling and the cline in colour-phase ratios of breeding Parasitic Jaegers (*Stercorarius parasiticus*) in the eastern Atlantic. Apparently whatever selection maintains these two clines it is driven by the same phenomenon and temperature seems the most likely common factor.

The proportions of bridled birds throughout the eastern Atlantic appear to have been stable between 1959–60 and 1981–2, although some reduction in the proportion of bridled birds was observed at Bear Island between 1948 and 1970 (Brun 1971). The sharp step in frequencies between northern Scotland and the Shetlands and Faeroes, and between the west coast of Norway and Norwegian Barents Sea populations correspond to two of the subspecific boundaries (*aalge/spiloptera, aalge/hyperborea*). The situation in Iceland, where there is a much higher frequency of bridled birds in the south than in the north, appears very anomalous and deserves investigation.

Comparisons with pinnipeds

It is instructive to compare the distributions of the auks with those of their counterparts among mammals, the seals and sealions. These groups overlap widely with auks in their diet, and feeding areas. They also share the common disability (for marine animals) of having to come ashore to reproduce. Some seals avoid this problem by giving birth on ice flows, a strategy also adopted among birds by Emperor Penguins (*Aptenodytes forsteri*), but not by any auk.

There are 17 species of pinnipeds found in the northern hemisphere, of which three—the tropical and sub-tropical monk seals—belong to the otherwise southern hemisphere subfamily, the Monachinae, which also includes the Northern Elephant Seal (*Mirounga angustirostris*). Among those found in boreal and Arctic waters, seven species, of six genera and three subfamilies, are found only in the Pacific; three, all of different genera, but all belonging to the subfamily Phocinae, occur only in the

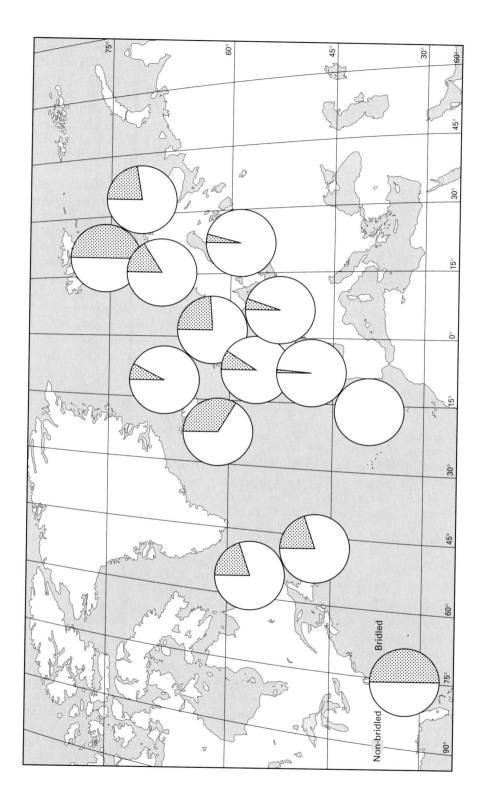

3.5 Proportions of bridled Common Murres in different populations (redrawn from Birkhead 1984).

Atlantic and four, of three genera (three Phocinae plus the Walrus, *Odobenus rosmarus*), are circumpolar or found in both oceans (Riedman 1990). Arctic waters are dominated by Phocinae, the subfamily of primarily Atlantic origin, presumably because of the permanent connection that persisted between Atlantic and Arctic oceans during the Pleistocene.

The greater diversity of Pacific auks is mirrored by their pinniped counterparts at all taxonomic levels; as in the auks, the pinnipeds of the Pacific include temperate and sub-tropical populations associated with the California current zone (California Sea Lion, *Zalophus californianus*, Guadalupe Fur Seal, *Arctocephalus townsendi*, Northern Elephant Seal). In contrast, there is no concentration of taxonomic diversity among pinnipeds corresponding to the richness of auks in the Bering Sea region. The diversity of pinnipeds in eastern North America, where six species occur, is very similar to that found in the Bering Sea (seven species). The corresponding figures for auks are seven (including the extinct Great Auk) and 13.

Numbers of large (more than 500 g) auks are similar, with five in the Atlantic and five in the Pacific. Hence, diversity of large auks is similar to that of pinnipeds. It is the great diversity of smaller, mainly planktivorous, auks in the North Pacific that begs an explanation.

Auks in marine bird communities

The marine avifauna of temperate and Arctic waters of the northern hemisphere is very different from that of the southern oceans. In the north, penguins are absent, albatrosses are confined to tropical and sub-tropical waters of the Pacific, and there are relatively few breeding species of Procellariiformes, principally a few storm-petrels and shearwaters and the Northern Fulmar. Numbers of Procellariiformes are enormously augmented in summer by non-breeding visitors from the southern hemisphere, but in sub-Arctic and Arctic waters, where most auks live, this influx involves relatively few

species, basically the Short-tailed (*Puffinus tenuirostris*) (Pacific only), Sooty (*P. griseus*) (both oceans), and Great (*P. gravis*) (Atlantic only) Shearwaters. The breeding seabird community is dominated by Charadriiformes: auks, gulls, terns, jaegers, and phalaropes. Pelecaniformes, in the shape of cormorants and gannets are also important, but much less so in Arctic than in temperate waters.

To assess the importance of auks in the marine bird communities to which they belong we used population estimates for selected areas and average weights from the literature to estimate their contribution to overall seabird biomass. Data are unreliable for some areas, so these figures should be taken as rough estimates only. For communities in boreal and sub-Arctic waters the proportion of biomass made up of auks ranged, in the Atlantic, from 38 per cent in the Faeroes (where there is a huge population of Northern Fulmars) to 97 per cent in northern Russia (Novaya Zemlya, White Sea coast, and Franz Joseph Land), and in the Pacific from 28 per cent in Japan, where biomass is dominated by the Streaked Shearwater (*Calonectris leucomelas*) to 77 per cent in the northern part of the Sea of Okhotsk (Fig. 3.6). Overall, auks make up 69 per cent of the seabird biomass of the areas considered, which probably include more than 90 per cent of the breeding seabird population of Northern Hemisphere boreal and arctic zones. The rest of the biomass consists of Procellariiformes (petrels, 14 per cent), larids (gulls and terns, 13 per cent), and cormorants (4 per cent). The preponderance of auks is greatest at high latitudes; they comprise 75–97 per cent among areas of predominantly Arctic waters.

Breaking the auks into 'fish-eaters' (murres, guillemots, and the puffin tribe) and plankton eaters (Dovekie, murrelets, and auklets; the *Brachyramphus* murrelets are ignored here) shows that the distribution of plankton eaters is highly irregular. They comprise 59 per cent of all seabird biomass in Arctic waters of the western North Atlantic, thanks to the millions of Dovekies that breed in North-west Greenland, and 22 per cent in the northern

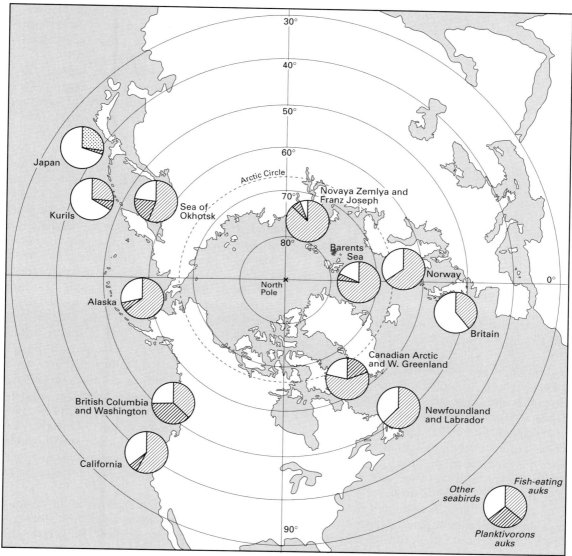

3.6   Auks as a proportion of total seabird biomass among breeding populations of the northern hemisphere.

Sea of Okhotsk where there is a large concentration of auklets. Uncertainty about the size of the huge Least Auklet populations in the Aleutian Islands makes figures for Alaska suspect, but planktivorous auks are unlikely to exceed 15 per cent of seabird biomass there and elsewhere they do not exceed 7 per cent, and their representation is greatly exceeded in all areas by that of the fish-eating auks.

Competition and the structure of seabird communities

The role of competition in determining the populations and ranges of species and hence the structure of animal communities is a topic that has exercised ecologists since the origin of the science (Wiens 1989). Classical ecological theory predicts that competition for food should lead coexisting predators to specialize on different prey species. However, exactly how much competition is necessary to create specialization, and how long it needs to take effect, are imponderables that make proof or disproof of the generalization hard to achieve.

Competition among auks

Cody (1973) attempted to explain the coexistence of certain auk species in terms of segregation in diet and foraging range. His views were severely criticized by Bédard (1976). However, the question of how it is that many auk species coexist, frequently in large mixed colonies, remains unexplained.

Some instances where coexisting auks were found to exploit different prey during the breeding season have been observed. Birkhead and Nettleship (1982b) found that Thick-billed and Common Murres, Razorbills, and Atlantic Puffins, all of which can frequently be found feeding on similar species of schooling fishes, fed their young strikingly different diets at the Gannet Islands, Labrador. At Triangle Island, British Columbia, Rhinoceros Auklets fed their chicks mainly on Pacific saury (*Cololabis saura*), while Tufted Puffins delivered mainly sandlance (*Ammodytes*) in one year (Vermeer 1979), although in other years their diet was similar (D. Bertram, ILJ). Elsewhere in British Columbia, sandlance is the main food fed to nestling Rhinoceros Auklets (Bertram 1988). The best instance of interspecific segregation in diet is provided by the three species of auklets breeding on St Lawrence Island, Alaska, where Bédard (1969c) found that, although foraging in the same general area, they had relatively little overlap in diet. In particular, the Crested and Parakeet auklets, although similar in body size, and hence presumably in diving performance, and taking similar sized prey, took a different spectrum of organisms: Crested Auklets concentrated mainly on euphausiids, while Parakeet Auklets took predominantly amphipods.

In contrast to the impressive examples of diet segregation, there are also striking examples of overlap. Common Murres, Razorbills, and Atlantic puffins have been shown to feed on similar prey in several areas, although they tend to deliver a different size range to their chicks (Furness and Barrett 1985; Harris and Wanless 1986b; Barrett and Furness 1990). The Ancient Murrelet and Cassin's Auklet, very similar in size, which coexist in large numbers in the Queen Charlotte Islands, British Columbia, both feed mainly on euphausiids and juvenile sandlance (Sealy 1975b; Vermeer *et al.* 1985). The former is the main food of Rhinoceros Auklets (adults, not chicks) and the latter is the main food of Marbled Murrelets in the same area (Sealy 1975b). At Buldir Island, Alaska, three species of *Aethia* auklets fed their nestlings mainly on calanoid copepods (Day and Byrd 1989). Wehle (1982) found little difference in diet between Horned and Tufted Puffins breeding together in the Aleutians, although the Tufted Puffins fed further offshore, while at the Farallon Islands in normal years all the local auks—Common Murres, Pigeon Guillemots, and Tufted Puffins—fed on juvenile rockfish (*Sebastes* spp.). However, in years when upwelling failed and juvenile rockfish were scarce, the auks showed much more segregation in diet and feeding areas (Ainley 1990).

The two Pacific puffins make an interesting case, because on a large scale their distributions appear almost identical: both have the centre of their populations in the eastern Aleutians and south of the Alaska Peninsula. However, looked at in detail, the two distributions have less resemblance. The bulk of the Tufted Puffins breed on islands associated with the major Aleutian 'passes', the channels through which water funnels between the islands, often at a ferocious pace. Horned Puffins are not nearly as abundant in that area, but outnumber Tufted Puffins a short distance to the east, among the tangle of rocks, reefs, and small islands dotted to the south of the Alaska Peninsula. The detailed differences in distribution suggest that there may be important differences in foraging requirements.

To some extent, the similarity in diet among coexisting auks is simply a manifestation of the general observation that most seabirds in a particular area tend to overlap broadly in diet. So, sandlance, which form the basis for the summer diet of Black Guillemots, Common Murrres, and Atlantic Puffins in Shetland, are also a principal food of Black-legged Kittiwakes (*Rissa tridactyla*) and Northern Fulmars in the same area (Heubeck 1989). Likewise, euphausiids in British Columbia, which form an important component of the diet for Ancient Murrelets, and Cassin's Auklets, are also taken in large quantities by Sooty Shearwaters and Black-legged Kittiwakes, as well as by several species of large whales (Gaston 1992a). Juvenile rockfish, the mainstay of Common Murres and Tufted Puffins at the Farallon Islands, are very important for the local Brandt's and Pelagic Cormorants (*Phalacrocorax penicillatus* and *P. pelagicus*) (Ainley and Boekelheide 1990).

Competiton with other seabirds

The only pursuit-diving competitors of the auks over most of their range are the cormorants. In parts of the Arctic, Red-throated Loons (*Gavia stellata*) feed commonly in inshore marine waters and, while non-breeding, other loons/divers (Gaviiformes) may also occur in the same marine areas. Probably the only significant competition occurs in continental shelf waters of the eastern Pacific, where the Pacific Loon (*Gavia pacifica*) is an abundant winter visitor (Briggs *et al.* 1987). In the western Atlantic, where many Common Loons (*Gavia immer*) winter offshore, most occur south of the area in which the fish-eating auks occur. At the Farallon Islands, there is a considerable overlap in diet between the resident cormorants and murres (Ainley *et al.* 1990b). However, cormorants tend to be inshore feeders to a much greater extent than murres and appear to be most likely to compete with guillemots. In British Columbia, the feeding areas of cormorants and guillemots are quite distinct, with Pigeon Guillemots feeding mainly in areas of rocky bottom and cormorants preferring sandy bottoms.

There is no obvious avian replacement for the plankton-eating auks in areas such as the temperate and sub-Arctic of the North Atlantic, where they do not occur during the breeding season. In Britain, there is a large breeding population of Manx Shearwaters (Lloyd *et al.* 1991) that undertake wing-propelled diving to depths probably similar to those of the small auklets. However, during the breeding season they feed mainly on fish and would therefore appear more likely to compete with the fish-eating auks (Brooke 1990). Numbers of breeding shearwaters in the western North Atlantic are negligible. It seems that whatever role is played by auklets in the North Pacific boreal and sub-Arctic food webs must, in the Atlantic, be filled by some non-avian planktivore.

Thus far, we have considered only the possible effects of locally breeding seabirds. This is because approximate numbers are available for breeding populations of most species in the northern hemisphere. However, non-breeding summer visitors from the southern hemisphere, mainly shearwaters, constitute an enormous additional influx of seabird biomass throughout the boreal and sub-Arctic waters of both oceans, from May to September (Sanger 1972; Wiens and Scott 1975). As their numbers are much harder to estimate, it is difficult to calculate even a rough contribution to overall seabird biomass. For Oregon waters, Wiens and Scott (1975) estimated that shearwaters took 47 per cent of food eaten by seabirds, compared with 32 per cent taken by Common Murres, the only abundant auk, while for the waters off eastern Newfoundland, Diamond *et al.* (1993) estimated that shearwaters consumed 76 per cent of food taken by seabirds in August, compared with only 4 per cent taken by auks. This contrasts with the 40 per cent share taken by auks in January–March. The shearwaters involved are all medium sized birds (500–1000 g) that dive to the sort of depths reached by the smallest auks (20–30 m) and feed on large zooplankton or small schooling fish. Large flocks of Sooty Shearwaters are known to exclude auks from certain prey aggregations (Hoffman *et al.* 1981). The influx of southern hemisphere seabirds may have a real impact on some auk populations, especially in sub-Arctic waters off Newfoundland and in the Gulf of Alaska and the Aleutians. Their prevalence may explain, to some extent, the concentration of auks in Arctic waters only thinly penetrated by the non-breeding shearwaters. In the high Arctic, murres, Dovekie, and puffins share colonies and feeding areas with large numbers of Black-legged Kittiwakes and Northern Fulmars, but both of these species feed exclusively close to the sea surface. Otherwise, potential competitors are either local or sparse and unlikely to have any great impact on the availability of prey for auks.

Biogeographical evidence for competition among auks

If species of the same genus are similar to one another in diet and foraging behaviour then they should be more likely to compete with one another than with species of other genera (a common assumption). Consequently if competition is occurring, we expect birds breeding in the same area to belong to different genera more often than expected by chance, especially if the congeners are similar in size. However, evidence from the present distributions of auks is equivocal regarding the influence of competition.

In the Atlantic, where only *Uria* is represented by more than one species, most communities necessarily consist of non-congeners. This is also true in the boreal zone of the Pacific; for instance, from California to British Columbia, we find one each of *Uria*, *Fratercula*, *Cerorhinca*, *Brachyramphus*, *Ptychoramphus*, *Synthliboramphus*, and *Cepphus* (excluding very small numbers of a second *Uria* and *Fratercula* species in British Columbia). Japan, likewise, supports one each of *Uria*, *Fratercula*,

Cerorhinca, *Brachyramphus*, *Synthliboramphus*, and *Cepphus* (two *Synthliboramphus* species occur in both areas, but are nowhere sympatric).

The situation in the Aleutians and the Bering Sea is much less clear, with congeneric *Aethia*, *Fratercula*, and *Uria* species breeding together at many colonies and two *Brachyramphus* species occurring side by side in the same waters for much of the year. The overlap between the two very similar murres is especially striking (Fig. 3.7), although there is a tendency for Thick-billed Murres to predominate in Arctic and offshore waters. Moreover, considering distributions at sea, there seems to be little tendency for congeners to use different water areas, although species of different genera may do so (Elphick and Hunt 1993).

Taking genera individually, only species of *Cepphus* and *Synthliboramphus* replace one another without major overlap. It may be no coincidence that the guillemots are rather sedentary and forage mainly on sedentary benthic fishes (Cairns 1987*b*; Ewins 1993), a resource potentially easier to deplete than the pelagic schooling fish and zooplankton taken by most other auks. On the basis of current distributions, the auks seem to provide little evidence for competitive exclusion. A similar conclusion, based on a lot more maths, was reached by Whittam and Siegel-Causey (1981) and Haney and Schauer (1994).

Aside from the historical effects of recent sea-level changes, there are other good reasons why we might not expect to see much overt competition among auks. The diets of seabirds of several different families, varying in size and foraging habits, overlap broadly within the temperate and Arctic waters of the Northern Hemisphere. Such overlap is no doubt magnified by the small range of prey items available in some areas. In a situation where any effect of the auks on their prey may be much diluted by the greater impact of other seabirds, especially shearwaters, and where even the combined impact of the seabird community may have little impact on the dynamics of their prey (Chapter 4), we might not anti-

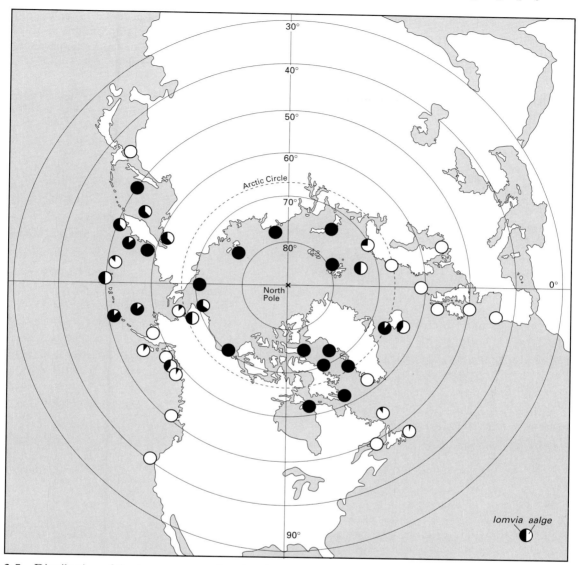

3.7   Distribution of the two murre species, showing colonies where they breed side by side.

cipate competition to be a major factor in structuring seabird communities. The evidence available for other seabirds suggests that specialization in diet is not typical of sympatric seabird communities, at least while breeding, while overlap of diet among coexisting seabirds is very common among tropical species (Diamond 1983).

In the high Arctic, where auks are the dominant seabirds and where food webs tend to be very simple, only a single planktivore (the Dovekie), a single offshore piscivore (Thickbilled Murre), and a single inshore piscivore (Black Guillemot) are present, along with a few Horned and Atlantic Puffins in their respective oceans. This could be seen as evidence for competition. However, the great changes in marine food webs occurring continually through large-scale processes affecting the world's oceans, and the more recent, rapid changes brought about by human activities such as commercial fisheries, make it unprofitable to dwell too much on present evidence for interspecific competition among the auks.

4

Auks in ecosystems

The activities of auks on land, especially at their major breeding colonies, are their most visible manifestations. Breeding colonies are like toadstools, whose bright presence on the forest floor is all that betrays the network of fungal tissues that permeates the soil and constitutes the true organism. Like toadstools, auk colonies sit at the centre of a network of transportation corridors, with birds ceaselessly shuttling between their breeding sites and their distant feeding grounds. The colony erupts annually, like the fructification of the fungus, discharges its spores, in the form of the year's crop of juveniles, and then subsides. However, the activities of the auks go on, largely invisible to the human eye, pursuing and capturing carbon, in the form of fish and invertebrates, processing it into fuel and replacement parts, and then burning it in millions of little metabolic furnaces, returning the carbon to the atmosphere, or through the death of individual birds, directly to predators and scavengers, or to entombment in deep ocean sediments.

If we sit on the cliff top at Digges Island, or the Pribilofs, or any other major auk colony, we can watch tens of thousands of birds hourly hurrying towards or away from their breeding place. At dawn, when there is scarcely light to read the time on your watch, birds will be delivering fish to their chicks. Even when the sea is being whipped to foam and the wind is hurling spray half way up the cliffs, this ceaseless traffic persists. A bird leaves its breeding site, where it has dozed for a day over its downy chick. It disappears over the horizon at a speed that a Peregrine Falcon would have difficulty sustaining in level flight. Eight hours later it returns after having swum to a depth of 200 metres; much deeper than any SCUBA diver cares to go, where light in the soupy, plankton-rich waters of the Arctic summer can scarcely penetrate. In doing so, it held its breath for longer than you or I could, although we have lungs hundreds of times larger. When a bird dives, there is, literally and figuratively, an area of darkness. If there have to be reasons to study auks, these demonstrations of their physical prowess and the mystery of their underwater behaviour are surely powerful arguments.

As researchers develop a greater appreciation for how the biological world functions we become concerned not only about individual organisms, but also about how whole systems function. On the scale of the planet, auks cannot be said to be important organisms, because they do not appear to dominate any trophic level, except perhaps in local areas around large colonies. The total biomass of auks and other seabirds is small compared with that of other groups of animals in the same marine ecosystems. However, their small size and resultant high metabolic rate mean that seabirds turn over much more energy than their biomass representation would suggest. They probably have an important role to play as consumers in certain northern hemisphere seas and a very localized, but striking, role in the ecosystems of offshore islands and other colony sites.

The organisms on which a bird feeds, and those for which it provides prey, define its place in the food web. Although birds affect ecosystems in other ways (nutrient enrichment through defecation, dispersal of seed and other propagules, for example) their main role tends to be either as predators on other ecosystem components, or as prey. In this chapter, we deal with their role as predators within the marine ecosystem: what they eat and where, how they locate it, pursue it and capture it, and how they perform in their chosen medium, in terms of locomotion and physiology. We also review their impact on terrestrial ecosystems at their breeding sites and their role as prey for terrestrial and marine predators.

Auks as predators

What we know about auks is very much determined by the methods at our disposal for studying them. With most terrestrial birds it is possible to follow them to their feeding areas and observe them capturing and handling their prey. For auks, even if we can follow them to their feeding areas out at sea, we have very limited means of studying what they eat and how they capture it. Consequently, some consideration of the methods at our disposal is essential to understanding the limitations on current information.

Comments on methods: diet

Among the topics discussed above, some have received much more attention than others from naturalists and researchers. There is a great deal of information available on some topics. For instance, among semi-precocial and intermediate species, the diet of the young is well known, because they are fed at the nest where the food can be easily sampled. For some species, sufficient studies are available to enable regional comparisons, such as those made for Atlantic Puffins by Harris and Hislop (1978), for Rhinoceros Auklets by Vermeer (1979) and Bertram and Kaiser (1993), and for Tufted Puffins by Hatch and Sanger (1992).

Information on adult diets is also available for most species, although these data need to be judged cautiously. Many studies have been based on the analysis of stomach contents from birds collected at sea or at the breeding colony. This method is subject to considerable, and so far more or less unquantified, biases, owing to variation in the rates at which different types of prey are digested (Bradstreet 1980; Gaston and Noble 1985). This is especially true of birds collected at their colony, or while flying towards it. These usually contain mainly persistent remains, such as fish otoliths, squid beaks, and annelid jaws, suggesting that birds may spend some time resting after feeding before returning to the colony, perhaps in order to lighten themselves for the journey. Collections made under these circumstances probably present a very biased view of the species' dietary habits.

To overcome biases introduced by digestion, several researchers have used the remains found in stomachs to reconstruct the diet at ingestion (Bradstreet 1980, 1982a). In this method, fragments such as fish otoliths are measured and the known relationship of otolith size to body size used to determine the size of fish eaten. The sizes of the intact organisms are then summed to provide a picture of the total diet. The drawback to this method is that organisms that leave persistent fragments, such as otoliths, and especially chitinous parts, such as squid beaks and annelid jaws, tend to be over-represented (Gaston and Noble 1985). As a generalization, we find that the reconstruction method tends to exaggerate the importance of fish and squid in the diet, compared with descriptions based on the wet weight or volume of remains. Conversely, descriptions based on numbers of organisms tend to exaggerate the importance of the smaller prey. Any evaluation of what constitutes the major prey based on published values of percentage wet weight needs to take into account the method used to arrive at the figures.

Recently, Hobson and co-workers (Hobson 1990, 1991; Hobson and Welch 1992; Hobson et al. 1994) have analysed the proportion of different stable isotopes of carbon and nitrogen in tissues to investigate the trophic level of the prey species on which different seabirds have fed. They have shown that the diets of several species of auk probably contain a greater proportion of invertebrates and a smaller proportion of fish than has been indicated by more traditional methods, something that was widely suspected, but previously unproven.

Comments on methods: ecology

Information on pelagic ecology has been obtained from numerous ship and aircraft surveys carried out over large areas of marine waters, mainly during the past three decades. Unfortunately, many of these surveys have had to be based on ships performing other functions (oceanographic and fisheries research vessels, ferries, coastguard patrols, etc.) and hence were not ideally designed for systematic surveys of seabird distributions (e.g. Brown 1986). Other surveys, especially those funded as part of the environmental assessment programme for offshore oil and gas exploration, were intensive and specifically designed to describe seabird distributions (e.g. Johnson et al. 1976; McLaren 1982). However, these were rarely continued for more than one or two years, and hence lacked any indication of year-to-year variability. A notable exception is the comprehensive series of surveys repeated over several years off California by Briggs et al. (1987), which emphasized the extent to which results can vary from year to year.

Much has been discovered over the past several decades about the effects of oceanographic processes in determining the distribution of seabirds at sea. In Alaska, especially, George Hunt and John Piatt and their associates (Hunt et al. 1990, 1993; Schneider et al. 1990a; Piatt et al. 1991b; Elphick and Hunt 1993) have made great progress in identifying the relationship between different water masses and the auks that inhabit them. However, much of this work has involved identifying areas and processes that concentrate birds. The larger goal, of estimating the relative importance of particular oceanographic areas and processes to species' populations, and hence assessing which are most important to their survival, will take much longer to achieve.

The behaviour of auks at sea has not been much recorded. Most surveys from boats have concentrated on obtaining distributional information. Data on the proportion of time spent feeding and the length of time spent underwater have rarely been collected. Although information on numbers of birds seen flying and on the water has been routinely collected as part of most surveys at sea, it has not often been presented. Anecdotal information on the duration of dives has been collected for a number of species, but many of these data have been obtained in circumstances that made it easy to collect, but may not have been typical of the species in question (e.g. close to shore, or in calm conditions with only a few birds present). The application of radio-telemetry to auk diving behaviour by Wanless and co-workers (Wanless et al. 1988) solved this problem for species feeding within transmitter range of a breeding colony. A different approach was taken by Cairns and co-workers (Cairns et al. 1987, 1990), who attached timers that recorded the accumulated time spent underwater, so that the time devoted to diving could be calculated as a proportion of the overall time budget.

What the auks do once they submerge was, until recently, something of a mystery. There are surprisingly few, even anecdotal, observations of auks underwater. Although there have been some attempts to use SCUBA techniques to observe auks (e.g. Hunt et al. 1988), none seems to have yielded any substantial information on foraging behaviour. Video cameras have been used to record the feeding behaviour of auks preying on fish schools at shallow depths, but what goes on at more typical feeding depths, where light levels may be very low, has not been observed. As much has been learnt by perceptive photographers filming underwater (e.g. Boag and Alexander 1986).

Diving depths have been inferred from the retrieval of birds drowned in fixed nets set at known depths (Tuck 1961; Piatt and Nettleship 1985), but this type of evidence cannot be used to estimate average diving depth because nets are not set uniformly at all depths. However, Piatt and Nettleship (1985) demonstrated convincingly that different species foraging in the same area foraged at different depths.

In the last decade, several depth recording devices have been developed that can be attached to auks and later retrieved. The simplest of these devices record only the maximum dive depth (Burger and Simpson 1986; Burger and Powell 1990), some record the proportion of time spent at different depths (Wilson *et al.* 1989), while some use a microchip to record the depth at preset intervals after submersion, giving a detailed profile of each dive (Croll *et al.* 1992). The latter type of instrument has so far been used on only one auk, the Thick-billed Murre, although similar information is available for several species of penguins and cormorants. These, being larger, are thought capable of carrying larger devices without inconvenience.

Recording devices attached to birds have the advantage that they make it possible to take a more random sample than might be obtained from net drownings, or shore-based observations. However, the attachments almost certainly affect the performance of the birds selected, in most cases to an unknown degree (Wilson *et al.* 1986). Moreover, it has only proved feasible thus far to deploy them on breeding birds that return regularly to their colony. This applies especially to devices that have to be removed to be read. A consequence of all these restrictions is that, despite recent technological advances, much of what can be said about the foraging of auks has to be based on inferences, rather than on direct observations.

Two examples

To illustrate the position of auks as components of ecosystems, we shall take two, fairly well-known, examples from northern Hudson Bay: Thick-billed Murres and Black Guillemots (Table 4.1). The Thick-billed Murre breeds at only two sites in the region: Coats Island (30 000 breeding pairs), and Digges Sound (300 000 pairs). The foraging ranges of the two colonies do not overlap during breeding, but the environments in which they forage support similar food webs, the colonies are at the same latitude, enjoy similar climates, and suffer from similar predators (Fig. 4.1). Together, birds from the two murre colonies forage over most of the marine waters of north-east. Hudson Bay. Most birds leave the area of their colonies in September, not returning until May. Most adults from both colonies reach Newfoundland by February, although first and second year birds from Digges Sound appear to winter further north (Donaldson *et al.* 1997).

Black Guillemots breed at many sites throughout the region, with usually less than 100 breeding pairs at any one site. During the breeding season, they feed principally within 10 kilometres of land. In winter, they probably remain in the region, but offshore, among mobile pack-ice, or in major leads. Most feeding at that season probably involves taking prey on or close to the undersurface of the ice.

In both these species, their total biomass standing stock is small in comparison with the amount of biomass they consume over the year. Thick-billed Murres, present from early May to early September, represent about 1000 metric tons of biomass and consume approximately 42 000 tons of food; the resident Black Guillemots, another 4.5 tons of biomass, consume 650 tons annually. Thick-billed Murres forage over an approximate water volume of 10 000 km^3 (assuming normal maximum diving depth to 150 m) and Black Guillemots 140 km^3 (assuming maximum dives to 35 m and range confined to within 10 km of shore). The ratio of their population

Table 4.1   Comparison of Thick-billed Murre and Black Guillemot populations and ecology in northern Hudson Bay

	Thick-billed Murre	Black Guillemot
Number (birds)[1]	1 million	10 000
Colony size	> 50 000	< 1000
Foraging range (km)	10–150	< 15
Foraging water depth (m)	> 30	< 30
Maximum dive duration (sec)	220	147
Major diet	Amphipods, mysids, Arctic cod, capelin, sculpins, blennies	Blennies, Arctic cod, amphipods
Adult body mass (g)	1000	450
Average daily energy expenditure (kJ)	1500	943
Chick feeds/day	< 6	> 10
Chick meal size (g)	12	15
Chick food as proportion of all prey taken by breeders during chick rearing	6%	24%
Chick mass at departure (g)	125–250	380
Predators (adults)	Gyrfalcon	Peregrine Falcon
Predators (eggs, chicks)	Raven, Glaucous Gull, Red and Arctic Foxes	Ermine
Kleptoparasites	—	Parasitic Jaeger
Wintering area	Newfoundland and Labrador	Locally

Sources: Thick-billed Murres from Gaston (1985), Gaston *et al.* (1985, unpublished data) and Croll *et al.* (1992, unpublished data); Black Guillemots from Gaston *et al.* (1985) and Cairns (1987*a, b*).
[1] Includes non-breeders.

sizes (100:1) approximates the ratio of their foraging volumes (72:1). There is broad overlap in the range of prey species taken, but the emphasis is different, with Black Guillemots taking mainly benthic fishes (especially blennies), while the murres take mainly Arctic cod (*Boreogadus saida*) and large zooplankton.

Adult diet

Prey size among predators in general is related to body size (Vezina 1985). In seabirds, it seems to relate especially to the size of organism that can be swallowed whole. Few seabirds dismember their prey, possibly because of the difficulty of doing so in a medium where the prey cannot be pinned down, and possibly because of the danger of being robbed by other birds. The auks swallow most of their food underwater, making it possible to obtain more than one food item at each dive.

In auks, both the maximum and median sizes of prey recorded are rather closely related to body size (Fig. 4.2). At the lower end of the prey size spectrum are organisms such as copepod crustacea and small pteropod molluscs (less than 0.1 g), taken by auklets and Dovekies, while at the upper end are medium-sized fishes up to about 80 g, taken by murres. The Great Auk probably took fish considerably larger than that (Olson *et al.* 1979).

For many of the larger auks, the median size of prey taken is much smaller than the maximum, probably because the size distribution of prey taken by any individual tends to follow a distribution typical of many biological phenomena, with many small and few large items ('lognormal' or 'canonical' distribution, Preston 1962, Fig. 4.3). If we regard the size of the throat as being a major factor constraining the upper size limit for prey (depending on the cross-sectional area of the prey), then we

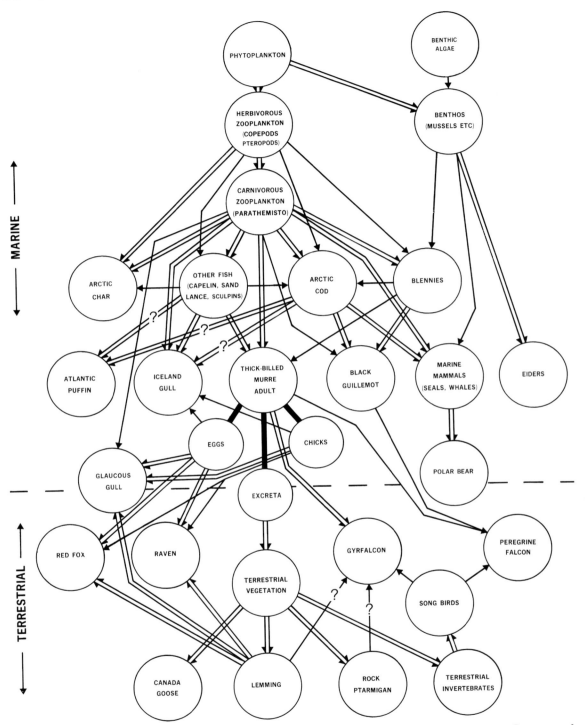

4.1 Food web of Thick-billed Murres and Black Guillemots in north-east Hudson Bay (after Gaston *et al.* 1985).

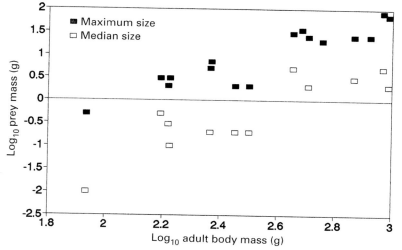

4.2   Maximum prey size and size of prey most commonly taken, in relation to body size among auks.

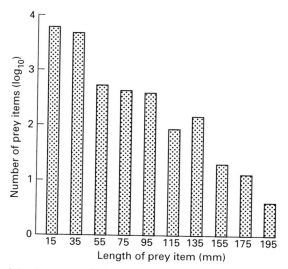

4.3   Lognormal distribution of prey size: data for Thick-billed Murres in eastern Canada (from Bradstreet and Gaston 1993).

intake (Fig. 4.2). We might anticipate this from the observation (Chapter 2) that larger predators can take a greater range of prey size than their smaller relatives. Hence, in considering the importance of different sizes of prey for each species it is more appropriate to make interspecific comparisons based on the size range making the greatest contribution to diet biomass.

The distinction drawn by Bédard (1969a), between those auks that take mainly fish and those taking mainly zooplankton, has been somewhat clouded by more recent findings. It appears that adult auks of all species, with the possible exception of the guillemots and, to some extent, the Common Murre, are more dependent on zooplankton, or squid, than on fish, outside the breeding season (Ogi 1980; Hobson 1991). Most auks seem to be opportunists that will take whatever they can swallow among a wide range of prey types.

The degree to which auks exercise selection in their choice of prey is hard to assess without detailed information on the abundance and accessibility of different prey organisms. Generally speaking, all species take a wide variety of the invertebrates and fishes available in their foraging habitat. However, certain types of prey that are common seem to be

might expect prey mass to be proportional to the body weight of the bird, as both prey and throat cross-sectional area are proportional to the square of linear dimensions. This appears to be approximately true when the maximum size of prey is considered, but less true when we consider the most important size for energy avoided. Jellyfish and comb jellies seem to be rare in the diets of most auks except the Parakeet Auklet (Harrison 1990), although comb jellies and medusae of a suitable size are common in most areas where auks forage (Bédard 1969a). Comb jellies were abundant in surface waters in an area of heavy foraging by murres investigated by Hunt et al. (1988), but the murres ignored them, swimming deeper to take euphausiids. Because these gelatinous animals are very fragile, we might expect them to be under-represented in diet studies, owing to their rapid disintegration after ingestion. However, the fact that they have rarely been reported at all from most auks suggests that they are generally ignored as prey. Likewise, pteropod molluscs and arrow worms, very abundant in Arctic waters, are rarely reported as dietary items for auks, and then usually only in small numbers, although they have a similar energy density to some crustacea commonly taken (Percy and Fife 1985). Norderhaug (1980) specifically noted the prevalence of comb jellies, pteropods, and arrow worms where Dovekies foraged in Spitzbergen, but did not find them in the birds he examined. These organisms may present difficulties in handling, or may be less digestible, distasteful, or more difficult to locate underwater.

Despite the wide variety of fish and zooplankton of a suitable size available in the boreal and Arctic waters of the northern hemisphere, a small number of prey species make up a large proportion of the diets of most auks throughout their range (Fig. 4.4, modified from Springer 1991). This reflects, in part, the predominance of a small number of dominant species in marine food webs, but this explanation is not sufficient to account for all of the uniformity of diet observed. It is likely that certain species of fish and crustacea behave in a way that makes them especially vulnerable to foraging by auks, either by forming dense swarms, or by approaching the surface. Capelin (*Mallotus villosus*), for instance, enter shallow, coastal waters in dense shoals to spawn in summer. The breeding cycle of auks

in Newfoundland is timed to coincide rather exactly so that chicks are being fed during this spawning period (Brown and Nettleship 1984). Likewise, the dense shoals formed by sandlance (sandeels), which may be corralled and herded by Rhinoceros Auklets (Grover and Olla 1983), Marbled Murrelets (Mahon et al. 1992), and probably by other auks, appear to provide especially suitable foraging targets. Surprisingly, the very large, dense schools of Arctic cod that form in shallow waters in the Arctic in late summer rarely attract auks, although they may be fed on by other seabirds and by marine mammals (Welch et al. 1993). Euphausiids (shrimp-like crustacea, usually less than 4 cm long) are an important constituent of diets for many Pacific auks, the species of euphausiid involved varying with region and time of year. Their habit of aggregating into dense swarms apparently makes them more attractive as prey than other common crustacea of a similar size; mysids fill a similar role in Arctic waters.

How prey is located

Warm-blooded marine animals, especially those inhabiting cold waters, need to find a great deal of food daily to satisfy their basic metabolic needs. Studies of energy expenditure by free living auks (field metabolic rates, FMR) have shown that they require from 358 kJ/day (Least Auklet, Roby and Ricklefs 1986) to 2200 kJ/day (Common Murre, Gabrielsen 1994) while rearing chicks. These figures, when corrected for the efficiency of digestion (80 per cent) translate roughly to 40 per cent of body weight in fresh weight of food for the Common Murre, 75 per cent for the Least Auklet. Moreover, compared with penguins and marine mammals, auks do not carry large energy reserves relative to their rate of expenditure. Thick-billed Murres wintering off Newfoundland were estimated to have only 3–4 days of energy reserves in peak condition (Gaston et al. 1983b); Common Murres during breeding carry only 1.5–2.5 days of

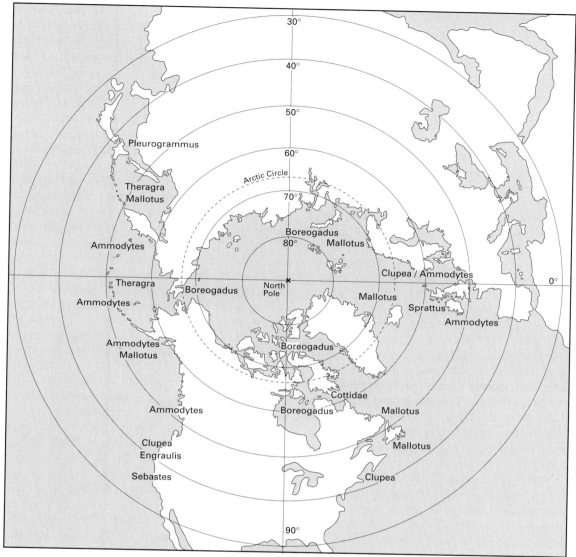

4.4   Major prey of auks by geographical regions (adapted from Springer 1991).

energy reserves (Gabrielsen 1994). Because they are larger and do not need to fly, penguins can store much more fat than the extant auks and frequently fast for weeks while moulting (Adams and Brown 1990). Accounts of Great Auks stress how 'excessively fat and oily' they were (Hardy 1888), suggesting that the elimination of flight in the species led to the maintainance of much larger fat reserves than those of their flying relatives.

If their prey was evenly dispersed at sea, auks would be hard pressed to survive. Simple calculations suggest that prey densities of at least 100 times the average are necessary for profitable foraging (i.e. energy gained in foraging exceeds energy expended, Brown 1980).

Consequently auks cannot forage everywhere in the ocean, but must discover areas where their prey is suitably clumped. Compared with terrestrial environments, the availability of prey in the marine environment varies much more over short periods of time and tends to be much less predictable (Haury *et al.* 1978; Gaston and Brown 1991). Tidal currents cause concentrations of food in a particular locality to appear and disappear in a matter of hours, while the herding of large predatory fishes and marine mammals may cause aggregations of forage fishes to form and disperse on a scale of minutes (Safina and Burger 1988). For marine predators this puts a premium on mobility.

Many specialized seabirds, especially the petrels and albatrosses, travel long distances in search of food. Their highly efficient soaring and gliding flight enables them to do so at minimum cost. The high wing-loading of the auks prevents them from imitating such mobility. Consequently, auks need to exploit more predictable and concentrated sources of food than those of albatrosses, petrels, terns, and tropicbirds. This is probably a major factor in confining most auks to continental shelf waters, where marine productivity is generally an order of magnitude higher than in deeper oceanic waters (Raymont 1980). It may also partly explain why auks have not colonized tropical waters. However, the absence from tropical waters of avian pursuit divers other than cormorants may also be related to competition with, and predation by, large predatory fishes, such as sharks. Even in winter, auks shun tropical and sub-tropical waters completely, with the exception of the southerly breeding Craveri's Murrelet.

Predictability

Inshore feeding auks, which include the three guillemot species, the *Brachyramphus* murrelets, and (probably) the Whiskered Auklet, generally forage in predictable feeding areas from day to day. Cairns (1987*b*), studying

Black Guillemots in north-eastern Hudson Bay, found that birds returned regularly to localized feeding areas throughout the chick-rearing period. In the Quoddy region of New Brunswick, the pattern of foraging by Black Guillemots was similarly predictable and determined by tidal cycles (Nol and Gaskin 1987), while in waters around Haida Gwaii, British Columbia, Marbled Murrelets occur in the same nearshore waters year after year (AJG).

The need to have predictable food supplies means that many auks take advantage of permanent food concentrations, in the form of current rips (turbulence, upwelling, convergence, etc.) that concentrate prey, or bring slow-swimming organisms closer to the surface than usual. Such features occur especially off headlands, among islands, and in areas of very uneven bottom topography, and in association with strong tidal currents (Schneider *et al.* 1990*b*). Examples of auks congregating at such permanent food sources are the huge multi-species concentrations among the Aleutian passes, especially Akutan and Baby passes, where up to eight species occur (Hoffman *et al.* 1981); aggregations of tens of thousands of Thick-billed Murres in the Nuvuk Islands of north-eastern Hudson Bay (Cairns and Schneider 1990) and off Cape Upright, St Mathews Island (Hunt *et al.* 1988); and of Ancient Murrelets in winter among the islands of the Straits of Juan de Fuca (Gaston *et al.* 1993*a*). The exact relationship of auk distributions to tidal currents may vary, depending on the favoured prey. Detailed investigations by Hunt *et al.* (1995) in Alaska, showed that Least Auklets fed on copepods concentrated in near-surface convergencies on the downstream side of a pass, whereas Crested Auklets fed on euphausiids where water upwelled at the upstream entrance. The relative distribution of the two species switched when the current reversed.

On a larger scale, the aggregation of Dovekies at the edge of the Newfoundland Banks in winter may result from similar upwelling processes (Brown 1985) and the large numbers of Common Murres breeding in California appear dependent on the enhanced productivity created by the cool upwelling of the California Current. In years when warm waters extend northwards and the upwelling fails, as a result of the El Niño phenomenon, reproductive success is generally lower than normal and occasionally fails completely (Boekelheide *et al.* 1990).

Feeding and flocking

The degree to which seabirds are aggregated while foraging depends very much on the scale of measurement (Haney and Solow 1992). All species will appear aggregated on the scale of the whole planet, because none is distributed evenly throughout the world's oceans. At the other end of the scale, we might find that individuals within a feeding flock are actually rather evenly dispersed on the scale of a few tens of metres. Deciding an appropriate scale on which to measure aggregation is important if we want to understand the role of habitat and prey abundance in determining the distribution of auks at sea. For Common Murres and Atlantic Puffins foraging off Newfoundland, Schneider *et al.* (1990*b*) found greater concentrations on a sampling scale of 5.5 km that corresponded to predictions based on oceanographic processes (fronts), than on the minimum scale of observations (20 m). However, it is possible that at the smallest scale (20 m), where neighbouring birds could see one another easily, social processes were also involved in determining spacing patterns. The idea of a universal scale governing seabird aggregations and based on physical processess is very appealing, but requires much more evidence to be substantiated. Thanks to the problem of defining an appropriate scale for measurement, current information on the dispersal of auks at sea is hard to interpret.

The distribution of auks may change very rapidly from day to day and even from hour to hour, depending on tides and weather condi-

tions (Cairns and Schneider 1990; Gaston *et al.* 1993*a*). Moreover, Hunt (1990) showed that seabird abundance and the availability of prey, as measured by various acoustic methods, are not necessarily correlated. This seems to be more true of planktivores than of fish-eaters (Russell *et al.* 1992), although Woodby (1984) found little relationship between concentrations of murres and of patches of their potential prey in the eastern Bering Sea. Given that people tend only to write about those studies that demonstrate some kind of effect (or journals only accept such papers), we may anticipate that the real relationship between seabird distributions and those of their prey is weaker than published data suggest. Despite much time and effort devoted to mapping offshore seabird distributions, our ability to predict where and when auks will form dense concentrations away from predictable tidal upwellings is still poor. Except in very limited areas, we may never know enough to be able to predict confidently in advance the consequences of an oil spill at a particular time and place.

Behavioural tactics

Despite the exploitation of some predictable food sources, many auks do not appear to rely on the type of food concentrations created by permanent tidal upwellings. Where they are confronted with finding an ephemeral and unpredictable prey their problems can be divided into two: (a) locating the whereabouts of a suitable underwater prey concentration and (b) actually pursuing and capturing the prey. During the breeding season the first problem is particularly acute, because birds have to leave the feeding area periodically to relieve mates or feed chicks. By the time their colony shift is over (as much as 4–5 days, in the Ancient Murrelet), the location of food may have changed, forcing them to search for a new feeding area.

Both of these problems pose challenges to auks. In some areas they apparently locate suitable prey concentrations by joining mixed species feeding flocks which form over fish schools or zooplankton swarms close to the surface. In such instances they may take cues from easily visible and wide-ranging seabirds, such as Black-legged Kittiwakes and other gulls (Hoffman *et al.* 1981; Porter and Sealy 1982). Murres, Rhinoceros Auklets, and Ancient Murrelets may all be involved in such flocks (Chilton and Sealy 1987). The presence of other species can be valuable to auks in enabling them to detect potential prey, but it can also be a deterrent where large concentrations of shearwaters occur (Hoffman *et al.* 1981). In some instances, it is the behaviour of the auks that leads to flock formation, as Rhinoceros Auklets and Ancient and Marbled Murrelets may herd sandlance towards the surface, making them available to gulls (Grover and Olla 1983; Litvinenko and Shibaev 1987; Mahon *et al.* 1992). Nor are auks attracted to all aggregations of surface feeders. It is not unusual to see murres and razorbills feeding within sight of flocks of kittiwakes and fulmars and taking no apparent notice of them. It is very rare to find auks following fishing boats, although Black Guillemots and Common Murres will do so occasionally (Ewins 1987; Camphuysen *et al.* 1995).

Species with the same ostensible feeding potential may adopt very different strategies towards food finding. Least, Crested, and Whiskered Auklets choose to feed mainly in limited areas of strong currents and tidal upwellings, necessitating feeding in dense aggregations. However, Parakeet Auklets breeding on the same islands feed solitarily, grazing on widely dispersed zooplankton (see species accounts). A similar offshore dispersal is characteristic of zooplankton-feeding Thick-billed Murres in many parts of the Arctic, whereas Common Murres, which concentrate on schooling fish, generally feed in much denser aggregations (Piatt 1990).

Despite Bedard's (1969a) statement that 'all alcids are solitary feeders', auks are often found in flocks on their feeding grounds. Marbled and Xantus' Murrelets are nearly always seen in pairs or small groups; apparently this was also characteristic of Great Auks. During the breeding season, the presence of other auks may be the main clue to the whereabouts of suitable feeding areas. Gaston and Nettleship (1981) suggested that a major factor causing the aggregation of Thick-billed Murres into very large (>10 000 pairs) colonies in the eastern Canadian Arctic was the need to have a large number of birds foraging from a single spot to track the movements of food souces that change from day to day in response to ice movements. The idea of colonies as 'information centres' will be dealt with further in Chapter 5.

Feeding underwater

How auks locate their prey once underwater is even more of a mystery. The role of 'head-dipping', periodic submersion of the head for 1–2 seconds while swimming on the surface, in locating underwater prey has been proposed, but not confirmed. An alternative explanation for this behaviour is that the birds are looking for potential predators, as several species of marine mammals, as well as some large fish, have been known to take auks. It is quite possible that such periodic underwater observations serve both functions.

No one has been able to detect any acoustic mechanism by which auks could locate their prey, and chemical detection (e.g. smell, taste) seems very unlikely, so we have to assume that vision is the universal detector. However, at the maximum depths to which auks, especially murres, typically dive very little light penetrates. Even in highly transparent water the intensity of light at 100 m is less than 5 per cent that at the surface and in typical coastal water, with high phytoplankton density it may be only 0.1 per cent (Valiela 1995). Moreover, some auks feed in water of high turbidity (e.g. Common Murres around the Farallon Islands, Briggs *et al.* 1987) where visibility may be obscured at much shallower depths. Kittlitz's Murrelets sometimes forage in fiords made turbid by glacial melt-water and although they may be feeding in saline water below the freshwater inflow, light must be very much reduced. Guillemots often forage among kelp and other seaweeds and on rocky or boulder-strewn bottoms where prey is likely to be extremely cryptic (Cairns 1987b). In addition, Black Guillemots may winter as far north as 80° N (Renaud and Bradstreet 1980), where, even at mid-day, there is little light at the surface, let alone under the ice where they feed. All this evidence suggests that auks have exceptionally good vision in low light conditions, although in some cases they may be aided by the bioluminescence of their prey.

In a general survey of the plumage colour of seabirds that pursue prey underwater, Cairns (1986) showed that while the majority are mid-water feeders and white below, benthic feeders are more likely to be dark below. This applied to cormorants, auks, and certain sea ducks. He suggested that a white belly made a bird descending on prey from above less conspicuous, but that benthic feeders generally came at their prey horizontally, while travelling along the bottom, and in that situation being black was less conspicuous than having white underparts. While concurring with the general idea that black-and-white plumage was a camouflage against detection, Wilson et al. (1987) proposed that the lateral stripes on fish-eating penguins of the genus Spheniscus assisted them in breaking up schools of fish. However, although several species of auks feed on schooling fishes similar to those pursued by the penguins, none has lateral stripes, unless the facial plumes of the Rhinoceros Auklet serve this function.

The guillemots, black below in summer, are white below in winter. In the case of the Black Guillemot, some populations at least are mid-water or under-ice feeders in winter (Gaston and McLaren 1990), which could be treated as evidence in support of Cairns' hypothesis. Boag and Alexander (1986), while confirming the lower visibility of puffins while seen from below, thanks to their white bellies, pointed out that their red feet were extremely conspicuous from the same position, apparently ruining the camouflage, at least as far as

colour-sensitive prey are concerned. However, red light is absorbed by seawater much more rapidly than shorter wavelengths, so that red coloration appears black at depths below 10–20 m and this may not be an important factor at typical foraging depths.

The brown breeding plumage of Kittlitz's and Marbled murrelets is presumably an adaptation to predator avoidance while on their inland nest-sites. Both become black above and white below outside the breeding season. It may be significant that both have a relatively short period in summer plumage, only attaining the brown breeding plumage just before breeding commences, although some individuals indulge in breeding activities, such as visiting the nest-site, throughout the winter. However, the all-dark plumage of the Tufted Puffin and Crested Auklet cannot adequately be explained by anything that we currently know about these species' ecology. It raises some questions about camouflage as a complete explanation for the prevalence among auks of plumage that is black above and white below.

A striking feature of the head feathering in murres and the Razorbill is the presence of a very narrow channel in the feathers leading back from the eye for several centimetres. In the bridled form of the Common Murre this channel is picked out in white feathers. Except in the bridled murre, the channel is hard to see and seems unlikely to have any significance in communication. It seems possible that it may function in some way to aid the flow of water over the eye while the birds are swimming rapidly, which would explain its occurrence only in the fast-swimming auks. The possibility that the strong development of the boney ridges above the eye sockets (supra-orbital ridges) in Alca and Uria may be associated with resisting deformation of the eyeball under the pressures experienced during deep dives deserves investigation.

Diving depth and behaviour

Our knowledge of the depths at which auks forage has increased considerably in the past decade, with the deployment of Maximum Depth Recorders (MDR) and Time Depth Recorders (TDR). Large birds, having a lower metabolism and cost of propulsion (because of lower drag per unit weight), but similar oxygen storage capacities, per unit mass, should be capable of diving for longer than small birds (Wilson 1991). Given similar or greater underwater swimming speeds, that should allow larger birds to dive to greater depths than small birds. Such a correlation has been demonstrated for the maximum diving depths of auks (Fig. 4.5), with the smallest species measured (Dovekie) diving to a maximum of 30 m, while the largest, the murres, dive to maxima of more than 150 m (Burger 1991). A similar correlation can be found in the average length of time spent submerged.

Considerations of morphology suggest that the murres, Razorbill, and murrelets should have the best underwater performance (Kuroda 1954, 1967), while auklets and puffins, being better adapted for walking, and having larger wing areas relative to weight, should be less efficient. Guillemots appear intermediate in structure. Adaptation for better underwater performance in murres makes sense because of their habit of pursuing schooling fish capable of taking evasive action. The typical prey of auklets (copepods,

euphausiids) is probably too slow to require high swimming speed. Evidence is currently insufficient to test this idea, but the maximum diving depths recorded for murres are strikingly greater than those of other auks (Burger 1991). Presumably the Great Auk was also a deep diver. Differences in preferred foraging depth could explain the coexistence of murres and puffins over most of their range, despite the fact that diets are often similar (Piatt and Nettleship 1985; Burger and Simpson 1986).

Detailed information on dive profiles is available only for Thick-billed Murres feeding chicks at Coats Island, Canada, and then only for six individuals and for one foraging trip each (Croll et al. 1992). Wilson (1991), writing before Croll et al.'s results were published, suggested that pelagic-feeding alcids probably spent similar amounts of time throughout their depth range ('bounce dive' trajectory), as appears to be the case for some penguins. However, Croll's results showed that Thick-billed Murres dived repeatedly and rapidly (1–2 m/sec) to a fairly uniform depth and then spent some time there before rising equally rapidly to the surface ('flat-bottom dive'). As the population at Coats Island was believed to be feeding mainly on mid-water fishes and zooplankton in water more than 200 m deep, the depth at which they foraged

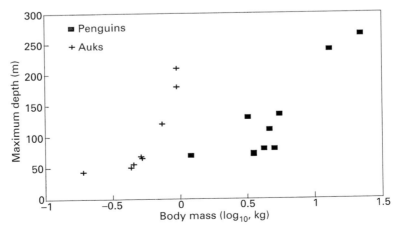

4.5   Maximum dive depth in relation to body mass for auks and penguins (adapted from Burger 1991).

presumably coincided with a layer of abundant prey, rather than representing the sea floor. Daytime dives were generally much deeper, and much less frequent, than those at night, indicating that the prey concentration may have moved up and down with a 24-h-cycle. Such behaviour is characteristic of the large zooplankton that make up much of the summer diet of Thick-billed Murres in that region (Gaston and Bradstreet 1993). King Penguins (*Aptenodytes patagonicus*) show similar alternation in diving depth with time of day (Kooyman *et al.* 1992).

In addition to the study by Croll *et al.* (1992), TDRs have also been attached to Rhinoceros and Cassin's auklets (Burger and Powell 1990; Burger *et al.* 1993). These studies, and some using MDRs (Barrett and Furness 1990), suggested that the normal diving depths of auks tend to be much less than the maximum depths recorded. Thick-billed Murres at Coats Island mainly foraged between 20 and 50 m, while Rhinoceros Auklets concentrated between 10 and 30 m and Cassin's Auklets at even shallower depths.

Information on the diving performance of bottom feeding guillemots can be obtained from plotting the position of diving birds and measuring the water depth, or reading it off a chart. Cairns (1987*b*) found that Black Guillemots in Arctic Canada fed mainly in water less than 30 m deep, while Clowater and Burger (1994) showed that Pigeon Guillemots avoided depths greater than 40 m and showed a marked preference for waters about 20 m deep.

While feeding, auks tend to make a series of dives fairly close together (a bout), followed by a prolonged rest. The length of inter-dive intervals (pauses) tells us something about the time needed by the bird for recovery and respiration and is often standardized in relation to the length of the dive by expressing it as the 'dive–pause ratio'. As this ratio decreases, the bird is taking longer to recover between each dive. This ratio varies with the length of the dive, being highest for brief immersions. This effect occurs because, as the length of dives increases, so does the proportion of the dive

during which respiration is anaerobic, causing the build up of lactate in the blood. The lactate must be cleared by aerobic respiration at the surface before the next dive. The diving performance of auks, in terms of duration and depth of dives, is rather good when compared with that of penguins. Such comparisons are inevitable for auk biologists because penguins, with their very long specialization in aquatic life, should be the best avian diving machines. Certainly, the performance of King and Emperor Penguins (*Aptenodyles forsteri*:) which regularly dive below 200 m, cannot be matched among auks. However, the statement by Kooyman *et al.* (1992) that for murres 'the vast majority of dives are <40 m' ignores the fact that most Thick-billed Murres reach greater depths at least daily, many diving below 100 m (Croll *et al.* 1992; AJG, personal observation). The normal diving depths of murres are greater than those measured for penguins several times as large (Burger 1991; Fig. 4.5) and may indicate exceptional physiological abilities relating to anaerobic respiration (Croll *et al.* 1992). However, this may not apply to other auks; anaerobic respiration appears to be rare in Rhinoceros Auklets (Burger *et al.* 1993).

Auks in marine ecosystems

There have been a number of attempts to assess the impact of marine birds on the food webs of which they form part, through estimating total prey consumption and comparing this with prey standing stocks or productivity. In some of these studies auks have been shown to take an important portion of the total amount of food being taken by the seabird community (Table 4.2). Overall, however, the share of auks in the biomass consumed tends to be lower than their representation in the resident seabird communites would suggest. The difference is accounted for in most cases (all except Foula) by the preponderance of wintering southern hemisphere visitors in the total seabird biomass.

Table 4.2   Energy use by auks in different marine areas, compared with energy use by other marine birds.

Area	Energy use (mgC/m²/year)		ref
	Auks	Other seabirds	
Oregon	151 (48%)	163 (52%)	Wiens and Scott 1975
Foula, Shetlands	89 (37%)	151 (63%)	Furness 1978
California	25 (27%)	65 (73%)	Briggs and Chu 1987
Bristol Bay, Bering Sea	43 (27%)	118 (73%)	Schneider *et al.* 1987
George's Bank	27 (20%)	110 (80%)	Powers 1983
Eastern Canada	47 (34%)	92 (66%)	Diamond *et al.* 1993

In eastern Canadian waters, Diamond *et al.* (1993) estimated the contribution of different seabird species to overall energy consumption by seabirds in different oceanographic areas. In winter (January–March), auks comprise 46 per cent of seabird energy demand on the New-foundland Banks, 41 per cent in the Gulf of St Lawrence and 22 per cent on the Scotian Shelf. These figures fall to 11, 17, and 0 per cent in July, when wintering Thick-billed Murres have moved northwards and non-breeding Great Shearwaters are present in large numbers, providing 59 per cent of seabird energy demand. In that month, auks comprise 89 per cent of seabird energy requirements in the high Arctic, 96 per cent in Hudson Strait and 89 per cent in Hudson Bay (Fig. 4.6). Overall, they were estimated to consume 34 per cent of the energy used by seabirds in the continental shelf waters of eastern Canada.

In the western Gulf of Alaska, Hatch and Sanger (1992) estimated the possible impact of puffins on stocks of late-larval/early juvenile walleye pollock. Being the basis of an important commercial fishery, pollock stocks are assessed intensively in that area. Their calculations suggested that puffins took about 10 per cent of that age-class of pollock. However, they assumed that adult puffins ate the same proportion of pollock that they fed to their chicks. As adult energy requirements made up more than 90 per cent of the total, this assumption had rather a large effect on the estimated impact.

Both biomass and energy estimates suggest that the ecosystems where auks should have the greatest impact are those of the high Arctic. These ecosystems are characterized by

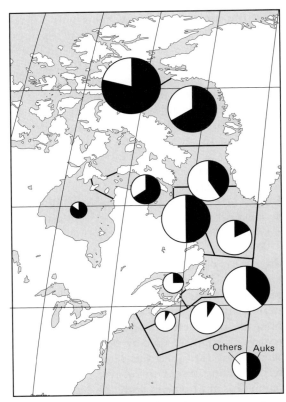

4.6   Energy consumption by auks as a proportion of total seabird energy consumption in the north-west Atlantic (data from Diamond *et al.* 1993).

a brief pulse of high productivity during the Arctic summer (starting with production by under-ice algae), and an even briefer window of opportunity between the break up of the consolidated pack-ice and the onset of refreezing in autumn. Production and consumption by different ecosystem components are probably best known for the Lancaster Sound–Barrow Strait area of the Canadian Arctic. There, consumption by seabirds was estimated to be 2.7 kJ/m²/year, of which Thick-billed Murres accounted for 66 per cent and Black Guillemots for a further 1 per cent (Welch *et al.* 1992). This compares with 10.6 kJ/m²/year taken by marine mammals in the same area (excluding the benthic feeding Walrus). Hence the impact of the seabirds on the pelagic food web appeared to be only a quarter that of marine mammals. Moreover, this calculation ignores the needs of Bowhead Whales (*Balaena mysticetus*), now much reduced from their earlier numbers, which might have played an important role in this ecosystem prior to commercial whaling. If we consider planktivorous auks, their impact on prey stocks is even smaller. Springer *et al.* (1987) estimated that all seabirds in the northern Bering Sea consumed less than 0.5 per cent of the copepods carried through Anadyr Strait in summer. The role of auks in Arctic marine food webs, although significant, should not be overemphasized.

Auks and terrestrial food chains

Vegetation

Although they visit land for only a short period, and generally only close to the shoreline, the very dense concentrations of birds at some auk colonies have appreciable affects on local soils and vegetation (Furness 1991). In the Arctic, manuring by Thick-billed Murres, Auklets, and Dovekies creates local pockets of vegetation in otherwise barren areas. The effects of Dovekies on terrestrial vegetation in Spitzbergen extends away from the colonies in peripheral areas where non-breeders circle and along the route taken by breeders to and from the sea (Stempniewicz 1990). Growth of vegetation as a result of manuring at the colony may make some breeding sites less attractive, resulting in a gradual shifting of the colony, and further extending the nutrient enriched area.

The moss *Dicranum groenlandicum* is a principal component of peat, laid down over thousands of years, on flat ground above Thick-billed Murre colonies on Coats, Digges, and Akpatok Islands in Hudson Strait (Gaston and Donaldson 1995). The peat provides a foothold for plants, such as cotton grass (*Eriophorum*), that are otherwise absent from the cliff-tops. Such peat does not occur on cliff-tops elsewhere in the region and presumably forms as a result of nutrients blown up the cliffs by strong winds (not uncommon in such places). On the colony cliffs themselves, where instability of the rocks does not encourage the formation of peat, scurvy grass (*Cochlearia officinalis*) often forms luxuriant clumps up to 40 cm in diameter. On the colony at the Minarets, Baffin Island, it grows up to an altitude of 800 m. Away from seabird colonies scurvy grass seldom forms clumps more than 10 cm across and does not grow above 200 m elevation in the Minarets area (AJG). In Greenland, several vascular plants occur only in the vicinity of seabird colonies at the northern limit of their range (Salomonsen 1979).

Seabird islands in low Arctic and boreal waters may also be affected by the manuring and burrowing of the auks. In the Queen Charlotte Islands of British Columbia, a distinctive tussock grass understory develops in forests subject to burrowing by Rhinoceros and Cassin's Auklets, but does not invade forest occupied only by Ancient Murrelets, which do not defecate in the burrow. Burrowing activities have resulted in the gradual erosion of some colonies, making the islands uninhabitable to the culprits. This seems to have occurred on Grassholm, Wales, where a colony of tens of thousands of Atlantic

Puffins destroyed most of the soil layer some time in the last century (Lockley 1953). Similar erosion has affected the puffin colony at the Farne Islands, England (Harris 1984).

Terrestrial predators

The dense aggregations of auks that form at colonies provide a tempting food source for predators. Adult birds of practically all species may be preyed on by terrestrial predators while breeding. Red foxes (*vulpes vulpes*) and Arctic Foxes are probably the most widespread and important predators. The concentration of Arctic Foxes associated with the Dovekie colonies in Thule District of Northwest Greenland forms the basis for an important local fur trapping industry, one that would probably not exist without the birds.

In Alaska and British Columbia, the local race of the Peregrine Falcon (*Falco peregrinus pealei*) specializes in feeding on marine birds and the smaller auks are its principal prey. Unlike most peregrines, but like Eleanora's Falcon (*Falco eleonorae*) (Walter 1979), Peale's peregrines hunt mainly over the sea, skipping low over the wave tops and snatching Ancient Murrelets and Cassin's Auklets either on the water, or in the act of taking off. The social aggregations of Ancient Murrelets that occur at sea several kilometres from their colonies are much denser and more active on calm nights than when the sea is choppy, perhaps because calm conditions allow them to detect peregrines more easily (Gaston 1992a). Peregrines nesting on coastal cliffs in northern Scotland also take many auks, apparently preferring Atlantic Puffins (Ratcliffe 1980), while Gyrfalcons (*Falco rusticolus*) may specialize on Thick-billed Murres at certain Arctic colonies (Gaston *et al.* 1985).

Conclusions

There is evidence from population studies of many auk species that populations track their food supplies fairly closely (Cairns 1992c).

The main factor controlling population size appears to be the survival of young birds to recruitment and this seems most likely to operate through winter mortality.

The fact that many populations are food-limited does not mean that density-dependent population adjustment necessarily occurs. As we have seen, we cannot expect changes in auk populations to have any strong impact on their prey organisms, because other, competing predators, such as marine mammals and large fishes, probably have a much greater impact. Some evidence of density-dependent effects on reproduction among auks has been presented by Gaston *et al.* (1983a), Gaston (1985c), and Hunt *et al.* (1986), who observed that certain measures of reproductive success decreased with increasing colony size. Cairns (1992c) dubbed this the 'hungry horde' effect, because the proposed explanation involves the depletion of food close to breeding colonies. This causes birds to forage further away, so that more time is spent on travel and less on prey capture. Some instances of density-dependent effects operating through breeding site limitation have also been observed (Atlantic Puffins, Nettleship 1972; Cassin's Auklet, Manuwal 1974a).

Despite this evidence of density dependence in some aspects of breeding, overall density-dependent population regulation may be absent if the main factor controlling populations occurs outside the breeding season, as proposed by Lack (1966). Over the sort of time scale that we can satisfactorily measure (decades) most auk populations exhibit linear trends and many sympatric species fluctuate in parallel. For example, throughout the period 1969–87 Black Guillemots, Common Murres, Razorbills, and Atlantic Puffins all increased in Britain, presumably because common prey species were abundant (Harris 1991; Lloyd *et al.* 1991). Similar parallel fluctuations among the same species in the Gulf of St Lawrence occurred over the same period (Chapdelaine and Brousseau 1991). Evidence for density dependence and interspecific competition are likely to be rather cryptic in the face of these larger trends. It seems best to view auks, along with the rest of the seabird community, as responding to larger ecosystem processes, rather than being keystone species driving important effects. Paradoxically, for marine organisms, their most obvious impact on ecosystems may be the local impact on terrestrial vegetation around Arctic breeding colonies.

5

Social behaviour

Introduction

Individual auks do not function independently; they interact with other individuals of their own and different species as competitors for food, for breeding sites, and for mates, and they co-ordinate with each other when foraging, breeding, and deterring predators. Social behaviour mediates all of these interactions, and thus is fundamental to the biology of the auks, as it is to most animals. As we have described in other chapters, a notable feature of alcid biology is the remarkable diversity of the ways they live: for example, ranging from extreme coloniality to solitary nesting, from displaying elaborate ornaments to hiding behind cryptic plumage, from rearing the chick to adult size at the nest to taking it to sea at 2 days of age, and from foraging close to shore within a few hundred metres of their breeding sites to making daily journeys of hundreds of kilometres (Bédard 1969a). Given this wide diversity of both social behaviour and natural history, it is worth considering the aspects of biology directly important to social behaviour that are shared by all auks. First, the form of their behaviour at their breeding sites has been strongly affected by natural selection related to their aquatic lifestyle. Second, like practically all seabirds, alcids are socially monogamous, meaning that breeding adults pair each year with a single opposite sex partner with whom they associate during the breeding season and co-operate in rearing the young. Social monogamy in alcids does not imply that courtship liaisons and even copulations outside the pair do not occur (see below), yet the basic reproductive unit is one male and one female. Because of the lengthy incubation period, and the demands of provisioning a growing chick with food that is obtainable only at a distance from the nest-site, both pair members are required for incubation of the egg and care of the young to independence. In some species, the chick is cared for exclusively by the male parent after it leaves the nest-site but before independence is achieved. The female remains at the colony and guards the nest-site from prospectors to ensure that it is available the following year.

Making sense of the enormous natural variety presented by auk behaviour is a challenging task. We have chosen to discuss selectively aspects of social behaviour that we believe are too important to omit, have been the subject of recent work, or are neglected topics that deserve more attention from seabird researchers. Much of our commentary is speculative—we hope to advance science if by no other means than by stimulating discussion of these complex aspects of auk natural history.

Coloniality

Like most seabirds, most alcids are colonial breeders. Many authors have explained their coloniality in terms of benefits such as avoiding predation (Lack 1968), a shortage of suitable breeding habitat (Ashmole 1971), and the potential for exchanging information about the location of feeding grounds (Ward and Zahavi 1973). Birkhead (1985) provided a detailed review of the various advantages and disadvantages of coloniality to breeding auks, while the adaptive significance of coloniality in birds in general was discussed by Wittenberger and Hunt (1985). We believe another detailed dissection is probably unnecessary. These reviews provide a catalogue of the potential advantages and disadvantages of coloniality, some supported by direct empirical corroboration. The evidence seldom permits us to assign a single factor as being responsible for coloniality in any one species. A simple explanation of auk coloniality is that auks inherited aspects of behaviour and morphology related to colonial breeding from their gull-like ancestors. Thus the tendency to nest in colonies is not an adaptation in the strict sense of the term (i.e. an evolutionary change related to a particular selective force) specific to the auks; it is part of their heritage shared with most of the gulls. Furthermore, virtually all the world's seabird species are colonial, so it is virtually certain that auks are colonial because they share certain ecological similarities with other seabirds. However, these explanations are unsatisfying because, although they tell us why auks in general tend to be colonial, they do not account for the variation in colonial behaviour among the living auks.

It is a fair generalization that those seabird species that feed the furthest offshore and at the greatest distances from their nest-sites are the most colonial breeders, while those that feed close to shore or close to their breeding sites tend to breed in small groups or solitarily (Lack 1968; Ashmole 1971). This suggests there is a basic link between foraging behaviour and social behaviour. Indeed, there is no seabird species that we know of that nests solitarily and feeds in large flocks at sea. Thus the key to understanding colonial tendencies among the auks may be a thorough understanding of the function and significance of their social interactions at sea. Unfortunately, biologists have as yet barely described the

worldwide distribution of auks at sea, the few detailed at-sea studies have concentrated on the distribution of birds and their prey, and a detailed understanding of their pelagic social behaviour is a long way off. Nevertheless, the most compelling of the functional hypotheses for coloniality suggests that colonies form because they bring individuals into close contact with large numbers of others of their species, allowing them to obtain information about the location of patchy and unpredictable food resources by watching the behaviour of other birds (i.e. in the terminology of Ward and Zahavi [1973] they act as information centres). The importance of this effect for foraging birds is still a matter of controversy, but for seabirds, and especially auks, the potential benefits of such information transfer are very great. Many auks travel long distances from their colonies to feed and expend a lot of energy on flight. Arriving at a previously productive feeding area to find that conditions have changed and food is no longer available could be a costly mistake and one that would be easily avoided by individuals observing the incoming flight directions of successful foragers.

Auks that breed at high latitudes, have diurnal colony attendance, and are offshore feeders on patchily distributed prey (e.g. Least and Crested Auklets, Dovekie, Thick-billed Murre) are likely to benefit the most from visual contact with conspecifics, at the colony as well as at sea (Birkhead 1985). For example, at Prince Leopold Island, Gaston and Nettleship (1981) noted that Thick-billed Murres departing from the colony immediately after delivering a fish generally headed straight for the horizon, alone. In contrast, those that had just completed a 12-h brooding shift departed first to the gathering area on the sea adjacent to the colony and then left to feed in flocks. They also noticed that outbound flocks often changed course in response to incoming birds, a clear indication that they were obtaining information about where arriving birds were coming from. Observations of other species of auks leaving their colony in co-ordinated groups to travel to the foraging area (Ashcroft 1976; Birkhead

1976b) also suggest that colonial auks may take cues from their neighbours when choosing where to feed. Clearly, the more predictable the prey, and the smaller the amount of time spent travelling to and from the feeding areas, the less useful the information centre effect is likely to be. This may partly explain why guillemots (*Cepphus*), which normally forage within sight of their breeding sites on predictably resident benthic fishes, do not form large colonies (Lack 1968; species accounts). The case of the Razorbill may be similar: among the Alcini it is the most likely to nest in small groups or even isolated pairs, and at the same time it forages the closest to shore and is the most likely to enter protected waters and estuaries to feed on sandlance, a demersal fish. Most of the small auklets nest in large, dense colonies, but Whiskered and Parakeet Auklets are exceptions. Two factors probably explain why Whiskered Auklets can nest at lower densities and take advantage of a greater range of breeding habitats than their congeners. First, this species feeds close to shore and to their breeding sites where their euphausiid prey is predictably concentrated (in tide rips in Aleutian passes; Hunt *et al.* 1993), reducing the 'information centre' benefit of coloniality. Second, Whiskered Auklets have taken to nocturnal colony attendance, providing protection from avian predators and reducing the possible 'predator swamping' advantage of coloniality (see below). Parakeet Auklets forage offshore, but utilize dispersed prey and are solitary or occur in small groups at sea (Harrison 1990; Hunt *et al.* 1993), again reducing the potential 'information centre' benefit of coloniality; how they avoid avian predation that seems to affect the similar sized Crested Auklet is unknown.

The role of predation in the evolution of coloniality among auks is unclear. Auks do not generally mob predators, in the way that gulls or terns may do, probably because of their limited manoeuvrability in flight, so this potential direct benefit is absent. However, there is considerable indirect evidence of the advantage of colonial breeding for predation avoidance from some auk species. Murres

combine in the 'Macedonian phalanx' defence, with bills pointing towards the predator, to ward off gulls and foxes (Birkhead 1977b; Gaston and Elliot 1996); this is necessitated by their choice of open ledges on which to nest. Murre colonies that have been decimated by hunting are subject to increased predation by gulls (G. Gilchrist, personal communication) because so many individuals are left stranded without the cover of conspecifics. These observations make successful solitary nesting by murres hard to imagine, considering predation alone. Individual auklets and Dovekies certainly reduce the chances of being taken by predatory gulls and raptors by nesting in huge colonies where the dense swarms of birds that fill the air confuse predators. Individuals that become separated from the wheeling flocks or attempt to land on unoccupied breeding habitat quickly become a predator's meal. However, such large aggregations may also act as a magnet for predators and no clear evidence exists about whether predation rates at small colonies are lower or higher than at large colonies. For the nocturnal species such as the *Synthliboramphus* murrelets and some auklets the predation-avoidance advantage of coloniality is diminished by the fact that individuals are normally invisible to predators when visiting the colony. Taken together, these observations suggest that predation is an important factor favouring coloniality in some species, but is probably not an important explanation for all species. In addition, the presence of predators may favour non-coloniality, such as in the case of the *Brachyramphus* murrelets (see below).

The species of auks with the largest colonies, and the ones least likely to occur in small colonies, include the Least and Crested Auklets of Beringia, the ecologically similar Dovekie of the North Atlantic, and the cliff-nesting murres. These species can be regarded as pretty much tied to colonial breeding. All breed at high latitudes, have diurnal colony attendance, and are offshore feeders that form dense aggregations over prey patches (Table 5.1, see species accounts). At the other end of

Table 5.1 Basic social behaviour characteristics of the Alcidae

	Degree of coloniality	Activity timing	Gathering ground	Ornaments	Copulation location	EPC frequency
Great Auk	+++	Diurnal	?	+	?	?
Razorbill	+++	Diurnal	++	+		
Common Murre	+++	Diurnal	+	+	On land	High
Thick-billed Murre	++++	Diurnal	+	+	On land	High
Dovekie	++++	Diurnal	++	−	On land	Low
Black Guillemot	++	Diurnal	++	++	On land	Occurs
Pigeon Guillemot	++	Diurnal	++	++	On land	Low?
Spectacled Guillemot	++	Diurnal	?	++	On land	Low?
Ancient Murrelet	+++	Nocturnal	+++	++	?	?
Japanese Murrelet	+++(?)	Nocturnal	+++(?)	++	?	?
Xantus' Murrelet	++	Nocturnal	?	−	?	?
Craveri's Murrelet	++	Nocturnal	?	−	?	?
Marbled Murrelet	+	Crepuscular	−	−	?	?
Kittlitz's Murrelet	+	Crepuscular	−	−	?	?
Cassin's Auklet	+++	Nocturnal	−	−	?	
Parakeet Auklet	++	Diurnal	++	+++	At sea	Low
Crested Auklet	++++	Diurnal	++	++++	At sea	High
Least Auklet	++++	Diurnal	++	++++	At sea	High
Whiskered Auklet	++	Nocturnal†	+++	++++	At sea	High
Rhinoceros Auklet	+++	Variable	+	++++	?	?
Atlantic Puffin	++++	Diurnal	++	+++	At sea	Low
Horned Puffin	+++	Diurnal	++	+++	At sea	?
Tufted Puffin	+++	Diurnal	++	++++	At sea	?

EPC, extra-pair copulation. Coloniality, + nests solitarily only; ++ usually nests in colonies of less than a thousand birds, occasionally nests solitarily; +++ usually nests in colonies of thousands of birds; ++++ usually nests in colonies of tens or hundreds of thousands; timing of colony activity, nocturnal if activity starts when darkness is complete, † Whiskered auklets are sometimes seen by day at one colony (see species account); gathering ground, − not present, + some social activity in small groups takes place at sea near colony, ++ intense social activity at sea, +++ intense activity and the only diurnal social activity takes place at gathering ground; ornaments include foot colour, mouth lining colour, bill colour, and accessory plates, presence of wattles, and number of crests and plumes; copulation location, marked as ? if there are few or no published observations.

the spectrum are the *Brachyramphus* murrelets, which stand out as the only truly solitary nesters among the auks. In between, the majority of auk species occur in colonies that cover a wide range of sizes. For these species, local population density is probably an important factor in determining colony size. The *Synthliboramphus* murrelets generally occur in colonies of a few hundred to a few thousand pairs. However, Ancient Murrelets in Haida Gwaii, the Queen Charlotte Islands of British Columbia where the species reaches its highest density, occur in colonies of up to 70 000 pairs, although most colonies are much

smaller. Likewise, Rhinoceros Auklets are almost entirely confined to colonies of more than 10 000 pairs in British Columbia and Washington but persist in much smaller aggregations in Oregon and California, on the periphery of the species' range.

There have been at least four cases of reduced colonial tendencies or solitary nesting in auks (for simplicity, we have assumed that the ancestral auk was colonial, so we suggest these changes involved the abandonment of colonial breeding by an ancestral species). These include the derivation of the *Brachyramphus* murrelets, the guillemots *Cepphus*, the

Whiskered and Parakeet Auklets, and possibly the sub-tropical *Synthliboramphus* murrelets (Table 5.1, see accounts). The *Brachyramphus* murrelets stand out as oddities among the alcids by choice of nest-sites, solitary habits, and cryptic appearance. These species forage in sheltered inshore waters, but growing evidence indicates they forage at considerable distances from their nest-sites on clumped unpredictable prey (see species accounts). The selective forces that led to the evolution of this group are unclear, but clearly, coloniality would have few advantages to small auks nesting in the open, and escaping predation must have been a factor. Predators are scarce in the bleak terrain where Kittlitz's Murrelets nest on alpine tundra and Aleutian barrens. Similarly, Marbled Murrelets' use of old-growth trees allows them to nest in areas where terrestrial predators, such as foxes, Raccoon, and Mink *Mustela* would preclude successful ground nesting.

Questions about the causes and consequences of colonial behaviour are likely to remain fodder for discussion among auk biologists for some time to come. One confounding factor in understanding the benefits of coloniality is that auk biologists seldom investigate situations in which their subject species breed in small groups. In the Pribilof Islands there are several tiny colonies of less than a dozen pairs of Least Auklets, raising the question: how can these isolated individuals of a typically highly colonial species persist? There are similar examples from other auk species that need attention. Perhaps because of our tendency to think of big groups of animals as the best examples of the nature of the species concerned, we potentially have an incomplete perspective. After all, every big colony must have started out as a small one.

Nocturnality

There are no truly nocturnal seabirds, as there are no species that restrict all their activities to the hours of darkness and roost immobile by day. However, many species nesting in burrows and crevices, particularly among the smaller members of the Alcidae and Procellariiformes, are active above ground at their breeding sites only at night. These species are present at breeding areas by day, incubating their eggs and brooding their offspring at concealed sites. However, they restrict their conspicuous social activity and their comings and goings to and from the nest-site to periods of darkness. These seabird species are commonly referred to as *nocturnal* and the remaining auk species that are active at nesting areas in daylight as *diurnal*, although their nocturnality refers only to the timing of social activity on land and may have little to do with the timing of foraging and other activities at sea during the breeding and non-breeding seasons.

We believe alcids evolved from a diurnal colonial ancestor, because most alcids are diurnal, as are virtually all of their closest relatives among the Charadriiformes. If this assumption is correct, nocturnality has evolved at least three and probably four times in the auks. All members of the murrelet genus *Synthliboramphus* and the Cassin's Auklet are extremely nocturnal at colonies, Whiskered Auklets are typically nocturnal at colonies, while Rhinoceros Auklets are nocturnal at most of their colonies and diurnal or crepuscular at others. Each case probably represents a separate evolutionary switch from diurnal to nocturnal social behaviour. For the purposes of this discussion, the *Brachyramphus* murrelets are considered to be crepuscular rather than nocturnal, because their social activity at nest-sites is focused around dawn and dusk, occurs partly during daytime, and is very reduced or absent during the hours of darkness (see species account). Thus nocturnality probably evolved once in the ancestral *Synthliboramphus* murrelet, possibly twice in the tribe Aethiiini, and once in the puffin tribe in the Rhinoceros Auklet.

The nocturnal alcids are much less agile on land than their diurnal relatives, with their legs set further back on their bodies. Consequently, their posture is more horizontal. This is most pronounced in the *Synthliboramphus* murrelets which rest on their sternum when perched. Presumably this awkwardness is due to their hind limb structure being better adapted to underwater swimming than to terrestrial locomotion. Nocturnality may have led to reduced selection for agility on land because with social activity at the colony occurring under cover of darkness, the risks from avian predators are low. On the other hand, extreme specialization to underwater swimming would lead to reduction of the hind limbs and could well have been a factor in the evolution of nocturnality! Whether their hind limb morphology was a cause or consequence of nocturnality, lack of agility has made nocturnal species extremely vulnerable to mammalian predators on land, and some of the most severe declines due to introduced mammalian predators have occurred in these species (e.g. due to Raccoons, Gaston 1994, and foxes, Bailey 1993).

Nocturnality probably has not led to development of exceptional adaptations of night vision, because the nocturnal alcids do blunder about in their colonies on dark nights, and do not consistently have larger eyes than their diurnal relatives. In any case, nocturnality might not be expected to add any additional selective pressure for vision in low-light situations, because most alcids hunt their prey deep under water where light levels are very low. For example, murres regularly forage at 60 m or more beneath the sea surface (Burger 1991), so they would be expected to have as acute vision in dim light as any of the nocturnal species.

One consequence of nocturnality in some avian species is an increased reliance on vocal rather than visual displays. Ancient Murrelets are a good example of this, because they are extremely vocal at night at their colonies, and have an unusually complex vocal repertoire which could be related to their nocturnal social behaviour. Their complex chattering calls provide information about individual identity, and gender, and are the basis for mate attraction and parental care (Jones *et al.* 1989; Fig. 5.1). The impressive vocal displays of males, often performed from the high branches of coniferous trees, are reminiscent of passerine songs to human listeners and almost certainly have a similar function in attracting makes. Ancient Murrelet vocal behaviour also has much in common with the

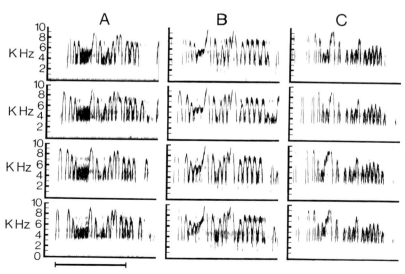

5.1   Sound spectrographs of the calls of three individual Ancient Murrelets, showing how the calls of individuals (in columns) are stereotyped while among individuals there is much more variation in structure.

nocturnal noise-making of many storm-petrel and shearwater species. However, other nocturnal auks such as Cassin's, Whiskered, and Rhinoceros Auklets do not seem to have more complex vocalizations than their diurnal relatives, although they are very vocal. Possibly, the *Synthliboramphus* murrelets have been nocturnal for a longer span of time and have thus managed to become more specialized. Taken together, these bits of evidence about vision and vocal behaviour suggest that the nocturnal alcids are not nearly as highly specialized as truly nocturnal bird species such as owls and nightjars.

What factors led to the evolution of nocturnal behaviour in alcids? The answer to this is certainly that nocturnality is an adaptation for avoiding avian predators. Small alcids are extremely vulnerable when exposed on land by day, are easily taken by a gamut of avian predators including corvids, gulls, and raptors ranging in size from Sharp-shinned Hawks (*Accipiter striatus*) to Bald Eagles *Haliaeetus leucocephalus*). Nocturnality is thus the rule among small alcids, except when they nest in remote regions at high latitudes. There, where daylight is continuous during the breeding season, they nest in enormous colonies (e.g. *Aethia* auklets, Dovekies) that swamp the few predators that are present. The other exception is the *Brachyramphus* murrelets that have taken to cryptic solitary nesting which reduces both mammalian and avian predation. Except for the Marbled Murrelet, all the small alcids nesting south of 53° latitude are nocturnal. Whiskered Auklets, the most southerly distributed of the genus *Aethia*, may have taken to nocturnality to avoid nest-site competition with Least and Crested Auklets, because under cover of darkness they could avoid avian predators and nest at low densities in a wider range of habitats. Rhinoceros Auklets are an interesting case, because they are nocturnal at some colonies and crepuscular or diurnal at others. Possibly, this species is facultatively nocturnal depending on the predators present. They are nearly fully nocturnal where Bald Eagles are common (i.e. British Columbia and South-east Alaska), but diurnal where eagles are scarce or absent or where they nest in sea caves that afford protection to individuals attending the colony (Aleutian Islands, Alaska; Sea Lion Caves, Oregon; Farallon Island and Channel Islands, California). The presence of kleptoparasitic gulls could also be a factor that drives Rhinoceros Auklets provisioning chicks to take cover in darkness.

The evolution of nocturnality probably did not have much to do with foraging requirements, because it offers no clear advantage or disadvantage to feeding. Nocturnal colony attendance probably does not restrict the nocturnal species to daytime feeding, as most have off-duty periods between incubation shifts lasting for a full day and night or more than 2 days. In any case, zooplankton, the preferred prey of most small alcids, tends to be closer to the sea surface and perhaps more available at night, when the nocturnal alcids are often busy socializing at their colonies. There is also little evidence that nocturnal species take a significantly different spectrum of prey than diurnal auks foraging in the same waters because of the timing of their activities on land versus the sea (e.g. Rhinoceros Auklets versus Tufted Puffins at Triangle Island, British Columbia (Vermeer 1979); Marbled versus Ancient Murrelets at Langara Island, British Columbia (Sealy 1975b); or Whiskered versus Least Auklets at Buldir Island, Alaska (Day and Byrd 1989)). We think that predation is a much more likely explanation for the occurrence of nocturnal colony attendance among auks.

Natural and sexual selection and the evolution of ornamentation

One of the attractive features of the Alcidae is the occurrence of unusual and even bizarre ornamentation displayed by some species during the breeding season. Although some species are very plain in appearance, more alcid species have breeding ornaments than expressed in any other Charadriiform group.

Two species, the Whiskered Auklet and Tufted Puffin, vie for the title of most elaborately ornamented seabird species. Natural selection explains many morphological, behavioural, and other traits of alcids, such as wing size and shape, egg size, patterns of colony attendance, and many aspects of social behaviour, because of the importance of these traits to survival and reproductive success. However, the question arises as to what selective forces have favoured ornamental traits such as the puffin's brilliantly coloured bill or the elaborate plumes of the Whiskered Auklet that seem to have no direct role in enhancing survival or reproduction. Darwin faced this question when formulating his evolutionary ideas, because his theory of natural selection did not explain adornments that had no obvious function in the rearing of offspring. If anything, these would seem to reduce survival by encumbering the individuals bearing them, or by making them more conspicuous to predators. The solution to this dilemma was his concept of sexual selection, by which he proposed individuals '... acquired their present structure, not from being better fitted to survive in the struggle for existence, but from having gained a (mating) advantage over other males ...' (Darwin 1871). In other words, sexual selection is the kind of selection that exclusively has to do with variation in mating success, rather than variation in other components of reproductive success (Andersson 1994). Stated another way, sexual selection can probably best be understood as a part of natural selection that considers only one component of reproductive success and fitness: the ability of individuals to obtain mates. Although relatively little attention has been paid to auks when thinking about sexual selection, we believe they provide an excellent example for study of this concept.

Monogamy, mutual mate choice, and sexual selection

Alcid ornamentation is displayed by courting birds during the breeding season, so is it likely that it is the result of sexual selection? For some birds and other animals it is easy to see that there is extreme variation in mating success, and that this variation has to do with ornamental traits. For example, Peacocks (*Pavo cristatus*) display their elaborate trains to females, and those males with the most elaborate adornments usually have the opportunity to mate with several females, while less ornamented males may obtain few or no matings (Petrie *et al.* 1991). This leads to variation in mating success among the males but not among the females, and explains the showy male appearance and dull female appearance of this and many other polygynous (i.e. males mating with multiple females) birds. However, most birds, and all alcids, are not only monogamous (i.e. pair with only one mate during a breeding season) but are also sexually monomorphic (i.e. males and females are identical or nearly identical in appearance).

It is less easy to see how sexual selection can work in a mating system based on monogamy, because by definition each individual gets only one partner, and in most cases both members of a pair must work together full time to care for the offspring, leaving no time for pairing with more than a single partner (Lack 1968). Given this, we might at first conclude that sexual selection cannot be important in the monogamous auks, because there is no variation in mating success to drive it. In fact, there are at least two important ways that variation in mating success arises in monogamous birds. The first is that mating success can be measured not only by considering the number of mates obtained, but also considering characteristics (i.e. quality) of the mate(s) obtained. The second is that mating success can vary in socially monogamous species if they are really cryptically polygamous because individuals engage in extra-pair copulations with other individuals in addition to their primary partner (Birkhead and Møller 1992; Wagner 1992c). In either case, research on alcids has provided some of the key insights on the related evolutionary theory.

Sexual selection and mate quality

Mating success can be measured by attributes of the mate(s) obtained. For example, on a nesting ledge in a seabird colony there might an equal number of males and females capable of breeding, and by the end of the pre-laying period each of the females will be mated to one of the males, so all individuals of both sexes will have the same mating success measured by numbers alone. However, some females will be capable of breeding earlier than others, and some males will be more capable of provisioning chicks with fish than others. That is to say, there is likely to be variation in many aspects of individual quality among members of both sexes, and there is likely to be competition and choice for the best mates. Some individuals will get high quality mates and other individuals will get low quality mates, producing variation in mating success. Traits that help individuals get higher quality mates will be favoured by sexual selection, whether these traits are preferences for certain types of mates, or physical characteristics that are attractive to potential partners.

An important aspect of sexual selection in the situation of monogamy is the possibility for mutual sexual selection (Darwin 1871), or inter-sexual selection simultaneously acting on both males and females. In most seabirds and all alcids, the individual characteristics that make a high quality male partner are likely to be the same that make a high quality female partner, because the sexes have such similar roles in reproduction, and sexual selection is likely to favour similar attractive traits in both males and females. Until recently there was no empirical evidence that similar mating preferences by males and females could be driving mutual sexual selection. However, Jones and Hunter (1993) performed experiments on Crested Auklets that demonstrated that males and females have a similar mating preference for the conspicuous forehead crest ornament, a preference that

5.2 A scrum of courting Crested Auklets, with three females surrounding a displaying male. Both sexes use their ornaments to attract mates.

favours individuals with large crests. Realistic male and female dummies were set up on display rocks on the colony at Buldir Island and the response of live birds was measured as they first looked at the fake auklets, which were equipped with either a crest of average size or a crest about the size of the largest naturally seen in the Crested Auklet population. Approaching birds acted as if the dummies were real and frequently tried to interact socially with them, and even more, both sexes reacted with more sexual displays to opposite sex dummies with larger crests, they approached closer, and they stayed interested longer. This was interpreted as indirect evidence that auklet crests evolved by sexual selection driven by a mating preference. Earlier work showed similar sexual preferences for bill colour and facial plumes by Least Auklets (Jones and Montgomerie 1992), but this study was less convincing because of the difficulty of sexing individuals of this species. Other alcid ornaments are probably favoured by similar mating preferences.

The evolutionary origin of auklet ornaments and preferences for them

It seems likely that most alcid ornamentation can be explained by inter-sexual selection, or in other words, that alcids have ornaments because these add-ons are preferred by both sexes when they choose mates. This may lead us to ask why they have these strange preferences in the first place. Darwin left open the question of why animals have exotic mating preferences, referring to an aesthetic sense, without giving an explanation (Darwin 1871). One explanation is that the ornaments and preferences function as species recognition or isolating mechanisms. The argument here is that the ornaments assist individuals in recognizing others of their own species, and in being recognized, and thus help avoid wasting time and energy on fruitless courtship with unsuitable mates. This hypothesis seems to make good sense, because ornaments differ a lot between species and could obviously aid in species recognition. However, there is only limited evidence that the origin of alcid ornaments and mating preferences has to do with their function as isolating mechanisms. Pierotti (1987) pointed out that no two sympatric North Pacific alcid species have exactly the same bill colour, consistent with the idea that the bill colour functions for species recognition. For example, Thick-billed and Common Murres, two congeneric species that breed in close association and sometimes side by side on the same cliff ledges, differ by the presence of a striking white line along the upper mandible of the bill in the former species. However, the spectacular adaptive radiation of these alcids could have resulted in a range of bill colours for other reasons. Furthermore, the different alcid species differ strikingly in size, vocalizations, display behaviour, and other non-ornamental aspects of appearance, leaving us wondering how ornaments could possibly add much to species identification. That alcid ornaments evolved as isolating mechanisms could be more firmly established if there was a case of repro-

5.3 An encounter between a Least Auklet and a Whiskered Auklet. The role of the ornaments in species recognition is unknown.

ductive character displacement (e.g. if two related species' ornaments were more different where they occur together than where they live apart), but there is no example of this from living alcids. Furthermore, evidence that Least Auklets, which have no crest ornament, have a mating preference for crests similar to the main ornament of the Crested Auklet (I. L. Jones and F. M. Hunter, unpublished data) is definitely inconsistent with the theory.

The isolating mechanism hypothesis has received little attention lately, but the origin of mating preferences for ornamental traits and elaborate displays has been a popular and controversial topic in evolutionary biology during the 1980s and 1990s. Currently there are three contending schools of thought on the origin of preferences for adornments: (1) the viability indicator or handicap mechanism (Zahavi 1975; Pomiankowski 1988); (2) the Fisherian runaway mechanism (Fisher 1930; O'Donald 1980; Lande 1981; Kirkpatrick 1982); and (3) sexual selection for sensory exploitation (Ryan 1990). The first mechanism is based on the idea that ornaments are favoured because they signal individual quality, the second on a positive feedback process that can favour arbitrary traits, and the third on the idea that biases built into the senses by natural selection can lead to the evolution of mating preferences for ornaments. Although testing these hypotheses has been fraught with difficulty (Balmford and Read 1991), the viability indicator mechanism is very compelling intellectually. It suggests that preferences for ornaments evolved because of the benefits to those choosing mates with well-developed ornaments, because well-ornamented mates are healthier or of higher quality. Attempts to test this hypothesis in auklet species have not provided confirmation, with little evidence that Least (Jones and Montgomerie 1992) or Crested Auklet (ILJ) ornaments signal quality, but with some suggestive evidence from Whiskered Auklets (I. L. Jones and F. M. Hunter, unpublished data). The negative evidence, and the observation that Least Auklets have a preference for the ornament of another species (I. L. Jones and F. M. Hunter, unpublished data), leave open the possibility that auklet ornaments are arbitrary traits that resulted from a Fisherian or sensory exploitation mechanism. Like many of nature's most spectacular phenomena, the origin of alcid adornments is likely to remain a mysterious concept for some time to come.

Understanding ornamental diversity among the alcids

The wide range of degree of ornamentation displayed in the auks, from the very cryptic *Brachyramphus* murrelets to the spectacular Tufted Puffin and Whiskered Auklet, is illustrated in the plates and summarized in Table 5.1. At first, the pattern seems bewildering, with both highly adorned and dull species occurring within the same tribes and genera (e.g. Aethiini and *Synthliboramphus*), and with no apparent relationship to nocturnal/diurnal colony activity. With such wide diversity among a small number of species, it is hard to apply the comparative method effectively and we are left merely to speculate about the origins of this confusing mosaic. Nevertheless, there may be some basic patterns worthy of discussion. The most recent phylogeny splits the alcids into two major groups, one that

includes the murres, guillemots, and all murrelets, and the other including the auklets and puffins (Friesen *et al.* 1996a; Fig. 2.7, page 31). This phylogenetic dichotomy concurs with much of the variation in showiness of ornamentation among the alcids (and this is reflected in even the earliest classifications of the auks, see Chapter 2). For example, among the 15 species in the murre–guillemot–murrelet group, not one species has a brightly coloured bill and only Japanese and Ancient murrelets have head plume ornaments (see Table 5.1). Among the nine species of the puffin–auklet group, all species except one have a brightly coloured bill with extra ornamental plates that are shed after breeding, and all species except (the same) one have complex facial ornaments including plumes, crests, etc. There is no obvious basic difference in ecology that could explain this dichotomy of ornamentation. The question remains: what is the basic idiosyncrasy of auklet–puffin social behaviour that requires them to have ornaments while the rest of the auks can get by without? The exception to the rule in the auklet–puffin group is the Cassin's Auklet which lacks ornaments, probably because it is extremely nocturnal and does not have social activity at a gathering ground, making visual displays next to useless. The degree of ornamentation seems to have more of an association with use of a gathering ground near the colony during daylight than with the diurnal versus nocturnal timing of activity on the colony itself. The nocturnal Ancient and Japanese murrelets and Whiskered Auklet all attend their colonies in pitch darkness, all have intense social activity at gathering grounds at sea near their colonies, and all have complex ornaments that are displayed in daylight.

One clear relationship between ecology and degree of ornamentation is found in the *Brachyramphus* murrelets which nest solitarily in the open on tree branches or on the ground. Not only do these species lack ornaments, which would make them more conspicuous, but they show the adaptation of extremely cryptic mottled brown plumage that must serve for concealment from predators. The strength of the phylogenetic effect on ornamentation is illustrated by the case of the Dovekie, which is similar ecologically and behaviourally to the *Aethia* auklets, yet has retained the appearance of its larger relatives among the Alcini and has no ornaments. With so much diversity and so little information on the function of ornaments, the living auks are certainly fertile ground for behavioural ecological studies of sexual selection.

Sperm competition

A potential source of sexual selection in socially monogamous birds such as auks occurs when individuals form a lasting pair-bond with only a single partner, but engage in extra-pair copulations (EPCs) and thus mate with multiple partners (Birkhead and Møller 1992). The resulting physiological process is known as sperm competition, but the term is often used to describe the behaviour and biology of this reproductive strategy in general. EPCs create the opportunity for sexual selection by increasing the variation in mating success of males and females as measured by number of partners mated. Mating success can be skewed in this way because some individuals may be more adept at obtaining extra-pair copulations (e.g. because they are more attractive) than others. The reproductive advantage of EPCs to males is obvious, since it immediately increases the number of offspring they potentially father. EPCs have generally been assumed to be costly to females, because they are constrained to having the same number of offspring regardless of how many males they mate with (Trivers 1972; Birkhead *et al.* 1985 but see Birkhead and Møller 1992). But female alcids do freely engage in EPCs, and several advantages of doing so have recently been suggested. By mating with more than one male, they may (1) ensure the fertility of their egg(s) in case their primary partner is infertile, (2) obtain a fertilization from a male that is of higher genetic quality, (3) increase the genetic

diversity of their offspring by having eggs fertilized by more than one male (summarized by Birkhead and Møller 1992), or (4) assess the quality of alternative male mating partners and increase the chances of acquiring them as mates in a subsequent breeding season (Wagner 1992*a*). For long-lived seabirds such as most auks where retention of a mate is a great advantage in successful chick-rearing, the tactic of increasing the diversity of offspring by engaging in EPCs may be especially appealing to females.

With a few notable exceptions, little is known about the copulation behaviour of auks. Based on what is known, it appears that copulation on land near or at the nest-site is the rule for species in the murre–guillemot–murrelet clade, while members of the auklet–puffin clade copulate at sea, usually at the gathering ground near the colony (Table 5.1). The reason for this apparent dichotomy is not clear and as more information becomes available we are quite likely to find exceptions to this rule. For all the murrelets and for Cassin's and Rhinoceros auklets, copulation behaviour has been rarely seen and published accounts are vague, fragmentary, or contradictory. Even for the relatively well-studied Ancient Murrelet, it is not clear where or how frequently they copulate (ILJ). However, EPCs have been recorded in all species in which copulation behaviour has been well observed (Table 5.1).

Indirect evidence concerning sperm competition is available in the form of comparative data on testis size among the auks (J. V. Briskie, S. Sealy, and J. Piatt, unpublished MS). These indicate that some highly social species, such as Razorbills, *Cepphus* guillemots, *Aethia* auklets, and Ancient Murrelets have relatively large testes controlling for the effect of differing body size (Least Auklets have testes larger than their brain size!), while puffins and solitary nesting species like Marbled Murrelets have relatively small testes. Interestingly, both Cassin's Auklet and Dovekie have small testes although they are both highly colonial. Since testes size is generally agreed to relate directly to the intensity of sperm competition, this suggests that EPCs and sperm competition are an important part of the social behaviour of Razorbills, guillemots, *Aethia* auklets, and Ancient Murrelets.

Copulation behaviour is well studied for only two auk species, the Common Murre (Birkhead *et al.* 1985; Hatchwell 1988b) and the Razorbill (Wagner 1991a,c, 1992a,c), and there are a number of interesting differences between these two species. Common Murres copulate only on cliff ledges at or very close to their nest-site. Females apparently resist most attempts at EPCs and are subjected to forced EPCs, while males vigorously defend their mate and attempt EPCs with neighbouring females when possible. Thus the ecological situation of the Common Murre at least in some cases favoured sperm competition associated with a high degree of conflict and aggression. However, in some situations female Common Murres do accept unforced EPCs, and these may be more likely to be successful in terms of successful cloacal contact that normal pair copulations (Hatchwell 1988b). Successful forced EPCs are rare or do not occur in the closely related Thick-billed Murre, which nests in lower densities on narrower ledges.

The Razorbill was the subject of a very interesting and detailed study by Richard Wagner at Skomer Island, Wales (Wagner 1991a,c, 1992a,c). In this species, copulation occurred on land away from the nest-site and at special arenas which were frequented for mating, controversially referred to as a 'lek' (Wagner 1992c) because this area contained no resources except mates, and females that sought EPCs there obtained only sperm. Females entered this area, actively solicited EPCs, and were able to control and avoid male forced EPC attempts completely by either throwing the male off by standing up or by protecting their cloaca with their long stiff tails (Wagner 1991a). Based on the patterns observed, fertility insurance may provide the best explanation for female Razorbill EPC behaviour (Wagner 1992a).

Among the auklets (*Aethia* and *Cyclorrhynchus* spp.), copulation takes place only on the sea at the gathering ground offshore after the pairs leave the colony site in the morning (F. M. Hunter and I. L. J, unpublished data). Least Auklets have the highest rate of EPCs, EPC attempts, and disruptions of courting pairs at the gathering ground of any auklet, which concurs with the observation that they have the largest testes relative to body size of any alcid (Briskie *et al.*, unpublished MS). Auklets with more elaborate ornaments have higher copulation rates and may have larger testes, further supporting a relationship between sperm competition and sexual selection.

There is a lot of work to be done on the subject of sperm competition in alcids. Although two studies have examined EPC behaviour in detail, there are as yet no published DNA fingerprinting results that can tell us whether EPCs are normally successful.

6

Chick development and the transition from land to sea

The vast majority of birds lay several eggs in a clutch. This makes sense, because the eggs need to be incubated for a fixed period of time, largely determined by egg size. Hence, the time invested in laying and incubating a single egg is almost the same as the time invested for a multi-egg clutch. Only a few land birds lay single-egg clutches, mainly very large raptors. Seabirds are remarkable in having a high proportion of species that lay single-egg clutches. Ashmole (1963) and Lack (1968) argued that the reproductive output of seabirds, as measured by brood size and growth rate, is constrained by their choice of foraging habitat. Those feeding in coastal waters generally rear more than one chick, whereas those foraging in offshore waters rear only a single, often very slow-growing, chick. The evolution of this single-egg strategy can be seen as a process whereby foraging range was gradually extended, with a consequent reduction in rate of provisioning of the chick, leading to a reduction in brood size.

The foraging range hypothesis, as an explanation for single-chick broods in pelagic seabirds, has been generally accepted. Auks mainly conform to this generalization, with the inshore-feeding guillemots rearing up to two chicks at the nest to close to adult size, while most other species forage further offshore and rear only one chick (Lack 1968; Birkhead and Harris 1985), often to a weight well below that of the mature adult. The only offshore-feeding auks that rear more than one chick at a time are the *Synthliboramphus* murrelets, which do not feed their young in the nest at all.

The *Brachyramphus* murrelets form an interesting exception to the general rule. Although entirely inshore feeders and often very abundant on landlocked inlets and among island archipelagos, they lay only one egg. However, they nest in a dispersed fashion, sometimes at high elevations and as much as 70 kilometers from the sea (see species accounts); consequently their travel time from breeding to feeding areas can be just as great as for offshore species. Moreover, perhaps because of the danger from falcons, much of the nestling provisioning is done at dawn and dusk, limiting the number of trips made each day. These peculiarities may explain why they rear only one chick.

Young birds vary in their condition at hatching and their subsequent strategy for leaving the nest. Nestling strategies have been divided into the following categories (Nice 1962): (1) altricial: young born blind, naked and helpless, and fed to near adult size in the nest (e.g. songbirds); (2) semi-altricial: downy and with eyes open at birth, but initially helpless (e.g. hawks, herons); (3) semi-precocial: young born active but fed at the nest to near adult size (e.g. gulls); (4) precocial: young born active and leave the nest more or less immediately to forage for themselves, or to be fed elsewhere by their parents (e.g. ducks, shorebirds). Nestling strategies are generally constant within families of birds. Auks are the only family in which there is wide variation in the developmental strategy of the chicks (O'Connor 1984). Their strategies range from the semi-precocial development of puffins, auklets, and guillemots, through the 'intermediate' strategy of the murres and Razorbill, to precocial nest departure in the *Synthliboramphus* murrelets (Sealy 1973*b*). This chapter considers the reasons for such diversity.

Chick-rearing among Charadriiformes

The ancestral strategy

The lack of variation in nestling strategies within families of birds suggests that these tend to be very conservative characteristics, which is another way of saying that moving from one type to another is very difficult. As a starting point, it is worth considering what the ancestral development strategy of the auks may have been, in order to understand how they reached their present situation and what strategy can be regarded simply as an expression of the evolutionary status quo within the family.

Among the Charadriiformes, some families comprise entirely precocial species (sandpipers, plovers) and others entirely semi-precocial species (gulls, terns, jaegers/skuas). The gulls are normally considered the closest relatives of the auks (Chapter 2) and on this ground we might consider the rearing of semi-precocial chicks to be the ancestral condition. This hypothesis is strengthened by the fact that all subfamilies of auks contain semi-precocial species, whereas intermediate and precocial species occur in only one subfamily each. However, this 'evidence' requires some qualification (see below). That semi-precocial chicks are the ancestral condition is also suggested by the fact that all other seabirds have either semiprecocial or altricial chicks. If auks and gulls, both basically marine families at present, were marine before they diverged, then it makes sense that their ancestor had already developed a strategy of rearing chicks at the nest-site.

Chick provisioning

Intimately associated with the decision on when the chick should go to sea is the method by which parents provision their young. The diet of nestling auks is affected both by the diet of their parents and by the means that their parents use to transport food. In the auklets, *Cyclorhynchus*, *Aethia*, and *Ptychoramphus*, and in the Dovekie, breeding adults develop a pouch in the throat in which the food destined for the chick is carried (internal transporters). In all other auks, food destined for the chick is carried externally, dangling from the bill (external transporters). In the puffins, the *Brachyramphus* murrelets, the guillemots, the Razorbill, and the Rhinoceros Auklet, fish are held crosswise in the bill, usually grasped close behind the head, so that the body trails backwards in flight. In the murres, most fish are carried with the head held inside the mouth and the tail sticking forwards, although very small fish or large crustacean may be carried crosswise. It would clearly not be feasible to carry many very small food items, such as amphipods or euphausiids, in the beak. Consequently, the chicks of the external transporters are fed almost exclusively on fish. This constraint does not apply to the internal transporters and the principal food of the chicks in all these species is planktonic crustacea and/or larval fish.

One result of the division between internal and external transporters is that, while internal transporters probably tend to feed their chicks very much the same food that they are feeding on themselves (Bédard 1969*c*), external transporters tend to feed themselves on smaller prey, reserving the largest to feed to their chicks (Gaston and Noble 1985), a habit common to many birds. Consequently, in external transporters, the composition of the chick diet can be quite different from that of the parents. An example is Kittlitz's Murrelet, in which the chicks are fed on fish of up to 12 grams, whereas the adults feed themselves mainly on tiny zooplankton (see species account).

Among external transporters, the puffins, Razorbill, and Rhinoceros Auklet habitually deliver several fish to their chick at each visit, and Marbled Murrelets occasionally do so, whereas guillemots, and murres usually carry only one. In the puffins, the ability to carry more than one prey organism is enhanced by a flexible hinge at the back of the jaw, which allows the mandibles to be opened at the tip while retaining previously captured items held near the back. This ability allows the multiple loaders to make use of smaller prey than those required by single loaders. The ability to carry multiple prey items is best developed in the puffin tribe, where up to 60 small fish are sometimes carried together.

Auks appear to exercise some selection in the prey that they deliver to their chicks. For external transporters, especially single-prey loaders, the items delivered to chicks are usually at the top end of the size spectrum of prey. Studies of murres, Razorbills, and Atlantic Puffins breeding on the same colonies show that the single-loading murres usually deliver a fish at the upper end of the size range of those delivered by the multiple-loading Rozorbill and Atlantic Puffin (Furness and Barrett 1985; Harris and Wanless 1986b; Barrett and Furness 1990). It has also been suggested that some species select fish that have a particularly high energy content. Brekke and Gabrielsen (1994) showed that the efficiency with which Thick-billed Murres assimilated the energy content of their food varied with the type of prey, so some selection among prey species could be useful to the birds. Atlantic Puffins may take sprats (*Sprattus* or capelin *sprattus*) preferentially over other pelagic schooling fishes (Harris and Hislop 1978; Brown and Nettleship 1984). However, the evidence for such selection is not conclusive, because of the difficulty of being certain that the chosen fish were not easier to obtain for other reasons (e.g. closer inshore, at shallower depths, slower swimming).

Internal transport has several advantages: it means that the food is carried close to the bird's centre of gravity, hence interfering as little as possible with flight dynamics,it reduces the chance that the food-carrying bird will be identified by potential kleptoparasites, and it leaves the bill free for other activities, such as preening and fighting. Among the gulls, the closest relatives of the auks, all food is delivered to the chicks by regurgitation. Among other Charadriiformes, only the terns and certain rather specialized shorebirds (e.g. oystercatchers, *Haematopus* spp.) are external transporters. The advantages of internal transport, and its universal use among their closest relatives, the gulls, makes the use of external transport by the majority of auks rather surprising. The facts that internal transporters among the auks have a specialized diverticulum for carrying the food, that such an organ developed independently in the auklets and the Dovekie, and that external transport occurs in all subfamilies except the auklets, seem to indicate that external transport is likely to be the ancestral condition for auks. Why should this be the case?

Not all birds that feed their chicks by regurgitation develop specialized organs for carrying the food. Petrels, penguins, cormorants, and gulls transport it in the proventriculus, the extensible area of the foregut where food is held before being processed in the stomach. The fact that auks do not carry food to their chicks in the foregut cannot be because they are unable to store food there. Birds collected while feeding frequently have large volumes of undigested food in the proventriculus. As auks certainly can transport food internally, and as this seems to be the most advantageous way to carry food, we have to assume that there is something about being an auk that makes it difficult to regurgitate food from the foregut. At present, we have no idea why this is. The answer may relate in some way to the requirements of swallowing food underwater, or to the possibility that storage in the proventriculus could lead to premature digestion. However, penguins and cormorants must have found some other way around the problem. An alternative idea is that the narrow, fusiform body cavity of the auks does not provide space for the proventriculus to expand sufficiently to accommodate the necessary amount of food.

However, this seems unlikely, because the proventriculus of birds collected while feeding frequently contains more food than the throat pouch can carry. That the pouch has evolved specifically for transporting food to the chick is demonstrated by the fact that it is absent in pre-breeding birds, and in breeders it regresses outside the breeding season, although in Cassin's Auklet it does not disappear entirely (Speich and Manuwal 1974). In any event, the throat pouch is not present in gulls, or in other Charadriiformes with semi-precocial young. The common use of external food transport among auks seems to be a major reason why they deliver relatively small meals to their chicks. That, in turn, must be a major factor in the evolution of auk development strategies.

Evolution of departure strategies

The following factors have been identified as potential contributors to the variation in brood size and departure strategies found among the auks:

(1) distance to the feeding area and the consequent expenditure of time and energy in travelling back and forth (Lack 1968; Birkhead and Harris 1985);

(2) the low load-carrying capacity of the auks, especially the larger, intermediate, species (Lack 1968; Sealy 1973b);

(3) the choice of nest-site and the relative vulnerability of the chick to predation (Cody 1971; Ydenberg 1989);

(4) the consequent need, at open sites, for one parent to guard the chick constantly, so that only one parent can forage at a time (Birkhead and Harris 1985);

(5) the risk to adults of predation associated with coming ashore to breed (Ydenberg 1989; Gaston 1992a);

(6) the effect on chick energy requirements of ambient temperature and parental brooding (Gaston 1992a).

The effect of body size on aspects of flight performance has been implicated in the evolution of the intermediate departure strategy (chick leaves when partially grown and parental care continues at sea). Consequently, we begin by describing the effects of size within the family; where it appears to be an important determinant and where it appears to be neutral. We then discuss the contribution of various potential environmental and developmental constraints in determining different growth and departure strategies. Finally, we examine the role of a formal model developed by Ydenberg (1989) and Ydenberg et al. (1995) to explain variation in departure age and weight of auk chicks.

Allometry

If we take a taxonomic group with fairly uniform ground plan, in terms of structure and behaviour, and a reasonable large spread of body size, then practically all species-specific phenomena are found to vary rather closely with body size (Calder 1984). The slope of the log:log regression of characters (egg size, wing length, life span, etc.) on body size is seldom unity. Consequently, the ratio of the character to body size changes, giving rise to the term 'allometry'. Allometry occurs because many physical properties have an allometric relationship with body size (e.g. surface/volume ratio, wing-area/weight).

Living auks exhibit about a tenfold range in body mass, but inclusion of the Great Auk makes it 50-fold. We have already noted that size is related to diving ability in the family (Chapter 4). Certain allometric relationships are likely to have an impact on chick-rearing strategy: (1) the effect of body size on egg size, which determines the initial size and energy reserves of the chick, and (2) the effect of body size on food-carrying ability, determining the maximum rate at which the chick can grow.

In common with other semi-precocial seabirds, auks lay comparatively large eggs, and somewhat larger than those laid by Charadriiformes in general (Fig. 6.1). Their

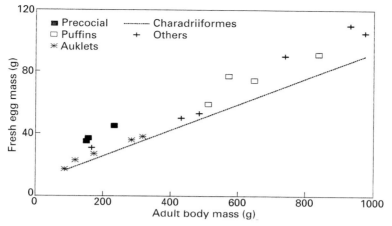

6.1 Fresh egg mass in relation to female body mass in auks. Solid line shows the regression for all Charadriiform birds (Rahn *et al.* 1975).

eggs are also much bigger, relative to adult body size, than those of penguins (Fig. 6.2). Those of the precocial murrelets are as large, as a proportion of female body mass (20–25 per cent), as any flying birds, comparable with those of storm-petrels (Hydrobatidae). It seems obvious that the size of eggs laid by *Synthliboramphus* species is related to their precocial departure strategy (Sealy 1975*a*). However, it is worth noting that the largest egg among the shorebirds is laid by one of the few

nidicolous species, the Crab Plover (*Dromas ardeola*), which incidentally, is similar to many auks in having two brood patches, but only laying one egg. Like the Kiwi (*Apteryx*), another bird that lays a very large egg, it nests in a burrow.

Incubation periods are closely correlated with egg size across birds as a whole (Rahn and Ar 1974). However, auks do not adhere to this generalization very well and it is worth examining the relationship to see what it can

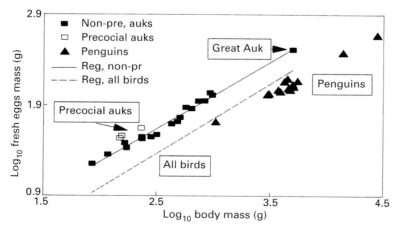

6.2 A comparison of fresh egg mass in relation to adult size in auks, penguins (data from Williams 1995, using female mass pre-laying), and all birds (Rahn *et al.* 1975).

tell us (Fig. 6.3). Most auks have longer incubation periods than their egg size suggests, but puffins and auklets have particularly prolonged incubation periods. When we compare auks and penguins, we find that murres, murrelets, and guillemots fall more or less on the same line as penguins, but that the auklets and especially the puffins have much longer incubation periods (Fig. 6.4).

Among non-precocial birds, there is a tendency for the incubation and nestling periods to be correlated and Lack (1968) proposed that this correlation is brought about by the need for a uniform growth rate both before and after hatching. For instance, petrels, which have longer incubation periods than auks, generally have slower growing chicks (Croxall and Gaston 1988). Among auks, the length of the

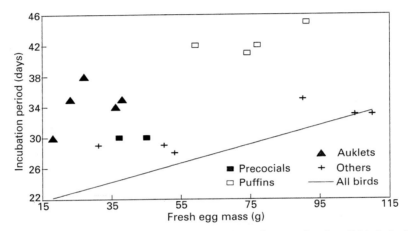

6.3   Incubation period in relation to fresh egg mass for auks and regression for all birds (solid line, Rahn and Ar 1974).

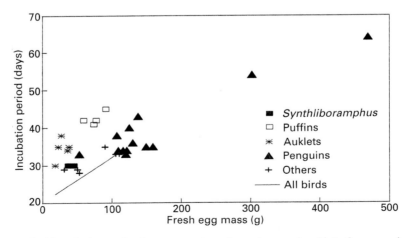

6.4   Incubation period in relation to fresh egg mass in auks and penguins (data for penguins from Williams 1995, for auks, from species accounts).

nestling period is strongly influenced by the weight of chicks at departure as a proportion of adult weight; consequently comparisons of nestling periods would be misleading. However, if we compare the maximum growth rate (g/day) of chicks (expressed as a proportion of adult weight) with their incubation period, we find that there is support for Lack's generalization; the chicks of species with long incubation periods tend to grow relatively slowly (Fig. 6.5).

The relationship between growth rate and incubation period looks a lot stronger if we ignore the murres and Razorbill, but is it legitimate to treat them separately, or do these species contradict Lack's generalization? In fact, these 'intermediate' species are growing in a very different way from puffins and guillemots. They have to be completely feathered and ready for life at sea by the time they leave the colony, which means that they have to undergo all the tissue and metabolic maturation by 20 days of age that the other auks undertake by 30 days or more. Whether it is this fact that determines their relatively slow weight increase while at the colony we do not know. At least there appears to be some

justifiable reason for considering them an exception when comparing growth rates and incubation periods.

Chick provisioning and departure strategies

Normally, big birds carry larger food loads than smaller birds, but there is little relationship between adult body size and the size of meals delivered by auks to their chicks (Fig. 6.6). The heaviest meals are delivered by the multiple-loading Rhinoceros Auklet (average about 30 g, but sometimes as much as 100 g, Vermeer 1979; Bertram 1988; W. J. Sydeman, personal communication), and Tufted Puffin (mean 22 g, Vermeer 1979), while the heaviest in relation to size are carried by Cassin's Auklet, which transports loads of up to 20 g, approximately 12 per cent of body weight (Burger and Powell 1990) and Crested Auklets which carry up to 14 per cent body weight (Jones 1993c). Among single-loaders, the guillemots generally bring larger meals (12–15 g, Oakley 1981; Cairns 1987a) than the murres (8–15 g, Gaston 1985a; Harris and Wanless 1985), although murres weigh

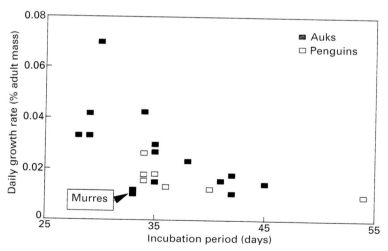

6.5 Daily maximum weight increase of auk and penguin chicks during the linear growth phase, expressed as a proportion of adult weight (daily growth rate) in relation to incubation period (data for penguins from Williams 1995, for auks, from species accounts).

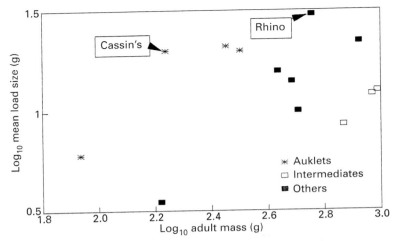

6.6   Meal size delivered to chicks in relation to adult body mass (log:log plot).

approximately twice as much as guillemots. The two nocturnal species (Cassin's Auklet, Rhinoceros Auklet), confined to delivering only one meal per parent daily (with rare exceptions), both deliver unusually large loads, relative to their body weight.

Murres and Razorbills may carry peculiarly small food loads for their size because a trade off between the needs of underwater and aerial locomotion (Birkhead 1977a; Gaston 1985c) has led to a reduction in their load-carrying capacity relative to those of other auks. This has been considered the main factor leading to the 'intermediate' departure strategy adopted by their chicks. It is certainly true that the mass to wing area ratio in auks is correlated with body size, with large auks being heavier in relation to their wing area than smaller species (Fig. 6.7), and it is also true that the species with an intermediate strategy (murres and Razarbill) have a higher wing loading than other auks. Moreover, breeding Thick-billed Murres and Least Auklets (and probably other auks) undergo an abrupt loss of mass at about the time of hatching that has been considered an adaptation to facilitate transporting loads to the chick (Croll et al. 1991; Gaston and Perin 1993; Jones 1994). This adjustment suggests that losing a little weight helps when transporting loads.

Mathematical modelling by Houston et al. (1996) suggests that, despite their poor load-carrying abilities, murres are potentially capable of rearing a fully grown chick at the breeding site, provided that they can find food without having to travel far from the colony, enabling them to make 10 or more feeding trips daily, as guillemots may do in some areas (species acounts). In fact, they could do so quite comfortably if both parents foraged simultaneously, leaving the chick unguarded. Consequently, we need to invoke a number of different selection pressures to explain the evolution of the intermediate departure strategy. Some possible candidates are outlined in Table 6.1. The consequences of these factors combined is that chick growth is slow and asymptotic weight is low, leading to the evolution of a strategy in which development at the breeding site involves the replacement of the down, and chicks leave well below adult size. The importance of colony size in the evolution of this strategy should not be overlooked. The deep diving of murres, and to a lesser extent Razorbills, allows them access to a greater volume of feeding habitat than other auks, and Arctic waters provide high densities of potential prey allied with low competition. A combination of these factors has made murres the

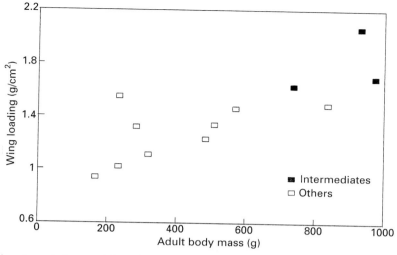

6.7 Wing loading (mass/wing area) in auks, in relation to adult body mass (wing loading data from Livezey 1988, body mass data from species accounts).

Table 6.1   Factors selecting for intermediate chick departure.

Factor	Result
Arctic breeding	No soil for burrows, terrestrial predators (e.g. foxes) ubiquitous
Deep diving	Large size, small wings, poor landing/take-off ability
Cliff nesting	Precocial departure impossible
Low temperature	Energy saved by brooding
Open nest-site	Chick guarding required
Large colonies	Distant foraging

dominant fish-eating auks in Arctic waters, creating a situation where, if colonies occur, they are likely to be large.

The great similarities between chick development in murres and the Razorbill suggest that this strategy was also developed in their common ancestor. If, as some taxonomists consider, the Great Auk was congeneric with the Razorbill, this suggests that either its chicks had a similar departure strategy, or at least that whatever behaviour they adopted must have been a modification of the intermediate strategy. In fact, early chick departure may have been a factor predisposing the species to flightlessness. The much slower rate of travel imposed by the loss of flight would

have created problems in provisioning a semi-precocial chick that could have been solved only by carrying a large volume of food in the crop, as penguins do. Hobson and Montevecchi (1991), on the basis of a single diet determination made from stable isotope ratios in the subfossil bone of a supposed nestling Great Auk, suggested that the young were fed on crustacea, presumably carried in the adult's crop and regurgitated. However, the total lack of historical accounts or museum specimens supporting the idea of chick development at the colony makes it more likely that the chicks left soon after hatching. Fisher and Lockley (1954) and Bengtson (1984) suggested that chicks left the colony at a very premature stage while still requiring only a relatively small daily food intake, perhaps as young as 8 days. This hypothesis is supported by accounts of downy chicks seen at sea and the apparently very contracted breeding season of the Great Auk (species account). We cannot rule out the possibility that the young of the Great Auk were completely precocial and were never fed on land.

The precocial murrelets

None of the features listed above as leading to the intermediate departure strategy apply to the *Synthliboramphus* murrelets. They are actually the most southerly genus of auks, with three of the four species breeding in, or at the edge of, sub-tropical waters. They are fairly small, without a particularly high wing loading, they nest in burrows, and, before people began to travel in ships, they bred on islands where nest predators were absent. Clearly, the arguments that we have advanced for the intermediate strategy will not work for these birds.

Sealy (1976) considered that the precocial strategy was an adaptation to a peculiarly unpredictable food supply that precluded commuting from the breeding site. However, Gaston (1992a) argued that if the food supply was unpredictable, taking the chicks to sea would be the worst strategy, because it would reduce searching mobility to the swimming speed of the chicks. Precociality seems to require that food be more, rather than less, predictable because family parties have very poor mobility until the chicks can fly. Predictable food supplies within swimming distance of the colony may be a condition for precociality in auks, but it hardly seems to provide a compelling cause. It is possible that the food is indeed predictable, but that the only suitable feeding areas are far from the colony, making precocial departure a more efficient strategy than commuting to the nest-site. The only argument against this hypothesis, proposed by Lack (1968), is that it applies equally to all pelagic seabirds, so it does not explain why these murrelets and no other

seabirds adopted this particular strategy. Cassin's Auklets take very similar prey to Ancient Murrelets living in the same area, but retain a semi-precocial chick that is fed at the nest for about 6 weeks. There is also the intriguing fact that the precocial murrelets rear two chicks, whereas all other pelagic seabirds that fly have adapted to the constraint on food delivery rates imposed by their lengthy foraging trips by reducing their brood size to one.

If multi-egg clutches are ancestral in the auks, single-egg clutches must have evolved as species sought to increase their foraging range. This transition can evolve very easily. A species laying two eggs, but only managing to rear one chick, as a result of poor provisioning ability, may benefit by laying a single egg and not wasting food initially feeding a young chick that will not survive. Variation in clutch size, which occurs in all guillemots, allows natural selection for single-egg clutches to proceed rapidly when food availability near the colony is reduced. The situation once evolution has fixed the clutch size at one is rather different. There is no variation on which selection can operate. Moreover, if a two-egg mutation arises it must do so in an individual that has the necessary behaviour (responds to increased food demands by the chick, accepts more than one chick) before it can be selected for. Single-egg clutches seem to be an evolutionary cul-de-sac, making it very likely that precocial murrelets evolved from an ancestor that laid more than one egg.

Why did the ancestor of *Synthliboramphus* not reduce its clutch to one before adopting the precocial departure strategy? The answer to this question can only be speculation, but the simplest explanation is that the ancestor was an inshore species which evolved a reduced nestling period, not because it was incapable of finding enough food to rear its chicks at the nest, but because the constant colony visits required by rearing chicks at the colony exposed the parents to a very high risk of predation (Gaston 1992a). This hypothesis has been criticized by Hatch (1994) for suggesting that clutch size has been adjusted to adult mortality, but he appears to have misinterpreted the idea. All organisms strive to maximize their fecundity within the constraints of their particular ecology. However, their subsequent survival is one of those constraints. The existence of deferred breeding in any organism illustrates a trade off between current reproduction and future survival. No organism will adapt by reducing its clutch size *because* it has achieved a high adult survival, but equally no organism can adapt to a reduced clutch size *unless* it already enjoys high adult survival.

Breeding adults of all *Synthliboramphus* species suffer heavy mortality at their colonies, despite being nocturnal, nesting in concealed sites, and taking other precautions against it. Predation on adults is clearly a strong selection pressure. If the ancestral murrelet occupied breeding areas with a similar risk of predation, selection for a reduction in the length of the chick-rearing period could have come about without any constraints on the parent's ability to find food. Under this hypothesis, the precocial strategy in auks is not an extreme modification of the intermediate strategy, but an independent strategy driven by completely different selection pressures. It is also worth noting that the advantage that accrues in saved travelling time by taking two chicks to sea is double the advantage that would accrue to a bird rearing a single chick (R. Ydenberg, personal communication). In fact, it seems possible that the precocial murrelets might be capable of rearing more than two chicks. The main thing preventing them from doing so may be the difficulty of laying and incubating a clutch of three such enormous eggs. The evolution of the precocial strategy would surely have been easier if, as seems likely, the ancestral murrelet inhabited fairly warm seas, reducing the difficulty that precocial chicks might otherwise have encountered in maintaining their body temperature (Gaston 1992a).

Growth rate variation

Lack (1968) contended that the prolonged nestling periods of many oceanic seabirds were an adaptation to periodic food shortages. Considering the auks specifically, he contrasted the slow growth of nestling puffins, presumed offshore feeders, with the faster growth of inshore-feeding guillemots. Although this comparison is true, a more general comparison of growth rates and foraging ranges among auks shows that growth rate varies widely among offshore feeders. In fact, foraging ranges within species are highly variable, and relatively poorly known. Hence, any generalization relating to the effects of foraging range among auks, other than contrasts between the inshore *Cepphus* and *Brachyramphus* spp. and the rest, must remain highly speculative.

Lack's idea concerning foraging range and growth rate has been criticized by Ricklefs *et al.* (1980) and Shea and Ricklefs (1985) on the grounds that slower growth has little effect on the average rate at which parents must supply their chicks with food. In the auks, the growth rate of puffin chicks (corrected for size) is substantially slower than that of auklets (Fig. 6.8). As the distribution of puffins and auklets, and the types of breeding site that they occupy, overlap broadly, the difference cannot be attributed to climatic variation, or differences in the vulnerability of their nest-sites. Kitaysky (1995) has argued that plankton, on which the auklets feed their chicks, is an inherently more predictable food source than the schooling fish on which puffins feed their young. In experimental tests, he showed that young puffins are much more resistant than young auklets to the effects of fasting. The puffin chicks responded to starvation by lowering metabolic rate and reducing the growth of tissues, bones, and feathers, whereas auklet chicks continued normal growth and metabolism, succumbing to starvation much earlier. It is possible that the variable growth of puffin chicks is a result of metabolic adaptations that allow them to 'shut down' in the face of reduced food deliveries, an option that is not open to the faster-growing auklets. These observations seem to support Lack's idea, and suggest that the long incubation and chick-rearing periods of the puffins are an

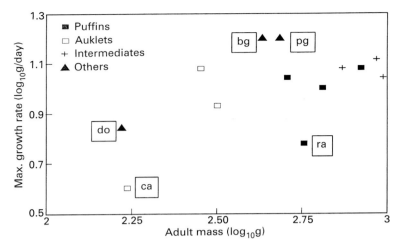

6.8   Maximum growth rate of chicks (g/day) as percentage of adult weight, in relation to adult weight. Species indicated: do = Dovekie, ca = Cassin's Auklet, bg = Black Guillemot, pg = Pigeon Guillemot, ra = Razorbill.

adaptation to unpredictable feeding conditions. Although most puffins rear chicks successfully each year, there are several accounts of partial or total breeding failure for puffins (Vermeer *et al.* 1979; Anker-Nilssen 1987).

Another feature of nestling growth in puffins is a decrease in weight in the last week before fledging, associated with a marked drop in feeding rate by the parents (Harris 1984; Bertram 1988). A weight recession also occurs in Dovekies in the last few days before fledging, when the female parent ceases to feed the chick (Konarzewski *et al.* 1993), and in auklets (Sealy 1968). Weight recessions have been identified in other auks, but sometimes by the dubious method of comparing departure weights with maximum weights, a procedure that, by definition, suggests a recession (Gaston 1985*a*); among the subfamilies of auks, weight recession is definitely most prominent in puffins (Fig. 6.9). Similar weight recessions are seen in all petrels and albatrosses and many penguins. It appears that the nestlings do not develop their tissues at a rate commensurate with the food supplied by their parents, so part of the energy supplied is stored as fat which is used to fuel the latter part of the development period. This strategy

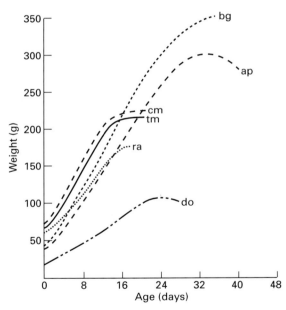

6.9   Representative growth curves for auks (adapted from Gaston 1985). Species indicated: do = Dovekie, ra = Razorbill, tm = Thick-billed Murre, cm = Common Murre, ap = Atlantic Puffin, bg = Black Guillemot.

enables adults to reduce, or even terminate, their provisioning before the nestling fledges.

Ricklefs (1979) and Ricklefs *et al.* (1985) argued that, in the storm-petrels, large fat stores are laid down because growth is limited by nutrients other than energy, which is provided in excess in the early part of the growth period. The fat deposits are therefore simply a by-product of the storm-petrel's diet. This explanation has been adopted by Taylor and Konarzewski (1989) to account for the large fat reserves built up by Dovekies during the nestling period, although experiments by Ricklefs *et al.* (1987) did not support the hypothesis for storm-petrels. Taylor and Konarzewski found that very few Dovekie chicks (less than 1 per cent) were ever in danger of starving at the colony where they worked. Consequently they rejected the idea that the fat deposits were an adaptation to periodic fasts.

Puffins are fed on whole fish, the same diet that allows guillemot chicks to grow rapidly and depart without weight recession. Hence, weight recession in puffins is probably not accounted for by the composition of their diet. It is more likely that weight recession in nestling puffins is linked to their slow growth rate and that these are both linked to adaptations related to unpredictability in their food supply. This is not necessarily contradicted by observations that some populations exhibiting weight recession show no signs of food stress. Even a small proportion of poor years in which survival was restricted to a few starvation-resistant chicks would constitute a very powerful selective bottleneck. Moreover, the survivors of such a selection event would enjoy better than average breeding opportunities, as a result of the lack of competition for pairing with experienced breeders. Consequently, lack of evidence for selection causing slow growth in average years does not preclude the possibility that food shortages are an important selection pressure maintaining prolonged nestling periods in puffins. Despite the findings of Taylor and Konarzewski (1989), the same could be true of Dovekies, because very few studies are available from their main breeding area in north-western Greenland. There is no

relevant information for auklets, but Kitaysky's (1995) findings suggest that auklet chicks are not well adapted to food shortages, so the cause of their weight recession is obscure.

The role of trade-offs between growth and survival

An important contribution to thinking on the departure strategies of the auks was made by Ydenberg (1989; Ydenberg *et al.* 1995) who developed a mathematical model that predicted the inter-relationship between the growth rate of chicks and their age at departure. His model depends on two key assumptions: that chicks can grow (increase in weight) faster at sea than at the colony, and that they are safer (have a better survival rate) at the colony than at sea. The model predicts chick age at departure from the colony on the assumption that they leave when the disadvantage caused by lower survival at sea is outweighed by the advantage gained from the higher growth rate achieved there. One of the most robust predictions of the model is that fast-growing chicks will leave the colony at a younger age and at a higher weight than slower chicks (the 'negative fledging boundary' Fig. 6.10). This prediction has been found to hold in several field situations (Harfenist 1995; Morbey 1995) and seems to be the normal condition in the family.

The general fit of the model's predictions to observations in the wild is impressive. However, there may be other explanations for the observations. If chicks are required to reach a certain condition, in terms of physical development, before they leave the colony, and if slower weight increase leads to slower development (e.g. slower growth of feathers) we might anticipate that slower growing chicks would remain longer at the colony. As nutrition is worse for these chicks, and as most chicks cease to increase in weight before they depart, the slower growing and later leaving chicks will also be lighter at departure. This would be true irrespective of their relative survival rates at sea and at the colony. Unfortunately, some crucial input parameters

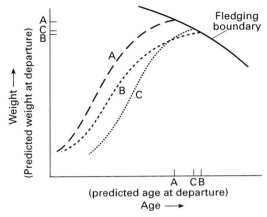

6.10 Illustration of the Ydenberg hypothesis for the relationship between weight and age at fledging. The fledging boundary is generated by the model. The hypothetical growth curves represent: A, an early chick growing at normal speed; B, an early chick growing more slowly; C, a later chick growing at normal speed. The age of the chick where its growth curve intersects the fledging boundary is the predicted age at colony departure. The negative slope of the fledging boundary means that late chicks and slow-growing chicks should depart at lighter weight than normal, early chicks.

used in the Ydenberg model have been measured only very crudely to date, and it will be hard to measure them with sufficient accuracy to establish unequivocally whether trade offs between growth and survival, or developmental constraints, are the main cause of the negative fledging boundary.

Although Ydenberg's model has so far been applied to predictions about what will happen within species' populations, it could equally be applied to interspecific comparisons. When this is done, it can be viewed as a more explicit formulation of Lack's generalization: fledging period is a function of nestling growth rate, which is determined by the rate of parental provisioning. Notwithstanding the criticisms of Ricklefs (1979) and Taylor and Konarzewski (1989), it is hard to believe that prolonged nestling periods are not related to provisioning rates, although selection for this trait may occur only periodically. Ricklefs and Schew (1994) came to a similar conclusion regarding weight recession in storm petrels.

7

Populations and conservation

Keeping their study species alive has become a major preoccupation with biologists. This is understandable as the ever-increasing impact that people are having on the biosphere threatens many organisms with reduction and disruption of their habitat and food supplies. We are at a stage when, for some observers, including many scientists, all organisms must be assumed to be threatened unless proved otherwise. At the same time, decision makers in government and industry continue to assume that any activity promoting employment or profit is acceptable unless proven highly deleterious. Comparing the influence of these two views, that of biologists appears relatively inconsequential, perhaps because of a failure to communicate with the public. In this chapter, we review the information available on the past and current status of auks, the possible causes of population change, especially human activities, and the prospects for maintaining, or even increasing, current numbers of auks.

There are many animals for which the changes in population brought about by human activities are irreversible. We shall never again (well, not for a very long time) see floods of bison churning the plains of Saskatchewan, lions in the hills of Greece, or tigers on the banks of the Ganges, because we have taken their habitat and converted it into farms, cities, golf courses, and airports. In that respect, the sea is different. Despite millenia of marine navigation, we are still transients on the seas; we have not colonized them as we have the land and, even where heavily exploited, many major

marine ecosystems remain similar in structure to those existing under natural conditions. Consequently, the opportunity exists for us to manage the oceans so that they maintain tolerably natural ecosystems and that means 'natural' populations of auks. Unfortunately, marine ecosystems are far from being in a natural state at present.

Fisheries impacts

Marine fisheries practices, although festooned with national and international regulations, are still based on the idea of an 'exploitable surplus' (also known as a sustainable yield—although the word sustainable is not necessarily appropriate): fish that are somehow in excess of what nature requires and therefore available for our consumption. In practice, this is taken to mean the maximum amount of fish that, if removed, allows a similar amount to be removed the following year (maximum sustainable yield, MSY).

Fisheries quotas generally make no allowance for the needs of the other fish eaters in marine ecosystems, such as seabirds. It is worth noting that maximum sustainable yield models predict lower standing stocks of fish than in an unharvested situation, because stocks are held at the level that promotes the optimum rate of growth. At the same time, the age distribution of stocks harvested under MSY is altered towards younger age-classes. This effect could benefit auks, which tend to take the youngest age-classes. The problem with managing commercial fisheries is that marine ecosystems are extremely dynamic, so that it is very hard for even the best-regulated fishery to anticipate and allow for naturally occurring changes in recruitment and mortality. The consequence has been that many large fisheries have crashed, with unfortunate effects on dependent fishermen and seabirds. Furthermore, with the disappearance of traditional commercial species, those lower in the food chain, such as sandlance and krill, are now being targeted. This brings fisheries into direct competition with seabirds.

Because auks live primarily in continental shelf waters of the northern hemisphere, and because these are the scene of some of the world's most intensive fisheries, the activities of commercial fishing operations form a constant background to the lives of the auks. Even if we never touched or disturbed the birds themselves, their lives would be strongly affected by our activities.

Because human alterations to the biosphere are all-pervasive, it is not possible to pigeon-hole trends in seabird populations neatly into 'our fault' and 'acts of God'. In 1969, tens of thousands of Common Murres and Razorbills died in the Irish Sea. Corpses that washed up were emaciated, suggesting that the primary cause was a failure of food supplies when the birds were moulting and unable to travel elsewhere. Laboratory analysis demonstrated that the dead birds had high levels of organochlorine compounds (PCBs) in their tissues, but healthy birds, shot in the same area, exhibited similar levels (Parslow and Jefferies 1973). Compounds stored harmlessly in fat deposits in healthy birds became mobilized in the bloodstream of starving birds. Did the mobilization of the PCBs in a situation where feeding conditions were bad contribute to the observed mortality? There is no rigorous scientific answer to this question, because it is impossible to replicate the conditions of wild birds at a period of physiological and nutritional stress when you do not know where they were, or what they were doing. However,

common sense suggests that, all other things being equal, birds with high levels of toxic compounds in their circulation are likely to be less able to survive periods of natural stress than those without. It is an unavoidable consequence of our current domination of the biosphere that our activities contribute to practically every event, yet science is unable to predict or explain many of the consequences.

Basic demography

Before considering what has happened to auk populations and the possible causes of change, we need to review some salient features of their demography. Compared with many other birds, but in common with most seabirds, auks are relatively long-lived. The lowest adult annual survival rates reported are 77 per cent for the Ancient Murrelet (Gaston 1990), and 79 per cent for the Least Auklet (species account), equivalent to a mean life expectancy of 5 years. Most auks have adult annual survival rates in excess of 90 per cent (Hudson 1985). Because most species lay only a single egg, their maximum annual reproductive output is only one chick per pair, although the guillemots and *Synthliboramphus* murrelets, which lay two eggs, achieve better than that. Like most birds with high adult annual survival, auks defer breeding for several years. A few Cassin's Auklets and Ancient Murrelets may breed at 2 years, and the majority by 4, but for the larger species (murres, puffins) the mean age at first breeding is 5 years or more (Croxall and Gaston 1988).

These demographic characteristics have predictable consequences for conservation:

(1) because of high adult survival and deferred breeding, population trends determined from counts of breeders may lag behind actual demographic changes affecting reproduction or pre-breeding survival;

(2) deferred breeding means that a substantial proportion of the population is not breeding in a given year. Some of the youngest

birds may spend the summer far from the breeding area, making them immune from disasters occurring close to breeding colonies;

(3) the low reproductive rate means that recovery from a population crash will be relatively slow;

(4) but (the other side of the coin) high adult survival makes small populations less liable to random (stochastic) extinction in the short term.

The last two points require a little thought. A lot of what has been written about the conservation of auks suggests that they are peculiarly vulnerable to extinction because of their low reproductive rate. However, it might be equally valid to say that they are peculiarly resistant to extinction because of their high adult survival rates. Populations go extinct because all their members die in a short space of time. The likelihood that this will happen to a population of murres is much lower than for an equivalent number of sparrows or thrushes, under natural conditions, because a population of murres can persist for a long time in the absence of recruitment.

There is some evidence that seabirds are more vulnerable to extinction than landbirds; they make up 30 per cent of species listed as threatened or near-threatened by Collar *et al.* (1994), compared with 20 per cent for all birds. However, this probably has more to do with the fact that they breed on limited numbers of rather small islands, sometimes only a single one, than it does with their reproductive rate. All of the seven penguins (44 per cent of all species) and 13 cormorants (48 per cent) considered threatened lay clutches of more than one egg. Only four auks qualify for inclusion among the threatened (18 per cent of species), of which three belong to the genus *Synthliboramphus*, which (along with *Cepphus*) has the highest potential reproductive rate in the family.

There is one catch to having a naturally high adult survival rate. Demographic models of long-lived populations are unambiguous in

demonstrating that, under such conditions, the key to population dynamics is adult survival. A species with a 95 per cent adult annual survival rate has a life expectancy of 20 years. If that annual survival rate falls by only 5 per cent (a doubling of the mortality rate), the future life expectancy is halved. Such species are relatively little affected by fluctuations in reproductive success, because they are frequently buffered by changes in juvenile survival rates that are influenced by population density. However, small changes in adult mortality rates can have very dire consequences. Thus auks are vulnerable to any human activities that increase adult mortality, such as oil spills (Piatt *et al.* 1991*b*), hunting (Shuntov 1986; Kampp *et al.* 1995), and drowning in gill-nets (Takekawa *et al.* 1990). There is evidence that some populations can adjust to increased mortality by reducing the age at first breeding, or through increased reproductive success (Gaston *et al.* 1994), but their ability to do so appears to be very limited. Consequently, auks are very vulnerable to changes affecting adult survival, less so to those affecting reproduction and recruitment.

Auk populations

Our knowledge of auk populations is very fragmentary, despite several decades of intensive surveys and censuses. We do not know total world populations accurately (within 25 per cent) for any species. Populations of most species are well known in certain parts of their range (usually north-western Europe, North America except for Alaska, Japan, parts of the Russian far east), but poorly known in others, making any consideration of global trends impossible. The Atlantic auks tend to be better known than those of the Pacific, but even there, the Atlantic Puffin and Razorbill suffer from great uncertainties about the very large populations in Iceland, while the Black Guillemot, because of its dispersed nesting habits, and other difficulties attendant on censusing the species, has not been censused at all over much of its range. The immense size of Dovekie colonies, especially those in north-west. Greenland, compounded by their remoteness, has prevented proper censuses. However, the status of at least one Atlantic species is known with certainty: the Great Auk is definitely extinct.

Turning to the Pacific, the two guillemots suffer from the same problem as the Black Guillemot, exacerbated by the remoteness of many of their breeding stations; their total populations are essentially unknown. The Marbled Murrelet has been the subject of very intensive surveys in parts of North America since 1980, especially from California to British Columbia. Because of its very scattered nesting dispersion, it is most successfully censused by counts at sea. Current estimates are probably better than for many other small auks. However, almost nothing is known of the Asian race. The number of Kittlitz's Murrelets, which breed in even remoter areas and have not been subjected to such intense scrutiny, is very poorly known. Ancient Murrelets, likewise, while tolerably well censused in British Columbia, are poorly known in Alaska, where there could be quite a lot; their world population must be treated as uncertain.

The less said about auklet numbers, the better. The diurnal species suffer, like Dovekies, from the immense size of their colonies, and their habit of nesting in talus which provides no convenient 'countable unit', such as a nest-site or burrow entrance. When the birds fly off, the would-be census taker is left with nothing but a pile of rocks streaked with guano to base a census on. Counts of birds on the surface of the colony may be anywhere from 10 to 80 per cent of the local breeding population and vary from day to day, seasonally, and between years in a chaotic manner (Jones 1992b). Of the two nocturnal auklets, Cassin's breeds in countable burrows and has been fairly well censused in its main breeding area in British Columbia. Assuming that numbers are fairly low in Alaska, the world population of this species must be one of the better known. The Whiskered Auklet breeds in

rock crevices, is rather widely scattered, and must be one of the worst known; most colony sites probably have not been identified.

The Pacific puffins are abundant, breed in large, but not vast colonies and visit them during the day, making them more amenable to censusing than the auklets. The main imponderable for them is the population of remote areas such as the Kuril Islands, where current estimates are only 'order of magnitude' guesses. The Rhinoceros Auklet, a nocturnal colony visitor, should be hard to count, but because the bulk of the world population breeds in Japan and British Columbia, two fairly accessible areas, its total population is as well known as that of any other of the abundant Pacific auks.

Given the above caveats, Table 7.1, which gives our best estimates of current auk populations, should be treated with great scepticism. We also include estimates for the Atlantic by Nettleship and Evans (1985) and for the Pacific, by Ewins et al. (1993) for *Cepphus* and *Brachyramphus*, Byrd et al. (1993) for *Uria*, *Cerorhinca*, and *Fratercula*, and Springer et al. (1993) for *Synthliboramphus*, *Ptychoramphus*, *Aethia*, and *Cyclorhynchus*. Footnotes give references where we dissent markedly from the earlier estimates.

Changes in populations

Several decades of intensive monitoring have brought us to the realization that there is probably no such thing as a stable population of auks, or indeed of any other seabirds. Most populations that have been accurately monitored showed steady upward or downward trends, or fluctuations over a period of decades. The exceptions to this generalization are populations that have suffered periodic catastrophic mortality (see Common Murre, below). If stable populations are rare (and they are probably rarer than they seem, because our ability to detect statistically significant trends for many species is rather poor), then in a balanced world, there must be as many declining

Table 7.1   Global population estimates for the auks (numbers of individual breeders).

Species	Previous estimate (breeders × 1000)		Our estimate
	Atlantic	Pacific	
Common Murre	6000–9000	6000	12 000–15 000
Thick-billed Murre	10 000–15 000	4700	15 000–20 000
Razorbill	600–2400	0	500–700[1]
Dovekie	7000–15 000	<1	7000–15 000
Black Guillemot	200–350	<5	250–500[2]
Pigeon Guillemot	0	85+	200–300[2]
Spectacled Guillemot	0	77+	100
Ancient Murrelet	0	710	1000–2000[3]
Japanese Murrelet	0	2	5–10[2]
Xantus' Murrelet	0	26–50	6–10[4]
Craveri's Murrelet	0	10	7–87[5]
Marbled Murrelet	0	300–1000	300–1000
Kittlitz's Murrelet	0	25–100	25–100
Cassin's Auklet	0	3350	3500+
Least Auklet	0	11 500	20 000+[2]
Whiskered Auklet	0	22+	100–250[2]
Crested Auklet	0	4400	5000–10 000[2]
Parakeet Auklet	0	875	1000–2000[2]
Rhinoceros Auklet	0	973	1250[6]
Tufted Puffin	0	3502	2500–5000
Horned Puffin	0	1220	1000–2000
Atlantic Puffin	7600–16 400	0	5000–10 000[2]

[1] Lloyd et al. (1991), [2] see species account, [3] Gaston (1992a), [4] Drost and Lewis (1995), [5] Pitman et al. (1995), [6] Gaston and Dechesne (1996)

populations as there are increasing ones. This is not necessarily a consequence of our malevolent influence on the biosphere, but a working out of natural processes that, through their complexity and the superimposition of numerous cyclical, chaotic, and catastrophic physical phenomena, are at present beyond our ability to predict.

Population change is one of the symptoms that we look at as an indicator of whether a population is likely to persist. If numbers are going up, we tend to be perfectly happy with the situation. If they are declining, we may be tempted to predict extinction. If we have insufficient information to detect a trend, it is tempting to assume that things are stable, although, in reality, that is seldom the case. It is a truism that any downward trend, if extrapolated, leads to extinction. In a given situation involving a declining population, we need to ask the following: what is the cause of the trend (natural, or human); if human, then what is the cost or likelihood of reversing it; and what is the likelihood that it will be halted through natural processes? All persistent declines are worth examining for their cause and seriousness, but we need to know when to become concerned.

Clearly, we need to be concerned about very small populations, especially where concentrated in a small area, because these can be driven to extinction by relatively local events. Hence total population size and range are important. We also need to be concerned where trends relate to large changes in adult survival; this is not the same as observing heavy mortality, although they are likely to be connected. Piles of corpses covered in oil, boatloads of birds being brought ashore by hunters, or flung overboard by disgruntled fishermen after extracting them from their nets, are certainly symptoms that need examination. However, of themselves, they do not tell us anything about the likelihood that human actions are driving the birds towards extinction. That can be judged only when we know the answer to many additional questions: how often does this happen; what is the age and breeding status of the birds involved; how big an area have they come from; how big is the breeding population of that area; and what is its capacity for recovery? Most importantly, in practical circumstances, how often does this happen without our knowledge, and how many corpses sink, decompose, are covered by beach sediments, or are removed by scavengers without ever coming to anyone's attention. To answer many of these questions takes a lot more trouble and ingenuity than counting corpses. It also takes a lot of time and it cannot be done after the event.

None of these questions, or their answers, should deny the importance of rectifying needless mortality caused by human activities. However, they can enable us to understand what is at stake in a given situation, and this may inform our actions in terms of where to apply limited resources. For example, an oil spill that killed a thousand first year Common Murres in the North Sea during the late 1980s could have had a negligible effect on the population, because during that period very few first year murres were surviving to recruit as breeders (for reasons that are unknown but probably did not relate to oil spills, Halley 1992). In contrast, if the same spill had killed a thousand breeders, it might have had a fairly striking effect on breeding populations the following year, as there were no pre-breeding birds ready to replace those that had died. Furthermore, an oil spill in the Izu Islands that killed a thousand breeding Japanese Murrelets could put an entire species in jeopardy of extinction. Cleaning live oiled birds for re-

introduction to the wild in the first case is justifiable mainly by our emotional concern for the unfortunate individuals involved, whereas cleaning oiled Japanese Murrelets, while improving animal welfare, could also have a significant impact on the survival of the species.

Population trends

For most auk populations, our knowledge of population trends extends only a few decades. There is much detailed information about trends in populations for limited geographical areas. Changes are easiest to detect in small, peripheral populations and these tend to be the best documented. Some are very dramatic. In this section, we try to sketch in regional trends, rather than trends at specific sites.

Little can be said about a number of species, because there is no firm evidence of trends except for limited or peripheral areas: Dovekie, Black Guillemot, Kittlitz's Murrelet, Craveri's Murrelet, Tufted and Horned Puffin. Lack of evidence for trends is not evidence of stable populations, but merely of our present ignorance. The other species are considered below.

RAZORBILL: after declining in the Gulf of St Lawrence during the 1970s, the species increased in the 1980s and is probably now as abundant as it has been in this century. The population of Britain and Ireland increased from 1970 to 1985, but the magnitude of the increase is not known.

COMMON MURRE: several populations in the North Atlantic increased steadily during the period 1950–90, especially those in Newfoundland and Britain, and to a lesser extent, the Gulf of St Lawrence (Nettleship and Evans 1985; Lloyd et al. 1991). Some British populations have declined since 1980 (Harris 1991), while the small Iberian population decreased from 20 000 to 300 between 1950 and 1980 (Barcena et al. 1984). The population of the southern Barents Sea, parts

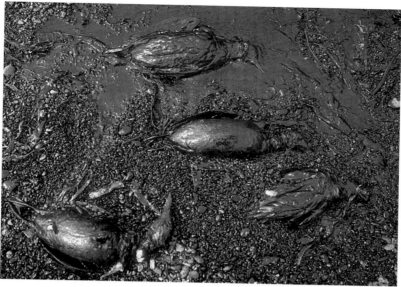

7.1   Murres and a Pigeon Guillemot killed in the Exxon Valdez oil spill, Ushagat Island, Alaska (photo ILJ).

of which had been declining since 1965, suffered a catastrophic decline in 1985, when a crash in capelin stocks led to an 85 per cent reduction in breeders in northern Norway and Bear Island (Vader *et al.* 1990). Numbers in central Norway have also declined—this population is seriously depleted. Populations in California decreased from 1850 to 1950, increased steadily by almost an order of magnitude until the 1980s (Ainley and Boekelheide 1990) and then decreased abruptly by about 50 per cent between 1982 and 1986 (Takekawa *et al.* 1990); those in Alaska showed a patchwork of increases and decreases (Byrd *et al.* 1993). The Common Murre is a classic example of steady increase punctuated by periodic crashes. The abrupt decline in California in the 1980s seems to have resulted primarily from gill-net mortality. The role of human agencies in the crash of Barents Sea capelin and the consequent decline of murres in that area, like most fisheries-related phenomena, cannot be accurately assessed. Overall, despite being by far the most commonly affected bird in marine oil spills, Common Murres seem to be holding their own in most areas.

7.2   Murre skeletons in a fragment of monofilament nylon driftnet at Buldir Island, Alaska (photo ILJ).

THICK-BILLED MURRE: the only well-documented population changes relate to direct harvesting of adult birds, or eggs at colonies. The population of western Greenland, south of Upernavik District, decreased from half a million in the 1920s to a quarter of a million in the 1940s and a few tens of thousands in the 1980s (Evans and Kampp 1991), mainly because of shooting of breeding birds at colonies (Kampp *et al.* 1995). In Novaya Zemlya and at Tyulenii Island in the Sea of Okhotsk, numbers were reduced by very heavy egg harvesting (Shuntov 1986). However, in the Barents Sea, Thick-billed Murres were hardly affected by the capelin crash that had such a devastating effect on Common Murres (Vader *et al.* 1990).

The largest auk hunt in the world involves Thick-billed Murres shot at sea off Newfoundland, with an estimated kill of more than half a million birds annually in the 1980s (Elliot *et al.* 1991). Despite the magnitude of the hunt, there is little evidence for changes in the population of Thick-billed Murres in Hudson Strait, at least since the 1970s (Gaston *et al.* 1993*b*). The majority of these birds winter off Newfoundland.

PIGEON GUILLEMOT: the data available relate only to very small fragments of populations. At the Farallon Islands, numbers appear to have increased by an order of magnitude since the early 1900s (Ainley and Lewis 1974), while in Prince William Sound, Alaska, there may have been an 80 per cent reduction from the 1970s to 1993 (Hayes 1993). Surveys at sea off Kodiak Island also suggest a reduction in that area from 1980 to 1994 (Zwiefelhofer and Forsell 1995).

SPECTACLED GUILLEMOT: the population of Teuri Island, Japan's largest colony, shrank from 7000+ to about 200 between 1949 and 1987. Mortality in gill-nets may have been responsible (Watanuki *et al.* 1986). We should note that this resulted in a more than 5 per cent reduction in the global population.

MARBLED MURRELET: more time and money is currently being devoted to research on this species than on any other auk, because of its association with old growth forests in western North America. Lumber obtained by felling their typical nesting habitat in ancient temperate rainforests is worth as much as $100 000/ha. An ongoing controversy exists over how much of this habitat should be protected from logging. Because of the difficulty in obtaining accurate population estimates, and a lack of historical data, the power to detect population changes has been and will continue to be very low. Consequently, population trends are very hard to assess. Nevertheless, it is clear that Marbled Murrelet numbers are greatly reduced in areas that have been extensively logged. In Alaska, the population fell by about 50 per cent between 1972 and 1992 (Piatt and Naslund 1995). In addition to habitat destruction, the *Exxon Valdez* oil spill killed an estimated 3 per cent of the Alaskan population and about 1 per cent die annually in gill-nets. In British Columbia, there is evidence of major declines near heavily logged areas around the Straits of Georgia, and Barclay and Clayquot sounds (Burger 1995). Numbers have also decreased in Washington, Oregon, and California (Ralph *et al.* 1995).

The future of the Marbled Murrelet as a common, widespread species is uncertain. Populations in the northern and Aleutian parts of their range, where breeding habitat is not threatened, will probably persist, but this comprises only a few per cent of the current world population. South and east of Prince William Sound, where most breeding is in valuable old growth trees, more than three quarters of their original breeding habitat will probably have gone by 2020. This may force the murrelets into fragmented forest remnants, to higher altitudes, or into second growth forest. We do not know whether these habitats are likely to prove suitable, but the fact that they are seldom occupied at present suggests not. In summary, although Marbled Murrelets are not immediately threatened with extinction, major population declines over most of their range appear inevitable unless steps are taken to preserve their remaining breeding habitat.

ANCIENT MURRELET: the population of the Queen Charlotte Islands, at present about half a million breeders, has been approximately halved since 1950 by predation from Raccoons deliberately introduced for fur trapping, and rats (*Rattus rattus and R. norvegicus*) that were accidental accessories to human activities (Gaston 1994). This downward trend is probably continuing. Populations on Alaskan islands were much reduced by Arctic and Red Foxes introduced as a form of 'fur farming' (Bailey and Kaiser 1993), but although the reductions probably were very severe, we have no way of knowing just how severe. Springer *et al.* (1993) suggested that the original Alaskan population was over 1 million, but this was based on a current estimate of only 150 000. If recent informed guesses of a current population in the region of 800 000 are correct, the original population of Alaska could have been several million; we shall never know. Programmes underway for some time to remove foxes from Alaskan islands give cause for optimism and Alaskan populations may be restored to their former size eventually. Numbers in the Kuril Islands may have been much reduced by rats, but again, we shall never know the details.

JAPANESE MURRELET: although numbers are known to have decreased on certain islands recently (Ono 1993, see species account), there is no information on what the population size might have been before the 1980s.

XANTUS' MURRELET: several breeding colonies on islands off the coast of Baja California have been extirpated in the last few decades, and that on Los Coronados Islands has been much reduced, probably by cats (*Felis domesticus*), which also reduced populations in the California Channel Islands. After the removal of cats from Santa Barbara Island, the population rose from very few to about 1500 in the

1990s (Drost and Lewis 1995). Given the limited number of potential breeding islands within the species' range, it may never have been very numerous, but a reduction approaching an order of magnitude since the last century seems likely.

THE AUKLETS: *Aethia*, *Ptychoramphus*, and *Cyclorhynchus* species can be treated together. Except for Cassin's auklets at the Farallon Islands, where the population is known to have declined since the mid-1980s (Carter *et al.* 1990), information is anecdotal. However, such accounts as exist make it clear that enormous numbers occurred in the Aleutian Islands prior to the introduction of foxes. The subsequent disappearance of the auklets from some of these islands can be confidently blamed on the foxes (Bailey and Kaiser 1993). The species most affected appear to have been Cassin's, Crested, and Least Auklets. There are no accounts of huge Parakeet Auklet colonies that are now small or extirpated and there is simply no information for Whiskered Auklets (Springer *et al.* 1993).

RHINOCEROS AUKLET: after disappearing from California in the nineteenth century, the species has now returned. There is also evidence of substantial population increases in Washington and British Columbia since the 1970s and the small population at Middleton Island, in the Gulf of Alaska, is also increasing. The only negative trend is in the Queen Charlotte Islands, where Raccoons and rats have reduced or eliminated a few colonies, involving the disappearance of more than 20 000 birds (Gaston and Masselink, 1997).

ATLANTIC PUFFIN: historical accounts of this species suggest that populations in some areas were greatly reduced during the eighteenth and nineteenth centuries, but such information is hard to evaluate. From Audubon's (1827) descriptions of populations on the north shore of the Gulf of St Lawrence, we cannot doubt that they were severely reduced, but whether the original populations comprised hundreds of thousands, or perhaps millions, is impossible to know. More recently, a steady decrease in the population of the Lofoten Islands, Norway during the 1980s resulted from reproductive failures because of the disappearance of the local herring stock (Anker-Nilsson and Rostad 1993). The very large population of St Kilda was considerably reduced prior to the 1970s, but enjoyed a slight recovery thereafter, while rats greatly reduced the colonies at Lundy Island and Ailsa Craig (Harris 1984). The species has almost disappeared from the English Channel coasts. Conversely, North Sea colonies expanded throughout the 1970s and 1980s. Overall, there must be a lot fewer Atlantic Puffins now than there were in the nineteenth century.

Cause and effect

Looking at the data available on population trends in auks, it is evident that the main factor influencing their populations in the past has been direct destruction of breeding birds through human harvesting, or the activities of introduced mammalian predators on colonies. Also important are fisheries, which affect auks by drowning them in gill-nets, or by removing potential prey stocks. The possible role of fisheries in enhancing prey by removing large predatory fish that compete with auks has not been possible to measure. Similarly, the impact of the reduction or extirpation of certain stocks of large whales, which may have had a significant impact on zooplankton stocks, cannot be evaluated. In the southern hemisphere, reduction of whales has been suggested as a contributory cause in recent increases in penguin populations. The almost total eradication of Bowhead Whales in the North American Arctic during the last century (Bockstoce 1986) may have had some similar effect on Dovekie and Thick-billed Murre populations, but we shall never know for sure.

Surprisingly, there are few cases where oiling, that most obvious cause of auk mortality, can be linked to reductions or extirpations of breeding populations. The disappearance of Atlantic Puffins and Common Murres from the English Channel coast and reductions of the same species in the Bristol Channel during World War II probably relate to oil, as these are areas that suffered enormous pollution as a result of the sinking of oil-burning ships. Colonies in Brittany, reduced in the early twentieth century by hunting, were similarly hard hit by the wreck of two large tankers: the *Torrey Canyon* in 1967 and the *Amoco Cadiz* in 1978 (Bourne 1976; Guermeur and Monnat 1980). Common Murres in California have been affected by numerous oil spills in that area, especially that from the *Apex Houston* in 1986 (Takekawa *et al.* 1990). The immense oil spill from the supertanker *Exxon Valdez* which ran aground in Prince William Sound, Alaska in 1989 killed 100 000–300 000 birds, most of them murres (Piatt *et al.* 1990*a*). However, despite the heavy mortality, population effects proved difficult to detect because of very poor data on the situation before the spill. This difficulty illustrates the need for good monitoring of populations, especially in areas close to tanker routes. From a global perspective, the species most affected by the *Exxon Valdez* spill was probably Kittlitz's Murrelet which was killed in smaller numbers than murres, but may have lost as much as 10 per cent of the world population (Van Vliet and McAllister 1994). Fortunately, most auks live all their lives far away from tanker routes and the chances that any single spill will affect more than a small fraction of the population are negligible except for those species that occupy rather small breeding ranges (e.g. Craveri's and Xantus' Murrelets), or those where a large proportion of the population breeds at a single site (e.g. Cassin's Auklet at Triangle Island, British Columbia, Rhinoceros Auklet at Teuri Island, Japan). The attention that has been paid to the threat posed to auks by oil spills has more to do with the poignancy conjured up by images of live creatures helplessly encased in black sludge than with demonstrable benefits to populations.

The problem cases

Three species of auks stand out as vulnerable to extinction in the short-term: Japanese, Xantus', and Craveri's murrelets. All have global populations of less than 10 000 breeding birds and all are menaced by immediate threats.

JAPANESE MURRELET: these birds face a variety of different threats, but all stem from a single cause; their breeding grounds are adjacent to one of the most densely populated areas of the earth's surface, the southern islands of Japan. Breeding colonies have suffered from the introduction of rats and from a build up of predators (crows (Corvus spp.), kites (*Milvus migrans*)), encouraged by refuse left behind by recreational and commercial fishermen. The majority of the breeding islands are very small, no more than 50 hectares in extent, and the obvious solution appears to be the purchase of the islands by some conservation organization and the management of alien or excess predators. If that can be achieved, the threat of drowning in gill-nets remains. Evidence obtained by Piatt and Gould (1994) indicated that many Japanese Murrelets were being killed far offshore by the Korean squid fishery. Modifications to such an international fishery, conducted in international waters, will be very difficult. However, further evaluation of the impact of the fishery on murrelets, and the circumstances under which they are killed, is required, so that a clear case can be made for conservation measures.

XANTUS' MURRELET: like the Japanese Murrelet, many of these birds breed close to large population centres. They are a little more fortunate, in that some of their breeding islands are already within the California Channel Islands National Park. Eradication of cats and rats from former breeding islands is a good possibility although reintroductions will always be a threat. The fact that some breeding islands are close to large concentrations of ornithologists means that monitoring introductions should be easy. The southern race, *S. h. hypoleucus*, is less fortunate, because its Mexican breeding islands, although remote, are regularly resorted to by fishermen. Plans are underway to eradicate cats from some former colonies, but subsequent monitoring will be more difficult (Drost and Lewis 1995). The northern population, in the Channel Islands, is menaced by oil spills, with the huge ports of Los Angeles and Long Beach close by. A major spill in the vicinity of Santa Barbara Island during the breeding season could wipe out a substantial proportion of the population of *S. h. scrippsi*. This makes the rehabilitation of populations on other Channel Islands more urgent, in order to spread the risk. Outside the breeding season, the species is scattered far offshore and although some mortality in gill-nets and through collision with ships at night has been recorded, it seems unlikely that current fisheries threaten the population.

CRAVERI'S MURRELET: has the smallest breeding range of any auk species. If we ignore the possibility that a few breed on islands off the Pacific coast of Mexico, then the entire breeding range consists of islands in the Gulf of California, a narrow inlet with a maximum axis of less than 500 kilometres. Like Xantus' Murrelets, Craveri's are vulnerable to introduced predators. However, most of their nesting islands are completely without surface water, making them inhospitable for cats and, perhaps, rats. The lack of information on their present distribution and status makes it difficult to know whether they are immediately threatened, or merely vulnerable.

Monitoring the health of marine ecosystems

A substantial amount of writing has been devoted to assessing the virtue of seabirds as indicators to be used in monitoring the health (quality) of marine ecosystems. The use of the words 'indicator', 'monitoring' and 'health in this context needs some elaboration. One of the virtues of seabirds as study objects is their visibility, compared with many other elements of marine ecosystems. Breeding birds use widespread feeding grounds. Their performance during breeding, as measured by variables such as attendance times, egg size, frequencies of food deliveries to chicks, meal size, and chick growth rates, integrates information about the state of the marine ecosystem in which they are foraging and makes it accessible to land-based researchers at a conveniently centralized location. Hence, variation in certain aspects of breeding biology can be used as indicators of what is happening elsewhere.

The word monitoring, when invoked in the context of conservation, means determining trends over time. Thus, for example, we can count the number and identify the species composition of food deliveries made daily to puffin chicks. If we perform these observations in a standardized manner and repeat them annually, using the same protocol, we can monitor inter-year variation in the rate at which puffins provision their young and in what they feed them on.

The concept of 'health' (used interchangeably with 'quality'), when applied to ecosystems, is much more difficult to define than our other terms. In an extreme situation, where the marine environment is heavily polluted by toxic chemicals, we can use the term in the same way that it is applied to people; impaired functioning. However, the type of human disturbances caused so far to marine ecosystems mainly lead to changes in their structuring and species composition. The extent to which we regard these changes as unhealthy depends on our evaluation of how the ecosystem of a given area 'ought' to be. Our concern stems mainly from the idea that, given a particular physical configuration in the ocean (temperature, salinity, water depth, presence of upwellings, etc.) the ecosystem will proceed towards a fixed configuration of species and relative abundances. In that case, any deviation from the naturally persistent state can be regarded as unhealthy.

The degree to which marine ecosystem health is equated with human concerns can be seen from recent events in Newfoundland waters. Beginning in 1989, the Atlantic cod (*Gadus morhua*) and some other groundfish became so rare in continental shelf areas off Newfoundland that a complete moratorium was placed on fishing for them, both commercial and subsistence. This event was an enormous economic and social disaster for New foundlanders and the marine ecosystem of the area was therefore considered to be sick. However, at the same time, and to the concern of the same fishermen, Harp Seal (*Phoca groenlandica*) populations whelping in the same or nearby areas in winter continued to increase. In addition, snow crabs (*Chionoecetes* and *Opilio* spp.) increased in Newfoundland waters to the extent that the value of the commercial catch now exceeds that of the previous cod fishery. Clearly, whatever happened to the marine ecosystem was in no way 'unhealthy' for Harp Seals or snow crabs. Fortunately, it is not necessary to place any particular value on different ecosystem states to justify monitoring seabirds. In whatever way our collective institutions decide to manage marine ecosystems (and inevitably it will be mainly for commercial fish stocks), information about the type of changes occurring is useful for a full evaluation of our options.

Naturally, our ability to interpret year to year variations in the biology of breeding seabirds depends on a knowledge of the general natural history of the species and of the causal connections between food availability and various measures of breeding performance. This has been discussed by Cairns (1987c, 1992c), Furness and Nettleship (1991), and Montevecchi (1993a,b). In the context of auks, it is worth noting that murres, probably the most studied species, and the most often promoted in monitoring studies, are the least likely to provide useful data. Owing to their 'intermediate' chick departure strategy (Chapter 6), the food consumed by chicks is only a small proportion of the total food required by breeding murres. Consequently, murres have proved much more resistant to changes in stocks of their prey species than birds like Arctic Terns and Black-legged Kittiwakes that normally attempt to rear more than one chick to a much larger relative size (Heubeck 1989; Monaghan *et al.* 1989). Reproductive success of Common Murres in Britain varied from 0.70 to 0.84 chicks/pair (variation 0.14 chicks/pair) among different regions in 1994, compared with 0.34–0.54 (0.20) for Northern Fulmars, 1.21–2.13 (0.92) for Shags (*Phalacrocorax aristotelis*), 0.45–1.14 (0.69) for Black-legged Kittiwakes, and 0.39–0.99 (0.60) for Common Terns (*Sterna hirundo*) breeding in the same regions (Walsh *et al.* 1995). Species of auks with semi-precocial chicks are much more likely to be useful in monitoring marine ecosystem changes, especially those species, such as Tufted Puffins, that appear to be susceptible to periodic breeding failures.

The future

If we stand back and take a global view of auk populations, there is no need to be too gloomy at present. Several species seem to have done quite well in recent decades (Common Murre, Rhinoceros Auklet), even if they are experiencing problems in limited areas. Several others that suffered big reductions through the harvesting of eggs and adults and the introduction of foxes for 'fur farming' in the nineteenth century, or the first half of the twentieth, are showing signs of recovery (Atlantic Puffins in the Gulf of St Lawrence, Least and Crested Auklets in the Aleutian Islands), or at least are enjoying some relaxation of our attentions (Thick-billed Murres in Greenland, Ancient Murrelets in Alaska).

Foxes were introduced to 455 Alaskan islands altogether, but in 1992 they survived on only 46, having died out, or been eradi-cated on the rest. Recolonization of the islands by seabirds after the disappearance of the foxes is often quite swift (Bailey and Kaiser 1993). Further eradications are underway, some funded by money from the financial settlement of the *Exxon Valdez* oil spill. A programme to halt the spread of Raccoons in the Queen Charlotte Islands was begun in 1993 and aims to prevent further erosion of seabird colonies in that archipelago.

In 1995, rats were removed from Langara Island, British Columbia, formerly the largest colony of Ancient Murrelets. If this removal proves permanent, a successful recovery of the murrelet population is a real prospect. Rats have also been removed from several small, but important, European islands in the 1990s, Atlantic Puffins being the most likely beneficiaries. Further accidental introductions, especially of rats, seem likely, but government agencies have now developed plenty of experience in dealing with these problems and there seems no reason to think that future events will compare in scale with what preceded them. Following predator removal, artificial enhancement of the colonization process is seldom necessary where potential population sources are available nearby. However, the extirpation of entire regional populations may require intervention in the form of transplanted chicks, as has been practised for Atlantic Puffins in the Gulf of Maine (Kress and Nettleship 1988).

Like all the inhabitants of the biosphere, auks face continual and probably accelerating change in their environment due to our activities, including direct mortality, habitat loss, and global climate change. Despite these threats, it is clear that we have the means to maintain auk populations given the necessary determination to share the biosphere with them. We remain optimistic that the beauty and fascination of these ancient survivors will pursuade humanity to make room for their continued existence.

PART II

Species accounts

Great Auk *Pinguinus impennis*

Alca impennis (Linnaeus, 1758, *Syst. Nat.*, ed. 10,i: 130—Arctic Europe).

The Great Auk (Garefowl or Penguin) is the only alcid to become extinct during historic times, the last specimen taken having been killed in 1844. The last reliable sighting was in 1852 (Bourne 1993). Extinction has been commonly blamed on their flightlessness making them vulnerable to ground predators, especially people.

It never appears to have entered into the calculations of the earlier generations of Great Auks that, sooner or later, evolution would produce a race of sailors to whom no flat coasts would be impregnable. (Parkin 1894).

Description

From contemporary accounts, it appears that Great Auks stood almost upright on land, presumably balancing on their tarsi. At sea, apart from the tiny wings, they looked like giant Razorbills. They were said to be strong swimmers ('easily outswam a 6-oared boat', Newton 1865), the very short wings presumably being used like those of penguins, but there do not appear to be any references to 'porpoising' as a means of travelling at sea; they either flapped along the surface, or dived as soon as disturbed. Plumage description from Witherby *et al.* (1941).

ADULT SUMMER: dark blackish-brown on all upper surfaces, except for narrow white tips to the secondaries and a large oval, white patch on side of head just above and in front of eye, reaching almost to base of bill; white below, except for dark chin and throat, and hoary brown underwing coverts; bill, legs, and feet black; iris chestnut; gape yellow-orange; bill with 6–7 deep vertical grooves on the terminal half of the upper mandible and 8–9 on the lower mandible; tail wedge-shaped, 12 feathers.

ADULT WINTER: white patch in front of eye and dark throat disappeared in winter, when plumage seems to have been similar to that of a winter-plumage Razorbill.

JUVENILE AND FIRST WINTER: similar to winter-plumage adult, but with a shorter, narrower bill.

CHICK: down grey (Fabricius 1780, in Meldgaard 1988).

VARIATION

No subspecies recognized. Measurements of bones from Funk I., Newfoundland and from middens in Norway suggest that birds from western Atlantic may have been larger than those from the east. See Table (from Burness and Montevecchi 1992).

MEASUREMENTS

(Witherby *et al.* 1941), wing 160–178 mm, tail 83–95 mm, tarsus 55–62 mm, culmen 82–90 mm, maximum bill depth 40–46 mm. Length distribution of a large sample of femurs was bimodal, suggesting possible sexual size dimorphism (Lucas 1890). Sexual dimorphism in bill size is probable (Livezey 1988).

WEIGHTS

One taken in Faeroes in 1808 weighed about 4 kg (9 Danish pounds, Feilden 1872). From skeletal measurements, Livezey (1988) estimated mean body mass at 5 kg.

Table: Great Auk skeletal measurements (mm)

Bone	Funk I.			Norway		
	mean	*s.d.*	*n*	*mean*	*s.d.*	*n*
humerus	104.4	2.8	82	100.0	3.3	15
ulna	58.5	1.9	100	53.2	1.8	14

Range and status

RANGE

BREEDING: only eight definite breeding localities known on the basis of historical accounts: in the western Atlantic, Funk I., Newfoundland, by far the largest colony and presumably supporting hundreds of thousands at the time of the first European visitors, Penguin I., off S Newfoundland (recently confirmed by Montevecchi and Kirk 1996), and Bird Rocks, Gulf of St Lawrence, perhaps also off Cape Breton; in Iceland, Geirfuglasker and Eldey Is. in the SW and the Westmann Is.; and in the eastern Atlantic, the Faeroes, St Kilda, Papa Westray in Orkney (possibly not a regular site), and perhaps Calf of Man. A specimen in the Copenhagen museum was taken as a breeder in Greenland, but exact locality is unknown (Salomonsen 1945). Great Auk bones are known from midden sites in Florida, new England, and Labrador, several sites in W Greenland, Iceland, the whole coast of Norway, Denmark, Holland, Brittany, and Gibraltar. Fossils have also been found in Pleistocene deposits in southern Italy (Violani 1975). This distribution is amazingly similar to that of its closest relative, the Razorbill, with the exception of the Florida records. It was apparently confined to boreal and low Arctic waters while breeding, perhaps extending southwards in winter. The abundance of material from Norway suggests that it was probably a widespread breeder there. The birds would presumably have been difficult to capture away from their breeding sites, so the occurrence of abundant midden material may be taken as evidence of nearby breeding. However, it is hard to imagine it breeding anywhere near Florida. Midden material there dates from 1000 BC to 1300 AD (Brodkorb 1960) and may indicate either that Great Auks were regular winter visitors that far south, or that bones or corpses were traded with Indian bands to the north. Their vulnerability to terrestrial predators would probably have excluded Great Auks from breeding in ice-covered Arctic waters where they would have been easy prey for Polar Bears (*Ursus maritimus*) and where the formation of ice bridges would have made their colony islands accessible to Wolves (*Canis lupus*).

WINTER: little known, but records from Gibraltar, and from Cape Cod and Florida, probably involved wintering birds, as no auks are known to have bred in those areas in historic times. According to Fabricius (in Salomonsen 1950), they appeared off Greenland in Sept and stayed until Jan. They were seen regularly off Massachusetts in winter in the eighteenth century (Audubon 1827).

STATUS

Extinct. Of the colonies known to have existed in the seventeenth century, Bird Rocks seems to have disappeared by 1700, St Kilda between 1720 and 1760, and Funk I. and the Faeroes some time about 1800, although one was reported killed in the Faeroes as late as 1808 (Yarrell 1884). The last pair was killed on Papa Westray, Orkney in 1812 (Newton 1861). In Iceland, a small, but perhaps viable, population persisted until volcanic eruptions in 1830 drowned the main breeding island. The remaining birds switched to Eldey Stack, a more accessible island, and the last known pair was collected there in 1844. There were a few, probably authentic sightings subsequently, but the species must have been extinct by 1860 (Newton 1861).

Voice

Only calls on record are 'low groans' or 'croaks'; apparently rather silent, compared with murres.

Habitat, food, and feeding behaviour

Little is known of distributions at sea, but they appear to have been confined to continental shelf waters, as early sailors used their presence as an indication that they had arrived on the Newfoundland Grand Banks:

…you may know when you are upon the Bank … by the great quantities of fowls … none are to be minded so much as the Pengwins [Great Auk], for

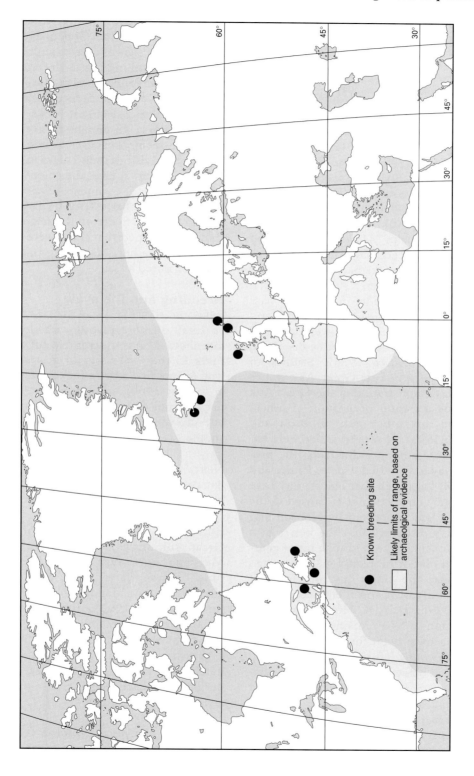

Known breeding site

Likely limits of range, based on archaeological evidence

these never go without the Bank, as the others do. (John Seller 1728, 'The English Pilot', reproduced by Christy 1894).

DIET

Food almost certainly largely fish, judging from bird's size and by analogy with its smaller relative, the Razorbill. Only one direct observation is available, of birds taken off Greenland, containing lumpsucker (*Cyclopterus lumpus*) and shorthorn sculpin (*Myoxocephalus scorpius*) (Fabricius, in Salomonsen 1950). Olson *et al.* (1979) identified fish remains found in soil attached to Great Auk bones from Funk I., which included capelin (*Mallotus villosus*), cod (*Gadus morhua*), striped bass (*Morone saxatilis*), shad (*Alosa* sp.), stickleback (*Gasterosteus aculeatus*), and menhaden (*Brevoortia* spp.). However, neither menhaden nor striped bass seems likely to have occurred off Funk I. and the possibility that the samples were contaminated from other sources cannot be dismissed (Bradstreet and Brown 1985). Studies of nitrogen isotope ratios in Great Auk bones from Funk I. demonstrate that the Great Auk probably took a greater proportion of fish in its diet than any other auk (Hobson and Montevecchi 1991). However, ratios were extremely variable among individuals; the preferred diet may have varied widely. Prince and Harris (1988) suggested that chicks may have been fed largely on zooplankton carried in the crop and this was supported by Hobson and Montevecchi's discovery that the bones of a juvenile Great Auk showed isotope ratios characteristic of a diet of zooplankton, rather than fish. However, the single juvenile in their sample was full-grown and identified only by incomplete ossification (K. Hobson, personal communication). No remains of partially grown chicks have been found and it seems likely that the young left the colony while still very small (see below). Hence, the bone analyzed by Hobson and Montevecchi may not have been that of a chick, but of a young of the year that had left the colony and somehow returned, presumably after feeding on zooplankton. K. Kampp (personal communica-tion) inspected the digestive tract of a Great Auk preserved in the Copenhagen Museum and found no sign of a diverticulum for carrying food, something that is present in all auks that feed their young on plankton. No diverticulum is visible in a preserved gullet illustrated in Salomonsen (1945); it looks identical to that of a murre. In the absence of further evidence, it seems best to assume that Great Auks carried fish to their chicks in their beaks, if they fed them at all at the colony.

Displays and breeding behaviour

Unknown. John Seller, in the reference quoted above, mentioned that there were never fewer than two together, suggesting that pairs may have consorted at sea.

Breeding and life cycle

BREEDING HABITAT AND NEST DISPERSION: bred on small offshore islands with sloping shores that allowed birds to scramble out of water. On Funk I., they were said to nest so close that 'you cannot find a place for your feet' (Thomas in Murray 1968), so presumably, like murres, they did not maintain any individual distance, but crowded together in contact with one another.

NEST: none was built, the egg being laid on flat rocks above high tide, in an area where birds had easy access from the sea.

EGG-LAYING: Martin (1698) reported that the Great Auk arrived at St Kilda, 'the first of May and goes away about the middle of June'. These dates must be adjusted for changes in the calendar to mid-May to late June (Bourne 1993). Eggs were taken in Iceland in late June, and eggs and chicks were said to have been present there in early Aug, suggesting that this was about the time of hatching. Assuming a 6-week incubation period (Birkhead 1993), this also suggests laying in mid–late June.

EGGS: one, elongated pyriform, very similar in shape to that of a Common Murre and with a range of ground colours also similar to murres: off-white through buff to pale bluish green (Witherby *et al.* 1941). Markings like those of Razorbill, many with blotches and scribbles, usually concentrated towards large end (Tompkinson and Tompkinson 1966). Dimensions: 123.4 (range 111–140) × 76.4 mm (70–83, $n = 36$, Witherby *et al.* 1941). Fresh weight estimated from volume as 327 g (Birkhead 1993), about 7 % of adult weight.

CHICKS: according to Salomonsen (1945) no specimens of chicks are known and none has been described. The only mention appears to be in the account of a raid on the Geirfuglasker colony by a British privateer in early Aug 1808, when eggs and chicks are said to have been present (Newton 1861, based on the researches of J. Wooley). Probably this visit would have been about the time of hatching. Both Bengtson (1984) and Birkhead (1993) argued that the nestling period must have been rather short and that, consequently, chicks must have left breeding site while still much below adult size, perhaps at an even earlier stage than those of murres and Razorbills. Downy chicks were reported at sea by Fabricius (1780, quoted in Meldgaard 1988), suggesting that chicks did not develop feathers before leaving colony, and went to sea still clothed in natal down, like *Synthliboramphus* murrelets (Chapter 6). This was also suggested by Duchaussoy (1897) and fits with accounts of chicks riding on their parents' backs (Denys 1672 quoted in Nettleship and Evans 1985). Lack of skeletal material of chicks among the large quantities of bones recovered from Funk I. seems to support this idea (W. A. Montevecchi, personal communication), although if all eggs were removed during the period of European raids, few chicks may ever have hatched. Departure from colony at a very young age seems to indicate prolonged parental care after leaving, especially if young were downy.

MOULT: primaries not shed synchronously, as in Razorbill, but periodically and in irregular order. This judgement is based on a single specimen, probably collected in October, and hence cannot be treated as definitely representative of the species (Stresemann and Stresemann 1966).

Population dynamics

Single egg laid; annual productivity was therefore not greater than that of a murre. Birkhead (1993) has argued that, given a correlation between size and adult annual survival rate in auks, survival of adult Great Auks under natural conditions was likely to have been very high. However, survival rate of unrelated, but ecologically similar penguins is lower than that of murres and puffins (Croxall and Gaston 1988), so we could plausibly argue for any survival rate. Given the late age at first breeding in razorbills and murres, and the fact that individuals continued to appear in Britain after the extermination of breeders, we can probably assume that the Great Auk also deferred breeding until it was several years old. Bengtson (1984) argued that the Great Auk was already a rare bird by the time that Europeans reached North America. This is supported by the lack of historical breeding sites and the fact that large numbers were reported from only two sites: Funk I. and Geirfuglasker. As both of these islands were rather small, it is hard to imagine that the world population in the seventeenth century could have exceeded a few hundred thousand, although they appear to have been common at sea off Newfoundland. The extinction of the Great Auk was brought about primarily by human activities, especially the killing of adult birds at Funk I. for their feathers and oil. The much smaller numbers breeding off Iceland were somewhat protected by the remoteness of their breeding islet until it was submerged during a volcanic eruption in 1830. The remaining birds apparently shifted to Eldey, where they were more vulnerable, and the demise of the last few was brought about by the availability of a market in skins and eggs for wealthy collectors.

Razorbill *Alca torda*

Alca torda (Linnaeus, 1758, *Syst. Nat.* 10, 1: 130—Type locality restricted to South Sweden).

PLATE 1

Description

A robust, heavy-billed auk with an unusually long, somewhat graduated tail. On the water and in flight behaves very much like a murre, although a little more agile, and upends more at the start of a dive. On land usually squats on the tarsi and shuffles about while maintaining a rather horizontal posture.

ADULT SUMMER: upperparts, head, and throat black, dark chocolate brown on face, but secondaries narrowly tipped with white; underparts, including underwing coverts, white; a narrow white line extends forward from the eye to the top of the bill; bill black, with prominent white vertical line just in front of nostril, gape bright yellow; iris dark brown; legs and feet black.

ADULT WINTER: as in breeding plumage, but throat, sides of neck, and face behind the eye white, brownish on the lores; white lines on head and bill less prominent, or absent; there may be a variable amount of dark brown mottling on the face, and the sides of the neck and breast.

JUVENILE AND FIRST WINTER: similar to winter-plumage adult, but with shorter, more slender, all black bill.

CHICK: down sooty brown with silvery streaking on head, which looks much paler than body, or grey-brown, without streaking; first plumage variable, similar to adult breeding, or non-breeding plumage or intermediate, as in Thick-billed Murre.

VARIATION

Two races generally recognized on the basis of size: *A. t. islandica* in Iceland, Faeroes, British Isles, and coasts of North Sea and English Channel; nominate race elsewhere. As with other Atlantic auks, the warmer water race, *islandica*, is smaller. Plumage of chicks

at colony departure may vary geographically: on Skomer I., Wales, Hudson (1984) found 51% in non-breeding type plumage, whereas only 2% at Gannet Is., Labrador were of non-breeding type (Birkhead and Nettleship 1985).

MEASUREMENTS

Very little difference between the sexes. Table: (1) *A. t. torda*, Hornøy, Norway, breeders (Barrett 1984); (2) Gulf of St Lawrence, breeders (G. Chapdelaine, personal communication), (3) *A. t. islandica*, Shetland, breeders (Furness 1983); (4) *A. t. torda*, Baltic (Salomonsen 1944, museum spec.); (5) *A. t. pica = torda*, Greenland, summer adults (Salomonsen 1944, museum spec.).

WEIGHTS

Little detailed information. Breeding adult *A. t. torda*, trapped at the colony on Hornøy, Norway (Barrett 1984) averaged 714 ± 52 g ($n = 73$); and in Gulf of St Lawrence (G. Chapdelaine, personal communication), during chick-rearing 727 ± 49 g ($n = 536$). *A. t. islandica* breeding in Shetland (Furness 1983) averaged 620 ± 41 g ($n = 35$).

Range and status

RANGE

Found in boreal and sub-Arctic waters of the Atlantic, where its range overlaps very broadly with that of Common Murre.

BREEDING: on cliffs and offshore islands from southern Britain and northern France to northern Norway and Iceland, and in western Atlantic from central Maine and the Bay of Fundy to Central Labrador. Scattered pockets are also found in E Canadian Arctic, from N Hudson Bay to SE Baffin I. and in W Greenland.

WINTER: more genuinely migratory than other Atlantic, low Arctic auks. Those breeding in Labrador and Gulf of St Lawrence mainly

Table: Razorbill measurements (mm)

♂♂	ref	mean	s.d.	range	n
wing length	(4)	209.4	5.1	201–216	10
	(5)	212.3	2.8	209–216	7
culmen	(4)	35.1	1.6	32–38	13
	(5)	36.5	1.6	34–39	9
bill depth	(4)	24.6	1.1	22–26	13
	(5)	25.1	1.1	24–27	9
♀♀	ref	mean	s.d.	range	n
wing length	(4)	205.8	3.7	201–213	15
	(5)	211.5	3.3	208–216	6
culmen	(4)	33.9	1.2	32–36	16
	(5)	35.9	1.0	34–37	7
bill depth	(4)	23.8	0.8	22–25	16
	(5)	23.6	0.4	23–24	7
UNSEXED	ref.	mean	s.d.	range	n
wing length	(1)	212	6		73
	(2)	208	5		457
	(3)	196.9	4.1		119
culmen	(1)	19.0	1.5		74
	(2)	32.8	1.4		461
	(3)	33.0	1.7		70
bill depth	(2)	23.0	1.1		461
	(3)	20.6	1.0		70
tarsus	(1)	36.7	1.9		6
	(2)	33.1	1.6		221

move to waters off Nova Scotia and E U.S. seaboard S to Massachusetts in winter. Main breeding areas are essentially deserted and few occur off Newfoundland in winter, although recoveries of marked birds indicate that they occur off Newfoundland in autumn. George's Bank, S of Nova Scotia, especially the shallow shoals region on the NW side, is a major wintering area (Powers 1983; Brown 1985). Christmas Bird Counts record high numbers off New Brunswick, Maine, and Massachusetts, with smaller numbers S to Long I. Highest counts are from Cape Cod, averaging 560 over 1961–92. In E Atlantic, birds from N Norway move S (Holgersen 1951) and some from Iceland move to the Faeroes (Gardarsson 1982). Some birds from the British Isles cross to the Norwegian coast in winter, and many, especially first years, shift to the Bay of Biscay and

S to the Mediterranean. Some birds remain in Biscay during their first summer (Steventon 1982; Teixeira 1986). An important concentration occurs off NE Scotland in Aug–Sept and presumably includes many moulting birds. Later, Razorbills scatter fairly evenly in the North Sea in winter, with no evidence of a concentration in the S (Tasker *et al.* 1988). In Oct large numbers arrive in the Dutch sector, presumably from Britain (Camphuysen and Leopold 1994).

STATUS
The majority of the world population breeds in Iceland, where numbers are very imperfectly known. Consequently, estimates of the world population are very approximate. Nettleship and Evans (1985) placed the total between 600 000 and 2.4 million breeding birds, of

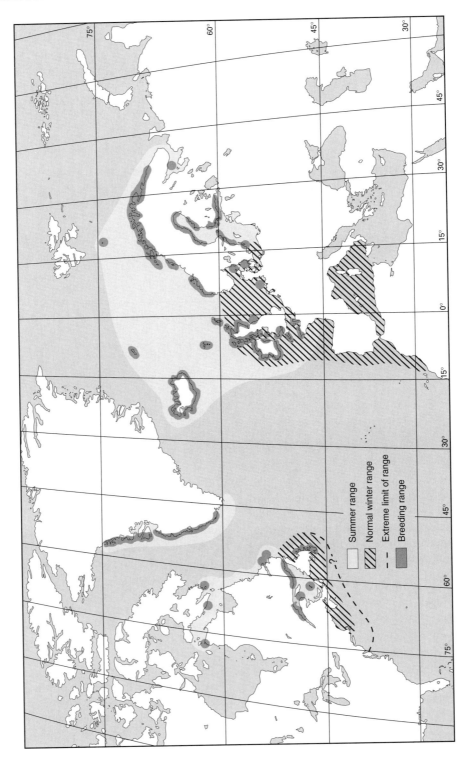

which about 70% bred in Iceland (estimated population 1 million, range 200 000– 2 million) and 20% in the British Isles, mainly in Scotland and Ireland. A further 4% bred in Norway, 2% in E Canada, and 1% each in Sweden and Finland. As many as three quarters of a million may breed on the immense cliffs of Latrabjarg in NW Iceland (Gardarsson 1985). The global estimate was revised downwards by Lloyd *et al.* (1991), who estimated the British population as 182 000, suggested that Iceland held 450 000 pairs, and that the world population was in the range 500 000– 700 000 pairs (of which half could breed at Latrabjarg!). Numbers in Britain may have increased between 1969 and 1987, although that is uncertain because of the imprecision of the earlier estimate. Some increase in Wales appears definite, but numbers on the Channel coasts have declined. Some increase occurred in the Gulf of St Lawrence between 1982 and 1993 (Chapdelaine and Brousseau 1992; Chapdelaine 1995).

Field characters

Very similar to murres in plumage, but the much thicker bill and longer tail distinguish Razorbills both on the water and in flight. Tail often held slightly cocked when on land, which, added to more horizontal stance, makes Razorbill posture very distinct from murres. Looking from above, the white line across tips of secondaries is longer and narrower in Razorbills. Head is blacker than in southern races of Common Murre (*albionis, aalge*). In winter, young birds might be mistaken for Thick-billed Murres, but differ in length of tail and extent of white on the face.

Voice

Comprises an almost undifferentiated series of growling or snoring sounds, rendered 'knorrrr', etc. Account based on Paludan (1947) and Cramp (1985).

ADULT: a variety of growls are given during agonistic interactions, or when disturbed or threatened. In alarm situations they tend to be short, coughing sounds, and in aggressive encounters more protracted growls. Cramp (1985) defines the following: 'alarm growl', <0.5 sec duration; 'attack call', a protracted growl lasting about 1.5 sec; 'ecstatic growl', 2–5 growls in a forward posture followed by a 2–3 sec growl in the head-vertical position, trailing off in growls of diminishing intensity; 'billing call', a series of growls diminishing in intensity; 'lure call', a series of growls 1.5–2 sec apart given in interactions with the chick.

CHICK: gives a soft 'piu-piu' when young, possibly a food-begging call. Older chicks give loud 'dui' or 'duiuiui' calls if separated from parents. Similar calls are given during departure.

Habitat, food, and feeding behaviour
MARINE HABITAT

Feeds in continental shelf waters, usually somewhat closer to shore than murres, but sometimes scattered among them. May make use of tidally induced upwellings in coastal areas. Enters and feeds in large estuaries with reduced salinity more frequently than other large auks (e.g. St Lawrence River).

FEEDING BEHAVIOUR

Rarely form dense flocks while feeding. Small numbers may associate on feeding areas, but while engaged in intensive diving bouts they usually spread out. Cayford (1981) noted that they tend to be more evenly spaced than Common Murres where feeding in association. They characteristically upend more than murres in submerging; 90% of those watched by Cayford oriented towards the sun while diving, suggesting that the direction of light could affect their ability to detect prey. Other birds noted feeding in association with Razorbills include Common Terns (*Sterna hirundo*), Herring Gulls (*Larus argentatus*), Little Gulls (*Larus minutus*), and Black-legged Kittiwakes (*Rissa tridactyla*) which apparently profit from prey driven towards the surface (Cantelo and Gregory 1975).

FORAGING RANGE: rather poorly known. Most studies suggest that Razorbills do not travel far from the colony during chick-rearing period, mainly feeding within 15 km of the colony. At Isle of May, Scotland, they travelled further than Common Murres breeding at the same colony, mainly >10 km, and fed mainly in shallow, inshore areas (Wanless *et al.* 1990). This was also true of Razorbills wintering in the Mediterranean Sea (Carboneras 1988). Relatively small food loads and relatively high delivery rates also suggest that foraging for chicks is commonly carried out close to the colony. However, most of the colonies studied have been small. It seems unlikely that the enormous number of breeders at Latrabjarg, Iceland, for instance, could all feed so close to their breeding sites without severely depleting local supplies.

DIVING BEHAVIOUR: dives to 120 m (Piatt and Nettleship 1985), but mean maximum dive depth of 18 birds breeding at Hornøy, Norway was 25 m (range 11–38 m, Barrett and Furness 1990). Average dive duration of birds feeding off the Isle of May, Scotland, was only 35 sec, implying, even at an underwater speed of 1.5 m/sec, a mean maximum feeding depth of about 25 m (Wanless *et al.* 1988). Data on diving times and feeding depths in shallow water (<10 m, e.g. Bradstreet and Brown 1985) are probably not characteristic of the species, because most feeding occurs in water more than 25 m deep.

DIET
ADULT: like that of Common Murres, consists mainly of mid-water schooling fishes: capelin, sandlance (*Ammodytes* spp.), herrings (*Clupea harengus*), sprats (*Sprattus sprattus*), juvenile cod. Take fish up to a maximum body depth of 23 mm (Swennen and Duiven 1977). In winter, off Portugal, pilchards (*Sardina pilchardias*) are important constituents of diet (Beja 1989), while in the Kattegat, gobies (Gobiidae), sprats, and sandlance were all common prey (Blake 1983).

CHICKS: are fed exclusively on fish, with capelin predominating in the diet in Greenland and the Barents Sea (Belopol'skii 1957;

Salomonsen 1967; Barrett and Barrett 1985) and sandlance in Britain (Furness 1983; Harris and Wanless 1986*a*; Swann *et al.* 1991). In the Gulf of St Lawrence capelin formed 53–69% of diet by weight, and sandlance the balance (Chapdelaine and Brousseau 1996). Sandlance were also important in chick diets in Labrador, where adults fed themselves on juvenile sculpins and euphausiids (Bradstreet in Bradstreet and Brown 1985).

Displays and breeding behaviour
COLONY ATTENDANCE: Colonies not attended as frequently as Common Murres, perhaps because most birds move further during the winter. In Scotland, most birds arrive at colonies in Mar, but some may visit as early as Oct, especially at Isle of May (Taylor and Reid 1981). At Skokholm I., they first appear in Mar and in Gulf of St Lawrence, in mid-Apr. Attendance during the pre-laying period is characterized by extreme fluctuations in numbers, from zero to most of the breeding population, with peaks occurring every 4–6 days. During this period there is much social behaviour on the sea, as well as at the colony.

SEXUAL BEHAVIOUR: Several behaviours that occur both on the cliff and at sea, where birds gather in rafts near the colony, are described by Conder (1950), Plumb (1965), and Hudson (1979*b*). Allopreening between pairs is common, usually directed at head, and neck and bouts may last for up to 6–7 min. Several related behaviours involve signalling with the bill: 'bill clicking', in which mandibles are clicked together while shaking head slowly from side to side; 'bill vibrating' ('ecstatic posture', Conder 1950), where head is thrown back and lower mandible vibrates, while a loud rattling noise is given, followed by sweeping the head forward and downward, and finishing with mutual billing; and 'bill touching', which also occurs in murres. Bill clicking and touching appear to be more frequent at sea than on land, while bill vibrating occurs mainly at the breeding site and appears to be a greeting display (Birkhead 1976*b*). Breeders gather in social groups on broad ledges or rocks adjacent to their breeding sites. These 'club' areas are attended mainly by ♂♂, but visited by ♀♀, both to secure copulations with ♂♂ other than their mate and to disrupt any attempts made by their mate to copulate with other ♀♀. This social arrangement has been described at Skomer I., Wales by Wagner (1992a,b,c), who equated it to a lek. There, 50% of ♀♀ received at least one copulation from a non-mate, although copulations outside the pair amounted to only 2.4% of all copulations. ♂♂ copulated with their mates on average 80 times in the 30 days before the laying of the first egg. Copulation is probably important in pair formation, which may span more than a year. Wagner (1991b) reported courtship and copulation between 4-year-old ♀ and a ♂ that apparently resulted in breeding together the following year. Initial contact was made at the club, followed by joint site prospecting; the ♀ then disappeared for the season, but the pair bred together the following year near where they had prospected, although the ♂ made no attempt to defend a site in the interim. ♀ may select the site.

Breeding and life cycle
Information mainly from Paludan (1947), Bédard (1969c), and Lloyd (1976).

BREEDING HABITAT AND NEST DISPERSION: in crevices on cliffs or among boulders and talus. Because chicks cannot fly when they leave the colony, the breeding site must give immediate access to the sea. Some breed on open ledges, in situations similar to those used by murres, while others use caves, crevices, and spaces under rocks. Sites are not immediately adjacent, as in murres, but are at least 10 cm, usually >30 cm, apart.

NEST: no nest is constructed, although a few pebbles or other debris may be collected at the site.

EGG-LAYING: mean laying dates at Skokholm, Wales were 17–29 May (3 seasons), with 80% of eggs generally laid within 10 days (Lloyd 1979). At Graesholm, in the Baltic, laying was between 10 May and 30 June, with median dates 22 May–19 June (6 years, Paludan 1947). At Hornøy, Norway, mean hatching was 1–7 July (2 seasons), implying mean laying about 27 May–2 June (Barrett 1984). In the Gulf of St Lawrence, median dates of hatching were 12–20 July, implying laying in early-mid June (3 years, Chapdelaine and Brousseau 1996).

EGGS: one, ovoid-pyramidal, not as pyramidal as murres, similar in ground colour and markings to those of Common Murre, but less variable. Ground colour mainly pale buff to tan, seldom bluish, with dark brown blotches and/or streaks of brown or black. Eggs from St Kilda and Faeroes more likely to have scribblings and finer markings than those from elsewhere. Eggs from Irish Sea colonies tend to have heavier blotches and darker ground colour (BMNH). Eggs laid by same ♀ are very similar, allowing maternity to be identified (Lloyd 1976). For measurements, see Table: (1) *A. t. torda*, North America (Bent 1919); (2) Greenland (BMNH); (3) Hornøy, Norway (Barrett 1984); (4) Baltic (Paludan 1947) (5) *A. t. islandica* Scotland (BMNH); (6) Skokholm I., Wales (Lloyd 1979); (7) Faeroes (BMNH). Fresh weights: *A. t. torda*, Labrador 96 ± 9 g (Birkhead and Nettleship 1984b); Baltic 95.6 ± 5.1 g, range 88.9–107.2, $n = 10$ (Paludan 1947); *A. t. islandica* 85 g (Witherby et al. 1941); about 13% of adult body weight. Egg size decreases with date of laying and increases with age (Lloyd 1979). Replacement eggs at Skokholm and in Labrador averaged 8% smaller than first eggs. Interval from loss to replacement was 15.1 ± 2.3 days ($n = 29$) at Skokholm, 14.2 ± 2.9 days at Hornøy, Norway (Barrett 1984). Composition of fresh eggs: shell 10%, yolk 35%, albumen 53%. Pipping eggs weigh 84% of fresh weight, of which neonate chick constitutes 62% and remaining yolk 16% (Birkhead and Nettleship 1984b). Eggshell thickness at Skokholm, 1970s, 0.38 ± 0.009 mm ($n = 38$, Lloyd 1976).

INCUBATION: no detailed studies. ♀ may incubate for 48 hrs immediately after laying.

Table: Razorbill egg dimensions (mm)

ref	length			breath			n
	mean	s.d.	range	mean	s.d.	range	
(1)	75.9		69.0–83.5	47.9		42 0–51.5	80
(2)	78.1	2.9		48.3	1.6		8
(3)	77.0	2.9		49.0	1.8		154
(4)	74.7	2.9	69.9–81.3	48.6	1.0	46.7–50.9	15
(5)	71.8	3.2		46.6	1.5		21
(6)	73.6	2.4		47.5	1.6		206
(7)	72.4	2.6	67.8–76.8	46.7	1.4	42.8–49.0	27

Partners exchange incubation duty several times daily. At Skokholm, 3–4 change-overs daily during early incubation period, falling to 2 near hatching (Lloyd 1976). Total duration in Sweden 35.5 days (*n* = 6, Paludan 1947) and at Skokholm, 35.1 ± 2.2 days (*n* = 239, Lloyd 1979). There are two lateral brood patches.

HATCHING AND CHICK-REARING: eggs take 2–4 days to hatch (Paludan 1947). Brooding is continuous for initial 3–4 days, but discontinuous thereafter. One adult is normally present at the site throughout nestling period, except at concealed sites (burrows or under large boulders) when attendance with older chicks may be intermittent (Lloyd 1977; Hudson 1979*b*). The timing of presence at the nest-site is not influenced by sex; both birds were together at the site 25% of the time at Isle of May, most frequently between 14:00 and 18:00 hrs (Wanless and Harris 1986). Chicks may be fed up to 11 fish at a time, with bill loads averaging 3.0–8.5 g (5 studies) and ranging up to 30 g. Most common number of fish delivered is 1 (about 40% of deliveries); <10% of deliveries involve more than 3 fish. Daily mean deliveries per chick were 2.9 at Isle of May (Harris and Wanless 1986*a*), 4.7 at Skokholm (Lloyd 1977), and 3.0–3.7 in the Gulf of St Lawrence (Chapdelaine and Brousseau 1996); rate of feeding was highest at dawn and decreased thereafter, with few deliveries in the last 4 hrs before dark. There is little consistent relationship between feeding rates and date or chick age, although Harris and Wanless (1986*a*) found that feeding rates were highest to 3–11 day chicks. Daily energy intake in the Gulf of St Lawrence was estimated at 251–357 kJ over three seasons (Chapdelaine and Brousseau 1996).

CHICK GROWTH AND DEVELOPMENT: 'intermediate', as in murres. Development of mesoptile plumage very rapid, with feather pins for primaries and contour feathers evident by 7 days, and all sheaths bursting by 10 days. Independent thermoregulation occurs by 9 days (Barrett 1984). Chick hatches at 63 ± 6 g; reaches maximum weight of 180–200 g at about 11–14 days, after which weight is more or less stable until departure (Paludan 1947; Lloyd 1979). Chick weights at departure were 195–207 g (3 years, Gulf of St Lawrence) and age at departure 17–18 days (3 years, Skokholm, Lloyd 1971), 20 days at Graesholmen, with departure weight 212 ± 24 g, wing length 82 ± 7 mm (Lyngs 1994), and 21 and 23 days in 2 years at Hornøy, Norway with departure weight 202 ± 30 g (*n* = 30) and 214 ± 37 g (*n* = 23) and wing lengths 85 ± 6 mm and 90 ± 6 mm (Barrett 1984). Chicks leave at dusk, on average slightly earlier in the evening than Common Murres. At Handa I., Scotland, 8% of departures resulted in failure because chick did not link up with its parent (Greenwood 1964). Captive chicks without parents began to exhibit departure behaviour at 16–21 days (Swennen 1977). The chick is accompanied by the ♂ parent (Wanless and Harris 1986), which continues to care for it, probably for at least a month (Lloyd 1977). A ♂ radio-tracked after departure with its chick travelled at 1.5 km/hr for the first 6 hrs (Wanless *et al.* 1988). First flight in captive chicks was at 61 days.

BREEDING SUCCESS: because most Razorbills use protected sites that must be visited to determine reproductive success, results may be affected by observer disturbance. Predation from Ravens (*Corvus corax*), Crows (*Corvus corone*), Jackdaws (*Corvus monedula*), and gulls (*Larus spp.*) can be high, especially at open sites (Hudson 1979b). At Skokholm, hatching success of first eggs was 74%, and fledging success 94%. Overall 71% of pairs that laid eggs reared a chick: proportion declined with date from 83% in the early part of the season to 56% at the end; 63% of egg loss was due to predation by Jackdaws (*n* = 735, Lloyd 1976). Success was higher for concealed sites (70%, *n* = 145) than for those on open ledges (53%, *n* = 252, Hudson 1982). Reproductive success at undisturbed plots in the Gulf of St Lawrence was 71%, 71%, and 76% in 3 years (Chapdelaine and Brousseau 1996); and at Graesholmen, Denmark in 2 years, 72%, and 76% (Lyngs 1994).

MOULT: complete post-breeding moult initiated immediately after breeding and involves a period of flightlessness. In North Sea, body moult begins Aug, at which time many birds are concentrated off NE Scotland; by Sept, practically all birds are in winter plumage. Feathers on chin and throat are moulted first, so moulting birds may appear to have black collar around neck. Primary moult is simultaneous, leading to period of flightlessness. Pre-breeding body moult begins Jan when 10% are in summer plumage, reaching 50% in Feb, 60% in Mar, and 90% by Apr (Tasker *et al.* 1988). First-year birds undergo this moult (Pre-alternate I) in May and proceed to post-breeding (Pre-basic) moult very soon after, hence are in breeding plumage only briefly (Bédard 1985).

Population dynamics

SURVIVAL: estimated at 89–92% from observations of ringed breeders at Skokholm (Lloyd and Perrins 1977). Analysis of British ringing recoveries suggested 91% adult survival (Mead 1974). In Denmark, survival of birds ringed as chicks to the fourth summer was 44% and to the sixth summer 40% (*n* = 68, Lyngs 1994). Recoveries of British ringed birds suggest some inter-colony movement, but this has not been quantified. G. Chapdelaine (personal communication) estimated adult survival among breeders on the N shore of the Gulf of St Lawrence at 90%.

AGE AT FIRST BREEDING: at Skokholm, 35% (*n* = 20) bred at 4 years, 60% at 5, and 5% at 6; 46% bred in the sub-colony in which they had been reared (Lloyd and Perrins 1977). Breeding at 3 years has been reported occasionally in Scotland.

PAIR- AND NEST-SITE FIDELITY: high (Lloyd 1976).

Common Murre *Uria aalge*

Guillemot, Common Guillemot (U.K.)

PLATE 1

Colymbus aalge (Pontoppidan, 1763, *Dansk Atlas*, vol 1: 621, pl. 26—Iceland)

Description

Pacific race is the largest living alcid (just); a robust bird but with rather a slender neck and narrow head, the profile of the crown forming nearly a straight line with the upper mandible. Its flight is rapid and direct but it manoeuvres poorly. On land, it stands nearly upright, but rests its weight on the full length of the tarsus. On the water, swims fairly buoyantly, with tail kept clear.

ADULT SUMMER: upperparts a uniform dark brown, in worn plumage showing pale tips to back and covert feathers; a small area of white shows on the trailing edge of the wing, formed by white tips to the inner secondaries; head, throat, and upper breast similar to upperparts, but head is more chocolate brown, especially in the race *albionis*; remainder of the underparts are white, sharply demarcated from the brown of the upper breast, with the flanks streaked to a varying extent with blackish brown; 'bridled' morph, which forms up to 50% of the population in some parts of the Atlantic, has a white ring around the eye and a narrow white line extending back behind the eye, highlighting a narrow groove in the feathers; underwing white, sometimes lightly mottled with brown, glossy blackish below primaries; bill, legs, and feet black, toes sometimes paler; irides dark brown or black.

ADULT WINTER: similar to summer, except that throat, sides to neck, and face are mostly white except for a smudgy dark streak extending from behind the eye towards the nape.

JUVENILE AND FIRST WINTER: similar to winter-plumage adult, but with shorter, more slender bill, and white streak behind eye obscure.

CHICK: down grey, occasionally medium chocolate brown, paler below, with silvery tips to down on head and neck, but without the brindling seen in Thick-billed Murres. The plumage developed at the nest-site resembles that of a winter-plumage adult, except that the prominent white post-ocular streak may or may not be present and a diffuse grey collar varies in prominence (Birkhead and Nettleship 1985)

VARIATION

Two races recognized in the Pacific, the smaller *U. a. californica* on the W U.S. coast, from Washington southwards, and the larger *U. a. inornata* elsewhere. No evidence of clinal variation within either race. Pacific birds generally larger than those from the Atlantic, with longer

bills and wings (Fig. 1). In the Atlantic, the pattern of geographical variation is very complex. The following description is taken mainly from de Wijs (1978). Five races have been described, but only three were accepted by Vaurie (1965): *U. a. aalge* on the Atlantic coast of North America, in Iceland, northern Britain, and Norway; *U. a. albionis*, a smaller bird found in England, Ireland, Brittany, Iberia, and Heligoland; and *U. a. hyperborea*, found in northern Norway, Bear I., Spitzbergen, and Novaya Zemlya. *U. a. spiloptera*, in Faeroe Is., characterized by heavy spotting on the underwing, is accepted by many authors, although its characters seem to be shared equally by many birds from Bear I. and the Shetlands. Birds from the Baltic, sometimes separated as *U. a. intermedius*, appear similar in size and plumage to *albionis*. Birds from the western Atlantic tend to have smaller bills than those from further

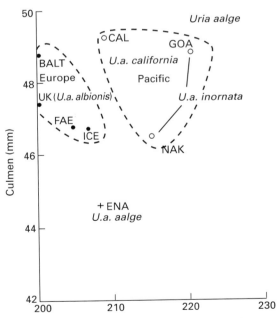

1   Plot of culmen length versus wing length for Common Murre populations. GOA = Gulf of Alaska, NAK = Northern Alaska, CAL = California, BALT = Baltic, UK = United Kingdom, FAE = Faeroe Is., ICE = Iceland, ENA = Eastern North America. All data from Storer (1952).

east. In the eastern Atlantic, de Wijs detected a continuous cline in wing and tarsus length, but not in bill length, increasing from the North Sea to N Norway and Spitzbergen. He suggested merging *hyperborea* with the nominate race. At the same time, he considered the dark backs, heavy streaking on the flanks, and dense underwing spotting of birds from Bear I., Faeroes, and Shetland to indicate that these populations had a separate origin from *albionis*. Those in northern Scotland and southern Norway that are intermediate in character may represent a secondary mixture of the other two populations. The frequency of the bridled morph increases from S to N and from warmer to colder water (Chapter 3). Bridled birds tend to be clumped together rather than randomly dispersed through the colony (Birkhead *et al.* 1980). No

apparent difference in behaviour between bridled and non-bridled birds, and mating between them is entirely random (Birkhead *et al.* 1985; Harris and Wanless 1986*a*).

MEASUREMENTS

Sexes very similar: ♂♂ slightly larger bill (average ♂:♀ ratios for culmen and bill depth from Table, 1.03), but measurements identical for wing length and tarsus. Wing length significantly longer in ♀♀ than in ♂♂ in a large sample measured at Isle of May (Harris and Wanless 1988*a*). Immature wing lengths (189.9) averaged smaller than those of adults (203.3) among birds collected in NW Scotland (Furness *et al.* 1994). Testes in Apr (maximum size) 33.0 ± 6.6 × 13.3 ± 3.7 mm (*n* = 18, Furness *et al.* 1994). See Table,

Table: Common Murre measurements (mm)

	ref	♂♂			♀♀		
		mean	*s.d.*	*n*	*mean*	*s.d.*	*n*
wing length	(1)	207	5.5	38	209	4.8	39
	(2)	215	6.2	9	214	2.7	4
	(3)	204	3.5	16	206	4.9	20
	(4)	197	2.9	11	196	5.1	13
	(5)	201	4.6	17	203	3.8	12
	(6)	218	5.7	45	217	4.3	25
	(7)	208	4.5	91	209	5.3	61
culmen	(1)	44.6	1.7	36	43.2	2.4	39
	(2)	42.1	2.2	9	41.7	1.2	4
	(3)	41.2	1.6	16	39.9	1.7	20
	(4)	46.7	2.7	11	45.4	1.5	13
	(5)	48.7	1.4	18	46.0	1.4	11
	(6)	47.7	2.2	42	45.8	2.0	24
	(7)	49.2	2.0	117	47.1	2.0	62
bill depth	(1)	13.8	0.6	35	13.3	0.5	38
	(2)	14.0	0.7	9	13.7	0.5	4
	(3)	13.2	0.5	16	12.8	0.5	20
	(4)	13.2	0.4	11	12.7	0.6	13
	(6)	14.5	0.9	42	13.6	0.9	19
	(7)	13.7	0.7	117	13.6	0.7	61
tarsus	(1)	38.4	1.4	38	37.9	1.3	41
	(2)	40.4	1.4	9	39.9	1.5	4
	(3)	38.4	1.1	16	38.2	1.4	20
	(6)	38.9	1.6	40	38.0	1.2	22
	(7)	38.3	1.3	122	37.8	1.3	66

(1) *U. a. aalge*, eastern North America (Storer 1952); (2) *U. a. hyperborea*, Bear I. (de Wijs 1978); (3) *U. a. spiloptera*, Faeroes (de Wijs 1978); (4) *U. a. albionis*, Skomer I., Wales, (Birkhead 1976*b*); (5) *U. a. intermedius*, Baltic (Salomonsen 1944, museum spec.); (6) *U. a. inornata*, Alaska (Storer 1952); (7) *U. a. californica*, California (Storer 1952).

WEIGHTS

Weight of breeders decreases during chick-rearing, but the loss is rapidly regained after departure (Harris and Wanless 1988*a*). At Gannet Is., Labrador (*U. a. aalge*, sexes combined), breeders during pre-laying averaged 968 ± 59 g (n = 42), incubation 975 ± 56 g (n = 31), and chick-rearing 960 ± 50 g (n = 30) (Birkhead and Nettleship 1987*a*). At Cape Thompson, Alaska (Swartz 1966), weights remained stable from May until July, then decreased, but rose again in Sept: means for the whole season, ♀ 985 ± 70 g (n = 37), ♂ 981 ± 67 g (*n* = 41).

Range and status

RANGE

Circumpolar, boreal, and low Arctic. In the Atlantic found between 43° and 56° N in the W and from 40° to 75° in the E; in the Pacific from 40 to 70° N in Asia and from 36° in California to 68° N in Alaska.

BREEDING: in the Atlantic, breeds commonly in eastern Canada from the Bay of Fundy (few) and the Gulf of St Lawrence to central Labrador and especially in E Newfoundland, in Iceland, Bear I., the British Isles, and Norway. In the Pacific, abundant throughout most of Gulf of Alaska from Prince William Sound westwards, the Bering Sea, and Sea of Okhotsk, also in California, Oregon, and Washington, but only a few thousand breed in British Columbia and SE Alaska.

WINTER: outside the breeding season some populations remain in the vicinity of their breeding colonies, while others disperse, sometimes moving S to avoid winter ice cover, especially in the Chukchi and Bering Seas and the Sea of Okhotsk. At this season confined to continental shelf waters. Initial dispersal after breeding is sometimes northwards, with birds from colonies in Oregon and Washington moving to British Columbia waters in large numbers in late summer (Campbell *et al.* 1990), those from eastern Newfoundland moving to the Labrador banks in Aug and Sept (Brown 1985) and first year birds from the Lofoten Is., Norway, moving to the Barents Sea (Holgersen 1951), where they probably remain over winter. By Dec, most Newfoundland Common Murres have moved S, being replaced on the Newfoundland banks by Thick-billed Murres. Common in winter off Nova Scotia and the New England states, uncommon S of Cape Cod, Massachusetts (CBC data). Likewise, off the W U.S. coast, numbers peak in California and Oregon in winter, presumably owing to an influx of birds from further N (Briggs *et al.* 1987), although large numbers also winter in the Aleutians and the Gulf of Alaska. In Asia, winter commonly as far S as the Sea of Japan (Ogi and Shiomi 1991) and the Pacific coast of Japan to Tokyo (Brazil 1991). Movements of the huge Bering Sea population outside the breeding season are more or less unknown. Birds from Scottish colonies disperse rapidly in July, but then concentrate off eastern Scotland in Aug dispersing gradually southward into the central and southern North Sea in winter (Tasker *et al.* 1988). Another big concentration of post-breeding birds, including many parent–chick pairs occurs in July at the Friese Front, near the Dogger bank, in the central North Sea (Camphuysen and Leopold 1994). Many pre-breeding birds from British colonies winter in the eastern North Sea, including the Skaggerak, where they are mixed with birds from northern Norway (Anker-Nilssen and Lorentsen 1995). Numbers wintering in the Dutch sector of the North Sea have increased in recent decades.

STATUS

Recent estimates suggest an Atlantic population of 6–9 million breeding birds, of which

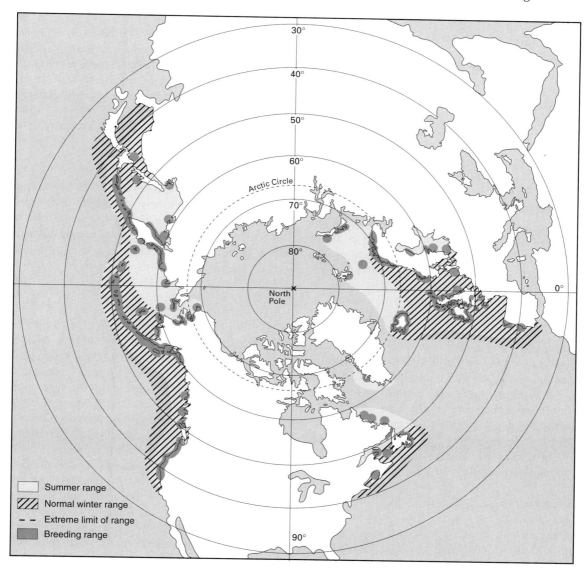

38% are in Iceland, 19% British Isles, 18% Bear I., 14% eastern Canada, 7% Faeroes, and 3% Norway (Nettleship and Evans 1985). In the Pacific, many large colonies of mixed Thick-billed and Common Murres in the sub-Arctic have not been differentiated to species. A recent estimate put the total Common Murres at 3.3 million breeding birds and the total unidentified murres at 5.1 million (Byrd *et al.* 1993). If we apportion the unidentified murres according to the ratio of the two species in each region, treating the Siberian Bering Sea and the Kuril Is. (where no ratios are available) as similar to the Alaskan Bering Sea, we estimate an additional 2.7 million Common Murres. Recent estimates for Oregon suggest about 300 000 more birds than given by Byrd *et al.* (1993) This gives a total of approximately 6.3 million breeders for the entire Pacific. Of those, about 34% breed in the Alaskan Bering Sea, 24% on the Alaska Peninsula, 10% in the Sea of Okhotsk, 9% on the Asian shores of the Bering Sea, 11% in Oregon and Washington, 6% in California, and about 5% in the Kuril Is. (assuming that Common Murres outnumber Thick-billed Murres there). Little is known of population trends in Pacific Common Murres, except in California, where the large population at the Farallon Is. (perhaps 400 000 in 1860) was decimated by overharvesting in the nineteenth century (Ainley and Lewis 1974), falling to a few thousand by 1920, then recovering through protection to reach 150 000 by 1982 and declining again owing to drowning in gillnets to less than 80 000 in 1986 (Takekawa *et al.* 1990). The population has subsequently begun to increase with the cessation of gillnetting in the area. Common Murres in the Bering Sea have declined gradually from the mid-1970s, perhaps because of changes in food supplies (Byrd *et al.* 1993). The history of Common Murre populations in the Atlantic is also one of recovery from human harvesting over most of the twentieth century (Nettleship and Evans 1985). In eastern Newfoundland the population increased by an order of magnitude since the early 1900s, the population of Funk I. going from very few in the late nineteenth century to 400 000 by 1950 (Tuck 1961) and in the British Isles the population increased from 550 000 in 1969–70 to 1.2 million in 1985–7 (Lloyd *et al.* 1991). In Norway, where harvesting continued to be legal until 1979, the population went on declining until the 1970s (Brun 1979; Barrett and Vader 1984). In northern Norway, a further decline occurred from 1965 to 1985, caused by drowning in gill-nets and, after 1985, a very sharp decline occurred in the Barents Sea colonies, including those on Bear I., coincident with a massive reduction in the Barents Sea capelin stock (Vader *et al.* 1990). The large populations in the Bering Sea and Gulf of Alaska have probably not been much affected by harvesting, although taken by native peoples in some areas. Numbers at the large colony at Bluff, on the Chukchi Sea, have declined continuously over the past 20 years for unknown reasons. This is the most

northerly colony in the northern Pacific to be dominated by Common Murres and changes in temperature have been implicated in the decline. At least one large colony was eliminated in the Aleutian Is. by introduced foxes. Otherwise, Arctic and sub-Arctic populations in the Pacific have shown variable and relatively small signs of population change (Byrd *et al.* 1993).

Field characters

A large black-and-white auk with very rapid flight. In winter best distinguished from similar Thick-billed Murre by whiter face, divided by a dark streak extending backwards from behind the eye. Thick-billed Murres are relatively dark faced, with black extending well below the eye. In summer, Thick-billed Murre has a prominent white line along lower edge of upper mandible, lacking in Common Murre, and the white of the breast meets the black of the throat in an inverted 'V' shaped wedge, whereas in Common Murre they meet in a smooth, unbroken horizontal line. In the right light, brown tint to upperparts of Common Murre is readily apparent. Bill shape usually not a good distinguishing character, especially in winter, when young Thick-billed and Common Murres have similar shaped bills (although shorter in Thick-billed) and the difference between adults is smaller. At all seasons thicker head and neck of Thick-billed Murre is noticeable, especially in side-by-side comparisons. In flight, at a distance, both species could be mistaken for Razorbill, but latter distinguished by much longer, graduated tail and deeper bill, which has a prominent vertical white line in summer.

Voice

Account mainly from Tschanz (1968), Cramp (1985), and personal observations. Very vocal at the colony and on the water adjacent.

ADULT: all calls are guttural and most are variants on a deep, growling, 'aargh', the 'crowing call', given by partner on arrival of a mate and on delivery of a fish ('greeting call'; terminated by a rattling crescendo); by ♂ prior to copulation ('♂ pre-copulation call', emphatic, but briefer than greeting call); and by adults calling to chicks during departure process ('luring call'), in shorter versions during other social interactions, given more softly during mutual footlooking (see below). A very intense version, during which there is noticeable tongue vibration, is given by ♂ to encourage copulation. Similar calls are given when chick is preparing to leave the colony, and during interactions between parent and chick on the water. 'Barking call': a sharp, dog-like bark given by both sexes immediately after greeting. The '♂ copulation call' is a continuous 'ah-ah-ah-ah' throughout mounting, given more rapidly than the simultaneous '♀ copulation call', a sharp, two-syllable 'uragh' given during solicitation and copulation. 'Alarm bowing call' is a short, nasal 'Arr', given during alarm bowing. A low, grunting call is given in response to the appearance of a potential threat, such as a large gull, and used in conjunction with bobbing the head. A soft, brief 'mm' ('contentment call') may be given by birds quietly resting or brooding.

CHICK: two calls: 'small chick call', a quiet peeping, given at low intensity before and during hatching and, more loudly, although still rather low, when being fed ('pecking call'), and from time to time during interactions with the brooding parent; 'departure call', a strident, 'wee-wee', or 'wee-wee-wee', given only by older chicks, either to locate an absent parent, or when preparing to leave the colony and afterwards until reunion with the parent is accomplished. The latter call is contagious, and nearby chicks may begin to call together.

Habitat, food and feeding behaviour
MARINE HABITAT
Found mainly in continental shelf waters in summer, ranging in surface temperature from 6 to 16°. May concentrate in areas of turbulence created by tidal currents, or where shallow banks occur offshore. Usually found offshore, but also occurs in sheltered inlets, not usually in estuaries, where turbidity may reduce prey visibility.

FEEDING BEHAVIOUR
Often feed in large aggregations, but there appears to be no co-ordinated herding of fish, as happens in some cormorants and pelicans. Off Newfoundland in summer, aggregations are mainly found over capelin concentrations (Piatt 1990). 20% of those recorded off California were in mixed species feeding flocks (Briggs *et al.* 1987). Feeding success may be affected by sea conditions (Birkhead 1976*a*) or tides (Slater 1976), However, in some studies the rate of delivery of food to chicks remained surprisingly constant even in high winds and rough seas (Harris and Wanless 1985).

FORAGING RANGE: seen carrying fish, presumably destined for chicks, up to 200 km from the nearest colony. At most colonies in the British Isles the majority of birds feed less than 30 km from the colony during chick-rearing period (Birkhead 1976*b*; Tasker *et al.* 1988; Wanless *et al.* 1990). At Hornøy in northern Norway, most feeding trips took less than 20 min, suggesting a foraging radius of not more than 12 km. At the Farallon Is., birds mostly fed within 60 km during the pre-laying period, and within 20 km during chick-rearing. At Witless Bay, Newfoundland, incubating breeders foraged up to 200 km, mainly 50–100 km from the colony (Cairns *et al.* 1990). During chick-rearing, their range was reduced to under 60 km, with most trips within 20 km. However, in the same area, large numbers of birds were seen carrying fish towards the colony from fishing banks 70 km away (Schneider *et al.* 1990*b*). Although capelin was the main prey (89% of fish delivered to chicks), local capelin densities had no effect on delivery rates, because adults spent less time resting when densities were low (Burger and Piatt 1990). In Shetland, Monaghan *et al.* (1994) found that mean foraging ranges were greater in a year of poor food availability than in a better year (7.1 versus 1.2 km). Individual birds do not usually return to the same feeding area on successive foraging trips (Wanless *et al.* 1990).

DIVING BEHAVIOUR: dives to maximum depths of more than 100 m (70 fathoms = 140 m, Holgersen 1951), but normal feeding depth is probably 20–50 m (Piatt and Nettleship 1985; Burger 1991). The diet suggests that prey is mostly taken in mid-water, rather than on the bottom. Dive times of post-breeding birds off the coast of Oregon averaged 101 ± 36 sec ($n = 137$), with a maximum of 153 sec for adults not surfacing with food, and 104 ± 33 sec for those feeding chicks. Inter-dive intervals averaged 44 sec, giving dive/pause ratio of 2.2. Birds feeding chicks surfaced with food after 80% of dives (Scott 1990). In Scotland, dives of breeding birds equipped with radio-transmitters averaged 67 sec, with 80% falling within 20–119 sec and a maximum of 202 sec. During diving bouts birds averaged 61–65% of the time underwater, giving a dive: pause ratio of just under 2 (Wanless *et al.* 1988). Close to the Farallon Is. dives averaged 58 sec (Ainley *et al.* 1990b). In Shetland Is., Scotland, Monaghan *et al.* (1994) measured dive times by telemetry in two consecutive years, one when food was abundant, one when it was scarce. Dive durations (mean 76.6 and 71.8 sec) and inter-bout periods (26.5, 21.2 sec) were similar, but dive: pause ratios and numbers of dives per bout were lower in the good year (2.2 and 3.5) than in the bad (3.7, 15.4) and surface pauses were longer in the good year (43.1 versus 24.6 sec). In the good year, birds spent 15% of their time underwater while away from the colony, but 34% in the bad year.

ENERGY REQUIREMENTS: adults require approximately 440 g fresh weight of fish daily to cover their energy expenditure while feeding chicks (mean 2200 kJ, Gabrielsen 1994). This contrasts with a mere 30 g delivered to the chick by each parent (see below). Hence, more than 90% of foraging during the chick period is devoted to supplying their own energy needs. The small additional amount of food required to feed the chick probably explains

why Common Murres are capable of breeding successfully in situations where other species dependent on the same prey (Shags, *Phalacrocorax aristotelis*, Arctic Terns, *Sterna paradisaea*) fail (Heubeck 1989).

DIET

The most important prey are midwater schooling fishes: in the Atlantic, sandlance, capelin tomcod (*Microgadus*), herring, and sprat and in the Pacific, sandlance, capelin, anchovy (*Engraulis mordax*), juvenile rockfishes (*Sebastes* spp.), and walleye pollock (*Theragra chalcogramma*) (see Table).

ADULT: stomach analysis suggests that the main food of adults throughout most of their range and at all times of year, is fish. Most exceptions to this generalization come from the Pacific: squid, especially *Loligo opalescens* in California, and large pelagic zooplankton, especially euphausiids and amphipods. In the NW Bering Sea in July, zooplankton made up 76% of the diet by volume, with the amphipod *Themisto libellula* the most abundant species (Ogi *et al.* 1985). Elsewhere in the Bering Sea, fish (saffron and Arctic cod, sculpins (*Triglops*, *Gymnocanthus*) sandlance, and capelin dominate the diet (Springer *et al.* 1984). Squid were also an important diet item in a few collections off western Scotland (Halley *et al.* 1995) Dramatic changes in diet can occur in response to changes in oceanographic conditions. In Monterey Bay, in two normal summers, rockfish dominated the diet, whereas in the El Niño year of 1983 market squid was the dominant prey (Croll 1990). When sampled in the same area Common Murres have generally been found to take more fish and less zooplankton than Thick-billed Murres (Swartz 1966; Springer *et al.* 1984; Vader *et al.* 1990; Piatt *et al.* 1991b). Representative diet information is given in the Table: (1) Monterey Bay, summer (Baltz and Morejohn 1977); (2) Oregon, summer (Mathews 1983); (3) Monterey Bay, non-El Niño summer (Croll 1990); (4) Cape Thompson, Alaska, summer (Swartz 1966);

Table: Common Murre adult diet

ref	(1)		(2)	(3)	(4)	(5)	(6)	(7)
method	presence	items	presence	IRI	presence	presence	presence	presence
n			503		66	702	250	109
INVERTEBRATES								
squid (*Loligo*)	33	19	32	2–15				
crustacea					6		1	
annelids						3	31	2
FISH								
herrings (Clupeidae)			14			29	23	
cod (Gadidae)			48		77	26	38	8
rockfishes (Scorpaenidae)	33	30	25	70–85				
smelts (Osmeridae)			29	8–10				
sandlance (Ammodytidae)			19		27	5	39	71
anchovies (Engraulidae)	54	32	14	5				
gobies (Gobiidae)						49	10	1

IRI = Index of relative importance.

(5) Skaggerak (Blake 1983); (6) Shetland, winter; (7) summer (Blake *et al.* 1985).

CHICKS: almost entirely fish, occasionally squid. Food fed to chicks tends to be extremely uniform at most colonies, being dominated by a single prey species: sandlance in Britain and Norway, capelin in the Barents Sea, Gulf of St Lawrence, and Newfoundland and Labrador, rockfish in California, sandlance and capelin in the Gulf of Alaska and the Bering Sea.

Displays and breeding behaviour

General information from Cramp 1985; Birkhead 1985; notes on posture displays are from Williams (1972) and Birkhead (1978*b*, 1985).

COLONY ATTENDANCE AND FLIGHT DISPLAY: at the Isle of May, Scotland, where chicks leave the colony in the second half of June or first half of July, breeders return to the colony in Oct, ♀♀ slightly earlier than ♂♂, attending on most days throughout the winter, except when there is prolonged bad weather, and usually only in the morning. Breeders attend more often than non-breeders (Harris and Wanless 1990*a,b*). Fierce fighting occurs in early winter (Greenwood 1972). A similar attendance schedule occurs at the Farallon I., with birds first arriving late Oct.; attendance is intermittent, and mainly occurs in the mornings during Nov and Dec. Breeding begins late Apr or May (Boekelheide *et al.* 1990). The first big arrival at the Farallons can be quite spectacular: 'The murres concentrate at dawn in large flocks, rapidly flying around the islands, then approach the cliffs warily, circling many times before landing. Once a few land, others pour from the sky until the colonies are again filled.' (Boekelheide *et al.* 1990). At Skomer I., Wales, breeders do not return until Dec. Attendance in the pre-laying period is not daily, but periodic, becoming cyclical in Apr, with peaks of 2–3 days of attendance, separated by periods of 3–6 days when the colony is deserted. Similar cyclical attendance occurs at St Kilda during Apr. During attendance periods, birds are present throughout the day, but leave for the night. Sites were occupied on only 34 of the last 90 days before first egg-laying (Birkhead 1978*a*; Hatchwell 1988*a*). During individual attendance peaks, ♂♂ tend to arrive slightly earlier than ♀♀ and leave slightly later, some ♂♂ losing 10 per cent of body weight during their 2–3 day sojourn at the colony (Hatchwell 1988*b*). Timing of breeding at Skomer is similar to the Isle of May. In Newfoundland, colony attendance does not begin until Apr and is irregular at first, becoming daily, but not cyclical (Mahoney 1979). Laying begins late May, peaking early June. Timing is similar in Labrador, where at Gannet Is. median laying ranged from 15 to 24 June (3 years, Birkhead and Nettleship 1987*a*). At the Semidi Is., in the Gulf of Alaska, earliest arrivals were on 10 Apr, after which attendance was intermittent until late May. In one year a 4–5 day cycle was apparent, but in others attendance was irregular. Laying began early June (Hatch and Hatch 1989). At Cape Thompson, in the Chukchi Sea, murres of both species do not arrive until mid-May. Attendance during the pre-laying period peaks in the evening. Laying begins late June and continues throughout July (Swartz 1966). Non-breeders in their third year begin to attend the colony after start of incubation, while second year birds, some of which do not visit the colony at all, appear on the colony only in mid–late incubation (Hudson 1985; Halley 1992). These two cohorts, along with some fourth years, form the bulk of birds loafing on rocks close to the sea, in what are known as 'clubs' (Birkhead and Hudson 1977). Many also socialize on the sea near the foot of the colony cliffs. Older pre-breeders spend more time on the actual breeding ledges, with some tendency to concentrate near the top of the colony, often on ledges giving a good view of many nearby breeding sites (Halley and Harris 1994). Pre-breeders desert the colony at about the same time as the majority of breeders, usually when about 90 per cent of chicks have departed (Piatt and McLagan 1987).

AGONISTIC BEHAVIOUR: aggression involves birds jabbing at one another's head, or, at high intensity, grappling bills, and beating one another with their wings, with feet spread wide apart and wings half open. Birds involved in fights may fall from ledges and continue fighting on the sea, including chasing both underwater and on the water surface. Such chases are sometimes ritualized, both birds beating along the surface without taking off. Other birds in the vicinity of a fight at a breeding site, especially those incubating, usually flatten themselves against the cliff and turn away. Sometimes fights move back and forth over the bodies of neighbouring birds. Fights between mated pairs sometimes end in token copulations, in which ♀ solicits, then stands up, throwing ♂ off. Neck stretching and side preening displays are used in appeasement situations. Aggression is frequently terminated by turning away and preening flanks or scapulars in a desultory and symbolic manner. Side preening also occurs frequently after arrival at the site, if the mate is absent. Non-breeding prospectors maintain upright stance with bill somewhat raised, and frequently keep their wings flapping after landing.

SOCIAL BEHAVIOUR: very sociable on the water close to breeding colonies, frequently forming flocks in line abreast. Forssgren and Sjolander (1978) and Cayford (1981) describe displays on the sea, including 'bill lifting', where the bill is tilted upwards, usually in response to the approach of another bird, 'billing' in pairs; gentle mutual pecking directed at the other bird's bill, and 'fencing', similar to billing, except more vigorous and aggressive. Groups, principally composed of non-breeders, may perform 'rushing', in which several individuals raise bills, lift breasts higher than normal, and swim rapidly and erratically away from one another, frequently followed by a communal dive. Mass dives sometimes occur in waves, spreading from a centre of initiation.

SEXUAL BEHAVIOUR: arrivals of paired birds of either sex during the pre-laying period are often followed by copulation, even as much as 5 months before laying. Frequency of copulations peaks in last 15 days before laying. Copulation is solicited by ♂ by giving a loud 'graaa' call, while standing beside ♀ which crouches, giving a distinctive 'a-ah' call, and tossing back head with mouth open. Calling and head-tossing continue while copulation takes place. ♂ paddles briefly on centre of her back, then shuffles backwards, standing up and balancing with wing tips drooped on ground. Tail is waggled from side to side before cloacal contact occurs. Each successful copulation averages two cloacal contacts (Hatchwell 1991). Copulation is generally terminated by ♀ standing up. Copulation is frequent during pre-laying, averaging about three copulations/pair/day, and increasing during last few days before laying, when it may reach more than six times/day (Hatchwell 1991). Extra-pair copulations occur, both from 'rapes' (forced extra-pair copulations) and, more rarely, through ♀ solicitation while her mate is absent. About 20 per cent of rapes involve more than one ♂ and some attempts involve other ♂♂ usually immediately after ♀ has landed. Only 6 per cent of rapes result in cloacal contact (Hatchwell 1988*b*). Courtship displays include mutual 'footlooking', in which both birds bend forward and take an intense interest in their toes and webs, often nibbling at them and giving 'aargh' calls, mutual fencing, in which the pair indulge in a kind of mock fight, fencing with open bills, and mutual allopreening, usually of the head and neck. Footlooking and mutual fencing are commoner among pairs without eggs or chicks, either before laying, or after the loss or departure of the chick; allopreening continues throughout the season, and may be directed at neighbours as well as mates. Footlooking is also performed by single birds occupying sites, often following comfort movements, such as defecation or wing-stretching. Egg-laying involves foot scraping and eventually standing to deliver the egg, small end first, towards the cliff. ♂, if present, inspects the egg, before ♀ initiates first incubation shift. Birds arriving at breeding site land on a common (i.e. unde-fended) area and approach their site on foot, keeping wings held out and back, and head and neck thrust forward ('landing display'). If the mate is present an emphatic 'aargh' call is

given. The mate usually stands, or at least stretches upwards while still crouching, neck bows, and also calls 'aargh'.

MAINTAINANCE BEHAVIOUR: preening and oiling have been described by Williams (1972). The chin and lower mandible are passed over the preen gland and then rubbed over the feathers of the back and scapulars and thoroughly ploughed through the breast feathers. Individual feathers may be passed through the mandibles. Inaccessible regions of the crown and nape are rubbed against the coverts, a considerable contortion.

Breeding and life cycle
Socially monogamous, with semi-precocial young, but with chick 'fledging' at only a quarter of adult weight ('intermediate' strategy, Chapter 6).

BREEDING HABITAT AND NEST DISPERSION: colonial, on cliff ledges, closely packed or scattered among boulders, on flat low-lying islands, or on the tops of offshore stacks. Where breeding on flat ground, birds do not spread out, but form discrete clumps (Johnson 1941), adjacent birds usually in contact with one another. On mixed cliff colonies, Common Murres occupy larger ledges, usually touching more neighbouring birds, than Thick-billed Murres. At the Gannet Is., Labrador 50% of Thick-billed Murres were on narrow ledges and only 8% of them had more than two neighbours, compared with only 9% of Common Murres on narrow ledges and 31 per cent with more than two neighbours (Birkhead *et al.* 1985). Dense nesting has a significant effect in deterring predation by gulls and density is positively associated with breeding success (Birkhead 1978*b*). Hatchwell 1991).

NEST: no nest is made, although a few pebbles may be placed around the laying site.

EGG-LAYING: varies considerably with latitude and climate, with median laying occurring approximately 5 days later for every 1 °C decrease in sea surface temperature (Fig. 2). This trend is similar in both oceans. Some anomalies occur: at Triangle I., BC, where July sea surface temperature is 15 °C, laying does not begin until June (Rodway *et al.* (1990b). Egg laying in a given year lasts for at least 35 days and does not usually exceed 50 days, irrespective of the population involved, so that eggs or chicks are present for about 100 days. Hence, although the timing of pre-laying arrival at the colony varies considerably among different populations, the length of the breeding season does not.

EGGS: one, ovoid-pyramidal, with a very thick, chalky shell. Ground colour off-white to deep turquoise, usually with some bluish tinge, occasionally unmarked, but normally with speckles, blotches, or elongated scribbling, ranging

from pale brown to black. At the Farallon Is., California, ground colour was thought to vary from year to year in response to changes in diet (Boekelheide *et al.* 1990). For measurements, see Table, (1) *U. a. aalge*, North America (Mahoney 1979); (2) *U. a. albionis*, British Isles (Witherby *et al.* 1941); (3) Portugal (BMNH); (4) *U. a. inornata*, Japan; (5) *U. a. californica*, North America (Bent 1919). Egg weight is approximately 11% of adult weight. Egg size varies among years at the same colony, usually being smaller in years of later laying. Eggs that are lost may be replaced; replacement eggs average 6% smaller than first eggs of same females, with interval between loss and replacement being 14–16 days (range 6–37). Second replacements, after the loss of two eggs, have been reported only at Isle of May, where intervals were 13–22 days ($n = 7$ in 6 years, Harris and Wanless 1988b). Yolk formation takes 14–16 days (Birkhead and Nettleship 1984b; Boekelheide *et al.* 1990; Hatchwell and Pellatt 1990). Composition (fresh weight) averages: shell 15%, yolk 36%, albumen 49–51% (Hatchwell and Pellatt 1990), with some inter-year variation. Embryo reaches 50% of hatching weight at 28 days (Mahoney 1979).

INCUBATION: normally continuous and lasts 33 days (s.d. 1.4–2.1). At the Farallons, the range was 26–39 days ($n = 1202$, Boekelheide *et al.* 1990). Adults take equal shares in incubation. At Isle of May there was a tendency for ♂♂ to incubate more often than ♀♀ in the middle of the day and for ♀♀ to incubate more often overnight. The off-duty bird always spent the night at sea; the pair spent an average of 5 hrs together at the site during the day (Wanless and Harris 1986). Incubation shifts averaged 17 hrs (range 1–38) at Gannet I., Labrador (Verspoor *et al.* 1987); at Skomer I., incubation shifts were 14–16 hrs, with change-overs being commonest at 06:00–10:00 and 18:00–22:00 hrs (Hatchwell 1988b). There is no tendency for adjacent pairs to change-over at the same time of day.

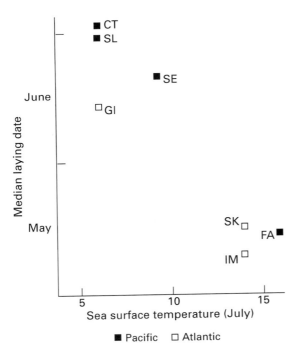

2   Relationship of median laying date to July sea surface temperature for Common Murres. IM = Isle of May, FA = Farallon Is., SK = Skomer I., GI = Gannet Is., SE = Semidi Is., SL = St Lawrence Is., CT = C. Thompson.

Table: Common Murre egg dimensions

ref	length (mm)			breadth (mm)			fresh weight (g)		n
	mean	s.d.	range	mean	s.d.	range	mean	s.d.	
(1)	82.7	3.6		50.9	2.2		108.4	9.2	30
(2)	81.5		72.5–92.7	49.7		44.0–53.6	105		100
(3)	82.4	1.8		49.4	1.4		—		21
(4)	82.5		77.5–90.5	52.2		47.0–58.0	112		41
(5)	82.2		69.5–90.0	50.2		42.5–54.0	—		74

HATCHING AND CHICK-REARING: hatching takes 1–3 days from first pipping, and during that period the incubating bird is often restless, looking at the egg frequently. Chicks are fed at the site for an average of 22–25 days (s.d. 1.1–2.4), after which chick-rearing is performed exclusively by one parent, normally (perhaps always) the ♂ (Scott 1990). Young chicks are brooded continuously and rarely seen. At Gannet I., Labrador, brooding shifts averaged 4 hrs (range 1–14) during daylight and 12 hrs (range 7–20) overnight (Verspoor et al. 1987). After about 2 weeks, chicks emerge regularly, especially in fine weather, to exercise their wings, a procedure that becomes more common as departure approaches. Chicks are normally fed 3–5 fish per day, up to about 160 mm in length, 30 g weight. Average daily food intake amounts to 30–60 g. There is some indication that the size of fish fed increases with chick age (Burger and Piatt 1990), but feeding rate does not, and often declines after 17 days, sometimes peaking as early as 7–8 days (Birkhead 1976b; Hatchwell 1988b). Relationship of feeding rate to chick age varies with food availability. In Shetland, in a year of poor food supply, feeding rate increased to beyond 20 days, but in a year when food was abundant it peaked at 10 days (Uttley et al. 1994). Some studies reported feeding rates were highest at dawn, while others found no strong diurnal variation (Harris and Wanless 1985; Birkhead and Nettleship 1987b; Hatchwell 1988b). Chick feeding occurs immediately after arrival of returning adult. Both parents stand, their partially spread wings forming a protective tent, and the fish is offered by the arriving bird, bowing forward and relaxing its grip on the fish, so that it slips out of the gullet and is held near the tip of the bill. Large chicks take the fish directly from parent's bill, while smaller chicks generally allow it to fall to the ground and pick it up from there, orienting it in relation to the position of the eye and the thickest part of the body before swallowing it head first (Oberholzer 1975).

CHICK GROWTH AND DEVELOPMENT: growth in weight is more or less linear during the first 10–14 days. Thereafter, the increase tapers off to reach a plateau of 200–260 g, varying somewhat among colonies, at about 21 days. Development of wing feathers (primary coverts) is more or less linear until departure (Hedgren 1981; Hatchwell 1991). After leaving the colony, chicks grow at 15–16 g/day; faster than during the colony period, reaching an asymptote of 90–95% adult weight by 45–60 days after leaving (Varoujean et al. 1979; Harris et al. 1991).

CHICK DEPARTURE: intense vocalization by parent and chick precedes departure, but decision is taken by chick (Tschanz 1959). In captivity, in the absence of parents, chicks initiate departure behaviour at 15–21 days old (Swennen 1977). On low-lying sites, chicks are led to the sea by ♂ parent, up to 50 m at some colonies. Departures occur at dusk, or throughout the night at high latitudes, where there is no complete darkness (e.g. northern Alaska). Chicks departing earliest in the evening suffer higher predation from gulls than others. At colonies where some chicks on cliff sites cannot glide from their ledge to the sea, predation on those chicks landing on the shore is much higher than on those reaching the sea. Overall, predation at departure may be up to 17 per cent (Williams 1975). At Middleton I., Alaska, where uplift in a recent earthquake has left the occupied cliff 0.5 km from the sea, all chicks land on the beach and are joined by their parents. Some may take several days to reach the sea (Hatch 1983). Care by ♂ continues for about 2 months after chicks have left the colony (Varoujean *et al.* 1979). During this time ♂ initiates its post-breeding moult. Captive chicks began to feed themselves 1 week after 'going to sea' (Swennen 1977), so self-feeding may begin fairly early; began flying at 94 days old. ♀♀ at Isle of May in 2 years continued to visit their breeding site for an average of 18 days ($n = 39$) and 13 days ($n = 63$) after their chick had left, averaging 11.5 hrs/day at the colony (Wanless and Harris 1986).

BREEDING SUCCESS: in the Pacific, 27–77% of pairs attempting to breed reared chicks (44 colony-years) with 7/10 colonies averaging >50% success (Byrd *et al.* 1993). In the Atlantic, success has generally been higher, except at certain Norwegian colonies. At Stora Karlso in the Baltic Sea mean success was 77–82% (4 years), at Isle of May, Scotland 71–82% (6 years), at Gannet I., Labrador, 80–85% (3 years), and at Skomer I., Wales 72% in 1973–5 and 79% in 1986–88 (Hedgren 1980; Harris and Wanless 1988*b*; Birkhead and Nettleship 1987*b*; Hatchwell and Birkhead 1991). Most failures occur at egg stage, with survival from hatching to fledging always greater than 90%. Among causes of egg loss at Isle of May, 43% rolled off ledge, 15% failed to hatch, 8% were lost in fights, 6% were washed off, and the rest were abandoned, cracked, poorly incubated,

or taken by gulls ($n = 298$). Birds in their first year of breeding are much less successful than experienced birds.

MOULT: breeders undergo a complete moult soon after leaving colony, beginning July–Sept, according to timing of breeding. ♂♂ with chicks probably begin moult towards end of period of chick dependency (Harris *et al.* 1991). Contour and flight feathers are moulted simultaneously, with primaries all being shed within a few days. Nine captive birds that had been cleaned after oiling took 42–90 days to renew their primaries (mean 63 days) and were probably flightless for an average period of 4–6 weeks (Birkhead and Taylor 1977). In most, but not all, individuals, body moult is completed within the period of primary moult. Pre-breeding moult involves only feathers of head and neck and may commence as early as Sept, with apparently only a short pause between post and pre-breeding feather replacements. Timing of pre-breeding moult varies among populations and also among age-classes. At Isle of May, birds in complete summer plumage begin to arrive at colony at end Oct and all are in summer plumage by early Jan (Harris and Wanless 1990*a*). However, birds seen offshore in North Sea still included 50 per cent in winter plumage in Jan, and 30 per cent in Feb (Tasker *et al.* 1988). This discrepancy presumably relates to different moult schedules of pre-breeding birds. The occurrence of birds in summer plumage follows a similar pattern in California to that seen in Scotland. Moult schedules elsewhere are poorly known because age and breeding status of birds seen at sea cannot normally be determined. Birds in summer plumage are not seen in Newfoundland until Mar, suggesting that population there moults much later. Attendance at colonies in Newfoundland does not begin until Apr.

Population dynamics

SURVIVAL: annual adult survival was estimated at 94% in Britain and California (Mead 1974;

Sydeman 1993 Harris and Wanless 1988*b*; Hatchwell and Birkhead 1991); Survival of young birds from departure to breeding age was estimated to be about 40% at Skomer Is. in the 1980s (Hatchwell and Birkhead 1991). Survival to 2 years or more varied from 14 to 39% (6 years) at Isle of May (Halley 1992). Survival of young birds in some years, or at some colonies, may be partly dependent on weight at (Hatch 1983), or date of (Harris *et al.* 1992) departure.

AGE AT FIRST BREEDING: most birds begin to breed at 5 or 6 years old at Isle of May, with ♀♀ breeding at a younger age, on average, than ♂♂ (Halley 1992).

PAIR- AND NEST-SITE FIDELITY: breeders normally return to same site each year (89% at Isle of May, Harris and Wanless 1988*b*) and hence pair-bonds persist, sometimes for many years. Inter-colony movement is restricted and most birds return to the colony where they were reared. However, some inter-colony movement occurs and 12% of recruitment at Graesholmen, in the Baltic, was estimated to come from Stora Karlso, a larger colony 280 km away (Lyngs 1994).

Thick-billed Murre or Brunnich's Guillemot   *Uria lomvia*

Alca lomvia (Linnaeus, 1758, *Syst. Nat.* 10th ed, I: 130—Northern Europe).   PLATE 1

Description
Large, chunky auk; more robust than Common Murre, with heavier head and neck, but usually slightly smaller. On land, stands upright, resting on the tarsus, or lies forward on the belly. On the water, rides high, with tail clear of the water. It takes off from water with difficulty, requiring a long taxi, except when there is a strong wind, frequently bouncing off wavetops. Never lands on flat ground.

ADULT SUMMER: upperparts brownish black, fading to dark brown, in worn plumage showing pale tips to back, coverts, and primaries; small area of white shows on the trailing edge of the inner secondaries; head, throat, and upper breast similar to upperparts, the remainder of the underparts white with sparse brown streaks on flanks; no 'bridled' form; meeting between black of throat and white of breast forms an inverted 'V'; underwing white, but blackish below primaries, bill dagger-shaped, black, but with a white streak 2–3 mm wide extending from the gape to just behind the tip of the upper mandible; interior of mouth and tongue yellow; legs and feet black or dark slate; tarsus and toes sometimes blotched with brownish yellow; iris dark brown to grey-brown.

ADULT WINTER: similar to summer, but bill less deep and with white streak reduced or absent, lacking black on throat and breast, with black of face extending to just below the eye.

JUVENILE AND FIRST WINTER: similar to winter adult, but bill distinctly thinner, somewhat shorter and without any trace of white line

CHICK: down colour ranges from silvery grey, through brown, to sooty black, usually paler on head and much paler on belly, paler chicks often brindled, 'pepper and salt'. Legs and feet dark slate grey, claws elongated and strongly curved. Pin feathers visible on the wing at 5 days, and begin to burst from their sheaths at 8–10 days, by 14 days most contour feathers are developing, and by 18 days all down has been replaced except on the head. The plumage at departure ('mesoptile') resembles adult plumage, but may be winter or summer, i.e. with or without black on the throat (Birkhead and Nettleship 1985), or intermediate; occasional albinos occur.

VARIATION

Pacific birds (*U.l. arra*) are generally larger than Atlantic specimens (*U.l. lomvia*; Fig. 1). Within the Atlantic, those in the North American sector appear smaller than eastern birds, but there is considerable overlap. Measurements vary significantly among birds from different colonies in Hudson Strait, with those from Akpatok being smaller than those from other colonies (Gaston *et al.* 1984). However, detailed investigation of DNA sequences in the cytochrome *b* gene by Birt-Friesen *et al.* (1992) did not reveal any genetic differentiation between colonies, either within the western Atlantic, or between E and W. Significant differentiation was found between Atlantic and Pacific populations, supporting their subspecific status. *U. l. eleonorae*, described from northern Siberia (Preobrazheniya I. and eastern Taimyr Peninsula) and *U.l. heckeri*, from Wrangel I. and the northern Chukotsk Peninsula, have a greyish hue to the upperparts, compared with other races; *heckeri* has larger bill than *eleonorae* (Golovkin 1990*b*). They appear poorly differentiated from one another and from populations to E and W, and their status as subspecies seems doubtful. Birds in this area could have resulted from a mixture of Atlantic and Pacific races. Occasional hybrids with Common Murres occur, exhibiting a mixture of characteristics from both species (Cairns and DeYoung 1981; Birkhead *et al.* 1986; Friesen *et al.* 1993).

MEASUREMENTS

♀♀ are generally slightly smaller than ♂♂ in most skeletal measurements, the difference being greatest for cranial measurements and least (or absent) for wing elements (Stewart 1993). Among external measurements, sexual dimorphism is greatest for culmen length and bill depth and least for wing length. In winter, bill depth is less than in summer; first winter birds are smaller in all dimensions than older birds, but especially in bill depth. See Table: (1) *U. l. lomvia*, Digges I., Canada, breeding (Gaston *et al.* 1984); (2) Akpatok I., Canada, breeding (Gaston *et al.* 1984); (3) Newfoundland, first winter; (4) older than 1 year (Gaston *et al.* 1983*b*); (5) *U. l. arra*, Cape Thompson, Alaska, breeding (Swartz 1966).

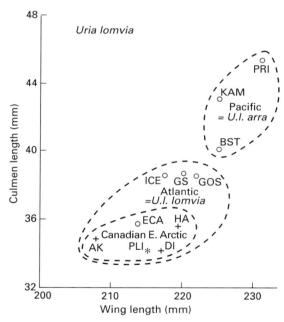

1 Plot of culmen length versus wing length for Thick-billed Murres. PRI = Pribilof Is., KAM = Kamchatka, BST = Bering Straits, ICE = Iceland, GS = Greenland Sea (Bear I., Spitzbergen), GOS = Gulf of St Lawrence, ECA = Eastern Canadian Arctic (including W. Greenland), AK = Akpatok I., HA = Hantzsch I., PLI = Prince Leopold I., DI = Digges I. Sources: Storer 1952 (o), Gaston *et al.* 1984 (+), Gaston and Nettleship 1981 (*).

WEIGHTS

Weight varies considerably during the year. In the breeding season, it tends to increase during pre-laying and incubation periods and declines sharply by about 50 g just after hatching, remaining static, or declining slightly, during chick-rearing (see Table). The weight decrease is triggered by hatching and can be altered by switching eggs to extend incubation (Croll *et al.* 1991; Gaston and Perin 1993). Winter weights generally similar to, or higher than, highest breeding season weights, although first year birds are lighter. See Table: (1) adult

Table: Thick-billed Murre measurements (mm)

♂♂	ref	mean	s.d.	n
wing length	(1)	217	5.7	122
	(2)	207	6.9	10
	(5)	224	6.2	80
culmen	(1)	34.8	1.9	127
	(2)	35.2	1.7	10
	(5)	38.3	1.8	79
bill depth	(1)	14.8	0.7	124
	(2)	14.0	0.9	10
tarsus	(2)	37.2	1.8	80

♀♀	ref	mean	s.d.	n
wing length	(1)	216.6	6.0	83
	(2)	209	5.0	13
	(5)	223	6.1	59
culmen	(1)	33.8	2.0	84
	(2)	34.6	2.0	11
	(5)	37.6	2.3	58
bill depth	(1)	14.5	1.0	78
	(2)	13.6	0.7	11
tarsus	(2)	36.4	1.5	59

SEXES COMBINED	ref	mean	s.d.	n
wing	(3)	204	5.6	6
	(4)	216	5.6	44
culmen	(3)	31.5	3.4	8
	(4)	34.2	1.8	43
bill depth	(3)	11.1	0.9	7
	(4)	13.0	0.5	42

Table: Thick-billed Murre weights (g)

BREEDING ♂♂	ref	mean	s.d.	range	n
June	(5)	961	69	870–1090	20
	(6)	933	54	836–1074	47
July	(5)	989	66	880–1180	40
	(6)	913	75	799–1033	24
August	(5)	962	59	820–1130	126
	(6)	903	68	802–1017	14
summer	(7)	972	92		94
	(8)	975	70		7

BREEDING ♀♀	ref	mean	s.d.	range	n
June	(4)	955	51	880–1050	9
	(5)	928	86	825–1101	20
July	(4)	934	86	740–1120	13
	(5)	855	66	734–960	11
August	(4)	947	64	810–1140	51
	(5)	841	50	765–936	19
Summer	(6)	939	99		53
	(7)	950	73		60

SEXES COMBINED	ref	mean	s.d.	range	n
winter	(1)	1286	144	1032–1481	11
	(2)	952	71	795–1150	44
	(3)	827	45	750–862	6
incubation	(4)	943	51		30
chick-rearing	(4)	899	57		30

U. l. arra, collected off Hokkaido, Japan, in winter (Hashimoto 1993); (2) *U. l. lomvia*, Newfoundland, Feb, >1 year old; (3) first winter (Gaston *et al.* 1983*b*); (4) Gannet I., Labrador (Birkhead and Nettleship 1987*a*); (5) Coats I. (AJG, unpublished data); (6) Prince Leopold I. (Gaston and Nettleship 1981); (7) Digges I. (Gaston *et al.* 1984); (8) *U. l. arra*, Cape Thompson, Alaska, breeding (Swartz 1966).

Range and status
RANGE
Circumpolar, arctic and sub-Arctic, breeding from 46° to 82° N in the Atlantic/Arctic oceans and from 50° to 72° N in the Pacific.

Abundant in waters subject to winter ice cover.

BREEDING: the largest colonies in the Atlantic sector are in Hudson Strait, the eastern Canadian high Arctic, W Greenland, Iceland, Spitzbergen, and Novaya Zemlya, and in the Pacific on the islands and coasts of the Bering and Chukchi seas, and the Sea of Okhotsk. In most of their Pacific range, and in Iceland and Bear I., share colonies with large numbers of Common Murres. Elsewhere colonies are predominantly of one or other species (Fig. 3.7). In Alaska, mixed colonies of the two murres

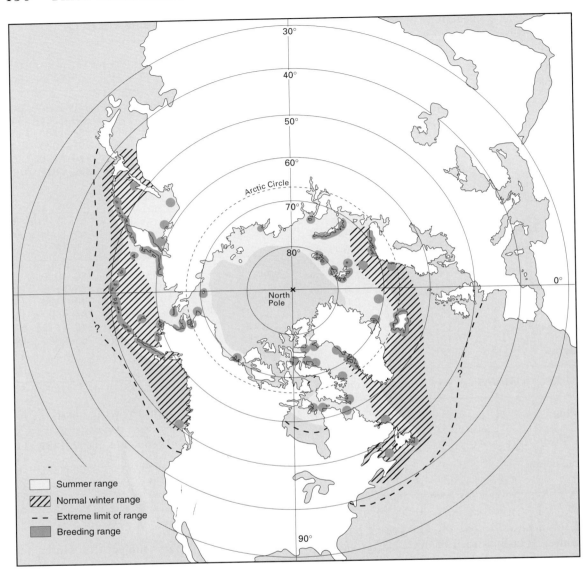

Summer range
Normal winter range
Extreme limit of range
Breeding range

comprise predominantly Common Murres in the Gulf of Alaska and Bering Sea coast, and predominantly Thick-billed Murres in the Aleutian I. and the Chukchi Sea. Relative numbers seem to relate to water temperature and bathymetry, with Thick-billed Murres favouring sites with lower water temperatures and closer to deep water (Springer 1991).

WINTER: northern parts of range are deserted in winter, and populations in some areas migrate long distances to escape dense pack-ice. In eastern Canada most, and in W Greenland about half, of the population winters off eastern Newfoundland (Fig. 2). Contrary to many earlier statements, the species does not winter in any numbers in Hudson Bay or Hudson Strait. Many birds from Spitzbergen and the Russian Arctic winter off SW Greenland, although they remain common among mobile pack in the Barents Sea throughout the winter (Hunt *et al.* 1996). The 'large flights' (Bird and Bird 1935) seen passing Jan Mayen in Oct may have comprised birds on passage from the

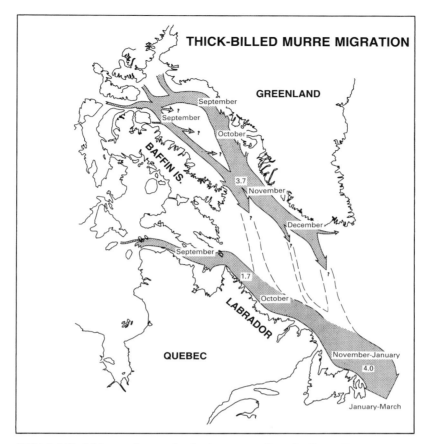

2 Migration of Thick-billed Murres from colonies in eastern Canada (from Gaston 1980).

European Arctic to SW Greenland. Young of the year apparently travel faster than older birds, arriving in Newfoundland in Oct–Nov (Donaldson *et al.* 1997). Breeding birds mainly arrive after Dec. Pacific populations winter in the southern Bering Sea, Sea of Okhotsk, and the northern Pacific S to northern Honshu, Japan in the W, and northern British Columbia in the E (Brazil 1991; Morgan *et al.* 1991), rarely to California (Monterey, Small 1994), but some persist as far N as Cape Thompson throughout the winter (Swartz 1966).

STATUS
Recent estimates are 10–15 million breeding birds in the Atlantic (Nettleship and Evans 1985) and 4.7 million breeding birds in the

Pacific (Byrd *et al.* 1993, using the same correction as for Common Murre: see p 137). Most of the uncertainty for the Atlantic is caused by the scanty information on the very large Icelandic population. The total world population of 15–20 million is distributed as follows: Iceland 22%, Bering and Chukchi Seas 22%, eastern Canadian Arctic 16%, Spitzbergen 10%, Novaya Zemlya 10%, Bear I. 8%, Greenland 6%, Sea of Okhotsk 2%. Populations in eastern Canada (probably) and W Greenland (certainly) have declined since the 1950s, in Greenland by more than 50%, with some large colonies entirely extirpated. Hunting, both at the colonies and on passage and wintering areas, is probably the primary cause, although drowning in salmon gill-nets was common in the 1960s and oiling has also been responsible for many deaths. Hunting at colonies in Greenland during the breeding season has been prohibited since 1978, with much stricter regulations on hunting elsewhere from 1988 (Evans and Nettleship 1985; Evans and Kampp 1991). Until 1993, the very large harvest in Newfoundland waters was unregulated between Sept and Mar and illegal sale of birds was widespread (Elliot 1991). In the Bering Sea there was a decline at the Pribilof I., and some other colonies, from the 1970s to late 1980s, although the magnitude is unclear and the cause is unknown (Springer 1991; Byrd *et al.* 1993). The colony at Tyulenii I., in the Sea of Okhotsk, declined from 650 000 to 140 000 between the late 1940s and 1960. A similar decline occurred at Guba Bezmyannaya, Novaya Zemlya where there were 1.6 million murres in 1933–4, but only 290 000 in 1948. In both cases, the cause was commercial hunting and egging, with 200 000–300 000 eggs being taken annually in Novaya Zemlya in the 1950s (Kozlova 1957). Following the cessation of commercial exploitation, there were increases at both these colonies, although neither has reached its former size.

Field characters

Similar to Common Murre in general appearance, in winter best distinguished by the face pattern; without a white streak behind the eye. In summer, deeper, shorter bill and white line along lower edge of upper mandible are distinctive, also the shape of the meeting between the white of the breast and the black of the throat, in an inverted 'V'. At all seasons has a distinctly larger head and thicker neck than Common Murre. In flight, at a distance, could be mistaken for Razorbill, but latter distinguished by its much longer tail. Juvenile and first winter birds are similar to winter-plumage adults, but have shorter, more slender bills. In the hand, can be aged as first year or older by inspection of primary coverts, which are pale and worn compared with the other coverts in first years, but uniform with the rest in older birds. On dead birds, take a half-inch (12.5 mm) crescent wrench (spanner) and

push it down vertically over the eye sockets. If the wrench slides down into the sockets the bird is a first year; if the wrench will not fit, it is older (can also be done with calipers). A simple ageing criterion used by fishermen in Newfoundland is to hold the (freshly dead) bird by the lower mandible, with the body hanging down. If the mandible bends, it is a first year (Gaston 1984).

Voice

Adults have a variety of guttural calls ranging from a faint 'urr' to a loud, emphatic 'aargh', depending on context. The following descriptions are from K. Lefevre (unpublished data)

ADULT: 'laughing'; a prolonged 'RAH-rah-rah-rah-rah', descending and diminishing, that causes whole body to shake when given loudly (Fig. 3); occurs periodically during incubation and chick-rearing, the bird standing up and pointing its head down towards the egg. It may be repeated by neighbouring birds and pass along the colony as a wave of calling, especially noticeable when other activity is low, as in the middle of the night. 'Nodding call'; a low 'urr', given in response to arrival of potential, but not immediate, danger (e.g. close approach of a Glaucous Gull, *Larus hyperboreus*). May be repeated over several minutes, as long as danger persists, accompanied by nodding action; the call is given while craning head downward and forward, sometimes bending low so that beak almost skims the ledge. 'Cawing'; loud 'RAH-ah' (rendered 'aarr-rrr-rr' by Pennycucik 1956), the first syllable harsh, ascending and longer than second, which is lower and descending; has at least five variants: (a) 'greeting call' given by a breeder on its site when its mate arrives; (b) 'copulation call', given loudly by ♂ while attempting to mount, (c) 'change-over call', given during change-over by the bird taking over, or occasionally by both (not used when chick is approaching fledging age); (d) 'fight call', given during agonistic interactions; (e) 'chick contact call' ('luring call' of Tschanz 1968), given by adults in situations where they have become

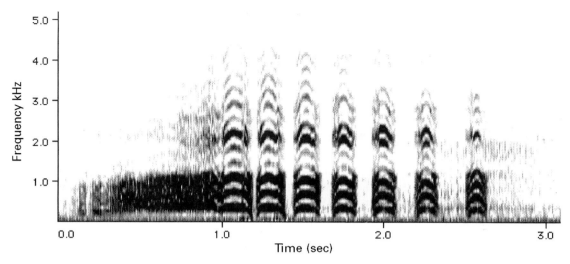

3   Laughing call, showing the rhythmic, descending, and fading structure of the call.

separated from their chick—the main adult–chick communication, less harsh and distinct from change-over call. The chick contact call is given very emphatically during colony departure. 'Adow call'; two short syllables, with the stress on either, given (a) by ♀ during copulation (presumably the origin of the Inuit name for murres, 'akpa'), (b) during agitation caused by predators or researchers, (c) by pair member arriving at site, as part of greeting ceremony. 'Growl'; a prolonged, low pitched, 'aarrr' delivered with bill closed and usually by birds incubating or brooding and not invoved in any special interactions (equivalent to the 'contentment call' of Tschanz 1968).

CHICK: similar to those of Common Murre: a low peeping during most of the chick-rearing period, and a louder 'wee', ascending then descending, used to attract parental attention. Close to departure, switch to a strident, di- or trisyllabic 'wee-wee-wee', given at frequent intervals in the last few minutes before jumping from the cliff and as soon as it reaches the water. Chicks separated from their parent at departure give this call continually until reunion is effected.

Habitat, food and feeding behaviour
MARINE HABITAT
Found in continental shelf waters ranging in surface temperature from 0 to 10 °C. Less often seen in coastal waters than Common Murre. Throughout the year, prefers areas with floating ice, up to 90% cover. In spring and early summer concentrates at seaward edge of land-fast ice, where it occurs over deep water. Dense aggregations form at ice edges and where upwelling and other oceanographic processes concentrate slow-moving prey, such as euphausiids (Piatt *et al.* 1992). Occurrence at ice edges may be affected by water depth, distance from land, wind speed and direction, and wave height (Bradstreet 1979). Birds in this habitat apparently forage under the ice. In winter in Newfoundland, frequents large bays, but usually in water more than 100 m deep. In northern Hudson Bay, large concentrations occurred on a 12-hr cycle in channels among islands, when tidal currents produced upwelling, resulting in concentrations of amphipods (Cairns and Schneider 1990). Elsewhere, and perhaps more typically, scattered while feeding, or consorting in small groups of fewer than 10 (Nettleship and Gaston 1978). Never occurs on fresh water unless storm-driven, but

large numbers of young birds appeared periodically on the Great Lakes in early winter during 1890–1950. This behaviour, although sometimes associated with high winds, appears to have been more akin to irruptions caused by lack of food in traditional wintering areas than to wrecks resulting from storms (Gaston 1988).

FEEDING BEHAVIOUR

FORAGING RANGE: may forage at considerable distances, with birds seen carrying fish towards the colony from up to 150 km (Swartz 1967; Brown 1980), although at some colonies birds feed much closer (Hornøy, Norway; 11 km, Furness and Barrett 1985). Birds commute to and from feeding areas in compact flocks flying within 2 m of the water (higher when assisted by a tailwind) at approximately 80 km/hr. Flocks often follow one another in distinct corridors, perhaps to enhance their ability to navigate in poor visibility (Gaston and Nettleship 1981). Leading birds periodically fly up several metres, apparently to get a better view, and changes of course often occur following such inspections.

DIVING BEHAVIOUR: dive down to more than 100 m, but normal feeding depths in northern Hudson Bay were 20–60 m, deepest at mid-day and shallowest at midnight (Burger 1991; Croll et al. 1991). Diet suggests that prey is mostly taken in mid-water, rather than on the bottom, but there is much variation among colonies.

DIET

Appears to be very catholic in its diet, taking a very wide range of fish and invertebrates. Information from stomach analysis and stable isotope ratios suggests that fish and large zooplankton both form important components of the diet throughout the year (Gaston and Bradstreet 1993; Hobson et al. 1994).

ADULT SUMMER: most important prey organisms in the Atlantic are the amphipod *Parathemisto libellula* and Arctic cod (*Boreogadus saida*) (Hobson and Welch 1992; Gaston and Bradstreet 1993). A variety of benthic and midwater fishes, squid (*Gonatus fabricii*), other amphipods (*Gammarus*, *Apherusa*), mysids (*Mysis*, *Boreomysis*), and euphausiids (*Thyssanoesa* spp.) are also taken. The latter are the most important diet item for birds feeding off the Pribilof I. in Aug (Schneider et al. 1990b; Coyle et al. 1992).

ADULT WINTER: off Newfoundland, where a large portion of the Atlantic population winters, Tuck (1961) reported that in the 1950s diet was dominated by capelin (93% weight). In the 1980s there was a marked change, with Arctic cod, and to a lesser extent capelin and squid, the main prey in Nov–Dec, and crustacea, especially euphausiids in Jan–Mar (Elliot et al. 1990). Off Greenland, the wintering population feeds mainly on capelin and euphausiids. As in Newfoundland, fish are predominant in the early part of winter (Falk and Durinck 1993). In the western Pacific, in winter, the diet is mainly squid (*Gonatopsis* and *Berryteuthis*) and euphausiids, with smaller quantities of fish, especially lanternfish (*Benthosema*) (Ogi 1980), although net-drowned birds off Hokkaido contained exclusively sandlance 186–253 mm in length (Hashimoto 1993). See Table: (1) W. Greenland, winter (Falk and Durinck 1993); (2) Newfoundland, Nov–Dec; (3) Jan–Mar (Elliot et al. 1990); (4) Canadian high Arctic; (5) Canadian low Arctic, summer (Gaston and Bradstreet 1993); (6) Chukchi Sea, summer (Piatt et al. 1991a);

CHICKS: most important prey delivered to chicks over much of the range are small cod, especially Arctic cod. Other fishes commonly fed to chicks include sculpins (Cottidae, especially *Triglops* spp.), blennies (Blennioidea), sandlance, capelin, and juvenile pollock (Gaston 1985c; Gaston and Bradstreet 1993). At the Gannet Is., Labrador, most important component of chick diets was the blenny (*Lumpenus maculatus*) (Birkhead and Nettleship 1987c), but this was in an area where Thick-billed Murres were greatly outnumbered by Common

Plates

Plate 1

Common and Thick-billed Murres, genus *Uria* and Razorbill, genus *Alca*

1. Common Murre

Uria aalge p. 133

Length about 40 cm; the largest living auk with a very high weight/wing area ratio, consequently takes off with difficulty; very clean-cut blackish-brown upperparts contrast with white breast, belly and underwing; throat black in summer, white in winter, when the face is also white except for the area in front of and a black line behind the eye; a very abundant bird, and the most widespread seabird in the northern hemisphere.

(a) Adult, summer.
(b) Adult, summer, 'bridled' form.
(c) Adult, winter.
(d) First winter.

2. Thick-billed Murre

Uria lomvia p. 147

Length about 40 cm; very similar in size and appearance to the Common Murre, but with a deeper, shorter, less dagger-like bill, sporting a white line along the lower edge to the upper mandible, most prominent in summer; in winter has a dark face; found almost exclusively in waters subject to winter ice cover and breeding only in Arctic waters; but in Pacific occurs in Low Arctic waters and overlaps widely with Common Murre: one of the most numerous seabirds in the northern hemisphere.

(a) Adult, summer.
(b) Adult, winter.
(c) First winter.

3. Razorbill

Alca torda p. 126

Length about 42 cm; longer than murres because of long tail, but actually smaller in body, with a much deeper, blunter bill with a vertical white stripe; similar to murres in plumage pattern; found mainly in inshore waters; breeding range overlapping broadly with Common Murre in the Atlantic, but found somewhat further south in winter.

(a) Adult, summer.
(b) Adult, winter.
(c) First winter.

Plate 2
Black, Pigeon, and Spectacled Guillemots, genus *Cepphus*

1. Black Guillemot
Cepphus grylle p. 168
Length about 35 cm; a plump-looking auk with rather broad wings; very distinctive in summer, with black plumage, white shoulder patches and bright vermillion feet; in winter mainly white except for black primaries, but races variable; found throughout boreal and Arctic waters of the Atlantic and the Arctic Ocean; mainly coastal, but offshore amid pack-ice; sedentary except in high Arctic.
(a) Adult, summer, ssp. *atlantis*.
(b) First summer
(c) Adult, winter, ssp. *mandtii*.
(d) Adult, winter, ssp. *atlantis*.
(e) Recently fledged juvenile.
(f) First winter.

2. Pigeon Guillemot
Cepphus columba p. 178
Length about 35 cm; very similar to Black Guillemot in plumage, but with a dark wedge in the white scapular patch; white almost obscured in the Kurils' race, *snowi*; found in coastal waters from California to Kamchatka and throughout Bering Sea; mainly sedentary.
(a) Adult, summer, ssp. *adianta*.
(b) Adult, summer, ssp. *snowi*.
(c) Adult, winter.
(d) Recently fledged juvenile.
(e) First winter.

3. Spectacled Guillemot
Cepphus carbo p. 186
Length about 38 cm; all-dark plumage except for a pale patch around and behind the eye; in winter plumage, pattern resembles murres; black above and white below, with a dark bill, but eye retains a distinctive pale ring; transitional plumage appears scaly on neck and flanks; endemic to Sea of Okhotsk and adjacent areas; behaviour similar to Pigeon Guillemot, with which range apparently overlaps in the Kuril Is.
(a) Adult, summer.
(b) Adult in transition from summer to winter.
(c) Adult, winter.

Plate 3

Marbled and Kittlitz's Murrelets, genus *Brachyramphus*, and Rhinoceros Auklet, genus *Cerorhinca*

1. Marbled Murrelet

Brachyramphus marmoratus p. 191
Length about 25 cm; a small, rather slender auk, with a narrow bill and pointed wings; flies very rapidly, often banking and veering; in summer, dark, mottled brown all over; in winter blackish above and white below, with a white partial collar and white sides to the rump, prominent when taking off; found in inshore waters from California through Aleutian Is. to Sea of Okhotsk and Japan (Asian form, *perdix*, now considered a separate species); usually occurs in pairs, or small groups.
(a)   Adult, summer.
(b)   Recently fledged juvenile.
(c)   Adult, winter.
(d)   Adult, summer, ssp./sp. *perdix*
(e)   Adult, winter, ssp./sp. *perdix*

2. Kittlitz's Murrelet

Brachyramphus brevirostris p. 200
Length about 25 cm; similar to Marbled Murrelet in *jizz*, but with a much shorter bill; in summer paler, more greyish brown; in winter, head and neck white except for crown and hind-neck; coastal waters from SE Alaska to Chukchi Sea and Aleutian Islands, also Sea of Okhotsk; often found in association with tidewater glaciers.
(a)   Adult, summer.
(b)   Recently fledged juvenile.
(c)   Adult, winter.

3. Rhinoceros Auklet

Cerorhinca monocerata p. 270
Length about 35 cm; a heavy-bodied grey-brown auk; in summer dark greyish brown all over except paler on the belly, with two white (not yellowish as shown) plumes curving downwards, one from behind the eye, the other from base of bill; bill orange, surmounted by a white 'horn'; in winter loses the horn; from California to N Japan, but mainly from Washington to Gulf of Alaska and in Hokkaido, coastal and offshore.
(a)   Adult, summer.
(b)   Recently fledged juvenile.
(c)   Adult, winter.

Plate 4
Ancient, Japanese, Craveri's, and Xantus' Murrelets, genus *Synthliboramphus*

1. Ancient Murrelet
Synthliboramphus antiquus p. 214
Length about 25 cm; a small auk with a blunt, sparrow-like bill; flies strongly; in summer, very distinctive with dove-grey upperparts, white underparts, black head and breast, and white plumes around crown and upper back; winter similar, but throat white and fringes on crown reduced or absent; common inshore and offshore British Columbia and less common through Gulf of Alaska to Yellow Sea.
(a) Adult, summer.
(b) Adult, winter.

2. Japanese Murrelet
Synthliboramphus wumizusume p. 222
Length about 22 cm; similar to Ancient Murrelet, but slightly smaller, with deeper bill and a broad white plume on either side of crown and neck; endemic to small islands offshore of Japan and found mainly in offshore waters.
(a) Adult, summer.
(b) Adult, winter.

3. Craveri's Murrelet
Synthliboramphus craveri p. 211
Length about 21 cm; a small, slender auk with a slim, sharp bill; black above and white below, except for black underwings; summer and winter plumage alike; breeding confined to Sea of Cortez; non-breeding distribution extends from Mexico to S California, mainly offshore.
(a) Adult, all seasons.

4. Xantus' Murrelet
Synthliboramphus hypoleucus p. 205
Length about 22 cm; very similar to Craveri's Murrelet, but slightly larger, with white underwing and white extending further up on face (race *scrippsi*, California), or in front of eye (*hypoleucus*, Guadalupe and San Benito Is.); offshore except when visiting breeding sites.
(a) Adult, ssp. *scrippsi*.
(b) Adult, ssp. *hypoleucus*.

Plate 5

Dovekie, genus *Alle*, Cassin's Auklet, genus *Ptychoramphus*, and Parakeet Auklet, genus *Cyclorhynchus*

1. Dovekie

Alle alle p. 161

Length about 20 cm; a small, chubby auk, not unlike a tiny murre, but with a very short, thick neck, a large rounded head and a small bulbous, finch-like bill; summer black above and on head and breast except for white streaks on scapulars and a white spot above eye; winter similar, but white below and on sides to neck, except for dusky breast band; throughout Arctic waters of Atlantic sector and small numbers in Chukchi Sea, mainly offshore.

(a)   Adult, summer.
(b)   Recently fledged juvenile.
(c)   Adult, winter.

2. Cassin's Auklet

Ptychoramphus aleuticus p. 227

Length about 24 cm; a plump, medium-sized auk with a distinctive, wedge-shaped bill; flight rather laboured; dull brown above and on throat and breast, otherwise pale below; white spot in front of eye and white iris; offshore from N Mexico to W Aleutians, commonest in British Columbia and Gulf of Alaska.

(a)   Adult, summer.
(b)   Recently fledged juvenile.
(c)   Adult, winter.

3. Parakeet Auklet

Cyclorhynchus psittacula p. 235

Length about 25 cm; chubby, medium-sized auk with a stout, red bill; plumage black in summer, except for white lower breast and belly, white plume behind the eye and prominent white iris; in winter chin, throat, and breast white; breeds from Gulf of Alaska through Bering and Chukchi seas to Sea of Okhotsk; in winter widespread offshore in North Pacific.

(a)   Adult, summer.
(b)   Recently fledged juvenile.
(c)   Adult, winter.

Plate 6
Least, Whiskered, and Crested Auklets, genus *Aethia*

1. Least Auklet
Aethia pusilla p. 252

Length about 15 cm; tiny auklet with stubby bill, surmounted by small rounded knob; plumage in summer varying from mainly black above and white below, through spotted below, to black on breast and upper belly with white throat; narrow white plumes on face, white streaks on scapulars, and prominent white iris in all morphs; winter black above and white below; Bering and Chukchi seas and Sea of Okhotsk; small southward movement to edge of pack-ice in winter.

(a) Adult, summer, pale morph.
(b) Adult, summer, intermediate morph, with food.
(c) Adult, summer, dark morph, with food.
(d) Adult, winter.
(e) Recently fledged juvenile.

2. Whiskered Auklet
Aethia pygmaea p. 262

Length about 18 cm; a small auk with extravagant head ornaments, plumage blackish in summer except for a paler grey vent, with narrow plumes on face and neck, a bright red bill and white irides; a long, narrow, black crest curves forward from the crown; in winter, plumes reduced; Aleutian Is., and Sea of Okhotsk, inshore and offshore, non-migratory.

(a) Adult, summer.
(b) Recently fledged juvenile.
(c) Adult, winter (November–January).

3. Crested Auklet
Aethia cristatella p. 242

Length about 25 cm; plump, medium-sized auk with short, blunt bill, surmounted by a substantial, forward-curling crest; plumage all sooty grey, with single narrow, white plume behind eye and white irides; bill bright orange in summer, brown in winter; breeding Bering and Chukchi seas and Sea of Okhotsk; non-migratory, but shifting to S edge of range in winter; mainly offshore.

(a) Adult, summer.
(b) Recently fledged juvenile.
(c) Adult, winter.

Plate 7
Tufted, Horned, and Atlantic Puffins, genus *Fratercula*

1. Tufted Puffin
Fratercula cirrhata p. 298

Length about 40 cm; unmistakable; a large, all-dark auk with a huge bill; plumage all sooty black except white face and long golden tufts from nape; bill bright orange, greenish at base; in winter, all dark except for pale stripe above and behind eye, bill dull orange, dark at base; from Central California to Japan and throughout Bering and Chukchi seas, moving S in winter and remaining well offshore.

(a) Adult, summer.
(b) Recently fledged juvenile.
(c) Adult, winter.

2. Horned Puffin
Fratercula corniculata p. 293

Length about 38 cm, a crisp, black-and-white auk with a large, rounded head and triangular bill; plumage black on crown and rest of upperparts, on neck and throat, otherwise white; bill yellow, with orange tip; eye surmounted by black, fleshy 'horn'; legs and feet bright orange; in winter, face dark grey, bill dull, constricted at base; from British Columbia to Sea of Okhotsk and throughout Bering and Chukchi seas, wintering south to California, usually well offshore.

(a) Adult, summer.
(b) Recently fledged juvenile.
(c) Adult, winter.

3. Atlantic Puffin
Fratercula arctica p. 282

Length about 35 cm; shape and plumage similar to Horned Puffin but face suffused grey; bill orange and yellow, dark blue-gray at base with orange rictal rosette; Arctic populations similar to Horned Puffin in size, but southern populations much smaller; Atlantic from English Channel and Gulf of Maine to Arctic Ocean, moving south in winter; offshore.

(a) Adult, summer, ssp. *grabae*.
(b) Adult, summer, ssp. *naumanni*.
(c) Recently fledged juvenile.
(d) Adult, winter.

Plate 8
Winter plumage auks in flight

1. Common Murre
2. Thick-billed Murre
3. Razorbill
4a. Black Guillemot, ssp. *atlantis*
4b. Black Guillemot, ssp. *atlantis*, recently fledged juvenile
4c. Black Guillemot, ssp. *mandtii*
5a. Pigeon Guillemot, ssp. *adianta*
5b. Pigeon Guillemot, ssp. *snowi*
5c. Pigeon Guillemot, ssp. *adianta*, recently fledged juvenile
6. Spectacled Guillemot
7. Ancient Murrelet
8. Japanese Murrelet
9. Cassin's Auklet
10. Least Auklet
11. Dovekie
12a. Marbled Murrelet, ssp. *marmoratus*
12b. Marbled Murrelet, ssp./sp. *perdix*
13. Kittlitz's Murrelet
14. Crested Auklet
15. Atlantic Puffin
16. Parakeet Auklet

Table: Thick-billed Murre adult diets

ref n	(1) 202	(2) 491	(3) 616	(4) 117	(5) 205	(6)[1] 46
INVERTEBRATES						
amphipods	20	14	23	63	35	0
euphausiids	51	20	72	0	1	0
decapods	0	2	2	2	1	0
squid	6	30	< 1	2	14	0
annelids	7	< 1	< 1	0	35	0
FISH						
Capelin (*Mallotus villosus*)	57	26	5	0	18	0
Arctic cod (*Boreogadus saida*)	0	49	10	41	29	94
saffron cod (*Eleginus gracilis*)	0	0	0	0	0	2
Atlantic cod (*Gadus morhua*)	0	13	7	0	0	0
sandlance (*Ammodytes* spp.)	7	0	0	0	0	0
sculpins (Cottidae)	0	0	0	1	29	3
snailfish (*Liparis* spp.)	0	0	0	0	15	0
other	21	23	6	0	39	0
					2	0

(% of stomachs containing food except [1] based on % weight

Murres. See Table: (1) Coats I., northern Hudson Bay (M.Hipfner and AJG, unpublished data); (2) Akpatok I., Ungava Bay (Tuck and Squires 1955); (3) Prince Leopold I., Barrow Strait, NWT (Gaston and Nettleship 1981); (4) Gannet Is., Labrador (Birkhead and Nettleship 1987*b*).

Displays and breeding behaviour

Based on personal observations.

BEHAVIOUR OF PRE-BREEDERS: birds in their first summer do not visit the colony. Some marked at colonies in Hudson Strait have been recovered off SW Greenland in their first and

Table: Thick-billed Murre chick diets

ref n	(1) 594	(2) 2630	(3) 178	(4) 650
INVERTEBRATES				
squid (*Gonatus fabricii*)	1	3	0	0
crustacea	3	4	3	0
annelids	< 1	< 1	0	0
FISH				
capelin (*Mallotus villosus*)	14	1	0	30
Arctic cod (*Boreogadus saida*)	53	35	78	< 1
blennies (Blennioidea)	13	20	0	69
sandlance (*Ammodytes* spp.)	3	1	0	0
sculpin (Cottidae)	12	28	18	0
lumpsucker (*Eumicrotremus* spp.)	< 1	1	0	0
snailfish (*Liparis* spp.)	0	2	< 1	0
Greenland halibut (*Reinhardtius hippoglossoides*)	0	4	0	0

(% of items delivered)

second summers, suggesting that many do not return to the vicinity of their colony. Pre-breeding birds that do attend the colony generally arrive later than breeders, with those in their second summer appearing only towards end of incubation. Counts of known-age birds suggest that most 3-year-olds attend the colony, but less than half of 2-year-olds do so. Birds attending the colony for the first time (second- and some third-summer) spend much time on the water at the foot of the cliffs, flying up periodically to circle repeatedly near the breeding ledges and make very short touch-downs, often only balancing on the edge of a ledge, with wings still flapping. Touch-downs by the same bird may occur in several parts of the colony, separated in some cases by hundreds of metres. More experienced pre-breeders (mainly third-year) land for longer periods, congregating mainly on broad 'loafing' ledges, especially near the top of the occupied area. Failed breeders may also visit the loafing ledges later in the season. After some time, pre-breeders begin to spend longer periods on the colony (>10 min), and by then they are usually in an area close to their rearing site. About 90 per cent of new recruits begin to breed on sites within 50 m of where they were reared. Non-breeders 4 years or older generally occupy potential breeding sites and behave very much as breeders do during pre-laying. Pre-breeders are much more easily disturbed than breeders. The enormous clouds of birds that leave large colonies in response to aircraft disturbance, or shooting, are mainly pre-breeders and off-duty breeders. Incubation is not usually abandoned except in the face of immediate danger, sometimes not even then; incubating and brooding birds can often be caught by hand and sometimes continue to sit while their egg is removed from the brood patch.

AGONISTIC BEHAVIOUR: fighting is frequent among neighbours, and between breeders and prospectors, and increases after hatching and among failed breeders. At high intensity, combatants grapple bills and beat one another with the carpus of the wing. Because ledges are

generally small, many fights end in both birds falling off the ledge into the sea and fights may continue there, the two birds spinning round in circles. Ritualized chases, with wings beating heavily along the surface, or diving underwater, also occur ('water dances' of Tuck 1961).

SOCIAL AND SEXUAL BEHAVIOUR: Posture and vocal displays are generally similar to Common Murres, but less ritualized appeasement is evident. More phlegmatic than Common Murres in all socializing (T. R. Birkhead, personal communication). Side preening, head up, and wing lifting all occur, especially among young birds prospecting for sites. Copulation is frequent in the pre-laying period, following every arrival of either pair member. Attempted extra-pair copulations are common, but rarely successful. Copulation attempts involving several ♂♂, as seen in Common Murres, are very rare. Pairs and neighbours frequently allopreen one another, usually nibbling at feathers around the head and neck, especially the base of the bill. 'Muck flicking' ('flinking' of Williams 1972), dipping the bill in mud or faeces on the ledge and then shaking the head so that the substance is sprayed about, is frequent during incubation, and manipulation with the bill of nearby objects, such as pebbles, feathers, and dropped fish is also common.

Breeding and life cycle

Chick departure strategy is 'intermediate' (Chapter 6). Breeding is colonial, nearly always on small ledges on precipitous cliffs, frequently intermixed with Common Murres, kittiwakes, sometimes Northern Fulmars (*Fulmarus glacialis*). Following notes are derived from Gaston and Nettleship (1981), Gaston *et al.* (1985), and personal observations, unless otherwise stated.

BREEDING HABITAT AND NEST DISPERSION: on cliffs immediately adjacent to the sea. On broad ledges, often forms one or two rows of birds, normally touching one another while incubating; seldom forms dense groups of the type seen in Common Murres. Some ledges are very small, so that tail of incubating bird hangs off edge, frequently also slope towards the sea by up to 15°.

NEST: egg is laid on bare rock and no nest is constructed. Birds incubate with breast propped against a cliff face or rock. Occasionally utilize rock crevices, or sites among boulders.

EGG-LAYING: first arrivals at the colony are usually in May, sometimes Apr (Alaska, SW. Greenland, Novaya Zemlya). Laying is in June or early July, continuing for 35–40 days. Median laying during 23 June–3 July at Coats I. (12 years), 29 June–19 July at Prince Leopold I. (6 years, Gaston and Nettleship 1981 and personal observation), 10 July at Akpatok I. (2 years), 28 June–3 July at Digges I. (3 years) and 19–26 June (3 years) at Gannet I. Labrador (Birkhead and Nettleship 1987a). Prior to egg-laying, ♀ spends up to 12 hrs at the site. As laying approaches, she crouches frequently, with wings partly spread, and scrapes with her claws, while intermittently contracting the abdomen and drawing the tail forward. To lay the egg she stands on tip-toe, with legs straight, leaning against some vertical surface and stretching the wings back. As egg emerges, small end first, tail is bent forward under the body, so that egg is kept away from edge of ledge. Within a minute, ♀ relaxes and steers egg under the single brood patch, oriented with long axis parallel to body, large end against the patch.

EGGS: clutch one; a replacement is usually laid if first egg is lost less than 2 weeks before cessation of laying. Occasionally, a second replacement may be laid. Interval between replacements is 13–18 days. Shape is elongated-pyriform, very variable, with ground colour dirty white to turquoise, spotted or blotched with brown or black. Markings may be rounded or elongate, with sharp or smudged edges, sometimes forming a ring, occasionally solid, around large end of egg. Eggs are frequently soiled with excreta and substrate, and sometimes caked with it so that original colour is hidden. Shell is rough, chalky, and very thick. Egg volume decreases with date of laying and replacement eggs are smaller than first eggs by c. 5% (Birkhead and Nettleship 1982, 1984b). The trend is not always linear and at Digges I. there was a sharp decrease in egg size after about 75% of first eggs had been laid (Gaston et al. 1985). First time breeders lay smaller eggs than experienced birds (de Forest 1993). For measurements, see Table: (1) U. l. arra, Commander and Kuril Is. (BMNH), (2) U.l. lomvia, Coats I., Canada, random sample; (3) experienced breeders; (4) first time breeders (M. Hipfner, unpublished data); (5) Prince Leopold I.; (6) Coburg I., Canada (AJG). Fresh egg weights: Gannet I., 107 g in both of 2 years (Birkhead and Nettleship 1987a); Prince Leopold I. 98.5 ± 8.1 g (n = 54); Coats I., first time breeders 97.1 ± 4.2 g (n = 9), females with >3 years of breeding, 110.8 ± 8.4 g (n = 10, M. Hipfner, unpublished data)

Table: Thick-billed Murre egg dimensions (mm)

ref	length			breadth			n
	mean	s.d.	range	mean	s.d.	range	
(1)	81.6	2.7	76.3–84.8	50.9	1.6	48.9–53.5	15
(2)	80.0	3.4	67.9–86.0	50.6	1.9	42.0–55.0	77
(3)	80.8	2.8		51.1	1.3		24
(4)	77.4	3.6		48.8	1.2		27
(5)	76.0	2.8	70.1–84.3	48.4	1.7	45.0–52.7	50
(6)	76.9	2.7	70.8–83.6	47.6	1.4	43.7–50.3	50

approximately 11% adult weight. Composition: shell 12%, yolk 35%, albumen 53% (Uspenski 1956). Chick weight at pipping 70% of fresh egg weight, of which 17% is yolk (Birkhead and Nettleship 1984b).

INCUBATION: continuous from laying onwards, with eggs left unattended only during periods of extreme food stress. Shifts average 12–48 hrs, varying from colony to colony and generally decreasing in duration as incubation proceeds. Change-overs frequently involve delicate manoeuvres to prevent the egg from rolling off the site. Parents recognize their own egg when given a choice, but will incubate any egg on their site if their own is removed. Birds occasionally steal eggs from neighbours after their own is lost (Gaston *et al.* 1993c). Incubation period 29–35 days (median 33).

HATCHING AND CHICK-REARING: the time elapsed from first starring of the shell to hatching is usually 3–4, occasionally up to 6, days, from pipping 2–3, occasionally 4 days. Chick is brooded against the flank, the adult adopting a characteristic pose with wing drooped on the side with the chick, and tip of chick's bill frequently visible poking through tertial feathers. Fish are sometimes delivered before chick has hatched. They are never used in display, as occurs in Common Murre, but occasional birds holding fish are seen at the colony, perhaps breeders that have recently lost a chick. Adults arriving with a fish land directly on the site, often on the back of their mate or neighbour, where the ledge is narrow. The brooding mate stands up instantly, giving the crowing call, and the arriving bird thrusts the fish under its mate's breast. The fish is released as the chick grasps it, securing it behind the head. With a series of jerks, the head is oriented towards the gape and the fish is swallowed. The tail of a large fish frequently protrudes from the chick's bill for some minutes. Dropped fish are frequently left lying on ledges, where they may be scavenged by gulls or kittiwakes. Chicks are fed an average of 2–5 fish per day, varying with season and colony. In the eastern Canadian Arctic, the mean weight of fish delivered at Digges I. ranged from 5.7 to 8.7 g (3 years) and at Coats I. from 12.8 to 18.0 g (8 years). Total daily intakes were 20–30 g at Digges I. and 50–65 g at Coats I. (5 years). Other published meal sizes and feeding rates fall between these limits, although chicks leaving the colonies at Cape Thompson, Alaska and the N colony at Akpatok I., Canada were smaller than those at Digges I., (Swartz 1966; G. Chapdelaine, personal communication) and hence presumably had received less food. Birds leaving the colony after an incubation or brooding shift usually bathe and preen on the sea before joining outgoing flocks. These birds may use deep, slow wing-beats in descent (termed 'butterfly flight' by Pennycuaik 1956). Birds returning to feed after delivering fish to a chick fly directly towards the feeding area without joining flocks (Gaston and Nettleship 1981).

CHICK GROWTH AND DEVELOPMENT: similar to Common Murres; weight increases steadily by approximately 10 g/day up to about 14 days, the rate of increase diminishing and sometimes ceasing thereafter. Growth of the primary coverts is more or less linear to about 22 days and provides a useful criterion for age, although varying somewhat with nutrition. Development of contour plumage is likewise affected by nutrition, with very slow growing chicks having shorter feathers. Chicks at most colonies reach 200 g by 18 days old and fledge at 200–250 g. In the eastern Canadian Arctic, growth of chicks appears to be related to colony size, with chicks at large colonies (>100 000 pairs) growing more slowly and leaving the colony lighter than chicks at small colonies (Gaston *et al.* 1983a). Departure weights of chicks at Digges Sound (300 000 pairs) and Coats I. (30 000 pairs), colonies only 300 km apart, were 141–162 g (3 years) and 224–239 g (5 years), respectively (Gaston *et al.* 1985; AJG).

CHICK DEPARTURE: chicks depart at 15–30 days (median 21–24), accompanied by the parent; the parent and chick keep together for at least 1 month. Departure takes place in low light conditions, at dusk, or at high latitudes in the middle of the night, apparently to minimize predation, usually under calm conditions. In Spitzbergen, most departures took place between 17:00 and 01:00 hrs (Cullen 1954). If several days of storms occur when many chicks are old enough to leave, this may result in a mass departure on the first fine night, with as many as 30 per cent of chicks leaving at once. Departure involves a complex interaction between chick and ♂ parent. Chicks usually exercise by wing-flapping periodically throughout the 48 hrs prior to departure. As departure approaches they move away from their parents, towards the edge of the breeding ledge and begin to give the shrill, di- or tri-syllabic departure call. If the parent (father) does not respond, the chick may return to be brooded several times. In some cases the parent may move with the chick and attempt to brood it away from the site, in others, it stands beside the chick, facing the sea and gives loud chick contact calls, or leads the chick for up to 4 m to reach a suitable jumping point. Calls from the parent appear to encourage the chick to jump, but in some cases it jumps without the parent calling or leaving the site and the parent scrambles to catch up. Chick calling is most frequent just prior to jumping. Chicks crouch, then launch themselves horizontally, with full extension of the legs. Wing-flapping begins as soon as they are airborne and continues until they are clear of the cliff, following which they glide at a steep angle, flattening out before hitting the water, and frequently skittering along the surface on landing. The parent follows very close behind, almost touching the chick and braking with wings and feet spread to avoid overtaking it, often swaying from side to side in the process. They land within 1 m and immediately call, swim together, and indulge in mutual billing, after which chick ceases calling. Within 1 min, the parent begins to swim steadily away from the colony and the chick follows, keeping very close to the parent's flank. Chicks landing on the shore may be killed on impact, but most survive and begin walking to the sea. Some chicks are either not accompanied by their parent at departure, or fail to reunite with it on the sea, in some cases because they are slowed in descent by striking lower ledges. A chick on the sea without its parent continues to call, attracting a mob of other adults that surround it, cawing loudly and frequently pecking at it, or diving and surfacing underneath it. Some chicks are injured, or even occasionally killed in these mêlées. Other may be adopted by non-parents, although how permanent such adoptions are is unknown (Gilchrist and Gaston 1997). ♀ breeders continue to visit the breeding site periodically for up to 15 days after their chick has departed. They are frequently visited by ♂♂ other than their mate and allopreening may occur. In the absence of the ♀, the site may also be occupied by other birds, including pre-breeders. However, changes of site from year to year are rare; most breeders return to the same site.

BREEDING SUCCESS: mean chicks reared/pairs laying in the Pacific is 36–72% (median 52%, 8 studies, Byrd *et al.* 1993) and in the Atlantic 48–79% (median 68%, Birkhead and Nettleship 1981; Gaston and Nettleship 1981; Gaston *et al.* 1994). Reproductive success, like chick weight, may be affected by colony size (Hunt *et al.* 1986). Reproductive success increases with age up to at least 9 years old (AJG). In pairs of mixed age, success is influenced more by age of younger bird. Young birds also lay later than experienced breeders and this effect accounts for most of the decline in breeding success typically observed with date of laying (e.g. Birkhead and Nettleship 1982; de Forest and Gaston 1996). Pairs breeding on sites with neighbours, especially those in the centre of large groups, normally have better reproductive success than those without (de Forest 1993).

POST-BREEDING: by early Sept, most colonies are deserted except for a few birds attending late chicks. In the case of birds from colonies in Hudson Bay, parent–chick pairs undertake a rapid swimming migration through Hudson Strait, covering 500 km in 2–3 weeks (Gaston 1982). Movement of birds away from high Arctic colonies is probably similarly rapid, as young of the year from the Canadian high Arctic reach central W Greenland by mid-Sept, moving southwards approximately 10–20 km offshore, in parties of up to 30 birds. Return migration off W Greenland occurs in Mar and early Apr (Salomonsen 1979).

PREDATION: eggs and young are taken by Glaucous Gulls and Ravens, adults by Gyrfalcons (*Falco rusticolus*), and less commonly, by Snowy Owls (*Nyctaea scandiaca*) and Peregrine Falcons (*Falco peregrinus*) Glaucous Gulls are present at all Arctic colonies and some pairs forage entirely on murre eggs and chicks throughout the season. On peripheral ledges breeders may be taken by Red or Arctic Foxes, which also remove and cache many eggs. Most breeding failure is caused by eggs rolling off ledges, either because of fighting among neighbours, or because of insufficient co-ordination between pair members during change-overs. Heavy mortality to adults and eggs is caused by rock falls during breeding at some colonies. Adults landing on shore-fast ice at the foot of the colony may be unable to take off. Occasional birds are taken by Polar Bears and Walruses (*Odobenus rosmarus*) (Donaldson *et al.* 1995).

MOULT: little known, but assumed to be similar in duration to that of Common Murres. Adult birds from Canadian high Arctic colonies have finished the complete post-breeding moult when they begin to arrive in Newfoundland in early Nov, hence moult probably takes place in Sept and Oct off Greenland. Birds in summer plumage are rarely seen in Newfoundland before most wintering birds leave in late Mar, but all those attending colonies in May are in summer plumage, so pre-breeding moult occurs in Apr.

Elsewhere, summer plumage may be attained by Feb (Kaftanovskii 1951). Juvenile birds presumably moult completely soon after leaving the colony, while growing the first set of primaries. Primary feathers on second year birds visiting the breeding colony in Aug usually show much greater wear than those of breeders, suggesting that more time had elapsed since replacement. The first complete moult may occur during the first summer, when birds do not attend the colony. Bédard (1985) suggested that this moult may pass direct from first winter (Basic I) to adult winter plumage (Definitive Basic), without any summer plumage intervening.

Population dynamics

SURVIVAL: mean annual adult survival is estimated as 0.88–0.89 (Kampp 1991; Gaston *et al.* 1994). Both of these studies relate to the heavily hunted populations of W Greenland and the eastern Canadian Arctic; survival is probably higher in populations not subject to hunting.

AGE AT FIRST BREEDING: 3–8 years or older (median 5), with ♀♀ generally breeding about 1 year before ♂♂. At Coats I, Canada, survival of young from colony departure to breeding (5 years) averaged 40 per cent, with some variation among years (Gaston *et al.* 1994).

PAIR- AND NEST-SITE FIDELITY: more than 90% of breeders still alive return to the same site and mate as the previous year (Gaston *et al.* 1994).

DISPERSAL: inter-colony dispersal appears to be rare, although genetic homogeneity among colonies, both between E and W Atlantic and among the western Atlantic colonies (Birt-Friesen *et al.* 1992), suggests that some interchange occurs. Two recoveries of Canadian-marked birds of breeding age in W Greenland in summer probably indicate dispersal (AJG).

Dovekie or Little Auk *Alle alle*

Alca alle (Linnaeus, 1758, *Syst. Nat.*, ed.10, i: 131—Scotland)   PLATE 5

Description

Tiny black-and-white auk with puffy cheeks and chin; the only miniature auk in the Atlantic, with a stubby, sparrow-like bill. Flies with whirring wings, like a bumblebee. Has a distinctive, compact body shape, appearing almost without a neck, with horizontal posture making it look as if permanently ducking to avoid icy Arctic winds. Bill is short, deep and wide at the base; blending with contour feathers of crown and chin, to give an unusual, extremely distinctive blunt head-shape. Agile on land, walking on toes, but usually rests on toes and tarsi with a nearly upright posture; jumps nimbly from boulder to boulder or takes off instantly at the approach of a predator.

ADULT SUMMER: upperparts, including head, throat, and upper breast, mantle, upper surfaces of wings, rump, and uppertail coverts uniform jet black (upper breast and face tinted brown); sable cloak is set off by a small white spot above each eye; eye dark brown and hard to see, although white superorbital spots resemble highlights; narrow, silvery white margins to outermost scapulars form contrasting irregular stripes on back and white-tipped secondaries form white bar on trailing edge of wing; underparts otherwise unmarked white; upperwing coverts, primaries, and tail feathers brownish when worn; bill black; tarsi, toes, and webs black or blackish-brown.

ADULT WINTER: similar to summer, except upper breast, throat, outer edges to nape, sides of neck and face, white; white on neck extends high up, forming incomplete collar; irides dark brown and eye can be hard to see against the dark face.

JUVENILE AND FIRST WINTER: similar to summer plumage adult, but with slimmer bill, less glossy plumage, white spot above eye smaller, and throat paler. This plumage converts into first winter plumage (Basic I) by moult of feathers on chin and face immediately after going to sea; by Oct first winter birds are similar to winter-plumage adults.

CHICK: down black, grey on belly, in Spitzbergen; variable light grey to black in Greenland (Evans 1981); bare parts black.

VARIATION

Two races have been described on the basis of size. Most population belong to the smaller, nominate *A. a. alle*. The larger (*A. a. polaris*) is known only from Franz Josef Land and Severnaya Zemlya (presumably; Flint and Golovkin 1990).

MEASUREMENTS

Little difference between sexes, except bill measurements. 'Subadults' (birds without bursae or brood patches in breeding season) significantly smaller. See Table: *A. a. alle* (1) collected at sea in Baffin Bay, adults; (2) subadults (Bradstreet 1982b); (3) breeding birds trapped on the colony, Thule District, Greenland; (4) 'brown-winged' (? first years, Roby *et al.* 1981); (5) breeders at colony, Upernavik District, Greenland (P. G. H. Evans, personal communication); (6) collected at the colony, Spitzbergen (Norderhaug 1980); (7) Spitzbergen (Lovenskiold 1954).

WEIGHTS

♂♂ average 2.4–16.6% heavier than ♀♀ among breeders (see Table). 'Brown-winged' birds (supposed first years) lighter than breeders (Roby *et al.* 1981).

Range and status

Circumpolar Arctic, but practically all in the Atlantic sector.

RANGE

BREEDING: colonies occur between 68° and 82° N, the main concentrations being in the

Table:　Dovekie measurements (mm)

♂♂	ref	mean	s.d.	range	n
wing	(1)	120.5	2.8		45
	(2)	113.8	1.8		5
	(3)	123.1	2.7		117
	(7)	123.7	1.2	119–130	16
culmen	(1)	12.0	0.5		43
	(2)	11.6	0.5		5
tarsus	(1)	21.3	0.8		43
	(3)	20.3	0.3		5
	(7)	21.1	0.9	19–22	16
bill depth	(1)	6.6	0.5		10
tail	(3)	41.5	2.2		117
	(7)	36.7	1.8	32–39	16

♀♀	ref	mean	s.d.	range	n
wing	(1)	120.7	3.0		42
	(2)	108.3	5.9		6
	(3)	122.3	2.9		57
	(7)	123.7	1.2	122–125	6
culmen	(1)	11.4	0.5		42
	(2)	10.5	0.3		8
tarsus	(1)	20.7	0.8		42
	(3)	20.1	0.4		8
	(7)	20.8	0.4	20–21	6
bill depth	(1)	6.6	0.5		10
tail	(3)	40.6	2.0		57
	(7)	36.7	1.5	35–39	6

UNSEXED	ref	mean	s.d.	range	n
wing	(4)	117.2	2.2		18
	(5)	118.0	7.4		17
	(6)	118.8		106–129	185
tarsus	(6)	18.3		16–21	185

Table:　Dovekie weights (g)

♂♂	ref	mean	s.d.	range	n
	(1)	174.7	8.5		45
	(2)	155.4	2.9		5
	(3)	152.7	7.8		57
	(5)	147.0	9.4		17
	(7)	173.8	18.1	148–200	10

♀♀	ref	mean	s.d.	range	n
	(1)	167.3	10.1		43
	(2)	151.8	10.1		8
	(3)	146.3	8.9		56
	(6)	149.0	3.6	145–152	3

UNSEXED	ref	mean	s.d.	range	n
	(4)	140.4	7.4		18
	(5)	163.5		136–204	74

Thule District of NW Greenland and in Spitzbergen, which together support more than 90% of world population. Other colonies are found in Upernavik, Disko, and Scoresbysund districts of Greenland, and in Franz Joseph, Novaya Zemlya, and Severnaya Zemlya. Small numbers breed in Home Bay, Baffin, I., Iceland, Jan Mayen and Bear I. Non-breeders penetrate into Canadian Arctic archipelago in summer and autumn, as far W as Barrow Strait, NW Hudson Bay, and N Foxe Basin. There is a small, isolated population in the Bering Strait region, where breeding has been observed at Diomede Is. (Day *et al.* 1988). Stragglers from this population have been observed regularly at auklet colonies and in nearshore waters of St Lawrence I. and southwards to Pribilof Is., but total population of Pacific sector may be less than 1000 birds.

Winter: occur in areas from the edge of the mobile pack-ice S to Nova Scotia in the western Atlantic and Britain in E, with stragglers as far S as the Azores and Florida. Widespread from Barents and Greenland seas (Brown 1985) S to the northern North Sea from Nov to Feb, but mainly gone from the North Sea by Mar (Stowe *et al.* 1995). Autumn migration from NW Greenland mainly occurs on the W side of Davis Strait. Birds from Spitzbergen reach southern Greenland by Oct (Salomonsen 1979). Most of those breeding in W Greenland are thought to winter off Newfoundland and Nova Scotia; those wintering off SW Greenland do not appear until Nov and probably originate from colonies further E (Salomonsen 1979). Small numbers occur on George's Bank, at edge of continental shelf E off Cape Cod from Dec to May (Powers 1983). Birds remain off Jan Mayen throughout the winter (Bird and Bird

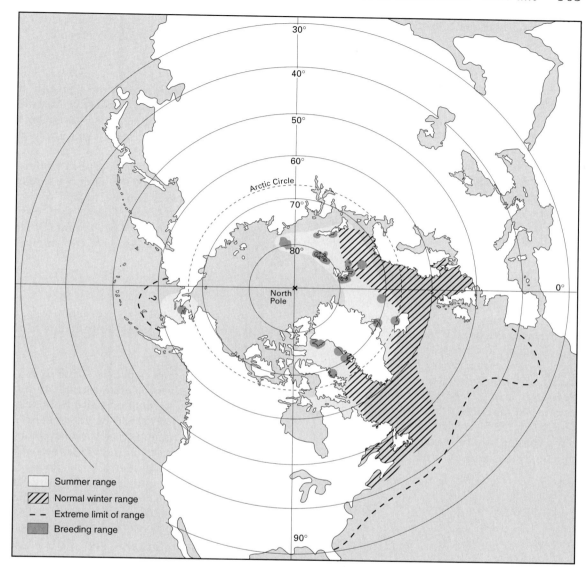

Summer range

Normal winter range

Extreme limit of range

Breeding range

1935). The wintering area of the Bering Sea population is unknown.

STATUS

Size of population in Thule, traditionally given as 30 million birds, is conjectural, but Renaud *et al.* (1982) estimated 14 million at sea in Baffin Bay in mid-May, presumably mainly breeders from Thule population. In Spitzbergen, population estimated in the order of 1 million birds (Mehlum, in Nettleship and Evans 1985); that of Scoresbysund, E Greenland, between 100 000 and 1 million. The population of Russian Arctic archipelagos are unknown, but apparently not as large as those in Greenland and Spitzbergen. No information on trends available for any population.

WRECKS: appear sporadically in large numbers in NE U.S. in winter, sometimes being found well inland, but rarely reaching the Great Lakes (unlike Thick-billed Murres).

Such inland wrecks occur mainly in Nov–Dec and tend to come in runs of several consecutive years: this was especially marked in the 1960s, when wrecks occured in 7 winters (Fig. 1). In contrast, no wrecks at all were recorded in four decades this century. Although inland wrecks are normally associated with periods of strong onshore winds, large numbers are sometimes seen from coastal observation points in New England without adverse weather conditions. Apparently large numbers occasionally move as far S as Massachusetts. Clumping of years with abundant Dovekie observations in New England suggests ultimate cause of these movements (like wrecks of Thick-billed Murres in the Great Lakes) is driven by changes in food supply or Dovekie populations occurring on a scale of decades. Similar wrecks occur in Britain, those of 1895 and 1912 being especially noteworthy. As in North America, inland occurrences are usually associated with storms, but conditions for wrecks require birds well S of their normal winter range. This was true in one case; large numbers were observed at sea off Scotland from Nov 1911, but the wreck occured in Jan 1912, after severe gales (Clarke 1912). In 1895 wreck, ♀♀ made up bulk of birds coming ashore at start of wreck, and ♂♂ preponderated later (Gurney 1895), suggesting

some difference in the winter ecology of the sexes; there is no subsequent information.

Field characters

In satisfactory viewing conditions, unlikely to be mistaken for any other auk in the Atlantic. Closest in size and appearance to juvenile Atlantic Puffin, but latter 50% larger and with different bill-and head-shape. In Bering and Chukchi seas, Dovekies may occur in summer among vast hordes of auklets. They are similar in size, but Dovekies can be picked out by their more compact body shape and crisp black-and-white plumage. They are more than twice the size of Least Auklets, the smallest alcids.

Voice

Birds on surface of colony, or in flight over it, are extremely vocal. Descriptions mainly from Ferdinand (1969).

ADULT: most common call is 'trilling call', which may be given while perched on the ground, or in flight, sometimes also from nest-site; lasts 1–3 sec with broad-band frequencies rising and falling rapidly; very variable among birds and may be used in individual recognition. 'Trilling calls' given by flying flocks in the distance described as 'something between the screams of gulls in the distance and a siren' (Ferdinand 1969), or 'screams of hysterical, witch-like, laughter' (Bateson 1961); they appear to fluctuate in tone, perhaps because of changes in the angle of the birds to the observer as they wheel. 'Aggressive calls' are a series of hoarse, unmusical notes given in pairs on ground and lasting a few seconds; start and end of sequence consist of shorter notes, and middle part involves a duet between the two birds. 'Clucking call' is audible only up to a few metres, given by pairs perched together on the surface and forming a duet of alternating notes. This call can apparently escalate into the 'aggressive call'. The 'snarling call' is a third form of duet, in which one bird con-

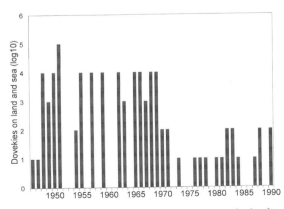

1   The occurrence of major Dovekie wrecks in the NE U.S. since 1945.

tinues 'clucking' and the other gives the 'snarling call', a rhythmic warble of very rapid, guttural sounds.

CHICK: 'a series of hoarse, wheezy whistles' (Cramp 1985). Begging call, a loud, monotonously repeated 'tick, tick,tick' is given once the chicks are half grown (Lovenskiold 1954). Chicks separated from their parent at sea give shrill peeping call (M. S. W. Bradstreet, personal communication).

Habitat, food, and feeding behaviour
MARINE HABITAT
Keeps to areas influenced by ice almost throughout the year and found from coastal to offshore waters. Non-breeding and post-breeding adults from the colonies in NW Greenland concentrate in the pack-ice zone off the E coast of Baffin I. in late summer (Brown 1980). In winter, mainly offshore, except when driven inshore by bad weather. In the W Atlantic, winter, associated especially with upwelling zones at the edge of the continental shelf and found especially along shelf edge off SE Grand Banks and Nova Scotia shelf (Bradstreet and Brown 1985). Not common away from continental shelf and slope.

FEEDING BEHAVIOUR
Usually found in flocks at sea. Follestad (1990) found birds widely scattered in offshore waters in the Norwegian Sea, but concentrated into dense patches in coastal waters. Feeds up to 100 km from colonies, but birds with chicks within 20 km, most within 5 km (Bradstreet and Brown 1985). Reported dive times rather short: mean 24.5 ± 9.0 sec (maximum 41 sec, $n = 35$, Brown in Bradstreet and Brown 1985). Distance travelled underwater up to 25 m.

DIET
Mainly zooplankton, ranging in size from copepods (<3 mm) to euphausiids and amphipods (<30 mm) and larval fish, especially Arctic cod. Little information available outside the breeding season.

ADULT SUMMER: in Baffin Bay, in May and July, mainly copepods (*Calanus* spp.), of which the commonest was *C. glacialis*, with some amphipods in July. In Aug 1978, mainly young-of-the-year Arctic cod, with some copepods and amphipods, especially *Parathemisto libellula*; in Aug 1979, mainly *Parathemisto* sp. Subadults (birds without bursae) collected in July–Aug 1978 had fed mainly on the amphipod *Apherusa glacialis*. Recently fledged juveniles, collected in late Aug 1979, contained mainly amphipods: proportions by dry weight, *Parathemisto* 60%, *Apherusa* 14%, *Onisimus* 6%, remainder mainly young cod, like the accompanying adults (Bradstreet 1982b). Some preference for *Parathemisto* evident, it being much commoner in diet samples than in the water column (Bradstreet and Brown 1985). In Spitzbergen diet appears variable: Lydersen *et al.* (1985) found 82% of adults examined at Hornsund (n = 11), in a sheltered fjord, contained small (<40 mm) Arctic cod, 64% *Parathemisto* spp., 45% snailfish (*Liparis* spp.), and 27% the pteropod *Helicina* sp.; in offshore waters, among ice, copepods and amphipods were the commonest prey, with 71% (dry wt) comprising the pelagic amphipod *Apherusa glacialis* (Mehlum and Giertz 1984). Lovenskiold (1954) and Norderhaug (1980) reported that adults and chicks in Spitzbergen fed mainly on euphausiids and copepods. Breeding adults consume 80% of their body mass in zooplankton each day (Gabrielsen *et al.* 1991).

ADULT WINTER: birds collected off southern Baffin I. in Mar and Apr contained principally the amphipod *Parathemisto libellula* (Smith and Hammil 1980).

CHICKS: copepod crustacea, especially *Calanus finmarchicus*, dominate diet everywhere. In five different studies, in Greenland, Spitzbergen, and Novaya Zemlya, the only other significant organisms (>10% by volume) were euphausiids (Novaya Zemlya) and amphipods (Thule District, Greenland) (Bradstreet and Brown 1985).

Displays and breeding behaviour
Account mainly from Ferdinand (1969) and Evans (1981).

COLONY ATTENDANCE AND FLIGHT DISPLAYS: mass flights over the breeding colony are a striking feature of Dovekie behaviour. Although they may be initiated by predator disturbance, they also occur periodically in the absence of predators, when they presumably have a social function. When Dovekies first arrive in Thule District, Greenland, they visit their colonies in huge flocks, arriving on calm days, just after midnight, and usually in mid-May (Salomonsen 1950) and circling first over the sea, at an altitude of 200–300 m and then above their breeding grounds, sometimes for hours (Ferdinand 1969). This behaviour is reminiscent of Least and Crested Auklets under similar circumstances. Circling flights over the colony continue throughout the breeding season and are usually accompanied by calling, but are sometimes carried out in complete silence, except for the rushing of wings. They are commonest in the early hours of the morning. The 'rushing flights' usually involved <50 birds which took off and landed together (Evans 1981). Larger flocks usually vocalized. Attendance at a colony in Spitzbergen was little affected by time of day, but in Upernavik District, Greenland, at a lower latitude, numbers on land peaked between 06:00 and 14:00 hrs (Evans 1981) and in Thule District from midnight to mid-day (Ferdinand 1969). Non-breeders visit colony during incubation and chick-rearing, but attendance of 'brown-winged' (? first years) ended by late July in NW Greenland, and older non-breeders were scarce during the latter part of chick-rearing (Roby *et al.* 1981).

SEXUAL BEHAVIOUR: before hatching, pairs frequently perform 'head-bowing', facing one another and bowing several times in quick succession, continuing for up to 1 min. After hatching this display is superseded by 'head-wagging', in which head is moved rapidly from side to side for up to 30 sec, accompanied by 'clucking', bill touching, and fluttering the wings. Birds hold themselves horizontally, with tail cocked. This display may also include allopreening, the bird doing the preening raising the feathers of the crown. In both head-bowing and head-wagging displays, the white feathers above the eye are very prominent. Head-wagging is sometimes preceded or followed by a 'butterfly flight' by one or both of the pair; flying with slow, deep wing-beats. Two other posture displays seen in social groups are an upright display, in which head is thrown back and bill pointed upwards, and a ritualized walk with back arched, and bill pointing down. The latter seems to be an appeasement display, given by subordinate birds.

Breeding and life cycle
Account mainly from Stempniewicz (1980), Norderhaug (1980), and Evans (1981).

BREEDING HABITAT AND NEST DISPERSION: breeding occurs in colonies ranging in size from about 1000 to several million pairs, usually situated in areas of extensive boulder screes on slopes of 25–35° either on the coast, or on nunataks (rocky outcrops surrounded by glaciers) or mountainsides up to 30 km inland. May nest up to 400 m above sea level. Nests tend to be aggregated into sub-colonies ranging in size from a few tens to many thousands of pairs on patches of suitable ground; birds from these sub-colonies tend to keep in flocks together, landing or taking off in synchrony and without reference to other sub-colonies.

NEST: most nests are among boulders and talus, with a density of 0.3–1 nest/m^2, and may be 0.3–1.0 m below the surface. Prefer boulders 0.5–1.0 m in diameter. Egg laid on bare rock in depression formed from collection of small pebbles which may be brought from outside the nest. A prominent rock, close to entrance, is used as display and launch pad for taking off, and is vigorously defended.

EGG-LAYING: may arrive at colonies as early as Mar (Novaya Zemlya); in Greenland in early May, when they are usually snow-covered. Egg-laying begins in second half June in all areas, continuing until early July, with peak laying 5–6 days after first eggs. Median laying in Upernavik District, Greenland 25 June ($n = 37$).

EGGS: one, somewhat bluish white, unmarked, rather spherical compared with most auks. Measurements: Spitzbergen (Norderhaug 1980), 49.0 (44.9–56.3) × 34.2 (30.9–39.2) mm ($n = 195$); W Greenland (Evans 1981) 48.2 ± 1.8 × 33.2 ± 1.1 mm ($n = 24$). Eggs lose 19% of initial weight during incubation.

INCUBATION AND CHICK-REARING: shared equally by the pair, with about four incubation exchanges daily. Duration 29 ± 0.8 days (range 28–31, $n = 38$, Stempniewicz 1980). Egg takes 2–4, sometimes as much as 7, days to hatch from first starring of the shell. Chicks brooded continuously for first 2 days, and intermittently up to 10 days, being kept against the parent's flank, with bill of larger chicks protruding under the wing (Stempniewicz 1980). Meals delivered to young chicks at Cape Atholl, NW Greenland by ♂♂ averaged 3.1 ± 1.2 g ($n = 57$) and by ♀♀ 3.3 ± 1.1 g ($n = 56$). At Siorapaluk, also in NW Greenland, ♂♂ delivered 3.9 ± 1.3 g ($n = 57$) and ♀♀ 3.7 ± 1.3 g ($n = 35$). Mean for all NW Greenland breeders 3.5 ± 1.2 g (range 1.0–6.5 g, $n = 204$, Roby *et al.* 1981). Chicks received an average 4–6 feeds/day in W Greenland; highest rates occurred in the darkest hours, between 20:00 and 08:00, peaking at 20:00–02:00 and with the pattern intensifying during the later part of the chick period (Evans 1981). In Spitzbergen, feeding rates averaged 8.5/chick (6–11, $n = 199$, Norderhaug 1980). Food is regurgitated from parent's throat pouch into mouth, from where it is taken by chick; both parents feed. ♀ contributes 64% of meals, but ceases to provision about 5 days before the chick fledges, after which ♂ does all feeding (Taylor and Konarzewski 1992).

CHICK GROWTH AND DEVELOPMENT: comparatively rapid. The vane of the longest primary

begins to develop at 5–6 days and contour feathers are growing by 9–11 days, when down begins to be lost from the face. Down has gone from the head, and most from back and wings, by 15 days, at which age the chick begins to exercise at the mouth of its nest cavity. All down has gone by 23 days, the earliest age at fledging. Chicks hatch at 21.5 g (range 16–25, $n = 84$), reach 125 g at 21 days (77% of adult weight), and fledge at 114.3 g (100–143), $n = 41$, 69% adult weight, Norderhaug 1980). Water content falls by 50% during growth period. At 23 days, they have the highest proportion of fat outside of the petrels. Their fat content falls steeply in the last 4 days before fledging (Taylor and Konarzewski 1989). Chicks are able to digest food with 81% efficiency at 20 days, but development may be constrained by level of calcium in the diet. The proportion of this element in their body tissues continues to increase right up to departure (Taylor and Konarzewski 1992).

COLONY DEPARTURE: duration of nestling period in Greenland averages 28.3 days (Evans 1981), in Spitzbergen, 27 ± 1.8 days (range 23–30, $n = 33$). Departure usually occurs during the darkest period, between 21:00 and 02:00 chicks fly from the nest-site, usually accompanied by one or more adults. As the fledglings at sea are usually accompanied by an adult ♂ (39/40 collected, Bradstreet 1982*b*), we presume that one of the adults departing with the chick is the ♂ parent, but this has not been conclusively observed. Departure is more synchronized than hatching, with most chicks leaving within 2–3 days. Contagious calling by the chicks appears to synchronize the departure. Predation of departing chicks by Glaucous Gulls and Parasitic Jaegers (*Stercorarius parasiticus*) can be heavy; in Spitzbergen, 90% ($n = 79$) of chicks attacked were killed (Stempniewicz 1980). Although chicks remain close to their attendant adults once at sea (M. S. W. Bradstreet, personal communication), it is not known for sure whether the adults feed them, or merely guide them to suitable feeding areas. Weight continues to increase for at least a month after leaving the nest. In 1976, a normal year, juvenile ♂♂ collected at sea in Baffin Bay weighed 150 ± 13 g (*n* = 20) in Aug and 160 ± 9 g (*n* = 31) in Sept. Corresponding weights for ♀♀ were 145 ± 13 g (*n* = 17) and 158 ± 12 (*n* = 22). Wing and culmen length also increased slightly after leaving the nest, but tarsus length did not (Bradstreet 1982*b*).

BREEDING SUCCESS: hatching success was 65% and fledging success 77% in W Greenland (Evans 1981).

PREDATION: adults are commonly taken on the colony by Arctic Foxes and Glaucous Gulls and less frequently by Gyrfalcons and Snowy Owls. Foxes excavate loose scree and eat eggs and chicks as well. Their predation is increased by their habit of caching food for use after the birds have left. Glaucous Gulls divide colonies among territory-holders; they switch almost entirely to fledging chicks as soon as their exodus begins (Stempniewicz 1980). Polar Bears have

been observed digging out breeders in Franz Joseph Land (Stempniewicz 1993).

MOULT: a complete post-breeding moult follows immediately after leaving the breeding colonies; it may start as early as Aug for non-breeders and is finished by Oct. Juveniles undergo a limited post-juvenile moult into winter plumage, involving feathers of chin and face at the same time. A pre-breeding moult (Pre-alternate) occurs in spring and involves replacement of the body plumage on neck and head. The timing of this moult is poorly known, but it probably occurs in Mar and Apr for most birds, later for first years (Bédard 1985). First years also begin post-breeding (Definitive Pre-basic) moult earlier than breeders; some collected in Aug had nearly completed primary moult (Bradstreet 1982*b*).

Population dynamics
Nothing is known about survival, site-fidelity, age at first breeding, or dispersal in this species. An excellent opportunity for research.

Black Guillemot or Tystie   *Cepphus grylle*

Alca grylle (Linnaeus, 1758, *Syst. Nat.*, ed.10, 1: 130—Sweden).   PLATE 2

Description
Medium-sized auk with a rather small, slender bill. The mainly black summer plumage is not found in any other Atlantic auk. It runs and takes off easily on land and is a strong flier, often performing aerobatic changes of direction and flying high into the air at steep angles during pair chases. On the water, it rides buoyantly, sometimes with tail cocked, and takes flight easily.

ADULT SUMMER: completely black except for a broad white wing patch on the greater and lesser coverts and with a white underwing, except for the tips and outer webs of primaries and secondaries; iris dark brown or black, legs and feet coral red, bill black but inside of gape crimson.

ADULT WINTER: mostly white mottled with sooty brown, especially on back, scapulars, and uppertail coverts; silvery grey-brown on head; greater and lesser coverts mainly white with some brown speckles, mostly white below, except for some pale brown flecking on upper breast and scattered brown flecks on belly. Intermediate plumage, with variable amounts of brown and white mottling on belly, occurs during primary moult and persists for some time after its completion, being seen in some individuals as late as Feb. Plumage very variable among different geographical areas (see Variation).

JUVENILE AND FIRST WINTER: similar to non-breeding adult, but feathers of upperwing coverts with brown tips, giving the covert patch brown/white mottling, and belly feathers also mottled with grey-brown. Feet and bill brownish and mouth lining a pale salmon.

CHICK: down entirely sooty black; bill and feet black, but with paler webs.

VARIATION

Up to seven subspecies varying in size and amount of white in the non-breeding plumage (Salomonsen 1944; Storer 1952). Variation does not appear to be clinal and, unlike most Atlantic auks, the Arctic populations are not consistently larger than those of boreal areas. Birds from the Baltic Sea (*C. g. grylle*) consti- tute a distinct population, having the longest wings. There is a sharp discontinuity (90% non-overlap in wing-length) between this population and that of Denmark and the Kattegat (Asbirk 1979c). Populations in the eastern Canadian Arctic and NW Greenland (*C. g. ultimus*) and the high Arctic of Spitz- bergen and the Arctic Ocean (*C. g. mandtii*) are all very white in winter (and are combined under *mandtii* by Cramp 1985), compared with those from the Gulf of St Lawrence and the Canadian maritime provinces, and the sub-Arctic of the eastern Atlantic (*C. g. arcti- cus*). Among the whiter group, those from the eastern Canadian Arctic have smaller bills, relative to their wing-length, than those from the high Arctic. Icelandic birds (*C. g. islandi- cus*) are small, and dark in winter (Ingolfsson 1961), but Bédard (1985) considered them similar to birds from eastern North America on the latter criterion. Those from the Faeroes (*C. g. faeroensis*) have the shortest wings and are sometimes considered distinct from, but more often included with, British birds as *C. g. atlantis*; British birds also have short wings and long bills. Alternatively, Cramp (1985) places British birds with those from the western Atlantic under *arcticus*, while keeping *faeroensis* distinct. A colour form that is all-black in summer plumage and exhibits variable melanism in winter has been reported from W Greenland (Salomonsen 1941).

MEASUREMENTS

No sexual dimorphism, except in bill depth. See Table: (1) *C. g. grylle*, eastern Sweden; (2) *C. g. atlantis*, Denmark; (3) Britain (Storer 1952); (4) *C. g. mandtii*, Spitzbergen (Salo- monsen 1944); (5) Alaska (G. Divoky, per- sonal communication); (6) *C. g. islandicus*, Iceland, adults; (7) immatures (Ingolfsson 1961); (8) *C. g. arcticus*, Gulf of St Lawrence (Storer 1952). First summer birds have shorter wings than older birds: Shetland, 154 $\pm$ 3.3 ($n = 14$), versus 162 $\pm$ 3.4 mm ($n = 68$, Ewins 1992).

WEIGHTS

As in most auks, body weights of incubating birds are higher than those of birds rearing chicks. Icelandic breeders, collected in June (pre-breeding or incubation), averaged 427 $\pm$ 39 g (378–489, $n = 13$) and immatures 431 $\pm$ 45 g (359–496, $n = 12$). Means for breeders in Denmark were 375 $\pm$ 23 g ($n = 158$) for incu- bating birds and 359 $\pm$ 23 g ($n = 32$) for birds rearing chicks (Asbirk 1979c). During incub- ation, there was no difference between $\male\male$ (376 $\pm$ 24 g, $n = 40$) and $\female\female$ (386 $\pm$ 18 g, $n = 24$). Breeders in Alaska (G. Divoky, personal communication) averaged: $\male\male$ 383 $\pm$ 22 g (316–455, $n = 421$); $\female\female$ 388 $\pm$ 28 g (315–525, $n = 450$).

Range and status
RANGE

Circumpolar in Arctic waters, boreal and sub- Arctic in Atlantic only, breeding from 43° to 78° N in the W Atlantic and from 50° to 80° N in Europe. Probably occurs further N than any other auk, with records to the 85th parallel (Kozlova 1957). Partially or mainly migratory in Arctic, Norway, and Baltic Sea, but largely sedentary in southern parts of its range, although post-fledging dispersal may involve movements of up to 500 km in U. K. (Ewins and Kirk 1985). In Shetland, adults move from exposed coasts in summer to sheltered inlets in winter. Those from Finland winter mainly on the Swedish coast of the Kattegat (Andersen-Harild 1969). Band returns indicate

Table: Black Guillemot measurements (mm)

♂♂	ref	mean	s.d.	range	n
wing	(1)	174	4	169–180	16
	(2)	163	4	156–168	29
	(3)	164	3	158–172	33
	(4)	167	4	159–171	10
	(5)	174	4	157–185	416
	(8)	164	3	161–170	17
culmen	(1)	32.8	1.1	30–35	16
	(2)	31.4	1.5	29–35	29
	(3)	33.3	2.0	29.5–37.5	26
	(4)	31.3	1.1	29–32.5	9
	(5)	29.6	1.6	25.2–38.0	412
	(8)	31.7	1.2	29.5–34.0	17
bill depth	(1)	11.5	0.5	11–12	17
	(2)	11.3	0.5	10.5–12	29
	(3)	9.8	0.4	9.1–10.6	25
	(4)	9.7	0.4	9–10.2	9
	(8)	9.8	0.4	9.1–10.6	16
tarsus	(3)	31.4	1.3	28.0–33.5	28
	(8)	32.0	1.1	30.0–33.5	18

♀♀	ref	mean	s.d.	range	n
wing	(1)	177	4	169–182	6
	(2)	162	4	158–173	22
	(3)	165	3	158–172	23
	(4)	168	4	161–174	9
	(5)	174	4	160–187	441
	(8)	166	4	158–173	29
culmen	(1)	33.2	1.0	32–35	6
	(2)	31.4	1.0	20–33	24
	(3)	33.2	1.6	31.0–36.5	24
	(4)	31.7	1.0	30.5–34.0	9
	(5)	29.2	1.4	25.1–36.6	433
	(8)	31.3	1.7	29.0–35.5	29
bill depth	(1)	11.2	0.5	10.5–12	6
	(2)	10.9	0.6	10–12	25
	(3)	9.4	0.6	8.4–10.7	22
	(4)	9.5	0.5	8.7–10.2	9
	(8)	8.5	0.4	7.5–10.3	29
tarsus	(3)	31.7	1.0	29.5–33.5	24
	(8)	31.7	1.1	30.0–34.0	30

SEXES COMBINED	ref	mean	s.d.	range	n
wing	(6)	160	3.3	152–169	36
	(7)	153	3.1	149–160	18
culmen	(6)	30.5	1.5	26.4–33.8	36
	(7)	30.2	1.4	26.0–31.7	18
bill depth	(6)	10.5	0.7	9.2–12.5	36
	(7)	9.5	0.8	8.2–10.6	18
tarsus	(6)	32.2	1.3	29.4–34.7	36
	(7)	32.5	1.1	30.8–35.0	18

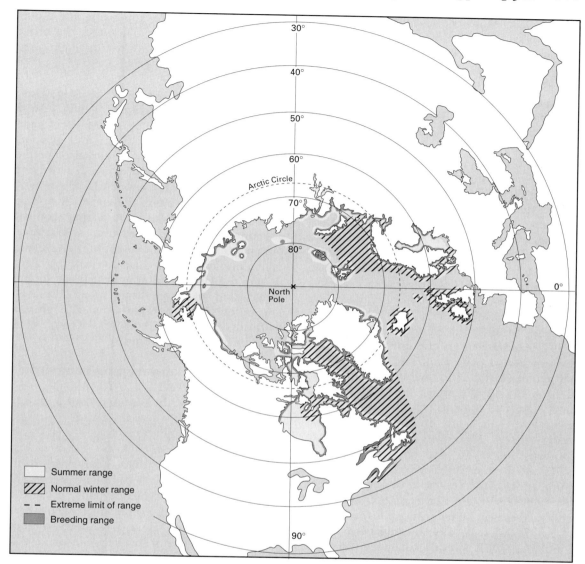

Summer range
Normal winter range
Extreme limit of range
Breeding range

some birds from the Gulf of St Lawrence move to Newfoundland waters in winter, while some pre-breeding birds from Iceland move to SW Greenland. Averages of >100 birds have been recorded on CBCs in Maine, Newfoundland, and offshore from Massachusetts (Stellwagen Bank). Numbers in all these areas suggest increases over local breeding populations. Some birds remain throughout the winter as far N as 78° N in Baffin Bay (Renaud and Bradstreet 1980) and off Novaya Zemlya, at 75° N (Kozlova 1957).

STATUS
Widespread and common throughout most of its range, but rarely seen in large numbers. Largest colonies (up to 10 000 pairs) are found in the high Arctic, especially in areas where land-fast ice or heavy pack-ice persists until late summer (Nettleship 1974). Nettleship and

Evans (1985) estimated the Atlantic population at 200 000–350 000 breeding pairs, with a best estimate of 266 000. There are probably no more than a few thousand in areas not covered by their surveys (Ewins *et al.* 1993). Numbers were distributed as follows: eastern Canadian Arctic 19%, Iceland 18%, Greenland 15%, Russian Arctic (Kola Peninsula, White Sea, Franz Joseph Land, Novaya Zemlya, and Arctic coast of Siberia) 15%, Denmark, Sweden, and the Baltic 8%, Norway 6%, Spitzbergen 6%, Newfoundland and Labrador 6%, eastern North America (New England, Canadian maritime provinces, Gulf of St Lawrence) 4%, British Isles 3%, Faeroes 1%. However, these numbers are very approximate, as most of the population breeds in small groups scattered along very extensive coasts and, in the Arctic especially, these have not been surveyed in detail. For Hudson Bay, Hudson Strait, and Ungava Bay, Gaston (1991) estimated a minimum of three times the number of pairs estimated by Nettleship and Evans (1985). The total world population is probably in the range 250 000–500 000 pairs. Because the species is hard to census accurately, there is little information on population trends. Moreover, non-breeders can comprise up to 50% of birds present at a colony in summer (G. Divoky, personal communication). Believed to have been more or less stable in Britain between 1969 and 1987 with increases in some areas and decreases in others (Lloyd *et al.* 1991). A decrease in the population of SE Sweden, following the introduction of Mink (*Mustela vison*), was accompanied by an increase in Denmark, where Mink were not established (Asbirk 1979*c*). The small (500–750 pairs) population of northern Alaska has probably doubled in the last 30 years owing to occupation of recent strandline litter.

Field characters

Because of the white patch on the upperwing coverts, not easily confused with other auks except, in the small area of overlap, with Pigeon Guillemot, which has a dark bar across the white patch on the upperwing and a dark underwing. Far off, in silhouette, the broad wings, fast, but rather fluttering flight, and small head and bill are good clues.

Voice

Taken mainly from Preston (1968), Asbirk (1979*b*), and Cramp (1985). A variety of high-pitched screaming, whistling, and peeping calls. On the approach of a human or predator, birds on land give a prolonged, screaming whistle, with bill wide open, which may also be given in flight, or on the water. A similar, more intense, whistle is given during agonistic encounters. During courtship, a plaintive, piping whistle is given, sometimes rising and falling in pitch. This is also given in the head-tossing display. A twittering whistle accompanies certain aggressive displays (see Displays). During the circling display, the male gives a loud, disyllabic peep. A short, staccato call is given during head-bobbing.

Habitat, food, and feeding behaviour
MARINE HABITAT

An inshore feeder in summer, found mainly along rocky coasts where it feeds in water less than 30 m deep and takes its prey mainly on the bottom. In boreal and sub-Arctic waters, not often seen more than 3 km away from land, but in the Arctic it commonly occurs far offshore among floating ice pans, or at the edge of land-fast ice, often over deep water (Bradstreet 1979). Found offshore in Sept even in the absence of ice (Sutton 1932; Nettleship and Gaston 1978). Occurs offshore in winter in areas of mobile pack-ice in Davis and Hudson Straits and northern Hudson Bay (Gaston and McLaren 1990). Population densities are highest in areas where ice persists into the summer; many birds feed in narrow leads in the ice in such situations. Elsewhere, favours areas with numerous small, rocky islands, submerged reefs, and sheltered waters, but may feed where there are strong tidal currents. May take advantage of eddies created by islands (Nol and Gaskin 1987). Frequently dives among kelp (*Laminaria*).

FEEDING BEHAVIOUR

FORAGING RANGE: in Shetland, and Denmark most breeders fed within 2–3 km of their nest-site during chick-rearing, but in northern Hudson Bay most fed within 5 km and some up to 13 km from the nest (Cairns 1987a).

DIVING BEHAVIOUR: dives to 50 m, but most benthic feeding is carried out in water <30 m deep. Dive bouts observed for breeding birds in northern Hudson Bay contained up to 21 dives, with a median duration of 73 sec (range 6–147 sec, $n = 47$, Cairns 1992b). Median inter-dive pause was 27.5 sec, but rose from about 10 sec for dives of <30 sec duration to more than 40 sec for dives in excess of 100 sec. Adult diet in the same area consisted of mid-water Arctic cod and benthic Arctic shanny (*Stichaeus punctatus*). Mean dive duration in Britain 45 sec (30–75, $n = 20$, Witherby *et al.* 1941).

DIET

In summer, mainly fishes, especially those living on rocky bottoms, including blennies and sculpins up to 30 g weight. Also takes sandlance and Arctic cod, large benthic invertebrates (annelids, *Pandallus* shrimps) and, especially in Arctic, midwater plankton (*Gammarus*, *Parathemisto*). Reported occasionally scavenging fish from fishing boats (Ewins 1987).

ADULT SUMMER: in northern Hudson Bay, 89% ($n = 46$) of stomachs collected in the breeding season contained fish: 52% Arctic cod, 24% Arctic shanny, and 9% four-lined snake-blenny (*Eumesogrammus praecisus*); 52% contained crustacea, including mysids, amphipods, and decapods; 17% contained annelid remains (Gaston *et al.* 1985). In Iceland, of adults collected near breeding sites in summer, 39% ($n = 38$, Petersen 1981) contained annelid remains, 42% shrimps, and 37% butterfish (*Pholis gunnellus*) (the most numerous item delivered to chicks). No other fish was present in more than 11% of birds. At Hornsund, Spitzbergen, 60% of birds ($n = 20$)

contained Arctic cod, mainly <50 mm long, 55% *Parathemisto*, 25% the shrimp *Margarites groenlandicus*, 35% *Mysis oculata*, and 35% the large benthic amphipod *Gamarellus homari* (Lydersen *et al.* 1985); the latter frequently taken in the high Arctic, and sometimes fed to chicks (AJG).

ADULT WINTER: little information. In the Arctic, may forage especially underneath sea-ice, taking amphipods on the under surface of the floes.

CHICKS: apart from a few large benthic crustacea (e.g. *Gammarellus homari*) they are fed entirely on fish. See Table: (1) Bay of Fundy (Preston 1968); (2) Gulf of St Lawrence (Cairns 1978); (3) Nuvuk Is., northern Hudson Bay (Cairns 1987b); (4) Coats. I., northern Hudson Bay (AJG); (5) Iceland (Petersen 1981); (6) Denmark (Asbirk 1979d); (7) Gulf of Finland (Bergman 1971).

Displays and breeding behaviour

Taken chiefly from Preston (1968) and Asbirk (1979b,d). Breeding birds during the pre-laying period and non-breeders throughout the breeding season, spend a substantial part of the day socializing, either on the water close to the breeding sites, or on rocky ridges or large boulders adjacent to sites. A perch close to the nest-site opening ('perch rock' of Preston) is used for pair displays and is frequently occupied by the off-duty pair member during incubation.

COLONY ATTENDANCE: Colony attendance occurs throughout the winter, provided weather is good, in some parts of the range (U.K., Greenwood 1987). In Shetland, birds come ashore at colonies from Dec onwards, but only about 20% of breeders attend until Feb, rising to 80% by Mar. During this period attendance is depressed by strong winds, but is not affected by tidal state. In Apr, peak attendance occurs between 04:00 and 10:00, with a smaller peak at 17:00–20:00 (Ewins 1985). At Kent I., New Brunswick, breeders arrive early Apr but do not come ashore until early May

Table: Black Guillemot chick diets

ref n	(1) 500	(2) 548	(3) 618	(4) 30	(5) 609	(6) 113	(7) ?
INVERTEBRATES	3	—	—	—	4	—	—
BENTHIC FISH	93	46	85	—	73	78	95
butterfish (*Pholis gunnellus*)	68	18	—	—	56	67	—
Arctic shanny (*Stichaeus punctatus*)	—	28	49	—	—	—	—
fourlined snake-blenny (*Eumesogrammus praecisus*)	—	—	24	—	—	—	—
fish doctor (*Gymnelus viridis*)	—	—	12	—	—	—	—
sculpin (*Myoxocephalus* spp.)	18	—	—	—	17	—	—
radiated shanny (*Ulvaria subbifurcata*)	9	—	—	—	—	—	—
eelpout (*Zoarces viviparus*)	—	—	—	—	—	4	95
MIDWATER FISH	5	40	—	100	23	22	5
Tomcod (*Microgadus tomcod*)	—	33	—	—	—	—	—
sandlance (*Ammodytes* spp.)	—	7	—	100	15	22	5
rockfish (*Sebastes* spp.)	—	—	—	—	4	—	—
herring (*Clupea harengus*)	2	—	—	—	—	< 1	—
snailfish (*Neoliparis* spp.)	3	—	—	—	—	—	—

% by number

and then mainly at high tide. In Iceland, breeders arrive off breeding islands in early May; peak attendance there occurs at 04:00–08:00 and 20:00–22:00, varying with weather, sea conditions, and tidal state (Petersen 1981). In northern Alaska, breeders are present near their colonies from Apr, but do not come ashore until snow clears in early June (G. Divoky, personal communication). Peak attendance there is from midnight to 04:00. Pre-breeding 1- and 2-year-olds begin to arrive in May.

AGONISTIC BEHAVIOUR: threat takes the form of the 'twitter-waggle' in which the bird stands obliquely to its antagonist, bill open and pointed downward, giving a low twittering whistle, while wagging head from side to side. Wings and back feathers may be raised and tail cocked. This display appears to be given by birds not confident of their superiority. Where greater confidence is warranted, a hunched posture is used, with neck withdrawn and bill open, pointing downwards; wings are held a little away from the body, in readiness. This threat may escalate to an attack in which the attacker runs at the other bird, head held low and bill pointed forward and agape, showing the scarlet lining, while uttering the screaming whistle. Such rushes may be followed by pecking, biting, wing-beating, and scratching with the feet, or both birds may take off and indulge in an aerobatic pursuit.

SEXUAL BEHAVIOUR: social groups congregate on the sea close to breeding sites, on rocks just above high tide, or on ledges or boulders close to sites. Pairing probably occurs at the nest-site. Pairs frequently sit or stand side by side. Both bob the head, keeping bills angled down, while facing and giving staccato calls. ♂ assumes an upright posture, with neck stretched and bill pointed slightly downward and struts around ♀ with a high-stepping gait, giving sharp calls, ♀ also executes a smaller circuit, in a hunched position. This may be followed by ♀ lying forward with neck held low and tail cocked, to solicit copulation. Following copulation, ♀ stands up and throws ♂ off.

Breeding and life cycle
Well known from studies in Denmark (Asbirk 1979*b*), Shetland (Ewins 1986), Iceland (Petersen 1981), New Brunswick (Preston

1968), the Gulf of St Lawrence (Cairns 1981), and northern Hudson Bay (Cairns 1987*b*).

BREEDING HABITAT AND NEST DISPERSION: usually breeds among boulders or talus, or on cliffs or rocky shores, within 50 m of sea; occasionally in burrows among sand dunes or soil, especially where exposed at the edge of eroding cliffs, or in cracks in stone walls. Will use beach debris, driftwood, and artificial boxes where natural sites are absent. Usually loosely colonial in aggregations of up to a few hundred pairs, with nest-sites 1–20 m apart. In Denmark, 37% of sites were more than 10 m from the nearest neighbour (Asbirk 1979*b*). Forms large colonies of up to 10 000 pairs in the Canadian high Arctic, where sites are mainly in crevices in cliffs.

NEST: in a cavity in a broad range of structures, the only requirement being overhead cover and a nest-chamber invisible from the entrance: usually in a crevice or under boulders. A shallow cup is formed from small pebbles and mollusc shell fragments, but no lining is used. At Kent I., New Brunswick, many sites were rearranged by winter storms, resulting in 32% of pairs relocating from one year to next (Preston 1968).

EGG-LAYING: usually the last auk to lay in a given area; laying somewhat more protracted than in other species. Mean date of first egg in clutch 24 May in Co. Down, Ireland (Greenwood and Marshall 1989), 24–26 May in Denmark (3 years, Asbirk 1979*b*), and 8–10 June in Shetland (2 years, Ewins 1986). Laying peaked 20 May–8 June in New Brunswick (means 28 May–8 June over 5 years) and estuary of St Lawrence, 14–22 June at St Mary's I., Gulf of St Lawrence. In Iceland, median laying 30 May–2 June (3 seasons), with 80% of laying taking place within 13–18 days. In Hudson Bay, laying begins about 24 June, with a peak in early July (Sutton 1932; Gaston *et al.* 1985); timing is similar in northern Alaska. Most chicks do not depart until first half of Sept. Mean interval between eggs in

two-egg clutches close to 3 days everywhere except Shetland, where 2.5 days (Ewins 1986). In Iceland 16% of intervals were 2 days, 71% 3 days, 9% 4 days (n = 245, Petersen 1981).

EGGS: clutch usually two, but one-egg clutches made up 14% in Hudson Bay, 15% in Denmark, 22% in New Brunswick. Occasional third, or even fourth, eggs indicate presence of two females, based on colour and markings. In New Brunswick, all clutches laid in the first week of laying were of two, but the proportion had fallen to 20% by the fourth and fifth weeks. Birds in their first year of breeding were more likely to lay one-egg clutches (27%, n = 15) than older birds (12%, n = 33, Preston 1968). Similar trends were observed in Denmark (Asbirk 1979*b*). Ground colour, cream, pale tan, pale grey, or pale bluish-green, usually with black spots, less commonly blotched and streaked with black or dark brown. Often show additional, incompletely expressed, markings. The range of colour and markings is smaller than that found among murres. Shape is elliptical-ovoid, much less pointed than murres. The two eggs are usually very similar, although not identical in ground colour and markings. In Iceland, first laid egg 2% and in Shetland 4% larger on average than second, but no difference in Denmark. Volume decreased with date from approximately 95 cm³ for early-laid eggs to 84 cm³ for those laid at end of season (Petersen 1981). Eggs in one-egg clutches smaller than those in two-egg clutches, and generally laid later. Replacement clutches are laid in a minority of cases where eggs are lost, the interval ranging from 14 to 28 days; mean in Iceland, 15 ± 4 days (n = 28), the eggs being smaller than those of the first clutch and similar to first clutches laid at the same date. Eggs lose 13% of initial mass by time of starring (Ewins 1986). See Table for measurements: (1) *C. g. mandtii* Spitzbergen (Makatsch 1974); (2) Alaska (G. Divoky, personal communication); (3) *C. g. arcticus* Sweden; (4) *C. g. islandicus* Iceland (Petersen 1981), (5) *C. g. atlantis* Denmark, two-egg clutches,

Table:　Black Guillemot egg dimensions

ref	length (mm)			breadth (mm)			weight (g)		n
	mean	s.d.	range	mean	s.d.	range	mean	s.d.	
(1)	58.6		56.1–63.5	39.3		38.0–40.9	51	3.9	18
(2)	57.5	2.3	50.9–67.8	38.8	1.1	34.9–42.3	47.2	3.6	347
(3)	59.7		53–68	40.6		36.6–47	52.3	4.2	102
(4)	58.2	2.2		39.2	1.1		49.7	3.9	607
(5)	57.5	2.1		38.9	1.1				204
(6)	57.7	2.0		38.9	1.1				200
(7)	56.9	2.4		38.3	1.6				57
(8)	58.3	2.2		39.6	1.2		49.9	3.9	51
(9)	57.4	2.0		39.1	0.9		47.9	3.2	44
(10)	56.7	2.5		38.3	1.0		46.4	3.8	6

first egg; (6) second egg; (7) one-egg clutches (Asbirk 1979*b*); (8) Shetland, two-egg, first egg; (9) second egg; (10) one-egg clutches (Ewins 1986). Egg weight is approximately 11% of adult weight and clutch weight approximately 22%.

INCUBATION: by both parents, initiated with laying of second egg; period 28 days (range 26–33 days, Preston 1968; Ewins 1986) or 29 ± 2 days (Petersen 1981). Most incubation shifts last less than 4 hrs, except overnight, when shifts are longer and either sex may incubate (Asbirk 1979*b*; Ewins 1986).

HATCHING AND CHICK-REARING: in New Brunswick and Iceland both chicks hatched on same day in more than half of broods, but where they did not, first laid egg usually hatched first (Preston 1968; Petersen 1981). In Shetland only 18% of brood-mates hatched on same day, 70% hatched 1 day, and 12% 2 days apart. Hatching order had no effect on departure order (Ewins 1992). Brooding is performed by both sexes and lasts less than 4 days in New Brunswick and Shetland, 5–6 days in high Arctic. Both parents feed the chicks. Rate of provisioning in Shetland rose from about 0.4 feeds/chick/hr during the first few days, to a peak of about 1 feed/chick/hr at 25–29 days, falling to 0.6 feeds/chick/hr by the

age of departure (Ewins 1992). In northern Hudson Bay, food intake likewise peaked at 25 days at 780 kJ/day (Cairns 1987*a*). Size of fish delivered increases with chick age. Some variation in feeding rate reported in relation to state of tide; may also be affected by the attentions of kleptoparasites. Kleptoparasitism of adults carrying food for chicks is common and involves Parasitic Jaegers, Herring and Lesser Black-backed Gulls (*Larus fuscus*), Black-legged Kittiwakes, and Arctic Terns. In northern Hudson Bay in 2 years, guillemots lost 8% and 22% of fish they attempted to deliver to attacks from Parasitic Jaegers (Birt and Cairns 1987). Guillemot chicks in Shetland at sites affected by Parasitic Jaegers, and large gulls grew more slowly than at sites where kleptoparasitism was absent (Ewins 1992).

CHICK GROWTH AND DEVELOPMENT: nestlings hatch at about 32 g. Weight increase becomes linear after 7 days, decreasing after 25 days, and reaching an asymptote at 30–35 days. Maximum and fledging weights reported vary widely, from 60 to 101% of adult weight. Pin feathers of primaries and coverts begin to burst at 15 days. By 30 days down remains only on mantle and head. Chicks fledge at 83% of adult wing length and grow to 96% of adult wing length within 1 month. Chicks in single chick broods reach higher asymptotic and fledging weights than those in two-chick broods (Shetland 428 g versus 396 g, Ewins 1992). First hatched chicks do not obtain any advantage except where the interval between hatchings is 48 hrs or more, in which case they reach higher weights. In Shetland, mean age at fledging is 33–36 days; chicks leave without parents, probably in the evening and either fly or flutter to the water. They quickly move away from the colony area and may disperse up to several hundred kilometres within the first month, mostly staying away from coastal waters. See Table for age and masses at fledging (broods of two only): (1) Shetland (Ewins 1986); (2) Iceland (Petersen 1981); (3) Denmark (Asbirk 1979*b*).

BREEDING SUCCESS: published figures are mainly in the range 0.5–1.0 chicks/pair, but may be biased by the effects of observer disturbance. At 'lightly disturbed' sites at St Mary's I. Gulf of St Lawrence, 1.03 chicks/pair fledged (Cairns 1981).

PREDATION: Predation of breeding adults by Peregrine Falcons, Gyrfalcons, Great Skuas (*Catharacta skua*), Snowy Owls, Ermine (*Mustela erminea*), and Mink reported.

MOULT: adults have a complete post-breeding moult beginning between July and Sept and with the flight feathers completely grown by early Nov. Primary moult may begin before body moult and includes a flightless period of about 5 weeks, during which birds may occur in flocks (Ewins 1988). In Shetland these flocks often occur on sheltered waters, but in northern Hudson Bay they tend to remain 10–25 km offshore (Sutton 1932). Post-breeding body moult may begin while still feeding chicks, and is sometimes arrested when partially complete; may overlap with the pre-breeding body moult that commences any time from Nov onwards in Britain. In Denmark, winter plumage uniformaly present only in Dec and Jan. Pre-breeding moult very rapid, in second half of Jan to late Feb. Flightless period for adults during Aug–mid-Oct (Asbirk 1979*b*). Regional variation occurs; Baltic birds begin pre-breeding moult 2 months later than Danish birds. Two Norwegian birds had fresh breeding plumage in Nov and Dec, whereas no British birds had breeding plumage before Feb. In Hudson Bay, pre-breeding moult does not begin before Apr and full summer plumage is not usually seen before mid-May (Sutton 1932). Juveniles undergo a post-juvenile (Pre-alternate I) body moult soon after leaving the nest. First year birds complete moult to first summer plumage only in May–June, retain worn wing and tail feathers.

Population dynamics

SURVIVAL: estimated as 87% in Iceland, based on the return of banded breeders. Survival to age at first breeding there was 46%, but the average of other banded populations is 24% (Petersen 1981). Annual survival in northern Alaska ranged from 75 to 95% (mean 85%, G. Divoky, personal communication): the population of that area was clearly limited by the availability of breeding sites. In Denmark, survival in an expanding population was 84%

Table: Black Guillemot age and weight at fledging

ref	age (days)				weight (g)			
	mean	s.d.	range	n	mean	s.d.	n	% adult
(1)	36.1	2.8	32–53	105	396	44	66	85%
(2)	33.1	2.0		108	356	25	86	89%
(3)	39.5	3.9		37	361	48	73	101%

(n = 204 bird-years) and in New Brunswick 81% (Preston 1968).

PAIR- AND NEST-SITE FIDELITY: in Maine and Denmark approximately 30% of breeders changed site from year to year; high for an auk. However, in northern Alaska, mate and breeding site fidelity was >95% (G. Divoky, personal communication). Changes were more frequent at unsuccessful sites; relocations were usually within 30 m of former site. Divorce was observed for 7% of surviving pairs (n = 42). Sexes appear equally likely to remain at a site in the absence of their previous mate.

Pigeon Guillemot   *Cepphus columba*

Cepphus columba (Pallas, 1811, Zoogr. Russo-Asiatica 2: 348—Kamchatka)   PLATE 2

Description

A Pacific species very similar to the Black Guillemot of North Atlantic and Arctic seas. Trim, graceful, and elegant birds with an upright posture, and with neck rather long and usually held extended. Stand on tarsi and toes or more rarely on toes alone. Bill rather long and slender, almost symmetrical and dagger-shaped, tapering to a fine point. Irides usually barely contrasting with the dark face but on some birds there is a hint of a lighter brown or whitish eye ring. Mouth lining bright red. Sexes apparently identical. Flight usually low near the sea surface, with shallow rapid wing-beats. Wings are rather rounded and broadest at the smooth division between secondaries and primaries.

ADULT SUMMER: sooty brownish-black overall, set off by flashy white upperwing patches formed by entirely white lesser coverts, and white and white-tipped greater coverts. Unlike Black Guillemots, only inner half of greater coverts are completely white; outermost greater coverts are black with white tips. On birds viewed in profile, this creates an oval white wing patch with a black wedge-shaped intrusion on its lower edge. Some birds, particularly in western Alaska, have additional dark feathers among the wing coverts, creating a second dark bar. In the subspecies *C. c. snowi* of the Kuril Is. only tips of greater coverts, lesser coverts, and outermost wing coverts are white, giving these birds a dark upperwing with several narrow white bars (Storer 1952). When flying, the eastern subspecies (*eureka, adianta*) upperwing patches appear almost divided into two white crescents, the larger formed by the lesser coverts and upper wing coverts. Flight feathers are blackish-brown all year, but bases of underside of primaries are glossy greyish or even whitish in some populations (Storer 1952). Underwing coverts blackish brown year round Bill black; irides dark brown; legs, feet, and webs bright coral red.

ADULT WINTER: underparts and head mostly white with some faint dark flecking; blackish-brown plumage retained around crown and forehead and around eye, which is usually narrowly ringed with white; fine brownish flecking extends from behind eye onto nape and down back of neck; lower back of neck and back blackish brown with pale edging; flight feathers and lesser coverts blackish brown as in summer, but wing coverts develop additional white edging in winter, making white patches less distinct. Legs, feet and webs pale red or pinkish.

JUVENILES AND FIRST WINTER: resemble winter plumage adults but upperwing patches are speckled with dusky spots, underparts washed or finely barred with dusky brown, feet dull coloured, and some show pale base to bill; their dirty looking underparts become whiter during the winter as dusky tips wear, and fresh white contour feathers replace juvenile feathering.

CHICK: covered with wooly black down after hatching, which is lost gradually until they are fully feathered a few days before fledging; feet blackish grey or brownish, lightening in colour somewhat towards fledging.

VARIATION

Five subspecies described: *C. c. eureka*, W coast of North America in California and Oregon; *C. c. adianta*, W coast of North America from Washington state to Alaska Peninsula, and westward to Atka I. (central Aleutian Is.); *C. c. kaiurka*, Adak I. (central Aleutian Is.) W through Andreanof, Delarof, Rat, and Near Is. to the Commander Is. *C. c. columba*, the Bering Sea coasts of Siberia from the southern tip of the Kamchatka Peninsula to the Bering Strait, coastal Alaska from Bristol Bay to the Seward Peninsula, and on St Matthew and St Lawrence Is. *C. c. snowi*, throughout the Kuril Is. Subspecies differ in body size and extent of white on the wing patches. There is a decreasing cline in body size from California to the Aleutian Is., with an increase in size further N and W on the Bering Sea Is. and coastal Siberia. Morphological variation has been reviewed in detail by Storer (1952) and Ewins (1993). Stotskaya (1990*b*) has questioned the validity of some subspecies, based on extensive inter-gradation. Pigeon and Spectacled Guillemots' ranges overlap in the Kuril Is. but it is not clear whether they breed in close proximity to one another. Their relationship in sympatry is not understood, although there is apparently some overlap in plumage traits such as white wing and eye patches in this region (Stotskaya 1990*b*).

MEASUREMENTS AND WEIGHTS

There is considerable geographical variation in mass and size—see Table, (1) *C. c. eureka*, S. E. Farallon Is., California (Nelson 1982); (2) *C. c. eureka*, California specimens (Storer 1952); (3) *C. c. snowi*, Kurils (Stotskaya 1990*b*); (4) *C. c. adianta*, British Columbia (Drent 1965*b*); (5) *C. c. adianta*, Alaska Peninsula (J. F. Piatt, unpublished data); (6) *C. c. kaiurka*, eastern Aleutians (J. F. Piatt,

unpublished data); (7) *C. c. adianta*, British Columbia (Storer 1952); (8) *C. c. kaiurka*, Aleutian and Commander Is. (Storer 1952); (9) *C. c. columba*, Bering Sea (Storer 1952).

Range and status

RANGE

Breed on the W coast of North America from central California through Oregon, Washington, British Columbia, SE Alaska, Kenai Peninsula, Kodiak I., Alaska Peninsula, and in Aleutians westwards to Near Is. Rare breeder in the Pribilofs, more common on St Matthew and St Lawrence Is. in the northern Bering Sea. Breeding colonies scattered along the Bering and Chukchi coasts of Alaska N to Cape Lisburne where suitable habitat is available (Sowls *et al.* 1978). In Asia breeds throughout the Kuril Is. northwards along the E coast of the Kamchatka Peninsula and along the Bering Sea coast N to the Bering Strait (Stotskaya 1990*b*). Mainly sedentary. Northern populations move S at least to the limit of continuous ice cover, but like Black Guillemots, winters among pack-ice and open leads in the Bering Sea. Much of the California and Oregon populations apparently move northwards in winter (Ainley and Boekelheide 1990).

STATUS

Populations apparently generally stable, although this based on speculation and minimal long-term quantitative data on populations. With widely dispersed population in hundreds of small colonies over a huge geographical area, censuses have been difficult and for most areas are non-existent. Population estimates are as follows: Russia, several tens of thousands (Stotskaya 1990*b*), Alaska, perhaps 200 000 (U. S. Fish and Wildlife Service 1993), British Columbia 10 200, Washington 6000, Oregon 3500, and California 15 470. The largest known breeding colony is located in the Farallon Is. California (Ainley and Boekelheide 1990). Locally affected by oil pollution (e.g. *Exxon Valdez* spill in Prince William Sound (Piatt *et al.* 1990*a*), where a long-term population decline was already underway) and gill-net

Table: Pigeon Guillemot measurements

ADULT ♂ ♂	ref	mean	s.d.	range	n
mass (g)	(1)	483	38	—	8
wing length (mm)	(2)	187.1	0.5	176–196	70
	(3)	183.8	—	180–190	7
tarsus (mm)	(2)	35.9	0.2	32.5–38.0	70
culmen (mm)	(2)	36.6	0.2	33.5–39.5	68
	(3)	34.1	—	31.7–36.8	7
bill depth (mm)	(2)	10.4	0.1	9.5–11.6	61

ADULT ♀ ♀	ref	mean	s.d.	range	n
mass (g)	(1)	487	20	—	9
wing length (mm)	(2)	188.1	0.4	179–197	88
	(3)	188.8	—	184–192	7
tarsus (mm)	(2)	36.3	0.1	34.0–39.0	87
culmen (mm)	(2)	36.5	0.2	33.5–39.0	80
	(3)	34.1	—	30.0–37.2	7
bill depth (mm)	(2)	10.1	0.1	8.8–11.1	83

UNSEXED BIRDS	ref	mean	s.d.	range	n
mass (g)	(4)	450	—	—	53
	(5)	531.5	110.8	—	42
	(6)	524.1	9.3	—	11
	(3)	417	—	389–446	5
wing length (mm)	(7)	180.2	4.2	173–191	85
	(3)	184.9	5.6	—	15
	(8)	177.3	3.8	169–183	19
	(6)	187.8	4.4	—	11
	(9)	186.2	4.0	178–194	25
tarsus (mm)	(7)	34.6	1.1	31.5–37.5	85
	(5)	34.8	2.4	—	13
	(8)	33.7	1.2	31.0–35.5	27
	(6)	37.2	—	—	1
	(9)	34.6	1.2	31.5–36.5	25
	(3)	—	—	32.0–36.5	—
culmen (mm)	(7)	34.5	1.4	32.0–38.5	91
	(5)	33.1	2.6	—	16
	(8)	31.2	1.3	29.0–33.0	19
	(6)	34.6	2.4	—	11
	(9)	32.4	1.3	29.0–34.5	21

fishing (King 1984), but their distribution at many small widely scattered colonies probably minimizes the likelihood of a catastrophic effect on the overall population. In the past, population declines probably occurred in the Aleutian Is. owing to the introduction of Arctic Foxes (Bailey 1993). Currently, the spread of introduced Raccoons through Haida Gwaii, BC is a major concern because of likely impacts to breeding populations.

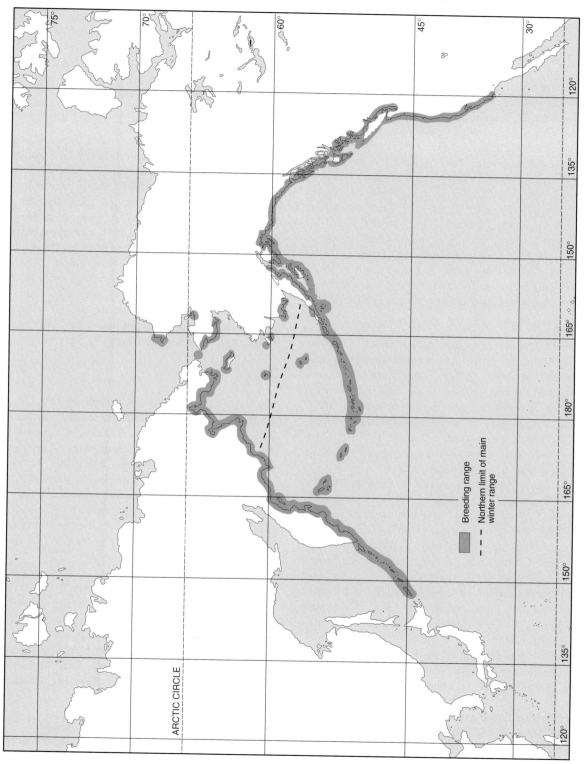

ARCTIC CIRCLE

Breeding range

Northern limit of main
winter range

Field characters

One of the easiest alcids to observe along Pacific shores during summer months and simple to identify, this species is the only mostly black diving bird with large easily visible white patches on the wing coverts and bright red feet. In the northern Bering Sea may rarely be seen together with Black Guillemots, which differ in summer in having oval white upper-wing patches unbroken by dark wedge(s), and gleaming white rather than dark sooty brown underwings. In winter as in summer, Black and Pigeon Guillemots differ in underwing colour and upperwing patch shape. In the Kuril Is. might be seen with Spectacled Guillemots; in typical breeding plumage the latter are uniformly sooty brown except for a splash of white around the eye. In winter Pigeon Guillemots are much whiter than Spectacled Guillemots.

Voice

Highly vocal at colonies, and have a variety of distinctive thin high-pitched twittering and whistling calls. Vocal repertoire was well studied Nelson (1982, 1984, 1985). Four main types of vocalization: 'hunch-whistles', 'trills' 'scream', and 'frequency-modulated notes'. There are three types of hunch-whistle (repeated short whistles that are usually slightly down-slurred or up-slurred in frequency): (1) the usual type given during courtship, (2) the 'neck-stretch'type given by territory intruders, and (3) 'low whistles' given by unmated birds. There also three types of trill (consisting of rapid 'pip-pip-pip' notes): (1) the usual type given by lone birds, (2) 'duet-trills' given by one pair member in the presence of its mate, and (3) 'trill-waggle' or 'twitter-waggle' calls given in agonistic interactions. 'Screams' are rasping whistles (1–2 sec duration) given as alarm calls in response to gulls, puffins, or human intruders. Frequency modulated notes (0.1–0.2 sec short notes) include 'peep', 'seep', 'chip', and 'FM notes' given in a variety of situations.

Habitat food and feeding behaviour

MARINE HABITAT

Pelagic range defined by mean sea surface temperatures in winter and summer of 4–16 and 0–15 °C. Normally forages close to shore in shallow protected waters where there are rocky reefs and shoals. Rare on remote offshore islands such as the Pribilof Is. and Buldir I. the large colony at SE Farallon I. California is apparently exceptional in this respect.

FEEDING BEHAVIOUR

Normally forages in shallow water by searching for benthic fish and invertebrates on rocky bottom and weedy areas, typically probing into crevices and cavities to find hidden prey. Most birds forage within 10 km of their breeding colony. Maximum dive depths of two birds were both about 30 m (Burger 1991), but apparently prefers water depths in the range of 10–20 m, to at most 45 m (Ewins 1993). Suspended schooling prey such as sandlance, herring, or invertebrates (*Euphausia pacifica*) are also taken opportunistically. In the Bering Sea in winter apparently frequent areas of heavy ice cover and forage under the ice like Black Guillemots.

DIET

ADULT: (summarized by Ewins 1993). fish: sandfish (*Trichodon*), capelin, cods, sculpins (*Myoxocephalus*), lingcod (*Ophiodon*), gunnels (*Pholis*), pricklebacks (*Lumpenus*), herring, sandlance and flounders (Pleuronectidae); crustacea: crab (*Cancer, Pangurus*), and shrimp (*Crangon, Pandalus, Heptacarpus*); other invertebrates: polychaetes, gastropods (*Lacuna*), and bivalves (*Musculus*). Adult diet in summer and winter contains a higher proportion of invertebrates than chick diet.

CHICKS: prey fed to chicks is easy to observe and identify at colonies, hence better known than adult diet—see Table; (1) se Farallon I., California (Ainley and Boekelheide 1990); (2) Strait of Georgia, BC (Drent 1965b); (3) Strait of Georgia, BC (Emms and Verbeek 1991); (4) Prince William Sound, Alaska (Kuletz 1983).

Displays and breeding behaviour

COLONY ATTENDANCE AND FLIGHT DISPLAYS: in general, colony attendance peaks early morning,

Table: Pigeon Guillemot chick diet

ref	California 1	British Columbia 2	British Columbia 3	Alaska 4
INVERTEBRATES				
squid (*Loligo* spp.)		trace		trace
octopus (*Octopus* sp.)	4			
FISHES				
lampreys (Petromyzontidae)		7		
herring (*Clupea harengus*)			9	7
smelt (Osmeridae)				8
cods (Gadidae)				5
lingcod (*Ophiodon elongatus*)			5	
sticklebacks (*Gasterosteus*)		trace		
blennies (Blenniidae)			1	
pricklebacks (Stichaeidae) & gunnels (Pholidae)	5	1	2	2
sandlance (*Ammodytes* spp.)		5	trace	1
rockfish (*Sebastes* spp.)	1	6	7	trace
sea-perch (Embiotocidae)		trace	8	
sculpins (Cottidae)	2	3	3	4
sea-poachers (Agonidae)		8		
flatfishes (Pleuronectidae)	3	4	6	6
UNIDENTIFIED		2	4	3

(by rank order of importance by numbers, recorded as trace if ≤ 1% of numbers, summarized from Ewins [1993]).

with activity declining steadily during day until evening when there is a brief resurgence of activity at dusk. However, in areas of high tidal range, activity is reduced during low tides and daily activity peaks are shifted to high tide (Vermeer *et al.* 1993a). Social activity is divided between the sea immediately adjacent to breeding colony, and rocky surface of colony site. Birds normally arrive and gather just offshore from colony at first light, immediately beginning intense social activity on water before going ashore. Their conspicuous social interactions on the sea near colonies has been described as a communal 'water-dance' or 'water-games' (Drent 1965b), involving birds circling each other, vocalizing, and engaging in frequent chases, both in flight and under water. Repeated nervous bill-dipping is a characteristic behaviour. Flight displays are not common, although mass flocks sometimes overfly the colony-site early in the season before birds have started coming ashore (Emms and Morgan 1989). Activity on land takes place on nearshore rocks and on boulders over nesting crevices.

AGONISTIC BEHAVIOUR: includes several threat displays described by Nelson (1984): 'hunch-whistle', 'neck-stretch', and 'trill-waggle' with accompanying vocalizations, aerial chasing, and also outright fighting with jabbing of bill and wing thrashing. Aggression results from conflicts over access to nesting crevices, which may sometimes be in short supply (Nelson 1982), or as intra-sexual contests over mating partners. Unlike most other alcid species, Pigeon Guillemot pairs and unmated ♂♂ appear to defend an exclusive space beyond the confines of their nest-site. This small 'territory' generally includes an area 1–4 m² around nest crevice entrance and also a perch or roosting site on a nearby boulder (Nelson 1982).

SEXUAL BEHAVIOUR: courtship displays take place on sea and on land, prospective pairs perhaps meeting at the gatherings on the sea first and moving to land later. Courtship of paired birds involves lengthy bouts of billing and 'duet-tril' vocal displays. Courtship displays are most frequent during prelaying period when pairs re-form, and during chick-rearing period when non-breeding 2- 3- and 4 year olds are most abundant. Copulation takes place on land, usually on a boulder or other perch near the nesting crevice. Copulations start at least 3 weeks before laying and peak about 12 days before the first egg is laid (Drent 1965b). Extra-pair copulation attempts (most apparently forced) occur frequently, but are rarely successful (Nelson 1982).

Breeding and life cycle

Relatively well-studied. Life history is based on a clutch size of two, which probably depends on foraging close to the nest-site, often on benthic fishes, allowing sufficient food to be delivered for two chicks to be raised to fledging.

BREEDING AND NEST DISPERSION: breeding colonies are located on islands and headlands with suitable rock crevices in rock piles, beach boulders, talus slopes, and cliff cracks, and with nearby shallow waters with rocky bottom for foraging. Often breeds on small low islands and skerries. Where mammalian predators such as rats, foxes, Mink, or Raccoons are present, nesting is limited to cliff crevices or other inaccessible sites. A colonial species, nesting in isolated groups of a dozen to a hundred pairs on some islands, with larger colonies rare. There are no quantitative data on nesting density; may be about 0.1 occupied crevice/m^2 in average colonies.

NEST: typically nests in rock crevices in natural rock piles and cliff cracks, with eggs laid on a shallow pebble-lined depression on the floor of the chamber. However, apparently will also excavate its own burrows in some situations, or use cavities under tree roots and driftwood piles, old puffin and rabbit burrows, hollow spaces under wharves, bridges, in shipwrecks, and even in abandoned buildings (Ewins 1993). The incubating bird is usually not visible through the crevice entrance. Nest-sites are usually within 30 m of shore.

EGG-LAYING: timing of breeding quite variable from year to year, varying as much as a month owing to variable oceanographic conditions. However, laying usually takes place early–late May in California, Oregon, Washington, late May to early July in British Columbia, June in northern Gulf of Alaska, mid-June or later in the northern Bering Sea. Flocks show up at colonies as much as 40 days or more before laying commences (Vermeer et al. 1993a).

EGGS: clutch size normally two, less commonly one. Mean clutch size: 1.8 in California, 1.7 in Oregon, 1.6 in Washington, 1.8 in British Columbia, 1.8 in Alaska (Ewins 1993). Relaying occurs when first eggs lost; time to replacement 13–18 days (Drent 1965a). At SE Farallon I. California, first clutch size was 1.8 ± 0.4 (1–2, n = 868), replacement clutch size 1.5 ± 0.5 (1–2, n = 26), both varying among years with fluctuating environmental conditions (Ainley and Boekelheide 1990). Laying interval about 3 days; 3.0 ± 0.6 days (1–4, n = 18; Drent 1965b), 1–5 days in the Farallons (Ainley and Boekelheide 1990). More than two eggs are occasionally found in a nesting crevice, owing to laying by a second female. Egg shape: pointed ovate to elongate ovate, smooth textured, greyish white or cream ground colour, sometimes with pale bluish or greenish tinge, with irregular brown, black, and dark grey spotting and flecking over entire shell, usually forming a slightly denser ring around larger end. Egg size: from Farallon Is., California, mass of fresh egg: 53.3 ± 6.0 (n = 15) (Nelson 1982); from Washington (Thoresen and Booth 1958), 61.0 (54.0–68.5) × 40.4 (36.5–43.5, n = 50) mm; from British Columbia (Vermeer et al. 1993b), 61.2 ± 2.1 (54.4–70.7) × 41.0 ± 1.5, (38.2–51.2, n = 130) mm. Egg weight approximately 12% of adult weight.

INCUBATION: breeding adults have a single brood patch wide enough to cover both eggs. The first egg is incubated intermittently by day until the second egg is laid; continuous incubation begins 1–3 days after laying of the second egg (Drent 1965a). Duration of incubation (days): first egg, 32.0 ± 1.1 (n = 11), second egg, 29.8 ± 1.4 (n = 11) in British Columbia; first egg 29.1 ± 1.8 (29–32, n = 369), second egg, 27.2 ± 1.9 (*n* = 282) in California (Ainley and Boekelheide 1990). Both sexes incubate, apparently in equal proportions, in shifts lasting from less than 1 hr to 17 hrs (Ewins 1993). One bird incubates continuously overnight, but the pair may exchange duties more than once during daylight hours. Eggs survive interruptions in incubation of up to about 24 hrs.

HATCHING AND CHICK-REARING: egg pips for about 48 hrs before hatching. The two chicks usually hatch 1–2 days apart. Chicks are semiprecocial, and are brooded continuously by parents until 3 days of age, regularly until about 7 days of age. Adults deliver single fish crosswise in bill, most frequently in early morning hours; delivery rate: 0.8–1.9 fish/hr/nest in British Columbia (Drent 1965*b*), 0.7–1.1 fish/hr/nest in Alaska (Kuletz 1983). At first, rather small fish are brought to chicks because all prey are swallowed whole. Provisioning rates increased with chick age, and parents delivered smaller sculpins and blennies to 1–2-week-old chicks than to 5–6-week-old chicks in British Columbia (Emms and Verbeek 1991). Adults delivering fish to older chicks often remain at crevice for only a few seconds before departing. Parents continue to deliver fish to crevice until after chicks fledge. No evidence of post-fledging parental care.

CHICK GROWTH AND DEVELOPMENT: chick mass at hatching 43.7 ± 4.5 g (*n* = 23; Drent 1965*b*). Mean growth rate 10–20 g/day, later hatching (second) chick sometimes grows slower. At Mitlenatch I., BC (Emms and Verbeek 1991) growth rates (g/day) of first chicks 15.9 ± 0.5 (*n* = 18), of second chicks 11.4 ± 0.9 (*n* = 17), and of solo chicks 16.1 ± 0.7 (*n* = 17). Duration of nestling period (days): 38.0 ± 3.8 (30–53, *n* = 368) for first solo chicks, 39.3 ± 4.1 (30–53, *n* = 140) for second chicks in California (Ainley and Boekelheide 1990); 35 (29–39) in the Strait of Georgia, BC (various studies, summarized by Ewins 1993); 38.0 ± 3.0 in Haida Gwaii, BC (Vermeer *et al.* 1993*b*); 42 ± 6 (35–54) in Gulf of Alaska (Lenhausen 1980). Most chicks

fledge at slightly (4–9%) lower than their maximum mass, although provisioning continues until day of departure (Ewins 1993). Fledging mass is about 300–400 g (65–85% of adult mass; Emms and Verbeek 1991). Chicks fledge by walking or fluttering down to the sea.

BREEDING SUCCESS: well studied (summarized here from various studies reviewed by Ewins 1993). Fledge about one chick per breeding pair per year. Mean hatching success: 73% in California, 66% in Oregon, 73% in Washington, 57% in British Columbia, 71% in Alaska. Mean fledging success: 67% in California, 69% in Oregon, 77% in Washington, 56% in British Columbia, 70.5% in Alaska. Mean number of chicks fledged per breeding pair per nesting attempt: 1.0 in California, 0.6 in Oregon, 0.9 in Washington, 0.7 in British Columbia, 0.9 in Alaska. These data probably do not reflect geographical variation, but do indicate range of productivity with varying environmental conditions and investigator disturbance.

PREDATION: Mammalian predators include rats which prey on eggs and nestlings in burrows, River Otters (*Lutra canadensis*) that visit some colonies (ILJ), Raccoons (Vermeer *et al.* 1993*b*), and Killer Whales (*Orcinus orca*; Stacey *et al.* 1990). Avian predators include Western (*Larus occidentalis*) and Glaucouswinged Gull (*L. glaucescens*; Emms and Morgan 1989), Great Horned Owl (*Bubo virginianus*), Bald Eagle (*Haliaeetus leucocephalus*), Peregrine Falcon and Northwestern Crow (*C. caurinus*; Emms and Morgan 1989; Vermeer *et al.* 1993*b*).

MOULT: moults and feather generations have not been studied in detail (but see Ewins 1993, summarized below). Juvenile plumage is acquired by 30 days of age and this is retained through their first winter, when they closely resemble winter-plumage adults except for dusky tipping to underpart feathers and coverts (Basic I plumage). First summer (Alternate I) birds attending colonies resemble breeding plumaged adults but their dark brown overall plumage is duller, lacking adult's slightly iridescent sheen, the wing patches are spotted with brown and the flight feathers are worn and tan coloured. These birds begin moulting into second winter (Basic II) plumage in July and Aug, earlier than adults, and by late autumn are indistinguishable from winter adults. This moult is complete and the flight feathers are dropped simultaneously, leaving the birds flightless for more than 1 month. Adult breeding (Definitive Alternate) plumage is attained before the third summer. All birds undergo a moult of underparts, back, head, and neck feathering during late winter and spring when overall plumage changes from white to dark brown; moulting birds have a peculiar piebald appearance (Definitive Pre-alternate moult, Feb–Apr).

Population dynamics

SURVIVAL: mean annual adult survival was 80%, varying from 62% to 89% at SE Farallon I., giving a predicted mean lifespan of about 6 years (based on resighting rates of colour-marked birds; Nelson 1991). Oldest known marked bird was 14 years old (Ewins *et al.* 1993).

AGE AT FIRST BREEDING: 1- and 2-year-old birds attend colonies and may both form pair-bonds and occupy nesting crevices, but do not breed. Earliest recorded breeding was at 3 years of age at SE Farallon I., California (Nelson 1991). Most bird probably begin breeding in fourth or fifth year.

PAIR- AND NEST-SITE FIDELITY: rather low divorce rate, with only four cases of divorce recorded in 95 pairs followed in more than 1 year (Nelson 1991). Pairs normally return and reuse same nesting crevice from one year to the next, if both pair members are alive and crevice is intact (87% of the time; Nelson 1991). Most birds (60%) whose mate disappeared retained their crevice, and ♂♂ and ♀♀ were equally likely to retain their nesting crevice after the loss of their mate (Nelson 1991). Factors leading to abandonment of a nest-site by a pair include alteration of site, death of one or both partners, or divorce.

Spectacled Guillemot　*Cepphus carbo*

Cepphus carbo (Pallas, 1811, Zoogeogr. Russo-Asiatica 2: 350, plate lxxix—Bering Sea)

PLATE 2

Description

Poorly known western Pacific alcid similar to other *Cepphus* species but considerably larger, and lacking pale wing patches. An elegant seabird with a graceful upright posture with long neck usually held extended. Legs strong enough to support the bird on the tarsi and toes or occasionally on toes alone. Bill, relatively somewhat heavier than bills of other *Cepphus* species, almost symmetrical and dagger-shaped, tapering to a fine point. Irides contrast strongly with the white face patches in all plumages.

ADULT SUMMER: uniform sooty greyish brown overall, except for the white spectacles (white patches around the eye, with a narrow whitish streak projecting backwards from the eye towards the nape) and small areas of white around base of upper and lower mandibles of bill; pale chin grades evenly into dark grey-brown throat; flight feathers and tail blackish-brown and wings rather rounded; upper and underwing coverts blackish brown year round; irides dark brown; bill black; mouth lining, legs, feet, and webs bright coral red.

ADULT WINTER: underparts and head turn to white with faint dark flecking, while upper-parts remain uniformly dark brown; blackish-brown feathering retained around crown and forehead and onto face, and reduced white eye and chin patches are maintained throughout the winter; back of neck and the back have blackish brown feathers with some faint pale edging; mouth lining, legs, feet, and webs pale red or pinkish.

JUVENILE AND FIRST WINTER: resemble winter adults but underparts heavily barred with brownish grey, particularly on throat and flanks, feet pinkish brown, and bill relatively short with pale tip.

CHICKS: covered with blackish grey down which is gradually lost until they attain complete feathering a few days before fledging.

VARIATION
Monotypic. No notable patterns of geographical variation in size or plumage have been described.

MEASUREMENTS AND WEIGHTS
From Talan I., Sea of Okhotsk (S. Kitaysky, unpublished data) (mm): wing 200.1 ± 10.2 ($n = 7$), culmen 39.4 ± 0.8, ($n = 7$), tarsus 38.1 ± 1.6 ($n = 7$), body mass 646.0 ± 23.5 g ($n = 7$). From the Sea of Okhotsk (Shibaev 1990c) sample sizes not given): wing, ♂ 189.3 (186–193), ♀ 188.5 (184–192); culmen, ♂ 38.5 (34.4–40.0), ♀ 35.7 (33.3–37.4); body mass, ♂ 598.5 (544–685), ♀ 633 (604–649). From Teuri I. (Minami *et al.* 1991); wing 204 ± 6 ($n = 5$), culmen 42.2 ± 2.8 ($n = 5$), tarsus 37.4 ± 1.6 ($n = 5$), mass 680 ± 67.7 ($n = 5$). From various localities, unsexed birds (Storer 1952): wing, 202.6 ± 6.7 (188–215, $n = 19$), culmen 42.4 ± 2.0 (38.5–45.5, $n = 18$), tarsus 37.2 ± 1.4 (35.0–39.5, $n = 19$).

Range and status
RANGE
Breeds most abundantly in the northern Sea of Okhotsk and along the NW coast of the Kamchatka Peninsula: Penzhina Bay, Taigonos Peninsula, Gizhiga Bay, and Shelekhov Gulf. Also breeds on the western shores of the Sea of Okhotsk, the Yama Is., Tauiskaya Gulf,

Shantar Is., and on Sakhalin I. on rocky headlands at Capes Terpeniye, Aniv, and Lamanon, and nearby Moneron I. On the Kurils breeds on Paramushir, Dym, Brouton, Chernye Brat'ya, Shikotan, Devyatyi val, and Srednii Is., and also Lis'ie, Storozhevaya, and Anuchin Is. Apparently breeds locally and uncommonly along the entire western coast of the Sea of Japan, S to the Russia–North Korea border. It is rather common in Peter the Great Bay (Nazarov and Labzyuk 1972; Shibaev 1990b). This species' range may be more extensive than previously suspected, since Dinetz (1992) observed several birds carrying fish at a small breeding colony in the Chukotka Peninsula, far to the NE of their recognized range. In Japan, most birds bred near the northern island of Hokkaido with historical nesting records from the Shiretoko and Nemuro Peninsulas and on Kushiro, Tomoshiri, Yururi, Moyururi, Kojima, and Teuri Is., the Daikoku Is., and at Capes Esan and Shakotan. Also nested locally on Honshu I. at Cape Shiriya, the Shimokita Peninsula, and on Sankan, Futago, and Totigani Is. (Austin and Kuroda 1953; Brazil 1991). In Korea, formerly nested on Kuk I. in Kangwon-do Province, and Ren I. in Hamgyon-pukdo Province (Austin 1948); current status there is unknown. Vagrants have reached the Commander Is. and the Chukotka Peninsula, so this species should be looked for at Attu I. in the western Aleutians and at St Lawrence I. Winters mainly in the sorthern part of the breeding range.

STATUS
Population status in Russia unknown, possibly numbers between 50 000 and 400 000 pairs (Shibaev 1990b). The largest colonies are apparently in the Shantar Is., where there are several colonies in the range of 5000–8000 pairs, and at Maty'kil I. in the Yama Is., 4000 pairs (Shibaev 1990b). Overhunting and disturbance of colonies in Peter the Great Bay has recently become a major threat to local populations (Shibaev 1990b). In Japan it is now considered to be rare, local, and declining on Hokkaido and northern Honshu (Brazil 1991).

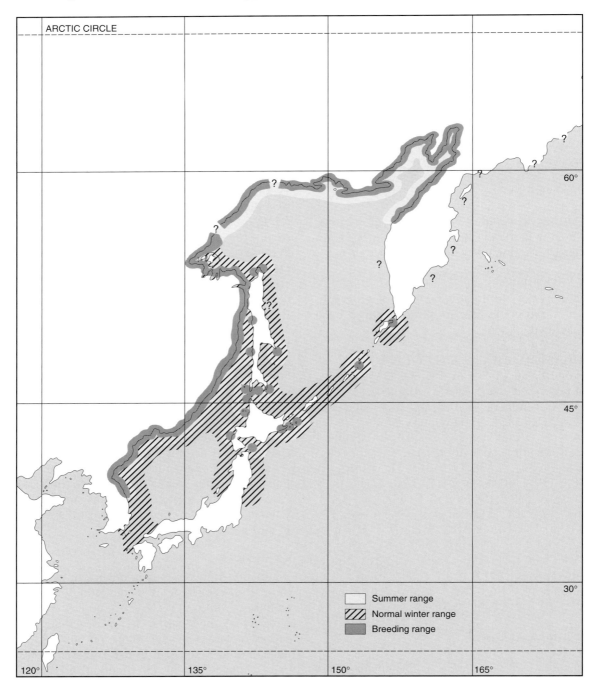

ARCTIC CIRCLE

60°

45°

30°

☐ Summer range
▨ Normal winter range
▓ Breeding range

120° 135° 150° 165°

Has undergone severe population declines in Japan owing to gill-net fishing and disturbance of breeding islands. For example, the population at Teuri I. declined from about 7000 in 1949 to only 213 in 1987, with the population still considered to be declining (Watanuki *et al.* 1988). Sadly, development plans for tiny Teuri I. include several tourist hotels and other commercial activities that are not likely to favour seabird populations.

Field characters

Although similar in shape and general appearance to other *Cepphus* species, differs considerably in plumage pattern, and thus is likely to be easy to identify. Range overlaps with Pigeon Guillemot only in the Kuril Is. and eastern Sea of Okhotsk, so Spectacled Guillemots are the only *Cepphus* species throughout most of their range. Confusion is possible with dark Pigeon Guillemots *C. c. snowi*, which have little white on the upperwings, but Spectacled Guillemots never show white wing coverts, always have a large white eye-patch, and are larger and stockier.

Voice

Vocal repertoire has not been studied or described in detail. Vocalizations apparently similar in tone and structure to those of other *Cepphus* species. One frequently performed vocalization described by Shibaev (1990*b*) closely resembles the 'trill' call given by Pigeon Guillemots: 'tsit-tsi-tsi-tsi-tsi-si-si-si-sisisi.' Chick calls undescribed.

Habitat food and feeding behaviour
MARINE HABITAT
Pelagic range defined by mean sea surface temperatures in summer and winter of 2–20 and 0–15 °C. Like the similar Pigeon Guillemot,

1 Adults at a breeding colony (photo Takaki Terasawa).

normally frequents and forages in shallow water close to shore, with breeding colonies situated near to protected bays and areas of extensive reefs. Apparently prefers warmer waters than other *Cepphus* species, since southern limit of breeding range lies in areas with summer maximum sea temperatures of 20 °C or more.

FEEDING BEHAVIOUR
Forages on benthic prey and opportunistically on pelagic schooling fishes, close to shore in water depths of 10–20 m, within a few kilometres of nesting colonies.

DIET
ADULT: not well studied but Austin and Kuroda (1953) reported that three birds shot in winter in Japan had small fish *Platycephalus indicus*, octopus, and crab in their stomachs, and Kitaysky (1994) found invertebrates including *Sabinea* and *Sclerocrangon* spp. and small fish including blennies (*Pholis*) and sculpins (*Triglops* spp.) in the stomachs of adults shot in summer near Talan I. Sea of Okhotsk.

CHICKS: polychaete worms, shrimp, and small fishes including Gobiidae, Hemiramphiidae, and Pholidae in Peter the Great Bay (Nazarov and Labzyuk 1972), sandlance (*Ammodytes hexapterus*) and anchovy (*Engraulis japonicus*) at Teuri I. (Austin and Kuroda 1953), rockfish (*Sebastes minor*), sandlance (*A. personatus*), and blennies (Blennioidea) at Teuri. (Minami *et al.* 1995), Arctic smelt (*Osmerus eperlanus*) Pacific sardine (*Sardinops sagax*) and Pacific herring at Moneran I. (Gizenko cited in Shibaev 1990*b*), and small fish including sculpins (Cottidae, 44% by number), the blenny *Pholis pictus* (27%), and small numbers of sandlance, *Alectrias alectrolophus*, and *Gimnelis* spp.

Displays and breeding behaviour
COLONY ATTENDANCE AND FLIGHT DISPLAYS: little information available. Flocks gather on the sea near colonies to court and socialize. No mass flight displays have been described.

AGONISTIC AND SEXUAL BEHAVIOUR: little information available. Aggressive behaviour on land is accompanied by a piercing call, head waving, and lunging at an opponent with half open wings (similar to the trill-waggle display of the Pigeon Guillemot; Nelson 1984). Courtship on the sea includes frequent flight chases involving pair members, trill calls, bill dipping, and circling behaviour. Copulation on land takes place near the nesting crevice (Nazarov and Labzyuk 1972).

Breeding and life cycle
Very poorly known, but based on the data available their natural history appears to be very similar to the other *Cepphus* species.

BREEDING HABITAT AND NEST DISPERSION: breeds on rocky islands and headlands free of mammalian predators where nesting crevices among piles of boulders and cliff cracks are available; from just above high tide to more than 120 m from shore. Some nests are located in cliff crevices as high as 100 m or more above the sea, and occasionally in rock crevices among bushes, or scrubby woodland on headlands (Nazarov and Labzyuk 1972). A colonial species. Usually nests on small islands in colonies of 20–40 pairs, but on larger islands in Peter the Great Bay, colonies of over 700 pairs have been described (Nazarov and Labzyuk 1972) and in the Shantar Is. there are several colonies with more than 1000 pairs. Based on the structure of breeding habitat, nesting density in colonies is probably about one breeding pair per 100 m^2.

NEST: typically nests in rock crevices in talus and in cliff cracks, with eggs laid in a shallow depression on the gravel floor of the chamber, or on bare rock which is sometimes wet from trickling water. Depth of nesting chamber varies from 0.2 to 1.5 m (Nazarov and Labzyuk 1972).

EGG-LAYING: in Peter the Great Bay, birds return to the vicinity of their colonies early Apr, laying begins early May and continues through early June, peak of hatching is between 10 and 20 June, and fledging occurs in July and early Aug (Nazarov and Labzyuk 1972), with a similar timing of breeding in Japan (Austin and Kuroda 1953). Egg-laying takes place a month to 6-weeks later farther N in the Sea of Okhotsk (Shibaev 1990b).

EGGS: clutch size normally two eggs, less commonly one (despite early reports that this species laid only a single egg). Among more than 60 nests examined before the end of the laying period, 60% had two eggs (Nazarov and Labzyuk 1972). The second egg is laid, on average, 7 days after the first (Kitaysky 1994). Egg shape: oval to subelliptical, smooth textured, whitish or pale bluish ground colour with numerous irregular brown, black, and dark spots, denser at the larger end. Egg size: from Peter the Great Bay: 61.1–66.8 × 41.1–45.2 mm (n = 14) (Nazarov and Labzyuk 1972); 64.2 (61.2–68.0) × 43.2 (41.1–45.5, n = 24) mm, mass of three fresh eggs: 62.5, 66.4, 66.3 g (Shibaev 1990b).

INCUBATION: breeding adults have two lateral brood patches. Incubation lasts for about 27 days (Kitaysky 1994). No information available on incubation shift length, or the relative effort of ♂♂ and ♀♀.

HATCHING AND CHICK-REARING: egg pips for 24–72 hrs before hatching. Parents provision chicks with invertebrates and small fish until they are about 75% of adult size and mass; duration of nestling period about 5–6 weeks.

CHICK GROWTH AND DEVELOPMENT: Chick mass at hatching 35.6 ± 4.5 g (n = 5; S. Kitaysky, unpublished data). Another 1-day-old chick weighed 37.7 g (Nazarov and Labzyuk 1972) Maximum growth rate, 22.1 g/day at 15 days of age (Minami *et al.* 1995). Mass at fledging: 466.3 ± 53.8 g (n = 7), S. Kitaysky, unpublished data, 620 g (Minami *et al.* 1995). Chicks fledge at night.

BREEDING SUCCESS: little quantitative information available. Kitaysky (1994) reported that both chicks survived to fledging in 45% of nests. Post-fledging mortality is often high in Peter the Great Bay (Nazarov and Labzyuk 1972).

PREDATION: Mammalian predators include Red Foxes, Ermine, and marten (*Martes* sp.) which take adults at colonies. Avian predators include Slaty-backed Gulls (*Larus schistisagus*) which take adults and fledglings near colonies, Black-tailed Gulls (*L. crassirostris*) which kleptoparasitize adults at colonies, Eagle Owls (*Bubo bubo*), probably Steller's Sea (*Haliaeetus pelagicus*) and White-tailed Eagles (*H. albicilla*), Peregrine Falcons which take birds at sea, and Jungle Crows (*Corvus macrorhynchos*) which feed on eggs and chicks (Shibaev 1990*b*).

MOULT: from (Shibaev 1990*b*). Juvenile plumage acquired by 30 days of age, and there is an incomplete moult of contour feathers from Aug to Oct when first winter (Basic I) plumage is attained. First winter birds typically have a mixture of fresh white feathers and brown-edged feathers retained from the juvenile plumage, but become very pale and almost adult like by mid-winter. First summer (Alternate I) birds attending colonies resemble breeding-plumaged adults but their dark brown summer plumage is not completely moulted until May, and the flight feathers are worn and tan coloured. The moult into second winter (Basic II) and adult winter (Definitive basic) plumage begins in Aug, is most intense in Oct, and by late autumn second-winter birds are indistinguishable from winter adults. This moult is complete and the flight feathers are dropped simultaneously, leaving the birds flightless for several weeks. Adult breeding (Definitive Alternate) plumage is attained before the third summer. All birds undergo a moult of underparts, back, head, and neck feathering during Dec–May when their underparts change from white to dark brown (Definitive Pre-alternate moult). Moulting birds have barred underparts by late Jan and this moult is complete and all adults have uniformly sooty underparts by Apr.

Population dynamics
No information available on survival and pair- and site-fidelity. (Shibaev (1990*b*) stated without supporting data that this species breeds at 3 or 4 years of age.

Marbled Murrelet *Brachyramphus marmoratus*

Colymbus marmoratus (Gmelin, 1789, Syst. Nat. 1,2: 583—Prince William Sound, SE coast of Alaska) PLATE 3

Description
Small fast-flying alcid with a cryptic speckled brown breeding plumage that complements its atypical solitary nesting habits. Typical posture when resting on the sea is with head uptilted and tail cocked nearly vertically. In alert posture, they show a long and thin-necked profile not found in most other small alcids, accentuated by their shallow sloping forehead. Adults' legs and feet are weak and set far back on the body, so this species has poor agility on land and cannot assume a fully upright posture, perhaps restricting nest-site selection. Bill long, slender, and pointed (longest in Siberian form). In flight, wings narrow and pointed, and seem almost to blur with the high wing-beat frequency. Siberian populations probably represent a separate species, the Long-billed Murrelet (Friesen *et al.* 1996*b*). This form is about 20% heavier, has a longer and thinner bill, is less rufous in summer plumage, and in winter has more extensive dark plumage on nape and lores, lacking a light band across nape, but with a larger white spot below eye (Konyukhov and Kitaysky 1995).

ADULT SUMMER: extensively mottled overall with dark chocolate brown on a whitish ground colour, with mantle feathers thinly edged rufous, and scapulars fringed white and rufous; flight feathers and tail, blackish brown; considerable variability in breeding plumage, with some individuals very uniform dark brown and others paler with extensive white spotting showing through, pairs often differ in plumage; in all plumages underwing coverts dark brownish grey; irides dark brown; bill, legs, and feet black;

ADULT WINTER: black-and-white winter plumage similar to murres' winter dress, recalling winter-plumage Common Murre, but has conspicuous white scapular patches contrasting strongly with otherwise dark upperparts; dark feathering forms contrasting blackish brown cap that extends over most of face to well below eye; nape, back, rump, and upper-wing-coverts dark grey to blackish grey in winter. Some birds (presumably immatures and sub-adults, 1- or 2-year-old birds) remain in a winter-like plumage during summer, others may attain a summer appearance intermediate between typical summer and winter plumage (ILJ).

JUVENILES AND FIRST WINTER: resemble winter adult but has dusky brownish barring on breast and flanks for more than a month after fledging.

CHICK: thick yellow or greyish yellow down spotted with brown and black on upperparts, and with grey underparts, retained until just before fledging.

VARIATION
Two distinct forms, described on the basis of size, shape and coloration: *Brachyramphus m. marmoratus*, Marbled Murrelet, Pacific coasts of North America W to Near Is. in the Aleutians; *Brachyramphus m. perdix*, Long-billed Murrelet, Pacific coast of Siberia S to Hokkaido. Recent studies of genetic variation suggest that the divergence between the Asian and North American populations of Marbled Murrelet is greater than that between the North American populations of Marbled and Kittlitz's Murrelets (Friesen *et al.* 1996b) supporting earlier classifications of the Long-billed Murrelet as a separate species.

MEASUREMENTS
♂♂ average slightly larger than ♀♀ in all measurements except body mass, the Asian form is significantly larger in all measurements and 30% larger in body mass. *B. m. marmoratus* from Haida Gwaii, B C (Sealy 1975a), adults in summer: (mm) wing, ♂ 134.2 ± 1.2 (128–140, $n = 25$), ♀ 132.6 ± 1.8 (122–139, $n = 23$); culmen, ♂ 15.5 ± 0.3 (13.2–17.4, $n = 36$); ♀ 15.3 ± 0.4 (13.7–17.6, $n = 32$); bill depth, ♂ 6.0 ± 0.1 (5.4–6.6, $n = 26$), ♀ 5.8 ± 0.1 (5.3–6.8, $n = 23$); tarsus, ♂ 16.2 ± 0.2 (15.1–17.6, $n = 37$), ♀ 15.9 ± 0.3 (13.9–17.3, $n = 39$). *B. m. perdix* Sea of Okhotsk (Shibaev 1990a); sample sizes and standard deviations not available): wing, ♂ 141.2 (136–147), ♀ 138.3 (130–145); bill length, ♂ 20.2 (18.9–22.2), ♀ 19.0 (18.0–21.0); tarsus, ♂ 18.1 (17.0–18.7), ♀ 18.0 (16.8–19.0).

WEIGHTS
B. m. marmoratus from Haida Gwaii, BC (Sealy 1975a), body mass (g) of adults in summer: ♂ 217.0 (196.2–252.5, $n = 37$), ♀ 222.7 (188.1–269.1, $n = 37$). From Desolation Sound, BC (Desrochers *et al.* 1994), unsexed birds mist-netted at sea: 199 ± 13.2 (164–234, $n = 168$); of unsexed birds shot at sea: from the Alaska Peninsula, 229.2 ± 21.8 ($n = 109$), from the Gulf of Alaska, 222.0 ± 20.1 ($n = 35$; J. F. Piatt, unpublished data). *B. m. perdix* Sea of Okhotsk (Shibaev 1990a); sample sizes and standard deviations not available): unsexed birds, 295.8 (258–357).

Range and status
RANGE
The North American form *B. m. marmoratus* breeds on the W coast of North America from S central California through Oregon (local), Washington, British Columbia (especially parts of Haida Gwaii, the mainland coast,

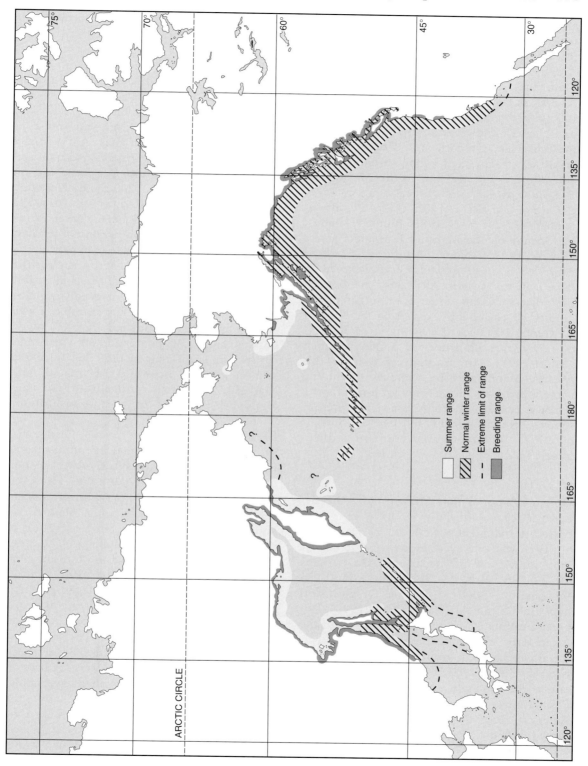

and the W coast of Vancouver I. where extensive stands of old-growth forest remain; Rodway *et al.* 1992*a*), SE Alaska (Alexander Archipelago), Prince William Sound, Kenai Peninsula, Barren Islands, Afognak, and Kodiak Is., on the Alaska Peninsula, and in Aleutians westwards locally to the Andreanof and Near Is. (Sowls *et al.* 1978). Populations in California, Oregon, and Washington are small and fragmented owing to habitat loss, and numbers are also low in the Aleutian Is. *Brachyramphus m. perdix* breeds along coasts of the Sea of Okhotsk and Kamchatka Peninsula, Sakhalin I. (Nechaev 1986; Shibaev 1990*a*), and locally on the island of Hokkaido, Japan (Brazil 1991). North American Marbled Murrelets occur around and certainly breed on Attu I. in the western Aleutians, but those of the adjacent Commander I. are of unknown affinity. Occasionally wanders as far N as St Lawrence Is. (Bédard 1966) and as far S as San Diego, California and southern Japan (Brazil 1991). The strong flying ability of this species is indicated by vagrants (including some individuals of the Siberian form) having reached eastern North America in Ontario, Québec, Newfoundland, and Florida (Sealy *et al.* 1982). Withdraws from northern and ice-affected parts of its range in winter.

STATUS

Only rough population estimates are available, because census techniques have not been perfected, and because most of range has not been well surveyed. However, 200 000 or more individuals may occur in Alaska, perhaps 50 000 in British Columbia, 5500 in Washington, 5000–15 000 in Oregon, and 6450 in California (Ralph *et al.* 1995). No precise estimates of the population size of the Asian form are available but may number in the low tens of thousands (Konyukhov and Kitaysky 1995). Marbled Murrelets are certainly the *hamadryads* among the alcids. Populations in southern parts of North American range have declined owing to widespread destruction of old-growth forests. The greatest historical decreases have occurred in

Washington, Oregon, and California, while the greatest current threat is from ongoing massive clear-cut felling of ancient forests in British Columbia, SE Alaska, and Siberia (Perry 1995). Marbled Murrelets have disappeared from cut-over areas; in the long-term this species is likely to be extirpated from regions without large old-growth forest reserves. Vulnerable to oil spills. An estimated 8400 were killed by the *Exxon Valdez* oil spill at Prince William Sound, Alaska (Piatt and Lensink 1989) and 170–200 by the Nestrucca oil spill in Washington (Rodway *et al.* 1990*a,b*), although these may be underestimates of real mortality owing to low probability of recovery of murrelet corpses. Considerable mortality in gill-nets has also occurred, and it is clear that use of nylon monofilament gill-nets in shallow waters where murrelets are present inevitably leads to heavy mortality. For example, Carter and Sealy (1984) estimated that 4% of the Barkley Sound, BC population drowned annually in salmon gill-nets in 1979–80. In Alaska, salmon gill-nets drown an estimated 3300 Marbled Murrelets annually (Piatt and Naslund 1995), 1.5% of the state population.

Field characters

In summer plumage, almost unmistakable from all other small seabirds owing to their overall dark brown coloration. However, where Marbled and Kittlitz's Murrelets occur together in S central Alaska, the Aleutian Is., and Sea of Okhotsk they can be difficult to discriminate in breeding plumage. The latter species has paler grey-brown mottling that does not extend onto the lower belly, vent, or undertail coverts. In all plumages Kittlitz's shows a paler face with a contrasting dark eye. Marbled Murrelets often first detected, and easily identified, by their loud penetrating *keer* (see voice below) calls, especially at dusk, dawn, or at night near suitable nesting habitat. Marbled and Kittlitz's Murrelets have similar black-and-white winter plumage, but the extent of the black cap is much greater in the former species, extending well below the eye, making Kittlitz's Murrelets look much whiter at a distance. Marbled and Ancient Murrelets are likely to be encountered together in winter, presenting possible identification difficulties in views of distant flying birds. However, Ancient Murrelets lack white scapulars, show more extensive black on head and neck, and have bright white underwing coverts.

Voice

Loud, distinctive vocalizations with a complex repertoire. It is the most vocal alcid at sea, unlike most species which are usually silent on the water. There are four general types of adult vocalizations (based on unpublished work by S. K. Nelson, S. W. Singer, J. Hardin, and B. P. O'Donnell): (1) 'keer-like calls', which are a class of variable piercing note sounding like the cry of a gull but shifted to a much higher pitch, composed of a harsh introductory element followed contiguously by a pure tone element in the range 2–4 Hz (Fig. 1); (2) 'whistle-like calls', which are similar to keer calls but emphasize the pure tone whistled component; (3) 'groan-like calls', which are broad-band vocalizations with a nasal quality, a frequency range spanning 0.3–4.5 kHz, and incorporation of multiple harmonics; and (4) chips, which are very short (*c.* 0.1 sec duration) calls.

Habitat food, and feeding behaviour
MARINE HABITAT
Summer and winter ranges lie within mean sea surface temperatures of 5–17 and 4–15 °C

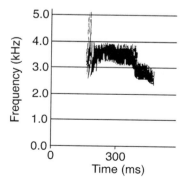

1 Sound spectrogram of Keer call (S. K. Nelson *et al.*, unpublished data)

(cf. Kittlitz's Murrelet). Normally occurs close to shore, and is the commonest alcid in sheltered waters of coastal British Columbia, SE Alaska, Prince William Sound, and around Kodiak I. in Alaska. Frequents areas with numerous small islands, and penetrates deep into bays, sounds, fiords, and estuaries where surface water is fresh or low in salinity (Hunt 1995). In some areas uses coastal brackish and freshwater lagoons for foraging and staging and regularly uses large inland lakes (as much as 75 km from the sea) in British Columbia (Carter 1986), the only alcid to do so regularly. Occasionally occurs far from shore (Morgan *et al.* 1991), but pelagic observations are exceptional. In southern parts of North American range, where breeding habitat has been mostly eliminated, concentrations occur on the sea adjacent to remnant fragments of old-growth forest.

FEEDING BEHAVIOUR
Usually forages in shallow nearshore water in sheltered bays and fiords, sometimes in tide rips between small islands, rarely offshore. Feeds on small schooling fish and invertebrates (the results of published and unpublished diet studies are reviewed in detail by Burkett 1995). Joins mixed species feeding flocks that include other alcids, gulls, and cormorants to feed on sandlance and herring schools. No estimates of dive depth available.

DIET
Not well known, compared with other aspects of this species' biology, because of the difficulty of sampling stomachs and observing adult's provisioning of chicks. Sandlance (reported in 62% of 26 studies) and herring (reported in 35% of 26 studies) are the major prey species of adults in summer. Like other alcids, they switch to a greater proportion of invertebrates during winter. See Table: (1) Langara I., Haida Gwaii (Sealy 1975*b*), (2) Vancouver I., BC (Carter 1984); (3) Vancouver I., BC (Vermeer 1992); (4) Kachemak Bay, Alaska (Sanger 1987*b*); (5) Kodiak I., Alaska (Krasnow and Sanger 1982). Chick diet includes sandlance, herring

Table: Adult Marbled Murrelet diet

ref	(1) summer	(2) summer	(3) winter	(4) winter	(5) winter
INVERTEBRATES					
Thysanoessa spinifera	2				
Euphausia pacifica	4				
unidentified euphausiids			2	3	2
mysids				4	3
FISH					
sandlance (*Ammodytes hexapterus*)	1	2		trace	
herring (*Clupea harengus*)	trace	1	1		
capelin (*Mallotus villosus*)				1	
rockfish (*Sebastes* sp.)	5	trace			
sticklebacks (*Gasterosteus aculeatus*)	trace				
seaperch (*Cymatogaster aggregata*)	3				
anchovy (*Engraulis mordax*)		trace			
unidentified smelt (osmeridae)	trace			2	1
gammarids				5	4
Loligo opalescens	trace				

By rank order of importance in numbers, trace if < 1%.

northern anchovy (*Engraulis mordax*) and possibly capelin, whitebait (*Allosmerus elongatus*) and surf smelt (*Hypomesus pretiosus*).

Displays and breeding behaviour

COLONY ATTENDANCE AND FLIGHT DISPLAYS: non-colonial, but social activities in nesting areas are a conspicuous aspect of behaviour; repeated overflights by individuals, pairs, and small groups of birds above the forest canopy, flights under the canopy near to potential nests-sites, and landings on tree limbs are common. Flights under the canopy and landings indicate the presence of suitable nesting habitat. Most inland activity occurs between 1 hr before and 1 hr after sunrise, peaking about 10–40 min before sunrise, with a less intense activity period at dusk (Naslund and O'Donnell 1995). Prospecting behaviour by breeding and non-breeding birds involves repeated circling over breeding habitat and flights through the canopy. Overflights and probably landings at potential breeding sites occur year round but peak during the breeding season (May–Aug). Two apparent flight displays have been described: (1) a noisy buzzing flight in which individuals produce a low-pitched buzzing sound, apparently with their wings; (2) jet-dives in which birds make a loud roar with their wings as they swoop over a nest-site (Nelson and Hamer 1995*b*), or at sea (AJG, ILJ).

AGONISTIC AND SEXUAL BEHAVIOUR: has been difficult to study because of the birds' solitary habits and the crepuscular timing of their activities. Apparently rather mild-mannered, since agonistic behaviour has not been frequently observed. Socially monogamous. Often seen in groups of two at all times of year, and 'pairs' shot on the sea have often included one ♂ and one ♀, leading to the suggestion that this species remains in year-round pair groups (Nelson and Hamer 1995*b*). Nevertheless, not known if these duos are indeed mated pairs. Courtship, including distinctive display postures, occurs on the sea from pre-laying period (even before breeding plumage has been attained) until well into summer. Courting pairs characteristically perform a mutual bill-up display with bills pointed vertically while circling and swimming side by side, this often accompanied by soft nasal calls (Nelson and Hamer 1995*b*). Pairs also chase one another in apparent flight displays. Courting pairs frequent the nest-site together up to several weeks before laying. Copulations have been observed on the sea (apparently the usual site) and in a tree (once). Extra-pair copulations have not been observed.

Breeding and life cycle

Until recently, Marbled Murrelets were among the most poorly known of any seabird species. Recently there has been an explosion of studies related to the economic importance of this species as an old-growth forest species (Ralph *et al.* 1995). Chicks are cared for at the nest-site until they fledge at about 70% of adult mass, parental care is apparently equally divided between the sexes, based on close observation of the activities of a few pairs at tree (Nelson and Hamer 1995*b*) and ground nests (Simons 1980). Murrelets visit their nesting areas year round in all parts of breeding range that have been checked. Their timing of breeding is less synchronous within localities compared with colonial alcid species.

BREEDING HABITAT AND NEST DISPERSION: typically nest in trees, but in sub-Arctic regions can nest on the ground. Along the Pacific coast of North America S of Prince William Sound, Alaska, nest in trees that form parts of old-growth stands (more than 200 years old and 81 cm diameter at breast height) (Ralph *et al.* 1995). Nest in cathedral-like ancient rainforests with extensive mosses and epiphytes; habitat characteristics for confirmed tree nest-sites in the Pacific NW, U. S. A. and Canada: elevation 332 ± 206 m a.s.l. (14–1097, $n = 33$); tree stand area 206 ± 351 ha (3–1100, $n = 16$); canopy height 64 ± 16 m, (38–88, $n = 20$); distance to ocean 16.8 ± 10.6 km (1.6–40.0, $n = 35$); and stand age 522 ± 570 years (180–1824, $n = 16$) (Hamer and Nelson 1995*a,b*). The oldest trees at a nesting site were from British Columbia—

1824 years, confirmed by dating stumps in a nearby clear-cut (Hamer and Nelson 1995*a,b*). Of 61 tree nests from North America, 46% were in Douglas Fir (*Pseudotsuga menzeisii*), 20% in Sitka Spruce (*Picea sitchensis*) 12% in Mountain Hemlock (*Tsuga mertensiana*), 10% in Western Hemlock (*Tsuga heterophylla*), 9% in Coastal Redwood (*Sequoia sempervirens*), and 1% each in Yellow Cedar (*Chamaecyparis nootkatensis*) and Red Cedar (*Thuja plicata*) (Hamer and Nelson 1995*a,b*). Size of nest trees: diameter at breast height 211 ± 84 cm (88–533, $n = 46$), tree height 66 ± 13 m (30–86, $n = 46$), nest branch height 45 ± 13 m (18–73, $n = 45$), and nest branch diameter 32 ± 16 cm (10–81, $n = 41$) (Hamer and Nelson 1995*a,b*). Despite the data, the requirement for old-growth forest habitat is still questioned by many government and forest industry officials, and most of their remaining nesting habitat is slated for logging. The most important nesting requirement is apparently the presence of natural platforms large enough to support an egg and incubating bird safely. These platforms result from dense growths of moss and other epiphytes, natural deformations, and exceptionally broad limbs. In sub-Arctic regions of S central and SW Alaska, small numbers nest in areas where trees are sparse or absent, and in these areas a few nests have been found on the ground on seaward-facing slopes, either partly protected in rock cavities (Simons 1980) or on open tundra (Day *et al.* 1983). Murrelets are rare, local, and sparsely distributed throughout most of their range where trees are entirely absent (Aleutian Is. and coastal SW Alaska), with these populations amounting to at most a few per cent of the overall North American population. About 3% of the Alaskan population was estimated to nest on the ground (Piatt and Ford 1993). Ground nesting is likely to be possible only on remote islands and alpine tundra areas where there are few or no mammalian and avian predators. The most perplexing area is Prince William Sound, Alaska, where murrelets are common, forest and tundra habitats are present, and both tree and ground nesting occur. In Siberia, nesting occurs in taiga larch forests with suitable platforms composed of mosses and lichens but in smaller trees than in North America (Nechaev 1986; Konyukhov and Kitaysky 1995). Nests are normally widely dispersed but occasionally may be clumped because young and eggs have been found in close proximity (S. K. Nelson, personal communication). Usually nest solitarily at low densities, although up to three previously used nesting platforms have been found in one tree where suitable habitat is limited (S. K. Nelson, personal communication). This species and Kittlitz's Murrelet nest the furthest from the sea (up to 70 km) and at the highest elevations (>1000 m) of any alcids.

NEST: normally located in depression in soft moss or detritus on branch of large tree (Hamer and Nelson 1995a), or in some parts of Alaska in similar depression on the ground (Simons 1980). Tree nests usually located 70–80% of the way to the treetop, about 1 m from the tree bole, in a mossy depression or nest cup on a level platform about 30 cm by 20 cm. Nearly all tree nests are invisible from above owing to cover from branches or boughs (87% of tree nest had more than 75% overhead cover; Hamer and Nelson 1995a), making incubating birds extremely difficult to detect from the air or ground.

EGG-LAYING: takes place from mid-Mar to mid-July in California, late-April to late-July in Oregon and Washington, early-May to early-July in British Columbia, and mid-May to early July in Alaska; chick-rearing period lasts until early Sept in all parts of North America (Hamer and Nelson 1995a,b).

EGGS: one egg, no records of relaying of lost first eggs. Egg shape: subelliptical, smooth textured, pale olive green to greenish yellow ground colour with numerous irregular brown, black, and purple spots. Egg size: from North America, *Brachyramphus m. marmoratus*, summarized by Day *et al.* (1983), 59.8 ± 2.1 (57.0–63.0) $\times 37.6 \pm 1.4$ (35.0–39.3, $n = 11$) mm;

mass of fresh egg, 36–41 g (Sealy 1975a; Simons 1980; Hirsch *et al.* 1981), about 16–19% of adult mass. From Sea of Okhotsk and Kamchatka, *Brachyramphus m. perdix*, (Konyukhov and Kitaysky 1995), $62.3 \pm 4.3 \times 39.2 \pm 0.2$ mm ($n = 4$).

INCUBATION: breeding adults have two lateral brood patches. Incubation shifts last 24 hrs, both sexes apparently contributing equally. Incubating bird remains motionless in a flattened posture, but may engage in nest defence behaviour (bill jabbing) in response to approach of predators such as Ravens (Nelson and Hamer 1995b). Duration of incubation period, 27–30 days (Nelson and Hamer 1995b). Incubation change-overs take place before dawn, immediately after arrival of relieving bird. Timing of incubation change-overs (minutes before dawn): 23.2 ± 13.9 (0–82, $n = 99$); duration of change-overs: 26.0 ± 39.2 sec ($n = 76$ exchanges; Nelson and Hamer 1995b). Egg is sometimes left unincubated for 3–4 hr periods during the day.

HATCHING AND CHICK-REARING: duration of pipping unrecorded. Chick is apparently brooded continuously for at least first 2 days after hatching, intermittently thereafter (Nelson and Hamer 1995b). Adults provision chick with single (sometimes two) fish held crosswise in bill. Most chick feeds take place within 1 hr of dawn and dusk, less frequently at other times of day. Chick provisioning rate: 3.2 ± 1.3 feeds/day (1–8, $n = 10$ nests; Nelson and Hamer 1995b); provisioning rate increases with chick age. Time spent at the nest during a feed: 12.6 ± 2.8 min (0.2–80, $n = 16$). The parents usually pause and remain motionless for several minutes after landing and before feeding chick, particularly at high altitude nests, apparently to rest from their exertions. Food passing is preceded by chick begging calls. The timing of the last feeding varies from 5 min to more than 2 days before the chick departs. Chicks fledge unaccompanied and unassisted by parents and there is no evidence of post-fledging parental care.

CHICK GROWTH AND DEVELOPMENT: semi-precocial. Mass at hatching: 32.0–34.5 g ($n = 2$) or about 15% of adult mass. Duration of nestling period: 27–40 days. Chick mass at fledging: 157.0 ± 28.5 g ($n = 9$; Sealy 1975a; Simons 1980; Hirsch *et al.* 1981; Nelson and Hamer 1995b) or about 71% of adult mass. Chicks remain motionless about 90% of the time on the nest until a few days before fledging, when they begin to exercise their wings frequently by flapping and stretching them, and they preen nearly constantly. Chicks remain about 80% down-covered until 48–24 hrs before departure, when they remove the down with their bill and feet (Simons 1980; Hirsch *et al.* 1981; Nelson and Hamer 1995b). Fledging takes place at dusk: among six nests from which fledging was observed, all fledged between 2020 and 2124 hrs, between 11 and 55 min after sunset (Nelson and Hamer 1995b). Successful fledgings apparently always involve a direct flight by the chick from nest-site to the sea (confirmed by radio-telemetry in one case). Fledglings have been found in watercourses but these chicks may be failed fledgers unlikely to survive (most were flightless). The mechanism(s) of seaward orientation are unknown, some chicks may be able to see the ocean soon after taking flight, but others must make long journeys before coming within line of sight of the sea. Possibly chicks observe their parents' incoming flight direction and use this for orientation.

BREEDING SUCCESS: it has been difficult to estimate productivity in this species because of the small number of nests that were usually found late in the season. Nevertheless, breeding success is apparently quite low, owing to predation of eggs and chicks by corvids, raptors, and squirrels. Only nine fledglings were produced in 32 nests monitored (28% reproductive success), although many of these nests were located in fragmented habitat and thus the sample may not reflect natural breeding performance (Nelson and Hamer 1995a). Probably suffer from increased predation rates owing to forest fragmentation, and the link between nesting success and habitat characteristics is a crucial issue. Another concern is that as birds are forced further inland and to higher elevations by habitat loss, fledging success will drop because fledglings will be more likely to get lost on the way to the sea.

PREDATION: Avian predators of eggs and nestlings include Raven, Northwestern Crow, and Steller's Jay (*Cyanocitta stelleri*; Nelson and Hamer 1995a). Western and Glaucous-winged Gulls, Bald Eagles, Peregrine Falcons, Sharp-shinned Hawks (*Accipiter striatus*), Goshawks (*A. gentilis*), and Great Horned Owls predate adults at the nest-site and at sea. No mammalian nest predation has been documented, but ground nests are likely to be vulnerable to predation by foxes and rodents where these are present. A Northern Fur Seal (*Callorhinus ursinus*) was observed capturing a Marbled Murrelet at sea (Nelson 1996).

MOULT: from Carter and Sealy (1995). Juvenile plumage acquired about the time chicks fledge and retained through their first winter. Recently fledged juveniles have white underparts evenly covered with fine speckling, this created by dusky tipping to contour feathers, sometimes with slightly darker band of speckling across upper breast. These birds have been identified and counted at sea to assess fledgling production. However, the speckling fades or wears quickly and within a month of fledging most juveniles closely resemble winter-plumage adults except for dusky tipping to their underpart feathers and coverts (Basic I plumage). Breeding (Definitive Alternate) plumage is attained in an incomplete moult (of contour but not flight feathers, Pre-alternate moult) that takes place in Mar, Apr, and May. First-summer (Alternate I) birds usually resemble breeding plumaged adults but their brown plumage is acquired later in the season. All birds undergo a complete moult in late summer and autumn (Pre-basic moult, Aug–Nov) when the flight feathers are dropped simultaneously, leaving the birds flightless for about 2 months. Adults retain their black-and-white winter (Definitive Basic) plumage from Oct to Mar.

Population dynamics

SURVIVAL AND AGE AT FIRST BREEDING: No measures of adult survival rate have been obtained. A mark–recapture study using floating mistnets was started in 1994 at Theodosia Inlet, BC, an area frequented by many murrelets but with minimal nesting habitat nearby (G. Kaiser, A. Desrochers, and F. Cooke); 176 birds were captured at sea, but given difficulty of recapturing individuals (seven recaptures out of 186 captures in 1995) it is likely to be difficult to obtain a useful survival estimate for this species. Age at first breeding unknown.

PAIR- AND NEST-SITE FIDELITY: No information on pair-fidelity. Tree nest-sites are sometimes used in successive years, but lack of marked individuals has hampered investigation of this question. For example, of 16 nest platforms checked in a year following occupancy, four were in use, although it was not clear whether the same individual birds had returned (S. K. Nelson, personal communication).

Kittlitz's Murrelet *Brachyramphus brevirostris*

Uria brevirostris (Vigors, 1829, Zool. J. IV no. XV: 354—San Blas, Mexico = North Pacific)

PLATE 3

Description

Like Marbled Murrelet, a small, slender-bodied alcid with a cryptic summer plumage that serves as camouflage in the bleak landscape of its alpine tundra nesting grounds. Bill very short, with culmen more strongly curved than Marbled Murrelet's, and with a nearly straight gonys. The small size of the bill is accentuated by the forward extension of feathers over the nostrils. This species and Marbled Murrelet have extremely short tarsi (Pitocchelli *et al.* 1995), shorter than all other alcids including even the tiny Least Auklet. Legs and feet weak and set far back on the body, even relative to other alcids, so this species has poor agility on land and shuffles awkwardly along the ground at the nest-site. Adopts a nearly horizontal posture on land, resting on tarsi and toes and often on sternum as well. Typical posture at sea has a short- or bull-necked profile with head uptilted and tail cocked. Flight appearance and style similar to Marbled Murrelets, with long narrow wings and fast wing-beat producing very fast direct flight. Primaries rather long, narrow, and pointed. In all plumages underwing coverts are dark brownish grey. Outer tail feathers mostly white, contrasting with grey-brown central tail feathers, like a small *Calidris* shorebird.

ADULT SUMMER: entire upperparts extensively mottled with greyish brown and white flecking resembling the summer plumage of a female ptarmigan or shorebird; mottling indistinct and pale on face, which serves to emphasize the large dark eyes; throat and breast heavily mottled with grey and brown flecking and indistinct barring that fades out on lower breast; belly and undertail coverts mostly white with some indistinct flecking; scapulars dark brown, fringed with buffy grey; flight feathers and upperwing coverts blackish brown, as are the upper and underwing coverts; bill black, irides, legs, and feet blackish brown.

ADULT WINTER: black-and-white winter plumage similar to murres' winter dress, but back is greyer. Back, rump, wings, and tail greyish with white scapular patches contrasting strongly with the otherwise uniform upperparts; scapular patches are visible when the bird is both resting on the sea and flying; darker feathering forms a small contrasting black cap covering the crown of the head and extending onto nape; sides of neck and nape are white, with a small pre-ocular triangle of dark feathering that probably functions as an anti-glare panel. Some birds (presumably immatures and sub-adults, 1- or 2-year-old birds) remain in winter-like plumage during summer, while others partly attain summer plumage.

JUVENILE AND FIRST WINTER: resembles winter adults but has indistinct fine greyish-barring on the underparts during autumn; probably nearly indistinguishable from adults by mid-winter.

CHICK: retains fluffy down until just before fledging. Down on the head is buffy yellow with black spotting, on back, grey with yellowish suffusion, on throat and breast, buffy yellow with black spotting grading to light grey belly. Bill is black with white egg tooth; feet pink with brownish webs.

VARIATION
Monotypic, no notable patterns of geographical variation have been described.

MEASUREMENTS AND WEIGHTS
From Sea of Okhotsk (Kozlova 1957), unsexed birds (mm): wing 126.0–142.6; culmen 7.9–10.5; tarsus 15.5–18.0. From Alaska (Sealy *et al.* 1982): wing 136.7 ± 4.6 (124.8–146.0, $n = 101$), culmen 10.7 ± 0.9 (8.5–14.0, $n = 100$), bill depth 5.0 ± 0.3 (4.1–5.8, $n = 90$), tarsus 15.1 ± 0.8 (13.5–16.5, $n = 36$), tail 30.7 ± 2.3 (26.8–36.0, $n = 52$). From Alaska (J. F. Piatt, unpublished data): wing 144.0 ± 4.1 ($n = 14$), culmen 12.1 ± 1.2 ($n = 14$), tarsus 17.1 ± 0.9 ($n = 14$). Body mass, from Alaska (Pitocchelli *et al.* 1995), 241.1 ± 25.3 g ($n = 28$).

Range and status
RANGE
In North America locally distributed from LeConte Bay in SE Alaska westwards through Prince William Sound, the Kenai Peninsula, Kodiak and Afognak Is., along the Alaska peninsula and sparsely along the Bering Sea coast of Alaska northwards beyond the Seward Peninsula to at least Cape Lisburne in the Chukchi Sea, and very locally through the Aleutian Is. (particularly the Fox, Andreanof, and Near Is.; Sowls *et al.* 1978; Day *et al.* 1983). Birds breeding on the Bering coasts must move southwards after breeding owing to sea-ice, but there is little southward movement out of the southern parts of the breeding range in North America, although vagrants have reached southern British Columbia, Washington, and California. The most accessible locations to see this obscure bird are Homer Spit in Katchemak Bay and Glacier Bay, Alaska. In Siberia, numerous only in the cold waters of the northern Sea of Okhotsk, with birds present during the breeding season in Shelikhov Gulf (nest found), and Babushkin Bay. Also present in Karaginskii and Korf Bays, and on the Bering and Chukchi coasts of the Chukotka Peninsula. Birds have also been seen as far S as the Kuril Is. (Paramushir Is. and Kolyuchin Bay), off Hokkaido, and as far N as Wrangel Is., where they may breed (Shibaev 1990c). Their distribution along the Siberian Bering Sea coast S of Chukotka is unclear.

STATUS
Common at only a few widely scattered locations, the best known of which are Glacier Bay and Prince William Sound, Alaska. van Vliet has speculated that the world population may be only about 20 000, of which 95% are in Alaska (van Vliet 1994). No population trends are apparent, owing to lack of quantitative data and only recent development of appropriate census techniques. However, large numbers were killed in the 1989 *Exxon Valdez* oil spill (Piatt *et al.* 1990a), possibly amounting to as much as 10% of the world population (van Vliet and McAllister 1994).

Field characters
Brachyramphus murrelets are distinguished from other alcids by their small size, slender bodies, and relatively long pointed wings. In flight at a distance, both Marbled and

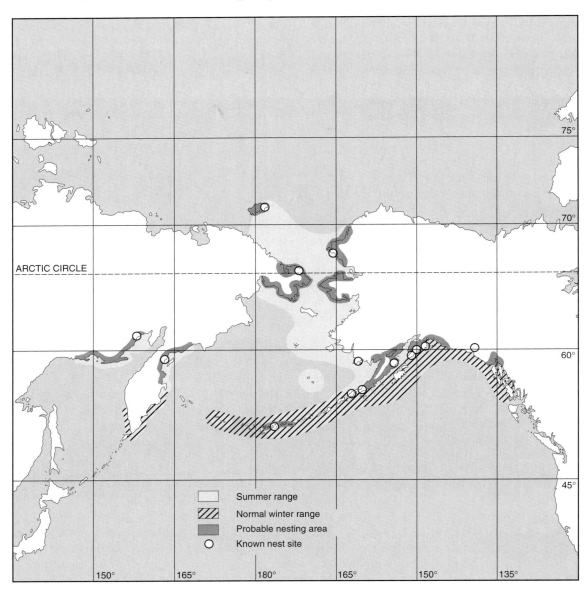

Summer range
Normal winter range
Probable nesting area
Known nest site

Kittlitz's Murrelets have dusky underwings and conspicuous white scapular patches that differ from Ancient Murrelet's uniformly dark grey back and white underwings. In moulting plumage, Marbled and Kittlitz's Murrelets can pose an identification challenge where they mingle together in Alaska and Siberia. However, breeding plumage Kittlitz's Murrelets have paler grey-brown mottling that does not extend on to lower belly, vent, or undertail coverts, and they have white outer tail feathers that flash conspicuously in flight. In all plumages Kittlitz's shows a distinctly paler face with a more contrasting and larger dark eye. Marbled and Kittlitz's Murrelets have similar black-and-white winter plumage, but in

Kittlitz's the face is mostly white with the eye isolated in a white field, and the back is greyer, while Marbled Murrelet's eye is hidden within their dark cap. In winter, Kittlitz's Murrelets differ from Ancient Murrelets by their white scapulars, harlequin-like black-and-white face pattern, and dark bill and underwing coverts.

Voice

Has not been described in the literature, but like Marbled Murrelet, this species has conspicuous flight calls. The call sounds like a harsh low Northwestern Crow or Oldsquaw, 'aah-aah-aah', usually repeated 3–5 times (R. H. Day, personal communication).

Habitat, food, and feeding behaviour
MARINE HABITAT

Summer and winter ranges defined by ocean surface temperatures of 2–10 and 1–8 °C, respectively. Like Marbled Murrelet, forages close to shore in sheltered waters of bays, inlets, and fiords; these two species are often seen together. Large concentrations occur in turbid water near tidewater glaciers at the heads of fiords in Prince William Sound and Glacier Bay, Alaska. Avoids areas of heavy ice cover. Absent from small isolated islands in Aleutians and Bering Sea, present around larger islands with extensive stony uplands and convoluted shorelines (e.g. Unalaska, Adak, and Attu Is., Alaska). In Beringia ranges as far S as the sub-Arctic waters around Adak Is. (54 °N) N to the high Arctic waters of the Chukchi Sea (70 °N).

FEEDING BEHAVIOUR

Without a sublingual pouch, cannot efficiently carry plankton to the nest, so chicks are provisioned with small fish. Feeding flocks gather at the heads of fiords, in constricted channels where there are strong tidal currents, and in silt-laden brackish water near tidewater glaciers (at Prince William Sound and Glacier Bay, Alaska) and where prey is concentrated. There is no information on dive depths or foraging behaviour.

DIET

Based on anecdotal accounts and inferred from their tiny bill, Kittlitz's Murrelets preferred prey is larval fish and zooplankton such as amphipods, copepods, euphausiids, and polychaetes. However, adults do prey on other fishes. According to Bent (1919), small crustaceans (probably copepods and amphipods) and 'a slippery slug-like animal about an inch long' (polychaete worms or pteropods). J. F. Piatt (unpublished data) found juvenile sandlance and walleye in the stomachs of adults shot in Katchemak Bay, Alaska. Chicks are fed single whole fish carried externally crosswise in the bill; chick diet at one nest included sandlance (67%), capelin (18%), and unidentified fish (15%) (Naslund *et al.* 1994).

Displays and breeding behaviour

Little published information. Non-colonial. Mated pairs and courting birds engage in flight chases similar to those of Marbled Murrelet.

Breeding and life cycle

Probably the least-known alcid species.

BREEDING HABITAT AND NEST DISPERSION: nest-sites are on stony fell-fields, on both steep talus slopes and nearly level areas, with little vegetation other than lichens, usually with snowfields or permanent ice nearby, in alpine areas well above tree line (Day *et al.* 1983). van Vliet (1994) pointed out that concentrations of this species occur on mountainous coasts with extensive glaciers and recently deglaciated stony uplands, since major concentrations occur near glaciers in Glacier Bay, Prince William Sound, and near the Kenai Peninsula, Alaska. Distance of nests from sea: 20.5 ± 25.6 km (0.25–75.0, $n = 15$), altitude of nests (above sea level): 573.6 ± 275.5 m (140–1070, $n = 15$), usually on north-facing slopes (Day *et al.* 1983; Murphy *et al.* 1984; Shibaev 1990*c*; Naslund *et al.* 1994). On high mountains, nests usually face south and often occur in patches of bare ground among snowfields (Day 1995). Nests solitarily, average nearest neighbour inter-nest distances probably on the scale of kilometres.

NEST: on bare ground, sometimes in surface scrapes or slight depressions, often at high elevations, usually immediately down slope of a rock larger than the incubating bird. The egg is laid on collections of small pebbles with moss sometimes present, unclear if any material is added by the birds. The first nests were described by native people, who informed skeptical pioneering ornithologists that the bird nested high on mountain tops among snowfields (Bent 1919).

EGG-LAYING: Laying takes place in mid-May to mid-June in SE Alaska, early–to mid-June in the Aleutian Is. and Bering Sea coasts, and late June on the Chukchi Sea coasts (Day 1996), and its timing is affected by the timing of ice breakup (especially in Chukchi Sea) and snow-melt from the upland nesting areas. Fledging takes place between early July and early Aug in south and in early to late Aug in north (Day 1996).

EGGS: one egg, relaying of lost first eggs has not been noted. Eggs subelliptical, smooth textured, pale olive green, bluish green or yellow ground colour with numerous irregular brown spots, similar to Marbled Murrelet eggs and may not be distinguishable from them unless the incubating bird is seen (Day *et al.* 1983). Egg size: summarized by Day *et al.* (1983), 60.0 ± 2.0 (57.8–62.5) $\times$ 37.3 ± 1.1 (35.6–39.0, $n = 9$), very similar to Marbled Murrelet egg dimensions; no data on mass of fresh eggs.

INCUBATION: no information on duration of incubation. Incubating birds adopt a flattened posture with head low to the ground. Parents sit tight on nest at approach of humans, and presumably other predators, relying on their camouflage.

HATCHING AND CHICK-REARING: no information on duration of pipping. Chicks are fed single whole fish by both parents several times per day (totalling 4–6 feeds per day, equal effort by both parents at Kenai nest). Food de-liveries occurred at all times of day but were most frequent in the hour before sunrise and the hour after sunset. Parents remained at nest for an average of 11.4 min at each food delivery (Naslund *et al.* 1994).

CHICK GROWTH AND DEVELOPMENT: no measurements of chick growth are available. Duration of nestling period: 25 days at the Kenai nest (Naslund *et al.* 1994). Fledging mass is probably close to 80% of adult mass, suggesting that growth rate is high. Chick remains covered in dense down and nearly motionless at nest until it reaches fledging size. The chick at the Kenai nest retained all of its down until less than 12 hrs before departure (Naslund *et al.* 1994). Chicks probably fly from their nest-site to the sea, although it has been speculated that some chicks may use streams or river to get to the coast (Day *et al.* 1983). Both parents delivered fish to the nest-site after the chick fledged at the Kenai nest, suggesting that the parents were unaware of the chick's location after its departure from the nest, and that there is no post-fledging parental care. No quantitative data are available on breeding success.

PREDATION: Little information. However, Arctic and Red Foxes may occasionally take adults, nestlings, and eggs at nest-sites. Avian predators are likely to include Slaty-backed (*Larus schistisagus*), Glaucous, and Glaucous-winged Gulls, Gyrfalcons, and Peregrine Falcons.

MOULT: moults and feather generations have not been described in detail. However, it is clear that there are two moult periods, one incomplete moult in spring (Pre-alternate, mid-Apr–mid-June) when the black-and-white winter plumage is replaced with the mottled summer plumage, and one complete moult in autumn (Pre-basic, starting in late Aug; Sealy 1977). During the latter moult, adults drop their primaries simultaneously and remain flightless for several weeks.

Population dynamics

No data on survival are available. Age at first breeding unknown. There is little information on pair- and site-fidelity, but the Kenai Peninsula nest had traces of old eggshells, weathered fecal material, and chick down nearby at the time the bird was incubating, suggesting it had been used in a previous year (Naslund *et al.* 1994). Other nests showed no sign of multi-year use (R. H. Day, personal communication).

Xantus' Murrelet *Synthliboramphus hypoleucus*

Brachyramphus hypoleucus (Xantus, 1859, *Proc. Acad. Sci., Philadelphia,* (1860): 299—Cabo St. Lucas, Baja California)

PLATE 4

Description

A small, relatively slender auk, with clean black-and-white plumage, small head and sharp, needle-like bill. Ungainly on land, not usually standing up, except to walk, when it shuffles on tarsi. A good flier, it can take off from flat ground and rises from the water without taxiing. Flight direct with rapid, whirring wing-beats.

ADULT: upperparts entirely black, with a faint bluish cast in fresh plumage, becoming greyish with wear; underparts white, including underwing and undertail coverts, but with black strip along flank; there are small white spots just above and below eyelids; white on cheeks and face extends horizontally back from base of bill, passing just below eye (*S. h. scrippsi*); in *S. h. hypolecus*, white extends further up in a crescent in front of and sometimes above, eye and to variable extent behind eye; white spot on lower eyelid forms a broader white streak; ear-coverts grey, rather than slate, with white sometimes extending to ear-coverts and sides of neck; bill black, tarsus and toes light blue, webs black, iris brown; no difference between winter and summer plumage.

JUVENILE AND FIRST WINTER: similar to adult, but with indistinct fine barring along flanks and shorter bill.

CHICK: down black above and white below, mimicking adult plumage, including white eyelid spots; legs and feet similar to adult (and similar in size), pale blue with dark webs.

VARIATION

Suprisingly, for a species with a very limited range, there is distinctive geographical variation with two subspecies recognized: *S. h. hypoleucus*, breeding on islets off Guadalupe I., and San Benito I., Mexico and *S. h. scrippsi* found over the rest of the species' range. The two races differ in the plumage patterning of the head (see Description) and in bill measurements; longer and thinner in nominate race. Measurements of birds from San Benito I. are somewhat intermediate between the two races and it has been suggested that both breed together, suggesting either that they are sibling species, or that interbreeding is in progress (Jehl and Bond 1975). A bird showing characteristics of *S. h. hypoleucus* was found nesting at Santa Barbara I. once (Winnett *et al.* 1979). The relationship and distribution of the two races requires elucidation.

MEASUREMENTS

♀♀ consistently slightly larger than ♂♂ in all measurements except bill depth (Jehl and Bond 1975). See Table: (1) *S. h. scrippsi* adults from the California Channel Is.; (2) adults from Los Coronados Is.; (3) *S. h. hypoleucus* adults from San Benito I.; (4) adults from Guadalupe I.

WEIGHTS

Breeding adults trapped at Santa Barbara I. averaged 171 ± 8.8 g ($n = 171$) and non-breeders 163.5 ± 10.6 g ($n = 191$). Among breeding pairs, females averaged 11 g heavier than males ($n = 13$, Murray *et al.* 1983). Average for birds arriving to begin incubation

Table: Xantus' Murrelet measurements (mm)

ADULT ♂♂	ref	mean	s.d.	range	n
wing	(1)	120.3	0.9	114–123	10
	(2)	119.1	0.4	115–125	44
	(3)	118.5	0.9	111–123	13
	(4)	119.3	0.7	114–128	29
culmen	(1)	17.0	0.3	15.6–19.3	10
	(2)	18.0	0.1	16.3–19.2	41
	(3)	18.6	0.4	16.8–21.4	11
	(4)	19.3	0.2	17.4–21.2	27
bill depth	(1)	6.2	0.1	5.8–6.4	10
	(2)	6.1	0.1	5.6–6.5	41
	(3)	5.7	0.1	5.3–6.1	12
	(4)	5.8	0.1	5.2–6.1	28
tarsus	(1)	24.5	0.2	23.4–25.5	10
	(2)	24.2	0.1	22.9–25.3	40
	(3)	23.6	0.3	22.4–25.2	13
	(4)	23.5	0.1	21.2–25.4	29

ADULT ♀♀	ref	mean	s.d.	range	n
wing	(1)	122.5	0.7	119–126	8
	(2)	120.0	0.5	115–127	37
	(3)	120.5	0.8	115–125	12
	(4)	120.5	0.7	115–127	27
culmen	(1)	18.0	0.4	16.8–19.0	7
	(2)	18.2	0.1	16.5–20.3	36
	(3)	18.2	0.4	16.0–20.5	12
	(4)	19.8	0.2	18.4–21.3	22
bill depth	(1)	5.9	0.1	5.4–6.8	8
	(2)	6.0	0.1	5.5–6.4	37
	(3)	5.7	0.1	5.3–6.3	12
	(4)	5.6	0.1	5.2–6.3	22
tarsus	(1)	24.9	0.3	23.8–26.0	8
	(2)	24.4	0.1	22.3–25.5	37
	(3)	23.9	0.3	22.0–26.9	12
	(4)	23.8	0.3	21.8–27.5	23

172.1 ± 15.5 g (n = 60) and those departing after incubation 161.8 ± 11.2 g (n = 39, Murray *et al.* 1983).

Range and status

RANGE
Boreal/sub-tropical, confined to waters off W coast of North America, from Baja California to British Columbia, inshore while breeding but mainly offshore otherwise. Usually in waters warmer than 12 °C.

BREEDING: on the coast of California, south of Point Conception and the Pacific coast of Baja California, from the California Channel Is., S to at least San Benito and Guadalupe Is.

WINTER: post-breeding, mainly to N of breeding range, in waters affected by the California Current. Regular off central California from July to Nov, rare off Oregon (10 records, of which one was *S. h. hypoleucus*, Gilligan *et al.* 1993) and accidental off Washington and British Columbia, although 10 have been recovered drowned in fishing nets set off northern Washington and British Columbia in July and Aug (Drost and Lewis 1995), suggesting that substantial numbers may reach these latitudes in late summer. The two races appear to be intermingled outside the breeding season, although *S. h. scrippsi* is commoner than *S. h. hypoleucus* off Monterey (Small 1994). Breeding birds return to the vicinity of their colonies by late Dec, but a few persist further N; seen regularly off Monterey on Christmas Bird Counts.

STATUS
Information from Drost and Lewis (1995). The majority of the population breeds on Santa Barbara I., in the California Channel Is. (about 1500 birds), on Los Coronados Is. Off Tijuana, Mexico (about 750), on San Benito I. (several hundred, perhaps over 1000), on small islets off Guadalupe Island, Mexico, and perhaps on the main island itself (W. T. Everett, personal communication 2400–3500 birds). Small numbers also occur on San Miguel, Santa Cruz, Anacapa, Santa Catalina, and San Clemente Is. in the Channel Is. (none >250). Breeding populations on Todos Santos, San Martin, and San Geronimo Is., off W coast of Baja California, are believed to have been extirpated (Jehl and Bond 1975), but additional sites may still be found. The dispersed nesting and unobtrusive habits of the species make it very hard to census and current estimates are very approximate. Based on current colony information, the total breeding population is almost certainly less than

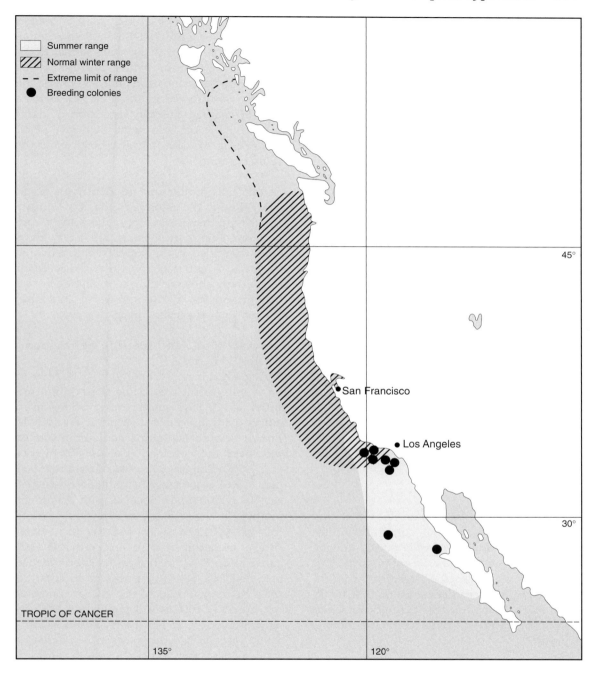

10 000 birds and this agrees with data obtained from offshore surveys. Introduced mammalian predators pose the main threat to Xantus' Murrelet populations. Cats are implicated in their disappearance from some islands off Baja California, and a considerable reduction in the population of Los Coronados Is. Cats also brought the population of Santa Barbara I. close to extinction in the early twentieth century, before they were eliminated by the

U. S. National Park Service in 1978. Ship rats (*Rattus rattus*) also pose a problem on some islands and may be responsible for the low population on Anacapa I. (Drost and Lewis 1995). Given the position of the main breeding colonies, adjacent to busy shipping lanes serving the southern California ports, the possibility of significant losses from oil pollution poses a constant threat. In addition, birds seem to be readily attracted to lights on ships and sometimes die when they collide with them. Lights on vessels moored close to breeding colonies could cause significant disruption of breeding activities and mortality to breeders.

Field characters

A small, unobtrusive auk, generally in pairs or small groups. Its size and clean black-and-white plumage distinguish it from all other species in the same area except for Craveri's Murrelet and winter-plumage Marbled Murrelet. The clean white underwing distinguishes it from both of these, which have grey underwing coverts and grey-brown flanks. Marbled Murrelets have conspicuous white scapulars and a shorter, less dagger-like, bill. Ancient Murrelets in winter plumage have white underwings, but are grey above, more dusky on the face, and have shorter, light-coloured bills.

Voice

Information from Dechesne (in Drost and Lewis 1995 and personal communication). Context and function of calls poorly known. Like other *Synthliboramphus* spp., calls frequently at night, on the breeding grounds, and especially on gathering grounds just offshore from the colony. Strident vocalizations by parents and chicks accompany colony departure.

ADULT: most frequent call at night on gathering grounds is the 'twitter', a series of 3–8 high-pitched 'seep' notes, descending slightly in pitch, also described as 'a trilling whistle given on one pitch' (Jehl and Bond 1975). This call is sometimes given twice in quick succession (about 0.2 sec apart); 'paired twitter'. Other adult calls include single note 'chips' and 'whistles'. 'Chips' are similar to the individual elements of 'twitter' call and may be given singly, in pairs, or in long sequences; the latter were observed in the presence of Barn Owls (*Tyto alba*) and may be a form of alarm call. Adults in the burrow at night give a soft 'twitter'; similar calls may also be given on the feeding grounds.

CHICK: calls loudly and frequently on sea at night, close to the colony, presumably to make contact with parents before making away from the shore. Commonest call is thin 'peer', given at sea, and also in the hand. When delivered urgently, the call may contain two discordant frequencies, showing that it originates in two distinct sources in the syrinx. At hatching, chicks also give a feeble 'peep'.

Habitat, food, and feeding behaviour

MARINE HABITAT

During breeding, presumably feed mainly in the vicinity of their breeding sites (Hunt and Butler 1980). Systematic observations from cruises off California between 35° and 43° N showed that during spring and summer most were over the continental shelf (Karnovsky *et al.* 1995). Similar cruises in autumn revealed them scattered offshore, with highest densities over the continental slope, but with maximum numbers in oceanic waters beyond, mainly >50 km offshore (Karnovsky *et al.* 1995). Total population of area surveyed was estimated to be 3862 in winter (95% confidence interval 2133–6895). As this is thought to cover a major portion of the species' range at that season, a total world population, including pre-breeders of less than 10 000 is indicated. Outside breeding season, scattered at low densities, often in pairs or groups of up to six, but never in larger flocks (Karnovsky *et al.* 1995). Nothing is known of diving behaviour.

DIET

Little known. During breeding season around Santa Barbara I., fed on larval anchovy

(*Engraulis mordax*, 36% of stomachs, *n* = 22), Pacific saury (*Cololabis saira*, 9%) and rockfish 9%). Another 36% of stomachs contained unidentified fish remains (Drost and Lewis 1995).

Displays and breeding behaviour

Virtually nothing is known. Colony attendance is entirely nocturnal. Vocal behaviour appears to be less complex than that of the Ancient Murrelet, but the prominence of gathering ground vocalizations (Voice) suggests that behaviour there may play an important role in pair formation. Birds are very abundant on the sea off their colonies at night during the late incubation period, as in Ancient Murrelets. These gathering ground aggregations appear to have an important social function. There seems to be very little above-ground behaviour on land, perhaps reflecting pressure from predators. Copulation has been observed on the ground, but if it occurs in nest-sites or at sea during the night, it would be unlikely to be seen.

Breeding and life cycle

Information mainly from Murray *et al.* (1983) and Drost and Lewis (1995).

BREEDING HABITAT AND NEST DISPERSION: nests on arid islands with sparse, thorny vegetation, usually within 200 m of the sea. Dispersion rather scattered, for an auk, probably because of a shortage of suitable breeding sites. Average distance to nearest nest at Santa Barbara I.: 5 m (range 0.15–40.0 m, *n* = 172, Murray *et al.* 1983).

NEST: mainly in small (<20 cm high) caves or crevices in rocks, or in cavities under boulders or roots. Sometimes on the surface below dense shrubs, occasionally in Rabbit or Burrowing Owl (*Athene cunicularia*) burrows. At Los Coronados I., sites used by Xantus' Murrelets early in the season are often used later in the same year by storm-petrels (Everett 1992).

EGG-LAYING: birds begin to call in vicinity of breeding colonies in Dec and come ashore in second half of Feb. Laying is rather unsynchronized, with 80% of eggs at Santa Barbara I. laid over 24–47 days (3 years, Murray *et al.* 1983). Earliest eggs laid 22 Feb, with peak laying during 21 Mar–21 Apr, and latest about mid-June. There is no evidence that timing differs between colonies, with the range of egg dates in the Channel I. similar to that at Guadalupe I.

EGGS: subelliptical to oval; variable in ground colour from off-white, through cream, to olive brown, with variable grey or brown blotches, spots, or streaks, similar to other *Synthliboramphus* spp. Clutch normally one (25%) or two (69% nests at Santa Barbara, Murray *et al.* 1983), but sometimes three or four, presumably the product of two females. Some one-egg clutches reported may have been incomplete, as the two eggs are laid 8 ± 0.8 days (5–12, *n* = 42, Murray *et al.* 1983) apart. More than 50% of intervals are 8 days. The two eggs are not as similar to one another in colour and pattern as is usually the case within clutches (Drost and Lewis 1995). The second egg averages about 1 g heavier than the first. Measurements: Anacapa I., 53.8 ± 2.1 (50.7–57.1) × 35.9 ± 0.9 (34.5–37.6, *n* = 17, Drost and Lewis 1995) mm; Los Coronados Is., 54.1 (50.0–57.2) × 35.8 (33.5–37.6, *n* = 85, Bent 1919) mm. Fresh mass at Santa Barbara I.: 37.2 ± 2.9 g (29.5–43.5, *n* = 100, Murray *et al.* 1983), approximately 22% of adult body weight. Pipped eggs average 31.4 g (Eppley 1984). Replacement clutches have been reported and there are 12 cases known where second clutches have been laid in a site after a brood had successfully departed. The need for prolonged parental care at sea in this species seems to preclude the possibility that these were second broods, unless the first brood was lost soon after departure (this possibility cannot be excluded).

INCUBATION: begins on average 2 days after laying of second egg; until then, first egg is unattended, at least during the day. Intermittent neglect for up to 4 days is common after start of incubation, with average of 2.9 days of neglect during the entire incubation period; 62% nests neglected for at least 1 day (n = 51, Murray *et al.* 1980). First shift usually by ♂ (81%, n = 11). Incubation shifts average 3 days (1–6, 82% 2–4, n = 408, Murray *et al.* 1983), although shorter towards the end of incubation. Some development of embryo continues despite neglect, because eggs with more neglect hatch after fewer days of incubation than those neglected less, although latter were heavier at hatching (Murray *et al.* 1980). Air temperatures during incubation: minimum 10–13°, maximum 18–24 °C. Mean number of days of incubation before hatching 30.6 ± 2.4 days (24–34, n = 37).

HATCHING AND CHICK-REARING: pipping and chick vocalizations (soft 'peeps') occur up to 5 days before hatching; chicks usually hatch within a few hours of one another (Murray *et al.* 1983). Chick hatches at 25.8 g (n = 22) and loses 8% of mass daily while in the burrow, departing at a mean of 23.8 g (n = 26). Chicks are brooded continuously until they leave the nest, but are not known to be fed there.

CHICK DEPARTURE: chicks and parents depart together at night, usually on second night after hatching. By this time, chicks are capable of great agility, running rapidly, jumping obstacles and swimming very fast; they can also thermoregulate in sea temperatures as low as 12 °C, although body temperature may be maintained at 2–4 °C below adult temperatures (Eppley 1984). Departure is accompanied by loud vocalizations by chicks and parents close to the nest, after which adults fly to sea and chicks follow rapidly on foot (Drost and Lewis 1995). Behaviour at departure may follow pattern seen in Ancient Murrelet, but has not been described in detail. Chicks reared in captivity exhibited a dramatic alteration in behaviour after 48 hrs, changing from snuggling docilely in their rearing boxes, to running excitedly all over the house (Z. Eppley, personal communica-

tion). Like Ancient Murrelets, family parties, usually two chicks and two parents, move rapidly away from the vicinity of the colony once they reach the sea.

CHICK GROWTH AND DEVELOPMENT: after departure, chicks are fed at sea until more-or-less fully grown, but nothing is known of this period. Captive-reared chicks grew very slowly (2.5 g/day, Eppley 1984) and feather development was not noticeable until 17 days. We do not know if this is typical of wild birds. Nothing is known of development in the wild, or when the young become independent.

BREEDING SUCCESS: production at Santa Barbara I. averaged 0.72 chicks departing per clutch laid; 57% of nests produced one or more young (Drost and Lewis 1995). Deer mice (*Peromyscus*) caused 44% of egg loss observed by Murray *et al.* (1983), although some of the eggs concerned may already have been deserted; an additional 14% of clutches were deserted in any case. The probability that reproductive success was influenced by observer disturbance seems quite high in this rather shy species.

PREDATION: Predation by introduced mammals and by Barn Owls *Tyto alba*, Western Gulls, and Peregrine Falcons accounts for many birds at breeding colonies; some adults also die through entanglement in or impalement on vegetation (e.g. cactus).

MOULT: no evidence of post-juvenile moult; the timing of the first moult is not known. Adults undergo a complete post-breeding moult in July and Aug, with body moult apparently lagging somewhat behind the replacement of remiges (simultaneous) and rectrices. This may be the only moult of the year, although some partial body moult may occur pre-breeding, as in most auks (Drost and Lewis 1995).

Population dynamics
Several birds ringed as adults have been re-trapped at more than 14 years old on Santa

Barbara I., suggesting that breeders may have a fairly high survival there, but there are no estimates of annual survival rates, or of age at first breeding. Nest-fidelity is high, with 59% ($n = 22$) of ringed breeders using the same site in 3 consecutive years.

Craveri's Murrelet *Synthliboramphus craveri*

Brachyramphus craveri (Salvadori, 1865, *Atti Soc. Ital. Sci. Nat., Milano,* 8: 387—Gulf of California = Raza Island).

PLATE 4

Description

A small, little-known murrelet, with clean black-and-white plumage: very similar to Xantus' Murrelet and thought at one time to be conspecific. It has pointed wings and a rather slender body, flying fast and low over the water. When sitting on the sea surface keeps low with an almost prone posture.

ADULT: dark blackish-brown on head, nape, neck, and back and on upper surfaces of wings and tail; gleaming white on lower chin, throat, breast, belly, and under tail; the black on the face extends to just below base of bill and backwards horizontally well below eye; black of the mantle extends onto sides of upper breast, forming half-collar; wide black panel along flanks, merging into dusky grey on axillaries and underwing coverts; bill is long, slender, pointed, and uniform black; irides dark brown; tarsus and toes light blue, with dark webs; no difference between summer and winter plumage.

JUVENILE AND FIRST WINTER: shorter wings and bills and blacker, less glossy plumage; otherwise similar to adults.

CHICK: black-and-white down pattern similar to adult plumage, but slightly reddish-brown in tone.

VARIATION

No subspecies recognized.

MEASUREMENTS

Slightly smaller than southern race of Xantus' Murrelet. ♀♀ seem to be slightly larger than ♂♂ all measurements except bill depth. See Table, adults in summer from throughout breeding range, based on museum skins (Jehl and Bond 1975).

WEIGHTS

A breeder weighed by Breese *et al.* (1993) during incubation weighed 152 g.

Range and status

RANGE

Found only in the Gulf of California (Sea of Cortez) and adjacent waters, inshore among islands and waters affected by powerful tidal currents before and during the breeding season; otherwise offshore in sub-tropical waters off the W coast of Mexico.

BREEDING: the only proven breeding is within the Gulf of California, where it breeds on small islands. There have been suggestions

Table: Craveri's Murrelet measurements (mm)

♂♂	mean	s.d.	range	n
wing	116.1	3.1	107–123	41
culmen	19.8	0.92	18.0–22.5	42
bill depth	5.4	0.03	4.9–5.9	43
tarsus	22.9	0.79	21.0–24.4	43

♀♀	mean	s.d.	range	n
wing	117.8	2.7	111–123	30
culmen	19.9	1.02	18.2–21.9	29
bill depth	5.3	0.16	4.9–5.7	28
tarsus	23.0	0.59	22.0–24.4	29

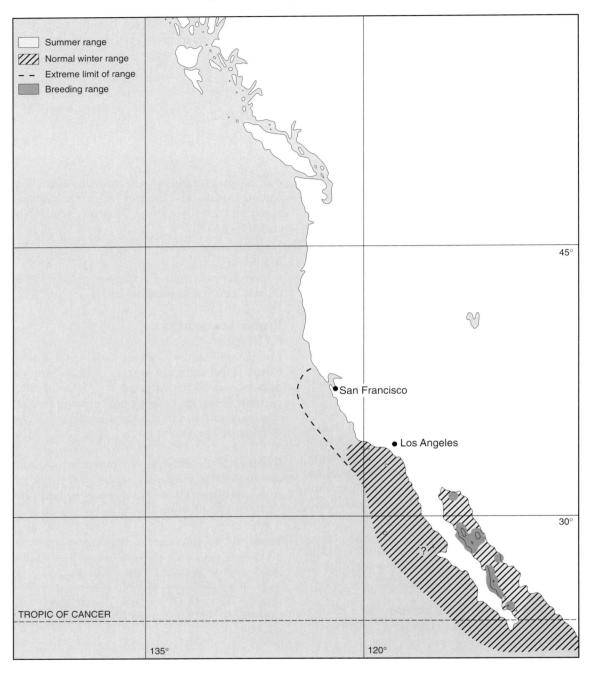

Summer range
Normal winter range
Extreme limit of range
Breeding range

45°

San Francisco

Los Angeles

30°

TROPIC OF CANCER

135°   120°

that the species may also breed on islands off the Pacific coast of Baja California (Jehl and Bond 1975). The type specimens were actu-ally taken off Isla Natividad, on the Pacific side of the peninsula (Violani and Boano 1990). Deweese and Anderson (1976) list 14 sites with proven breeding, extending practically the entire length of the Gulf, as far as islets off La Paz in the S, an axis of about 600 km. In the Canal de Ballenas, approximately half way up the Gulf, Tershy *et al.* (1993) found the species commonest in Nov– Feb, when they recorded about 1/hr on ship surveys. Very few seen on surveys in June–Oct.

WINTER: in autumn bulk of the population appears to be concentrated up to 100 km offshore off the coast of Mexico about the latitude of Puerto Vallarta, Mexico (21° N, Pitman *et al.* 1995). Some birds move northwards up the coast of California, as far as Monterey, where they are seen regularly from mid-July to mid-Oct, and especially in years when E1 Nino conditions bring warm water northwards. Apparently return to breeding areas in Nov.

STATUS
Everett and Anderson (1991) estimated about 5000 breeding pairs in the Gulf of California; if there are any breeding on the Pacific coast, they must be very few. This estimate suggests a total population, with pre-breeders, of 15 000–20 000 birds, which agrees remarkably well with an estimate from surveys at sea by Pitman *et al.* (1995) of 25 000 birds (95% confidence interval 7000–88 000).

Field characters
A small auk looking stark black and white at a distance. Best distinction from the very similar Xantus' Murrelet is the dusky, brownish-grey, rather than pure white, underwing. Bill is longer and thinner than Xantus' and head pattern resembles that of northern race of Xantus' Murrelet, not likely to be found off Baja California. However, both Craveri's and northern race of Xantus' may occur together off central California in autumn and there colour of underwing appears to be only reliable identification mark. A small black triangle extending down side of chin from just in front of eye has been proposed as a distinguishing

feature (Farrand 1983), but is not very apparent in the field.

Voice
At sea, gives short, high-pitched 'chip' calls resembling those of Ancient Murrelet. Bent (1919) refers to chasing and calling of adult birds ashore about 1 h before dawn and mentions three calls; one a harsh alarm call, the other two not described.

Habitat, food and feeding behaviour
During breeding, keeps in the mainly inshore waters of the Gulf of California. May make use of strong tidal currents among islands for feeding. Outside breeding, mainly offshore, at the edge of the continental shelf or beyond, although occasionally seen closer inshore (Small 1994). At all seasons, scattered in twos, or less commonly, in small flocks of up to eight birds while feeding. Only information on diet is from DeWeese and Anderson (1976), who collected five birds in spring in the Gulf of California. Four contained rockfish (17 items), herring (12 items), and lanternfish (*Benthosema panamense*) (8 items); species of jacks (*Carangus*), mackerel (*Scomber*), squid, and shrimp were found in one bird each. Fish ranged in length from 40 to 70 mm.

Displays and breeding behaviour
Presumably similar to Xantus' Murrelet, but almost nothing recorded. At sea, nearly always seen in twos, presumably pairs, which may remain together throughout the year.

Breeding and life cycle
Poorly known and requiring further study. Data mainly from DeWeese and Anderson (1976).

BREEDING HABITAT AND NEST DISPERSION: on arid islands with very low rainfall and sparse desert vegetation, usually close to the sea along steep rocky shorelines with abundant crevices and rock piles. Poisonous snakes are abundant on some islands, but terrestrial mammalian predators are generally absent.

NEST: under rocks or in caves and crevices on desert islands from 0.3 to 5.5 m above high tide, sometimes as much as 40 m above the sea (Breese *et al.* 1993); apparently rather low density in most places where it breeds, although a small islet off St Jose I. was said to be 'alive with them' (Bent 1919). Harder to discover than Xantus' Murrelet and more likely to be in a deep cavity where light cannot penetrate (Bancroft 1927).

EGG-LAYING: eggs laid Feb and Mar and hatch Mar and Apr. Start of breeding apparently rather unsynchronized, with young birds from a few days old to nearly fully grown reported in Apr (Bancroft 1927).

EGGS: clutch normally two (84%, $n = 63$), or one (11%), occasionally three (probably more than one female involved); subelliptical or oval, ground colour olive-buff, spotted or blotched, but not as heavily marked as Xantus' Murrelet. Measurements: 52.3 (48.5–55.0) × 34.9 (32.5–37.0, $n = 34$) mm. Mass 35 g, about 22% of adult body weight; shell 6.8% of mass.

INCUBATION: nothing reported, but presumably similar to Xantus' Murrelet.

CHICK-REARING: chicks precocial, leaving for sea with both parents 1–2 days after hatching. Reared entirely at sea. Nothing is known of their growth and development.

BREEDING SUCCESS: mean brood size observed at sea, mainly one quarter to two-thirds grown, was 1.47 chicks ($n = 62$). In one year 79% of pairs observed were accompanied by young, suggesting a minimum reproductive success of >1 chick/pair. In the following year, success was apparently much lower, with only 12% of adult pairs being accompanied by young (Tershy *et al.* 1993).

PREDATION: adults comprised 8% of remains at one Peregrine's nest. Regular predation by Barn Owls also occurs.

MOULT: not known.

Population dynamics
Nothing known.

Ancient Murrelet   *Synthliboramphus antiquus*

Alca antiqua (Gmelin, 1789, *Syst. Nat.* I. ii: 554—west of North America to Kamchatka and the Kuril Islands).   PLATE 4

Description

A small, comparatively slender auk with crisp black, white, and dove grey plumage, legs set well back on the body. Do not normally stand upright on land, like murres and puffins, but lie forward on belly. To run on land they flap their wings to keep upright, but fly strongly and can take off from flat ground. Takes off from water with little or no running; often lands by diving in head-first. Tarsus laterally compressed; bill short, laterally compressed (hence the generic name) and slightly decurved; external nostrils are circular with slightly raised edges.

ADULT SUMMER: back, upperwing, and upper-tail coverts bluish grey; head mostly black, with a prominent black bib extending to upper breast; hind neck black, and a black stripe runs at the base of the underwing; rest of underparts white, including bright white underwing coverts; flight and tail feathers dark slate. The crown is fringed by long, filamentous, white feathers, and similar feathers are scattered in the mantle. Long feathers on nape form an ill-defined crest, rarely visible; bill mainly pale pinkish to yellowish horn, with variable amounts of black at base and along ridge of culmen; legs and feet pale blue, some-

2 Summer plumage adults taking off (photo T. J. Lash).

times tending to flesh colour on webs, claws black, iris dark brown.

ADULT WINTER: similar to breeding, but lacking black bib on throat; chin and sides to upper neck have varying amounts of sooty grey suffusion, and white filamentous plumes are much reduced. Overall, upperparts have a slightly more sooty appearance.

JUVENILE AND FIRST WINTER: similar to winter-plumage adult, but with a shorter, darker, more slender bill and little trace of white trim to the crown.

CHICK: down pattern similar to adult plumage, grey above and white below, black on head, but white around eye and with grey spot behind ear.

1 Adult at breeding colony (Fiona M. Hunter photo).

VARIATION

Two races described: *S. a. antiquus* and *S. a. microrhynchos*, latter having smaller bill and less speckling on side of neck. *S. a. microrhynchos* is found in the Commander Is. the nominate race everywhere else (Shibaev 1990*d*). Measurements of specimens from Japan are similar to those from British Columbia but some wintering birds collected off British Columbia appear larger than local breeders; these may have originated in Alaska. The validity of *microrhynchos* appears doubtful.

MEASUREMENTS

No sexual dimorphism except in bill depth, but females tend to have slightly longer wings. Non-breeding birds trapped at the colony have shorter wings than breeders: 2-year-olds have shorter wings than 3-years-olds. See Table: (1) Queen Charlotte Is., breeders, freshly killed; (2) non-breeders, trapped; (3) 2-year-olds, trapped; (4) 3-year-olds, trapped (Gaston 1992*a*).

WEIGHTS

Weights of breeders vary among years at start of incubation, decrease during incubation, and

Table: Ancient Murrelet measurements (mm)

♂♂	ref	mean	s.d.	range	n
wing	(1)	139.7	2.3		15
culmen	(1)	13.5	0.7		15
bill depth	(1)	7.2	0.2		15
tarsus	(1)	27.2	0.7		15
♀♀	ref	mean	s.d.	range	n
wing	(1)	140.6	2.3		15
culmen	(1)	13.4	0.4		15
bill depth	(1)	6.7	0.2		15
tarsus	(1)	26.9	0.6		15
UNSEXED	ref	mean	s.d.	range	n
wing	(2)	138.1	3.2		859
	(3)	137.7	3.1		21
	(4)	139.3	2.8		6

usually attain a similar weight by the time of colony departure (Gaston and Jones 1989). Weights generally higher in winter than during breeding, although this comparison may be confounded by so far undescribed regional size variation (Gaston *et al.* 1993a). Breeders, Queen Charlotte Is., May (sexes combined) 213 ± 19 g (146–243, $n = 477$, AJG), ♀ with egg, just before laying 248 ± 10 g ($n = 87$); all non-breeders, 185 ± 12 g ($n = 924$, both sexes); 2-year olds, 180 ± 7 g ($n = 20$, both sexes); 3-year-olds, 197 ± 14 g ($n = 4$, both sexes). Adults in winter off British Columbia (Gaston *et al.* 1993a): ♂ 242 ± 25 g ($n = 49$), ♀ 251 ± 21 g ($n = 45$).

Range and status

RANGE
Pacific, boreal, and low Arctic, but extending to edge of sub-tropical waters off China. Substantial southward movement in winter, especially on American side of Pacific.

BREEDING: on offshore islands from 52° to 60° N in the eastern Pacific and from 35° to 62° N on the Asian shore. Uncommon in W Pacific, except in Peter the Great Bay, near Vladivostok and at scattered colonies in the Sea of Okhotsk; moderately common and widespread in the Aleutians and the Gulf of Alaska; abundant in SE Alaska and the Queen Charlotte Is. Largest colonies in Asia on Talan I. (10 000 birds, Springer *et al.* 1993) and Starichkov I., off Kamchatka (13 000, Vyatkin 1986), and in Alaska, Buldir I. (10 000, Byrd and Day 1986), the Sandman Reefs (10 000+, Bailey and Faust 1980), the Shumagin Is. (36 000+, Moe and Day 1977) and Forrester I. (60 000, DeGange *et al.* 1977). In Queen Charlotte Is., there are 13 colonies with more than 10 000 breeding birds, of which the largest are Frederick (134 000 birds), Hippa (80 000), and Rankine (52 000) (Rodway 1991).

WINTER: some birds remain within breeding range throughout year, except for a post-breeding dispersal, when most birds disappear from British Columbia waters, where they are otherwise abundant. Birds have been recorded moving N through Bering Straits at this season (Kessel 1989). There is a general dispersal southwards in winter, with birds reported on passage off Oregon in Oct–Nov and again in Mar (Gilligan *et al.* 1993) and many wintering as far as California on the American shore, where birds arrive in numbers by late Oct (Ainley 1976), and as far as Taiwan on the Asian side. Many winter off Japan and Korea, particularly from Dec to Mar (Austin and Kuroda 1953; Gore and Pyong-Oh 1971) and around Vancouver I. (Campbell *et al.* 1991). Christmas Bird Counts indicate highest numbers off Vancouver I., Washington, and Oregon, and counts in California are highest in N of the state. These numbers suggest centre of gravity of E Pacific wintering population is between 40 and 50° N.

STATUS
Population estimates very poor except in British Columbia, where about half a million birds breed, all in the Queen Charlotte Is. There are probably several hundred thousand in Alaska; Springer *et al.* (1993) indicated about 200 000, but this seems very conservative; V. Mendenhall (*in litt.*) suggested 800 000. There are several tens of thousands in Asia. The world population is probably between 1 and 2 million. Populations throughout the species range are much diminished as a result of the introduction of mammalian predators to colony islands, either accidentally (rats), or deliberately, for fur (foxes, raccoons, Bailey and Kaiser 1993). Springer *et al.* (1993) suggest that Aleutian populations may be only 20% of their former numbers, while Gaston (1992a) estimated that the population of the Queen Charlotte Is. may have fallen by as much as a half in the last few decades. The species is classified as vulnerable in Canada (Gaston 1994). It may well be in danger of extirpation from China, Korea, and Japan, where populations are small and unprotected.

Field characters
In summer, black head and throat distinguish Ancient Murrelets from Cassin's Auklet and

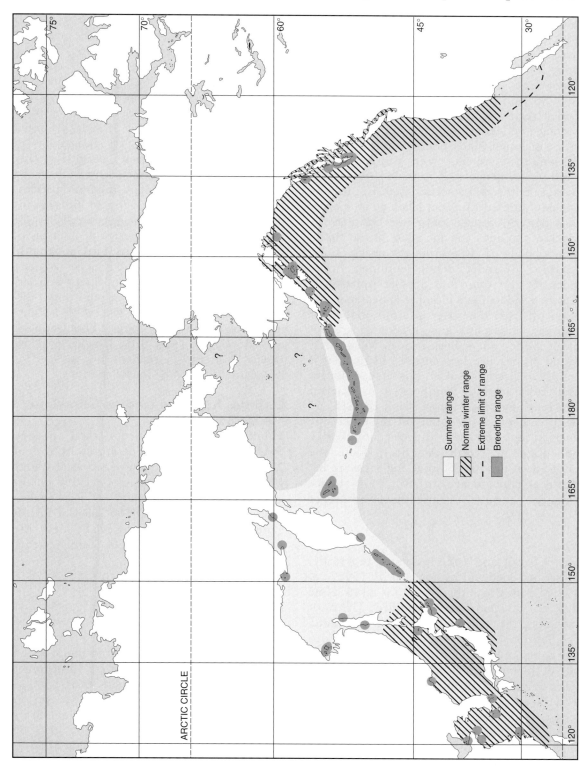

Summer range
Normal winter range
Extreme limit of range
Breeding range

ARCTIC CIRCLE

Marbled Murrelets. In flight, they show all-white underparts (except throat and axilliaries), head is often held a little above line of body and wing-beats appear shallower than those of auklets, with more banking from side to side. Non-breeding plumage is not maintained very long. Many birds are in breeding plumage by Dec, although some, perhaps first years, maintain winter plumage as late as May. In winter, confusion with winter-plumage Marbled Murrelets is possible, especially because both species occur together in inshore waters. Marbled Murrelet has a much cleaner face pattern, with a sharp line between the black of the head and the white of the cheek. When taking off away from the observer, the Marbled Murrelet shows prominent white sides to the rump and a paler upperwing. Cassin's Auklet has a plumper appearance and takes off from the water in a more laboured manner, underparts and underwing are not clean white, but sullied brown. Very rarely found on beached bird surveys. Leg colour is similar to Cassin's Auklet, but tarsus is longer and more laterally compressed in Ancient Murrelet. Dove grey upperwing coverts and back feathers are not found on any other auk, apart from the Japanese Murrelet (not likely to be found in North America). Note the plumage of head and throat and measure bill depth as clues to age. In Asia, look out for Japanese Murrelet, distinguishable by deeper bill and crest.

Voice

Has a large and complex vocal repertoire. Described in detail by Jones *et al.* (1987*a,b,* 1989), from which this account is taken. Nine different calls have been identified. The most common, the 'chirrup', is a short, emphatic, call, lasting <0.5 sec. These calls are individually recognizable from the number and relative size of the individual elements; each individual has a unique call. Elements of the 'chirrup' also occur in other calls; the 'chip', the 'bubble', the 'chatter', and the 'trill-rattle'. The 'chip' is a single, sharp note, shown on a sonogram as a single chevron, while the 'bubble' consists of an irregular series of chip-like notes, lasting up to several seconds. The 'trill-rattle' is a brief burst of chip notes, delivered very rapidly at evenly spaced intervals, and lasting <0.5 sec. The 'chatter' is a prolonged, high intensity burst of calling, which includes elements of all the other calls mixed up in variable ways, and including loud rasping notes and given by birds that are highly aroused. The 'song', performed by ♂♂ perched in trees or stumps, or on ground near nesting burrows, consists of one, or more, chirrups, connected by well-spaced 'chip' notes, and forming a rhythmic pattern. Like 'chirrup', 'songs' of different birds are individually recognizable. Songs generally last for several seconds, being repeated at intervals of 20–30 sec. Singing by one bird apparently stimulates others nearby, and bouts of countersinging can last up to 45 min. 'Long-whistles' and 'short-whistles' consist of pure tones, differing in length, while the 'wheeze' is a short, unstructured call which forms a broad smudge on a sonogram (Fig. 5.1). Chicks give low 'peeps' in burrow, a strident 'pee-pee-pee' at departure (see below, Chick departure).

Habitat, food and feeding behaviour

MARINE HABITAT

Found mainly in continental shelf or slope waters with surface temperatures from 4 to 20°. Also occurs commonly in coastal waters where tidal currents create concentrations of slow-swimming prey, such as euphausiids (Gaston *et al.* 1993*a*). Off W coast of British Columbia concentrates especially near the edge of the continental shelf (Vermeer *et al.* 1985).

FEEDING BEHAVIOUR

Frequently feeds in flocks of up to 50 that dive more or less simultaneously (Austin and Kuroda 1953; Gaston 1992*a*). Sometimes found in mixed species flocks with Black-legged Kittiwakes, Rhinoceros Auklets, and other species, usually associating with swarms of euphausiids. In these situations, usually occurs on the perimeter of the flock (Porter and Sealy 1981; Gaston 1992*a*). May attract gulls by driving sandlance and other schooling fishes to the surface (Litvinerko and Shibaev 1987).

Occasionally feeds around the bows of slow-moving ships (Stejnegar 1885; Bendire 1895).

DIVING BEHAVIOUR: diving depth unknown, but submergence times usually less than 45 sec, suggesting dives of 10–20 m (Gaston 1992a). Occasionally feed at the surface by 'head-bobbing'. Prey mostly taken in mid-water.

DIET
Not well known, in part because chicks not fed at nest-site.

ADULT SUMMER: Principal prey in Queen Charlotte Is. during breeding was adult euphausiids in Apr and May and juvenile sandlance and euphausiids in June, with smaller proportions of juvenile shiner perch (*Cymatogaster*) and rockfishes (Sealy 1975b). In the Gulf of Alaska the main prey was the euphausiid *Thysanoessa inermis* (Sanger 1987a) with smaller amounts of fish (capelin and walleye pollock), while in the Sea of Okhotsk fish, including sandlance, rainbow smelt (*Osmerus mordax*), and sculpins (*Triglops* spp.) predominated (A. Y. Kondratyev and A. Kitesky in Springer *et al.* 1993).

ADULT WINTER: the only evidence for winter food preference comes from waters off Vancouver I. where diet comprised almost entirely the euphausiid *Euphausia pacifica*, some stomachs containing over 1000 individuals (Gaston *et al.* 1993a).

CHICKS: those observed in Peter the Great Bay were fed juvenile herring and saury (*Hypoptychus dybowskii*) by their parents (Litvinenko and Shibaev 1987).

Displays and breeding behaviour
Information from Gaston (1992a) unless otherwise indicated.

COLONY ATTENDANCE: colony visits are nocturnal, the first arrivals occurring about 90 min after sunset and departures completed by 1 hr before dawn. Arrivals earlier, and usually more numerous, on overcast nights without moon (Jones *et al.* 1990). Attendance reduced during and after stormy weather. Arriving birds fly high over the colony, in forest habitat descending through the canopy, and frequently colliding with branches and twigs on the way. Most breeders land within 10 m of their burrow and run directly to it. Non-breeders may land anywhere and may spend extended periods sitting on the surface. Prospecting birds, mainly second years and some third years, occasionally first years, visit the colony in large numbers from mid-incubation until after the departure of all but the latest chicks (Gaston 1992a). Numbers generally peak midway through the departure period, very variable from night to night and lower during full moon periods (Jones *et al.* 1990). Visiting continues for a few weeks after the peak of chick departures, then ceases until the following spring.

GATHERINGS AT SEA: throughout the breeding season gatherings of birds waiting to visit the colony begin to assemble at discrete 'gathering grounds' 1–3 km offshore from colony area at up to 6 hrs before sunset. Similar post-departure gatherings occur at dawn, lasting for up to 5 hrs. Most birds seen on the gathering grounds after the start of incubation are probably non-breeders and numbers suggest that most of those prospecting the colony are present on peak nights. Gathering ground aggregation perform intensive social displays, including vocalizations, erection of the crown feathers, and 'flop displays', in which birds spring from the water to a height of 30 cm, banking to one side and falling back with lower wing trailing (Jones 1985, Fig. 2). Many birds are apparently paired. After dark, non-breeders approach close to the shore, continuing to socialize. Groups of loafing birds sometimes harass departing chicks by surfacing underneath them or pecking them.

AGONISTIC BEHAVIOUR: fighting, involving bill grappling, pecking, and scratching, sometimes occurs on surface at night, in one case continuing for 2 hrs. In Russia, has been described to defend territories 0.5–0.7 m in diameter (Shibaev 1990d).

2   'Flop display' on gathering ground.

SEXUAL BEHAVIOUR: non-breeding birds frequently land in trees when they first arrive, descending to ground later at night. Birds sing and 'chirrup' regularly from trees and stumps within the colony from soon after arrival. Song during incubation period is delivered by non-breeding ♂♂. Breeders may sing during pre-laying period, but this is not known for sure. Pairs exchange 'chatter' calls when they meet in the burrow, either as breeders exchanging incubation duty, or as non-breeders courting, and such vocalizations can go on intermittently for up to 30 min. Singing and chattering calls are also given in burrows by single birds during prospecting and these may attract other birds, possibly ♀♀ (Jones *et al.* 1987a). Copulation occurs on the colony surface, ♂ approaching ♀ form behind with head stretched forward; sometimes also at sea. After copulation ♀ tosses bill upwards several times (Shibaev 1990d).

Breeding and life cycle

Mainly from Bendire (1895), Ishizawa (1933), Sealy (1976), Jones *et al.* (1987b), Litvinenko and Shibaev (1987), and Gaston (1992a).

BREEDING HABITAT AND NEST DISPERSION: colonial, but colonies less dense than most auks, averaging less than 1 burrow/10 m² in the Queen Charlotte Is. (Rodway *et al.* 1988). Nests up to 400 m from sea, but more commonly less than 300 m. In temperate areas, breeds in mature forest, making burrows up to 2 m long under roots of trees, or fallen logs. Where forest is not available, as in the Aleutians, chooses densest vegetation available, often burrowing below grass tussocks. Also uses rock crevices or holes under boulders, occasionally artificial sites, such as under huts, or in walls. In British Columbia breeds in association with Cassin's and Rhinoceros Auklets, and storm-petrels.

NEST: cup usually sparsely lined with twigs, leaves, and dry grass. Most burrows curved, so that incubating bird is not visible through entrance. At short burrows, birds sometimes drag vegetation across entrance behind them, so that the burrow appears unused. Artificial boxes are sometimes used, but many birds prefer to dig new burrows when they begin to breed, rather than taking over one used previously. Consequently, most colonies contain many unused burrows. Burrow creation often makes use of natural cavities; excavation, by both sexes, does not involve expelling soil from the entrance, so it can be difficult to judge whether burrows are active. Non-breeders dig burrows at the end of the breeding season, presumably for occupation the following year.

EGG-LAYING: adults arrive in the vicinity of breeding colonies about 1 month before laying begins, coming ashore first 2–3 weeks before the first egg. Date of first laying in British Columbia 1–10 Apr (9 years). Median date of clutch completion varies from about 15 Mar in China (Zhao-qing 1988) to late June—early July in Kamchatka and the Sea of Okhotsk (Vyatkin 1986). Mean surface temperatures in surrounding oceanic waters in the month of clutch completion are 6–11°. Comparing sea-surface temperatures in Apr, there is a 6-day delay in median clutch completion for every 1° decrease in temperature (Gaston 1992a). Laying continues for about 45 days, with 50% of clutches initiated within 6–10 days (4 years). Interval between eggs, 6–10 days, normally 7–8 days, median 8 ($n = 23$).

EGGS: elliptical-ovoid, rather elongate compared with other auks, shell texture strikingly smooth, making them almost slippery. Ground colour pale buff to olive brown, sometimes bluish, with prominent brown speckles and blotches evenly distributed, very variable, 1–5 mm across. Most clutches very similar, but a few comprise two strikingly different eggs, presumably the product of two females. Measurements: Reef I., BC (range of means over 6 years, Gaston 1992a): 58.7–60.0 × 37.1–38.1 mm, fresh weight 43.8–47.6 g (n = 39–99); Buldir I., Alaska (G. V. Byrd in Gaston 1992a) 62.2 × 39.0 mm (n = 8); Kuril Is. (BMNH) 61.1 ± 1.9 (58.0–64.2) × 38.4 ± 0.8 (37.4–39.9, n = 7) mm. Egg 22–23% of ♀ body mass at laying. Shell 8%, yolk 45%, and albumen 47% of fresh weight (Birkhead and Gaston 1988). First and second eggs similar in size and appearance. Clutch is usually two (96%, n = 298), sometimes one (3%). A few clutches of three or four are probably the result of two ♀♀ using the same burrow; no replacement layings.

INCUBATION: first egg not incubated until second is laid; when clutch complete, incubation may commence immediately, or there may be a further delay of 1–2 days. Incubation period 29–37 days from clutch completion, but total days of incubation 28–33, usually 29–31 because 1–4 days of neglect are common (Sealy 1976; Gaston 1992a). Eggs very resistant to chilling and hatch successfully even if deserted for 24 hrs at 8 °C in late incubation (Gaston and Powell 1989). Incubation shifts 1–6 days (median 3), but attendance may be intermittent for first few days after laying of second egg. No defecation in burrow, either by adults or chicks, hence burrows are very clean and odourless (cf. Cassin's Auklet); however, adults disturbed in burrow during day may give a very large, smelly defecation, normally a prelude to nest desertion. Burrows never occupied in daytime except during incubation and brooding and more than one adult never remains during day.

HATCHING AND CHICK-REARING: eggs hatch within 24 hrs of one another and chicks leave for sea within 2 days. No provisioning in burrow. After leaving for sea, parents continue feeding until young are fully grown, probably at least 1 month after leaving colony (Litvinenko and Shibaev 1987). Family parties remain scattered, mostly in offshore waters, throughout brood rearing, during which the chicks do not feed themselves.

CHICK GROWTH AND DEVELOPMENT: precocial. Chicks are able to thermoregulate immediately. Chicks at Reef I., BC weighed 31 g at c. 12 hrs after hatching and usually spent 1–2 nights in the burrow without being fed, before making their way to the sea. While in the burrow they lost weight at 2 g/day and departed at 27 g, early chicks departing at higher weights than late chicks.

CHICK DEPARTURE: begins after arrival of off-duty adult, usually 90–150 min after sunset. Parents give distinctively emphatic rendition of 'chirrup' at burrow entrance and are followed within a few minutes by chicks which give a shrill 'pee-pee', or 'pee-pee-pee' call. Vocal interaction continues just outside burrow mouth for up to 2 min, then adults fly to sea where they call repeatedly using the same burrow departure call (Jones et al. 1987b); chicks run silently to shore, orienting towards light and downhill (Gaston et al. 1988b). During this period, chicks are hyperactive and if captured struggle constantly and vigorously: '... jumps off rocky ledges, climbs large rocks ... squeezes through narrow chinks and tirelessly seeks alternate paths when an obstacle is insurmountable...' (Shibaev 1990d). Some accounts of family departures assert that one or both parents accompany the chicks all the way to the sea. This may be the case on flat islands, where there is little topography to guide the chicks, but the evidence for continuous parental guidance is not explicit. Chicks begin to call again as soon as they reach sea; families reunite through mutual recognition of calls (Jones et al. 1987b) and swim directly away from land, continuing without pause for 12–18 hrs, by which time they are 20–40 km from the colony (Duncan and Gaston 1990). Chicks dive actively as soon as they reach the sea, successfully negotiating 1–2 m of surf and swimming underwater with wings as well as feet. They are occasionally taken by large predatory fishes. During travel, chicks stay within 0.5 m of parents, hydroplaning to keep up. They respond to parental alarm calls by diving for 10–15 sec.

BREEDING SUCCESS: up to departure of chicks at undisturbed, or minimally disturbed sites 1.44–1.69 chicks/pair over 5 seasons (Vermeer and Lemon 1986; Rodway *et al.* 1988; Gaston 1992*a*). Hatching success of eggs incubated at least 30 days, 96%. Most failure caused by desertion, often before incubation has begun (Gaston 1992*a*). Most family parties at sea include two chicks, so most of those leaving the burrow make a successful rendezvous with their parents. Deer Mice, which occur on many colony islands in British Columbia, eat pipping eggs or kill chicks if abandoned, but apparently do not affect broods or clutches in the presence of parents.

PREDATION: heavy on adult birds at most colonies by ravens, eagles, gulls, and Great Horned Owls, and where present, by rats, Raccoons and foxes; none of the mammals coexists naturally with Ancient Murrelets. On Talan I. some adults are killed by Tufted Puffins (Flint and Golovkin 1990). The Ancient Murrelet is an important prey species for the coastal race of Peregrine Falcon (*Falco peregrinus pealei*) (Beebe 1960), forming as much as 50% of the diet during the breeding season.

MOULT: very little known. Complete body moult apparently occurs in Aug and Sept, because adult birds collected in Oct are in fresh non-breeding plumage. In British Columbia, pre-nuptial body moult often follows very soon after the post-breeding moult and many birds are in full breeding plumage by Dec, most by Feb (Gaston 1992*a*). In the Russian Far East pre-nuptial moult occurs in Apr (Shibaev 1990*d*). Timing of moult in young birds is not known, but second summer birds trapped at breeding colonies in May and June frequently have very worn primaries, perhaps indicating that they are still wearing the first breeding plumage, adopted some time the previous year. A few birds trapped at their colonies at 1 year old were all in breeding plumage, but these individuals may not have been typical of the age group. Some non-breeding birds at colonies show variable traces of white in the throat, suggesting incomplete body moult.

Population dynamics

Mean adult annual survival 77%, age at first breeding 3–4 years. Population age structure at the start of the breeding season: 30% first years, 29% non-breeding second and third years, and 41% breeders in the third year or older. Inter-colony movements not well-known, but many birds visit colonies other than those at which they were hatched as pre-breeders (Gaston 1990).

Japanese Murrelet   *Synthliboramphus wumizusume*

Synthliboramphus wumizusume (C. T. Temminck, 1827–36, *Nouveau recueil de planches coloriées d'oiseaux.*, —Japan).

PLATE 4

Kanmuri umisuzume Japanese

Description

A smaller, more strikingly marked, relative of the Ancient Murrelet, with slender body, comparatively long, narrow wings and feet set far back, so that it moves clumsily on the ground.

The Japanese name, *Kanmuri*, refers to the crest, being the plume on a warrior's helmet. *Umisuzume* (mis-transliterated by Temmink in his scientific name), a name also used for the Ancient Murrelet, means 'sea sparrow'.

ADULT SUMMER: crown black, with long feathers forming crest extending back as far as bottom of nape and covering shorter, white feathers of crown; a few white filoplumes scatter base of hind-neck; feathers of crest floppy, central ones black and outers white; back, upperwing and uppertail coverts dove grey, head and throat black, flight and tail feathers dark slate; underwing white, edged with black, and a black band runs across base of wing, merging streakily into flanks; underparts white; bill bluish flesh, tipped with black; legs and feet bluish-tinted flesh pink, darker on webs; irides brown.

ADULT WINTER: as summer but crest and plumes absent, or reduced, head all dusky black and variable amount of dark smudging on chin.

JUVENILE AND FIRST WINTER: similar to winter-plumage adult, but with shorter, more slender bill.

CHICK: down pattern similar to adult, blue-grey above, white below; black cap on head to just below eye; narrow white ring around eye.

VARIATION
No races described, which is hardly surprising for a bird with such a small range.

MEASUREMENTS
Breeders from Biro I. (Ono and Nakamura 1993), sexes combined (mm): wing 123 (113–132, $n = 163$), tail 35 (19–42, $n = 152$), culmen (from base of skull) 15.9 (13.8–17.8, $n = 147$), bill depth 7.6 (6.7–8.6, $n = 124$), tarsus 25.8 (23.2–27.8, $n = 152$).

WEIGHTS
Breeding adults at Biro I. averaged 164 g (139–213, $n = 214$, Ono and Nakamura 1993)

Range and status
RANGE
Breeds only in Japan, a few islands off South Korea, and possibly in Peter the Great Bay, near Vladivostok. It is confined to warm-current areas, particularly islands off Kyushu and in the Izu Is. (Brazil 1991).

BREEDING: presently believed to occur on at least 12 islands around the coasts of southern Honshu and Kyushu. The main stronghold is believed to be in the Izu Is., where at least three islands are occupied and more remain to be confirmed (Ochikubo *et al.* 1995). Breeding is believed to occur on Gugul-do, SE of South Korea, and possibly on other Korean islands (Harrison *et al.* 1992). There have been several records in summer in Peter the Great Bay, near Vladivostock, including one fledgling female, leading to speculation that the species may breed there (Shibaev 1987). However, it is possible that these records refer to post-breeding dispersers.

WINTER: in the non-breeding season adults and young of the year disperse northwards as far as Sakhalin, the Kurils, and waters to the NE of Hokkaido. During Aug–Nov at least five were recovered drowned in gill-nets set for squid in an area bounded by 38–44° N and 142–147° E (Piatt and Gould 1994).

STATUS
Recent estimates suggest approximately 1000 birds in the Izu Is., where at least six sites are known, with the largest numbers on Tadanae I., Onbase Reef, and Sambondake Reef, with an estimated 100 pairs each in 1994 (Ochikubo *et al.* 1995). Otherwise, three sites are known off Honshu, of which the largest is Mimiana Is. on the S coast (150–200 birds) and the most northerly on the Nanatsujima Is., off the N coast; three off Kyushu, including the largest single island colony, 3000 birds, on Biro Is. (Ono *et al.* 1995), and one off Shikoku (Koshima Is.). They may also occur on the Danjo Is. The total world population in 1993 was estimated as 5–6000 birds (Ono 1996). Some breeding sites probably remain to be discovered, or rediscovered and some breeding islands are very rocky and inaccessible, so that actual breed-

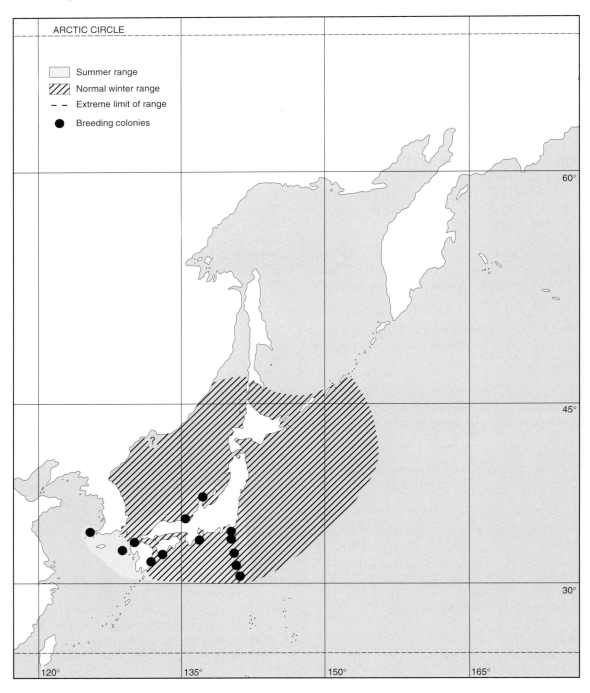

ing populations may be larger than estimated; however, the total world population cannot exceed 10 000 birds. The species has certainly declined in recent decades and this decline continues through the combined impact of introduced rats, human disturbance at breeding sites, and drowning in gill-nets. Very heavy pressure on breeding habitat from both commercial and recreational fishermen poses the main threat. Many breeding islands are used as periodic bases by commercial fishermen, while recreational fishermen use the islands to cast from the rocks. Both groups leave garbage that attracts Carrion and Jungle Crows (*Corvus corone* and *C. macrorhynchos*), and Black Kites (*Milvus migrans*) and may be instrumental in introducing rats which decimated breeders on Koyashima Is. in 1987. That population had comprised several hundred birds, but was reduced to less than 10 and had only slightly recovered by 1993, despite control of rats by poisoning (Takeishi *et al.* 1993). At Biro I., Jungle Crows cause heavy mortality of adults. In the past, eggs were harvested at some colonies, including Koyashima, by local people. This may still occur in a few places. Some birds drown in gill-nets set adjacent to colonies and larger numbers (relative to the presumed population size) drown in offshore gill-nets outside the breeding season (Piatt and Gould 1994). We consider the Japanese Murrelet to be the most likely auk to become extinct: probably fairly soon.

Field characters

In summer, can be easily distinguished from Ancient Murrelet, even at a distance, by broad white stripe curling down sides of neck from above and behind eye, also bulging appearance at back of head created by crest feathers. In winter probably hard to separate from Ancient Murrelet and even in the hand may be mistaken. Look for the deeper bill of the Japanese Murrelet and the paler legs, especially in beach-cast specimens.

Voice

Not fully described, but varied, with many calls reminiscent of Ancient Murrelet. 'Twitters', 'rattles' and 'whistles' (Fig. 1) are all given on

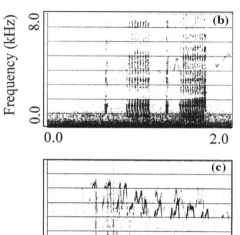

1 Sonograms of vocalizations recorded at Biro I., Japan (K. Ono, unpublished, sonograms courtesy S. Dechesne): (a) 'twitter', (b) 'rattle', (c) 'whistles'.

land. 'Whistles' resemble segments of the Ancient Murrelet 'song'.

Habitat food, and feeding behaviour

While breeding, seen in inshore waters around breeding sites, but otherwise found mainly away from coasts. Associated throughout year with sea surface temperatures 8–22 °C. Little is known of their ecology, but it appears essentially similar to that of the Ancient Murrelet. At sea, usually occurs in small flocks. Birds drowned in squid gill-nets were in the area where the cold Oyashio and warm Kuroshio currents converge and the resulting fronts may concentrate potential prey (Piatt and Gould 1994)

Displays and breeding behaviour

Rather poorly known, although much research was initiated in the 1990s. As far as published information goes, it appears to be similar to the Ancient Murrelet in all aspects of its life history and breeding behaviour. Monogamous.

Breeding and life cycle

Apparently resembles the Ancient Murrelet in most aspects of its natural history.

BREEDING HABITAT AND NEST DISPERSION: usually nests within a few hundred meters of the sea on small, rocky islands, either forest, bamboo brake, or turf-covered, and lacking mammalian predators. Apparently more dispersed than Ancient Murrelets, although this may be simply because the colonies are smaller. At Nanatsujima Is., the breeding area is shared with Streaked Shearwaters (*Calonectris leucomelas*) and at Koyashima Is. with Swinhoe's Storm-Petrels (*Oceanodroma monorhis*). It is not known whether these species interact, or compete for breeding sites.

NEST: may be placed in rock cracks, sometimes in cliffs, in crevices among boulders, or burrowed into grass tussocks or soil. On some inhabited islands, crevices in stone walls, or in the stairway leading to a lighthouse are used (Higuchi 1993*a*; Nakamura 1993). Burrowing under tree roots is not mentioned in any accounts. Most crevices and burrows used are relatively shallow and easily accessible compared with those of Ancient Murrelets.

EGG-LAYING: attendance at the breeding sites occurs in Feb–May, at sites close to the main islands of Japan, egg-laying occurs from late Feb to end Mar and family parties depart Apr or early May, but in the Izu Is. laying may be as late as early Apr (Moyer 1957).

EGGS: usually pale tan or buff with brown or liver-coloured spots or speckles, sometimes very heavily marked. Seldom warm buff or bluish. Clutch almost invariably two. Measurements: Kojine Reef, Japan (Ono 1993), $54.0 \pm 1.3 \times 35.4 \pm 0.3$ mm, mass 36.6 ± 1.7 g ($n = 5$); 'Japan' (BMNH), 53.2 ± 1.5 (51.8–56.0) $\times 34.8 \pm 0.6$ (33.5–35.5, $n = 10$) mm. Weight is approximately 22% of adult body weight.

INCUBATION AND CHICK-REARING: parents take equal shares in incubation, with shifts lasting 1–3 days, most often 2; periods of neglect of up to 5 days have been observed without effect on hatching success. Incubation period 31 days (Ono 1993). Chicks precocial, usually spend 1–2 nights in the burrow without being fed, before making their way to the sea after dark (Ono and Nakamura 1993). Departure at Kojine Reef occurred approximately 4 hrs after sunset (Ono 1993). Chick-rearing is undescribed, but presumably carried out by both parents, as in the Ancient Murrelet. Family parties disperse away from the breeding islands immediately.

PREDATION: like Ancient Murrelets, Japanese Murrelets suffer heavy predation while breeding from a wide variety of predators. These include Peregrine Falcon, Jungle and Carrion Crows, and Black Kite

MOULT: practically unknown. Complete body moult apparently occurs between June and Oct, because adult birds collected in Oct are in fresh non-breeding plumage. Pre-nuptial body moult presumably by Jan, as breeding activities begin by then.

Population dynamics

No information.

Cassin's Auklet *Ptychoramphus aleuticus*

Uria aleutica (Pallas, 1811, Zoogr. Russo-Asiatica 2: 340—North Pacific Ocean).   PLATE 5

Description

A small, short-winged, heavy-bodied auklet with dull plumage that remains virtually unchanged through the year. Rarely seen on land because of their strictly nocturnal habits. They are able to stand on their tarsi, but do not attain a fully upright posture like auklets of the genus *Aethia*, and move about clumsily on land, scuttling among the dense grass tussocks of their typical nesting habitat. Flight rather weak, often take off from the sea with considerable difficulty at the approach of a vessel, or merely paddle along on the sea surface to get out of the way.

ADULT: fairly uniform dark brownish grey, except for belly and undertail coverts, which are white or greyish white; face and crown slightly darker than rest of upperparts; lower breast and belly indistinctly mottled grey on white, so pale belly and vent grade evenly into darker breast plumage without contrasting borderline; flight feathers and tail blackish grey; upper- and underwing coverts dark brownish grey; wings short, broad, and rather rounded; irides white; small white or pale grey crescents above (larger) and below (smaller) each eye,

the only adornments possessed by this nondescript species; almost no difference between summer and winter appearance; mostly black bill large relative to other auklets' bills, and has pale grey base to lower mandible, triangular in profile with straight or even slightly concave culmen, an unhooked and sharp-pointed tip, and a broad base; legs short and placed relatively further back on body than in other auklets; feet blue-grey with blackish webs.

SUBADULTS: very similar to adults but have brown or brownish grey irides. 1-, 2- and 3-year-old subadults visiting breeding colonies have successively paler irides until they attain the white-eyed appearance of adults (Manuwal 1978).

JUVENILE AND FIRST WINTER: resemble adults but have dark brown eyes and relatively small all-black bill.

CHICK: soft medium grey down on upperparts and upper breast; white down on belly; short light grey down around eyes giving an eye-ringed appearance; bluish grey feet; brown eyes; all-black bill.

VARIATION
P. a. aleuticus Channel Is. of southern California, N and W to the Near Is., Aleutian Is., *P. aleuticus australe* Baja California (Manuwal and Thoresen 1993). There is a cline of increasing body size from Baja California, Mexico, N to the Farallon Is., California, with northern birds (British Columbia and Alaska) similar in size to the California population.

MEASUREMENTS
P. a. aleuticus from SE Farallon Is., California (Nelson 1981) (mm): wing, ♂ 125.8 ± 2.5 (120–129, $n = 14$), ♀ 125.7 ± 3.7 (119–130, $n = 18$); culmen, ♂ 19.8 ± 1.1 (17.9–21.6, $n = 14$), ♀ 19.5 ± 0.6 (18.4–20.5, $n = 19$); bill

1 Adult bird in colony, showing typical stance (ILJ photo).

depth, ♂ 10.9 ± 0.4 (10.4–11.7, *n* = 14), ♀ 9.5 ± 0.4 (8.9–10.3, *n* = 19); tarsus, ♂ 30.9 ± 1.0 (29.0–32.0, *n* = 14), ♀ 30.1 ± 0.9 (*n* = 18). *P. a. aleuticus* from Triangle Is. BC (I. L. Jones and H. Knechtel, unpublished data): unsexed adults (white-eyed), culmen, 19.7 ± 0.8 (17.2–22.5, *n* = 528); bill depth, 9.9 ± 0.7 (7.2–11.7, *n* = 760); tarsus, 25.6 ± 0.9 (22.5–28.9, *n* = 580); (possibly measured differently from Farallon study); unsexed subadults (brown-eyed), culmen, 19.7 ± 0.9 (16.7–21.8, *n* = 160); bill depth, 9.7 ± 0.7 (7.1–11.2, *n* = 214); tarsus, 25.5 ± 0.9 (22.8–28.5, *n* = 169).

WEIGHTS

P. a. aleuticus from Triangle, I., BC (I. L. Jones and H. Knechtel, unpublished data): unsexed adults (white-eyed), mass (g) of adults departing colony in early morning hours, pooled data for entire breeding season, 174.4 ± 12.3 (142–217, *n* = 778); unsexed subadults (brown-eyed), 171.8 ± 10.9 (134–200, *n* = 217). From Frederick I., Haida Gwaii (A. Harfenist, unpublished data), of birds departing colony in early morning, unsexed adults (white-eyed), 172.0 ± 16.4 (124–230, *n* = 46), unsexed subadults (brown-eyed), 170.9 ± 10.6 (144–190, *n* = 46). From the Alaska Peninsula (J. F. Piatt, unpublished data), body mass (g) of birds shot at sea, unsexed adults, 198.6 ± 12.9 (*n* = 28).

Range and status

RANGE

Widely distributed in the NE Pacific. Range along the W coast of North America from Baja California (San Geronimo, San Benitos, San Roque, and Asuncion Is.; Everett and Anderson 1991), northwards along the coast of California, Oregon, Washington, British Columbia, to SE Alaska. There is a notable gap in the breeding distribution in the Gulf of Alaska N of about 57° N. Further W and S, breeding colonies reappear along the southern coast of the Alaska Peninsula and Aleutian Is. W to the Near Is. Strays are occasionally observed as far N as the Pribilof Is. and Nunivak Is. in the Bering Sea. Northern range

limit is probably constrained by need for several hours of darkness for nocturnal activity at colonies. Unlike other auklet species, Cassin's are apparently absent as a breeding bird in Asia. The southernmost colony is located at Guadalupe Is., Mexico (Jehl and Everett 1985), while the north-westernmost outlying colonies are located at Buldir I. (Byrd and Day 1986) and near Agattu I. in the western Aleutians. There are two records from Siberia, one from Kamchatka, and one from the Kuril Is. (Stotskaya 1990*a*). There have been no records of vagrants inland, probably because of this species' poor flying ability (compared with Crested Auklet and Ancient Murrelet). There is little movement between winter and summer.

STATUS

Population estimates are: 1 000 000 in Alaska (U. S. Fish and Wildlife Service 1993), 2.8 million in British Columbia (about 70% of total world population (Springer *et al.* 1993), 43 800 pairs in Washington (Speich and Wahl 1989), 500 pairs in Oregon (Springer *et al.* 1993) 65 580 pairs in California (Springer *et al.* 1993) (declining), and 10–20 000 pairs nesting in Baja California, Mexico (Everett and Anderson 1991). The largest colony is located on Triangle Is. BC—370 000 pairs (Rodway *et al.* 1990). There have been few attempts to monitor populations quantitatively other than at the Farallons and Triangle I., so data are lacking on worldwide population trends, although the population may be at a historic low owing to the cumulative effects of introduced predators on many colonies. Like many nocturnal seabirds worldwide, Cassin's Auklets have suffered greatly from predation by introduced mammalian predators. In Baja California and California, cats, rats, and foxes have finished off many colonies (Everett and Anderson 1991; Springer *et al.* 1993*a*). In British Columbia were formerly abundant on Langara Is. but were extirpated by introduced rats; a recent attempt to remove rats from this island may restore seabird populations. However, populations in the rest of Haida Gwaii are now seriously threatened by expanding

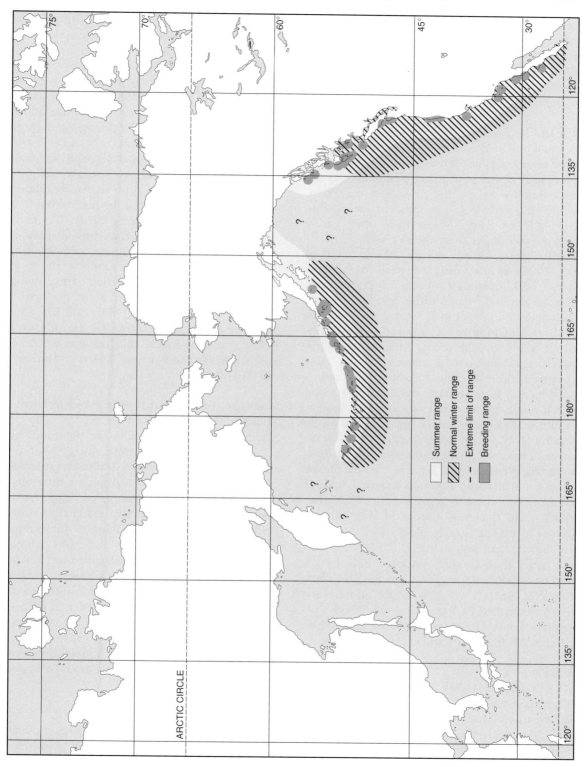

Summer range

Normal winter range

Extreme limit of range

Breeding range

ARCTIC CIRCLE

populations of introduced Raccoons. In the Aleutian Is. this species was probably the seabird most affected by Arctic Fox introductions, with large colonies extripated from numerous islands (Bailey 1993). Extremely vulnerable to oil pollution.

Field characters

A small pot-bellied alcid with short wings, dark grey upperparts and breast, and a white belly. Not easily observed ashore (see Colony attendance below). At sea, the pale lower mandible, white iris, and white eye crescents are visible only at close range. Weak fliers, often barely able to take flight from the sea surface, particularly after feeding heavily. Along much of the W coast of North America, sometimes occur with murrelets (*Synthliboramphus* and *Brachyramphus*), which differ greatly in shape (relatively long pointed wings and slender bodies) and plumage coloration. Also sometimes seen with Rhinoceros Auklets, which are much larger and have a massive usually light-coloured bill. Near the Aleutian Is., Cassin's Auklets may be seen with other auklets (*Aethia* spp.), which are sometimes superficially similar. Adult auklets in breeding plumage are easily separable by their conspicuous facial ornaments. However, identification of adults in non-breeding plumage and juveniles at sea can be more problematical, although Cassin's Auklet's chunky body shape is a good mark. Whiskered Auklets are the most similar in size, but are darker in colour (nearly black) and their bills are shorter and less pointed. Crested Auklets are larger than Cassin's Auklets, have longer and more pointed wings and lack the pale belly. Parakeet Auklets are also larger and have a more contrasting dark-grey-and-white body coloration. Least Auklets usually have large white patches on upperparts and are relatively tiny.

Voice

(ILJ; Thoresen 1964) A highly vocal species. Nighttime choruses at big colonies can be spectacularly loud. These choruses are socially facilitated and rise and fall in intensity, with seemingly the entire host breaking out in simultaneous song or just as suddenly falling silent. Vocal behaviour is most intense on dark foggy nights and much reduced or extinguished by moonlight. Their characteristic vocalization, the 'display kreek', is a harsh rhythmic call that is usually performed by birds concealed in burrows or vegetation, bouts can last for several minutes, and there are a number of variants. Manuwal and Thoresen (1993) suggested a group selection explanation for this call: to socially facilitate synchronized breeding. A more likely explanation is that the call is used for mate attraction by individual non-breeding birds. There is circumstantial evidence that this display is restricted to ♂♂ (I. L. Jones, unpublished data). Courting pairs, particularly those in burrows, perform the 'kreek duet'. Birds flying over the colony and sea at night perform a short harsh 'shriek' (or 'squeer') call. They are normally silent at sea during the day. Cassin's Auklet's 'display kreek' vocalizations are strikingly similar to the Parakeet Auklet's 'whinney' call (ILJ).

Habitat, food, and feeding behaviour

MARINE HABITAT

Summer and winter ranges are delimited by average ocean surface temperatures of 9–20 and 6–20 °C, thus Cassin's Auklets are tolerant of a wide range of oceanographic conditions, but individual birds must often experience cooler than average southern sea surface temperatures because of a preference for feeding in areas of cold upwelling. A highly pelagic species, normally feeding well offshore, or near the continental shelf break where it occurs near to the coast, and otherwise coming close to land only when nesting (and then at night), or when sick or injured. Although concentrations occur near continental shelf-breaks, Cassin's Auklets are normally widely dispersed in small groups or solitary individuals (Morgan *et al.* 1991).

FEEDING BEHAVIOUR

During the breeding season, feeding concentrations are normally found over the continental shelf break where their preferred prey occurs abundantly in upwelling water; less commonly they occur over shallow shelf waters (Hunt *et al.* 1993*b*). Rarely seen close to shore. In winter disperse more widely over deep offshore waters. Maximum dive depths averaged 28 m with a modal maximum dive depth of 40 m (Burger and Powell 1990).

DIET

Relatively well studied, compared with other auklets; see review by Hunt *et al.* (1993*b*). Diet variable, probably because this species is a generalist plankton feeder that exploits whatever prey is locally available. Major prey are copepods (mainly *Neocalanus cristatus* and *N. plumchrus*), euphausiids (mainly *Thysanoessa spinifera* and *Euphausia pacifica*), and larval fish (including *Sebastes*, *Citharichthys* and *Ammodytes*). It is not known whether adult and chick diets are different. Diet differences recorded in different areas may reflect geographical differences in zooplankton community structure or random noise in the data, since significant inter-annual variation has been recorded in replicated studies, and concurrent studies have never been attempted—see Table, (1) Channel Is., S California (Hunt *et al.* 1993; $n = 95$†); (2) SE Farallon Island, California (Manuwal 1974*b*; $n = 22$); (3) Triangle I.

(Vermeer 1984; $n = 259$); (4) Frederick I., BC (Vermeer 1984; $n = 237$); (5) Reef Is. BC (Burger and Powell 1990; $n = 58$); (6) Gulf of Alaska (G. A. Sanger, cited by Hunt *et al.* 1993; n; $= 8$†); (7) Buldir I., Aleutian Is. (J. C. Williams and others, unpublished data; $n = 14$). Samples of food loads being delivered to chicks, unless indicated by †.

Displays and breeding behaviour

COLONY ATTENDANCE AND FLIGHT DISPLAYS: the most nocturnal with respect to colony-attendance behaviour of any alcid. The first birds arrive at the colony only in nearly complete darkness, depart before dawn (although later than Ancient Murrelets at the same colony), and, activity at colonies is reduced on nights with bright moonlight (Thoresen 1964; Manuwal 1974*b*). Furthermore, they do not gather at sea near the colony at dusk like other auklets and Ancient Murrelets. Instead, like storm-petrels, they remain in small groups well offshore until darkness falls. Their moonlight avoidance, and nocturnality in general, seem to be adaptations to avoid predation by Bald Eagles, Peregrines, and large gulls. Landing and taking off at colonies involves frequent mishaps, with birds crashing into the ground, vegetation, buildings, people, and each other. Unlike other auklets with the

Table: Cassin's Auklet diet

ref	(1)	(2)	(3)	(4)	(5)	(6)	(7)
copepods (*Neocalanus* spp.)	—	—	1	1	trace	1	4
euphausiids (*Thysanoessa* spp.)	2	1	2	3	2	2	1
larval fish	1	4	3	2	1	4	—
squid	—	3	trace	5	—	—	—
amphipods	—	2	trace	trace	3	—	2
carideans	—	—	trace	4	4	—	—
scyphomedusae	—	—	trace	trace	trace	—	—
brachyurans	—	—	trace	trace	trace	—	—
cirripedia	—	—	trace	—	—	—	—
decapods	—	—	—	—	—	3	5
other	—	5	—	—	—	—	3

By rank order of importance by wet weight or volume, recorded as trace if less than 1% of wet weight or volume.

possible exception of Whiskered Auklet, Cassin's Auklets attend some colonies outside the breeding season, with activity in all months of the year at SE Farallon I., California (mostly Dec–July) (Manuwal 1974*a*); there is little information on this phenomenon at other colonies.

AGONISTIC BEHAVIOUR: social behaviour has not been well studied because of their nocturnality (but see Thoresen 1964—summarized below). Agonistic behaviour is frequent at colonies, especially where nesting densities are high. Individuals have a threat display in which the tail is raised, wings partly spread, and head feathers ruffled. Overt aggression takes the form of pecking, bill-jabbing, and outright battles between birds with locked bills and legs kicking. Manuwal (1974*a*) discussed territorial behaviour but there was little evidence of any exclusively defended space other than the actual burrow.

SEXUAL BEHAVIOUR: apparently socially monogamous. Sexual behaviour involves interactions at the breeding colony between vocally displaying birds (probably mostly ♂♂) and others visiting them (probably mostly ♀♀). Much of this species' social behaviour takes place underground in burrows or under cover of dense ground vegetation. Courtship by pairs includes head bobbing and bowing, head waggling, wing raising, circling, passing, and 'standing about face' movements (Thoresen 1964). It is unclear whether pairs associate anywhere other than in and near the nesting burrow, but birds caught leaving the colony in twos at Triangle and Frederick Is. could have been mated pairs (ILJ, A. Harfenist, personal observations). Copulations have seldom been observed and it is unclear whether matings take place mostly on land (as observed by Thoresen 1964), or at sea as in other auklet species.

Breeding and life cycle

The best-known auklet species (see reviews by Ainley and Boekelheide 1990; Stotskaya 1990*a*; Manuwal and Thoresen 1993). Under optimal conditions in the Farallon Is., pairs can have two broods in 1 year; unique among auks (Manuwal 1979).

BREEDING HABITAT AND NEST DISPERSION: nests on seaward facing slopes of islands lacking mammalian predators larger than mice (they coexist with small mammals such as *Sorex*, *Peromyscus*, and *Microtus*). Ideal breeding habitat must have sufficient soil for burrowing, and is often grassy and open, although this species nests in other situations. Typically, burrows are situated among dense grass tussocks on steep slopes within 500 m of the coast. In Haida Gwaii, many colonies are situated on mossy but otherwise bare ground under dense mature forest cover. Less commonly nests in sparsely vegetated rocky sites with little soil and in talus. Most colonies are situated on open grassy slopes, but many colonies in SE Alaska and British Columbia are at least partly forested (Campbell *et al.* 1990). On forested islands burrows under roots of living trees, and under fallen logs and stumps; in Baja California burrows under cacti and thorny shrubs. Some birds use wooden artificial nest boxes in the Farallon Is. where there is little soil (Ainley and Boekelheide 1990). A colonial species, nesting in isolated groups of a dozen to a hundred pairs on some islands (e.g. Haida Gwaii, BC), varying continuously up to dense colonies of hundreds of thousands of birds (e.g.) Triangle Is., BC, density of burrows was 1.36 ± 0.05 burrows/m^2 (Rodway *et al.* 1990)). In Haida Gwaii, nesting densities varied from 0.04 to 2.26 burrows/m^2, with 50% of the birds nesting at > 0.80 burrows/m^2 (Vermeer and Lemon 1986). Nesting density is not related to colony size.

NEST: nests in earth burrows excavated by the birds themselves, sometimes under driftwood piles, in dirt floored chambers with rock walls and roof, rarely in rock crevices on cliffs or in talus. Usually there are splashes of white and pink fecal material radiating from the burrow entrance, particularly after hatching. Nests

2   Cassin's Auklet breeding habitat at Triangle I., BC (ILJ photo).

become exceedingly malodorous as breeding season progresses, owing to chick's feces and spilled meals. Burrow dimensions: length 100.0 ± 27.0 cm (n = 40), height of entrance 13.0 ± 3.6 cm (n = 40), width 15.0 ± 4 cm (n = 40) in Haida Gwaii (Vermeer and Lemon 1986); length 75.0–100.0 cm in the Farallons (Manuwal 1974*b*).

EGG-LAYING: timing of breeding varies widely. In Baja California the rather asynchronous laying can begin as early as Nov and continues through to Mar (Jehl and Everett 1985). At the Farallons, laying occurs from Mar to May (rare second clutches in June and July; Ainley and Boekelheide 1990). In British Columbia, peak of laying is late Mar to late Apr, hatching late April to late May, and fledging early June to early July. At Buldir I. in the Aleutians laying takes place in early May, earlier than in *Aethia* species breeding at the same site (ILJ).

EGGS: clutch size is one egg, relaying of lost first eggs is rare, but about 10% of lost eggs replaced in Farallons, time to replacement

17.5 ± 8.0 days. At Frederick, 3 of 17 lost eggs were replaced, but it was uncertain if more than one female was involved (A. Harfenist, unpublished data). Two eggs are occasionally laid a few days apart in one burrow by two females (raising possibility of intraspecific brood parasitism), only one is incubated. Second eggs from same female are occasionally laid at Farallon Is. in a late season second brood (Ainley and Boekelheide 1990). Egg shape: elliptical ovate to ovate, smooth textured, white becoming lightly bluish or greenish tinged during incubation (Manuwal and Thoresen 1993), and often stained brownish from dirt in burrow. Egg size: from SE Farallon Is., California (Manuwal 1974*b*); 46.2 × 33.5 mm (n = 75); from Triangle Is., BC (Y. Morbey, unpublished data), 47.4 ± 1.8 (41.5–52.4, n = 221) × 34.1 ± 1.1 (31.1–38.2, n = 222) mm; from Haida Gwaii BC 47.6 ± 1.8; × 34.1 ± 1.0 mm (n = 75; A. Harfenist, unpublished data). Mass of egg within a few days of laying, 29.8 ± 2.4 g (24.0–34.0, n = 26) at Triangle I. (Y. Morbey, unpublished data), thus averaging about 17% of adult ♀ mass.

INCUBATION: breeding adults have two lateral brood patches 43 by 23 mm which defeather and become vascularized 1–2 days before laying, and begin refeathering at about mid-point of incubation (Manuwal 1974*b*). Notably, 60% of birds incubating the rare second clutch lack a brood patch. Duration of incubation at the Farallons: 37.8 days (37–42, $n = 86$; Manuwal 1974*b*); 39.0 days ($n = 493$; Ainley and Boekelheide 1990). Incubation is apparently shared equally between the pair members with change-overs at 24 hr intervals.

HATCHING AND CHICK-REARING: eggs pip for 24–72 hrs before hatching. Chicks semiprecocial, brooded continuously until day 4, thereafter left alone during the day but fed and brooded at night by one or both parents. Parents return once each night at dark with food in sublingual pouch, chick meal mass: at the Farallons, 27.8 ± 9.7 g (8.6–45.6, $n = 22$; Manuwal 1974*b*); at Frederick I. 18.7 ± 8.0 ($n = 30$; A. Harfenist, unpublished data).

CHICK GROWTH AND DEVELOPMENT: chick mass at hatching: 20.3 ± 1.6 g ($n = 23$, at Frederick I. (Vermeer and Lemon 1986). Maximum growth rate (g/day): 4.3 ± 0.7 ($n = 70$) at Triangle I. in 1994 (Y. E. Morbey, unpublished data), and 5.3 ± 0.7 (3.6–7.0 $n = 57$) at Frederick I. in 1994 (A. Harfenist, unpublished data). Chick mass at fledging: 162 ± 12 g ($n = 151$) at Triangle I in 1994; 167.1 ± 15.8 g ($n = 52$; Vermeer and Lemon 1986) at Frederick I. in 1980; 168.5 ± 17.4 g (129–201, $n = 46$) at Frederick I. in 1994, averaging about 90% of mean adult body mass. Duration of nestling period: in the Farallons, 41.1 days (35–46, $n = 16$; Manuwal 1974*b*); at Triangle I., 46 ± 3 days ($n = 147$; Y. E. Morbey, unpublished data), at Frederick I., 45 ± 3 days (41–54, $n = 46$; A. Harfenist, unpublished data). All measures of chick development fluctuate between years in response to changing environmental conditions.

BREEDING SUCCESS: productivity is probably about 0.5–0.7 chicks fledged per breeding pair per year. At SE Farallon I., mean number of chicks fledged per breeding pair per year was 0.6 in 14 years of monitoring, with the success of rare second brood about 0.1 chicks fledged per second nesting attempt (Ainley and Boekelheide 1990). At Frederick I., Haida Gwaii (2 years of monitoring; Vermeer and Lemon 1986) hatching success averaged 70% (although 22–29% of clutches were deserted, possibly in part due to investigator disturbance), fledging success 93%, and overall breeding success (proportion of eggs laid resulting in a fledged chick) 65%. Hatching and fledging success increased with breeding experience up to 3 years' experience for both sexes, and there was a weak trend of increased hatching success with number of years with the same mate (Emslie *et al.* 1992).

PREDATION: mammalian predators include Arctic and Red Foxes which were introduced to several breeding islands with disastrous results (Bailey 1993), and rats and mice which prey on eggs and nestlings in burrows. River Otters may visit some colonies and take a few birds (ILJ). In Haida Gwaii, this species is preyed upon and may be seriously threatened by introduced Racoons (AJG). Cassin's Auklets have been found dead after being ingested accidentally by Humpbacked Whales (*Megaptera novaeangeliae*) (Dolphin and McSweeney 1983). Avian predators include Western and Glaucous-winged Gull, Bald Eagle, Peregrine Falcon, Northern Raven, and Northwestern Crow (Vermeer and Lemon 1986; Ainley and Boekelheide 1990; Manuwal and Thoresen 1993).

MOULT: moults and feather generations have not been described in detail. There is considerable variation in timing of moult, probably related to age (Emslie *et al.* 1992). Adults apparently undergo moult of underparts, head, and neck feathering (but not flight feathers) during late winter (Definitive Pre-alternate moult, Feb–Mar). Gradual primary moult begins during breeding season (May–July); primaries are replaced sequentially starting with the innermost and working outward, followed by complete moult of body feathers (Definitive Pre-basic moult, July–Oct). Sub-adults (2- and 3-year-olds) attend colonies with very worn brownish or tan coloured primaries which are moulted during the breeding season, and they also begin body moult late in the breeding season (Definitive Pre-basic moult, July–Oct).

Population dynamics

SURVIVAL: Speich and Manuwal (1974) estimated annual adult survival to be about 83% at SE Farallon Is., based on rough calculations on 3 years of data. At Reef Island, Haida Gwaii, Gaston (1992b) estimated annual adult survival to be 88% (95% confidence limits: 0.78–0.93, 6 years of data), giving an estimated life expectancy of 7.6 years. A large scale demographic study of this species by Simon Fraser University is underway at Triangle and Frederick Is., BC, with estimates of survival based on mark–recapture expected soon.

AGE AT FIRST BREEDING: unusual among alcids in breeding at 2 years of age, although most birds probably begin breeding at age 3 (Emslie *et al.* 1992). Brown-eyed birds (all less than 4 years of age; Manuwal 1978) are often found breeding.

PAIR- AND NEST-SITE FIDELITY: little information on pair-fidelity. Pair-bonds are not 'permanent' (Manuwal 1974b; Manuwal and Thoresen 1993). For example, Emslie *et al.* (1992) noted six cases of divorce in a marked population inhabiting artificial burrows, but divorce frequency was unclear because of the difficulty of establishing whether both pair members of a split pair were still alive. Re-mating pairs normally return to the same burrow used in a previous year (Ainley and Boekelheide 1990). At Frederick I., of 40 burrows checked that had contained marked pairs in a previous year, 65% had same pair present, 15% had one of the original pair members mated to a new partner (owing to divorce or death of the previous partner), and the remaining burrows either had a new pair present or were empty (A. Harfenist, unpublished data).

Parakeet Auklet *Cyclorhynchus psittacula*

Alca psittacula (Pallas, 1769, Spicil Zool. I, V: plates ii and v, figs 4–6— Kamchatka)

PLATE 5

Description

A small alcid but one of the larger auklets, this species is vaguely coot-like, with a chunky body-shape, long neck, rounded wings, and relatively large feet. Bill is very unusually shaped, almost round in profile (hence Latin name) with an oval blunt-tipped upper mandible without a trace of overlap over the sharp-pointed and strongly upcurved lower mandible. The large pink tongue is usually visible when the bird calls. This strange bill morphology probably relates to feeding on jellyfish (see Diet below). On land, stand up fairly straight and move about with considerable agility, but they have an air of strangeness about them with their thin-necked, pot-bellied posture, large feet, round head, and vacant expression. Despite their rather short, broad, and rounded wings and slow wing-beat they appear to be quite strong fliers.

ADULT SUMMER: dark-grey upperparts and white underparts all year round; in summer-plumage adults demarcation between dark and light is right along the waterline, so birds resting on the water appear mostly dark grey; head and neck blackish grey, grading to dark grey on back, rump, and upperwing coverts; flight and tail feathers blackish grey (fresh) or brownish grey (worn), underwing coverts grey with white flecking; upper breast and flanks white, heavily flecked and indistinctly barred with dark grey; lower breast, belly, and under-tail coverts immaculate white; single narrow silvery-white plume tract (auricular plumes) extending backward from below and behind eye Bill deep red with small pale areas near gape; irides white these combine with the rather plain face to give the bird a pop-eyed look; legs and feet blue-grey (not yellow), occasionally tinged greenish, with black webs and nails. Subadults (2-year-olds) in summer have white flecking on the throat.

ADULT WINTER: similar to summer plumage but extent of white greater on lower neck, flanks, and upper breast, and white plumes are reduced or absent; bill dull grown.

JUVENILE AND FIRST WINTER: resembles winter-plumage adults but have smaller darker bills, greyish eyes, and dull blackish grey feet.

CHICK: uniform blackish grey down, bright blue eyes, and black bill and feet.

VARIATION
Monotypic. No notable pattern of geograph-ical variation in size or plumage have been described.

MEASUREMENTS
From St Lawrence I., Alaska (Sealy and Bédard 1973; Bédard and Sealy 1984 (mm): wing length, ♂ 147.4 ± 3.8 ($n = 20$), ♀ 145.2 ± 3.4 ($n = 20$), culmen, ♂ 15.8 ± 1.2 ($n = 20$), ♀ 15.6 ± 1.0 ($n = 20$), bill depth, ♂ 14.5 ± 0.6 ($n = 20$), ♀ 14.0 ± 0.5 ($n = 20$). From Buldir I., Alaska (ILJ): wing length, ♂ 149.3 ± 3.5 (142–157, $n = 20$), ♀ 146.6 ± 3.0 (141–152, $n = 11$), culmen, ♂ 16.1 ± 0.8 (14.1–17.0, $n = 32$), ♀ 15.7 ± 0.5 (15.1–16.5, $n = 11$), bill depth, ♂ 13.5 ± 0.5 (12.3–14.6, $n = 31$), ♀ 12.9 ± 0.4 (12.3–13.5, $n = 11$), tarsus, ♂ 30.6 ± 1.1 (29.0–34.0, $n = 32$), ♀ 30.3 ± 1.3 (28.2–31.8, $n = 11$), auricular plume length, ♂ 35.7 ± 3.5 (31.5–44.0, $n = 21$), ♀ 34.8 ± 4.3 (27.9–42.5, $n = 11$). From the Alaska Peninsula (J. F. Piatt, unpublished data): wing length, 155.8 ± 4.8 ($n = 53$), culmen, 15.4 ± 0.7 ($n = 53$), tarsus, 29.7 ± 1.1 ($n = 52$).

WEIGHTS
Body mass (g) of birds from St. Lawrence I., Alaska (Sealy and Bédard 1973; Bédard and Sealy 1984): ♂ 292.3 (245–347, $n = 79$), ♀ 285.7 (255–337, $n = 71$). From Buldir I., Alaska (ILJ), of birds shot at sea, ♂ 272.7 ± 19.9 (238–304, $n = 21$), ♀ 266.9 ± 16.8 (242–292, $n = 11$). From the Alaska Peninsula (J. F. Piatt, unpublished data):) of birds shot at sea, 256.7 ± 20.6 ($n = 109$).

Range and status
RANGE
BREEDING: in Asia, breeds on the Kuril Is. (Raykoke, Chirinkotan) and islands in the Sea of Okhotsk (Sakhalin, Iona, Talan, Matykil', Is.), Commander Is., and northwards locally along the coast of the Kamchatka Peninsula (Karaginski Is.), Cape Navarin, and on the Chukotka Peninsula (Konyukhov 1990*c*). In North America, breeds on headlands and islands around the northern Gulf of Alaska in-cluding Shumagin Is., Semidi Is., Chirikof I. near Kodiak, locally in Kenai Peninsula, and SE Alaska (St Lazaria I.), in the Aleutians (W to Buldir I.), and in Bering Sea (St Lawrence I. St Matthew I., Pribilof Is. and Nunivak I.) (Sowls *et al.* 1978). Although generally scat-tered among many small colonies, huge number of this species breed at St George I., Pribilof Is. (>100 000). The northernmost colony is located at Ratmanov (Big Diomede) I. in the Bering Strait, the south-easternmost possible colony is at St Lazaria I., SE Alaska where a few birds occur regularly, and the south-westernmost near Sakhalin I. in the Sea of Okhotsk (Konyukhov 1990*c*).

WINTER: regularly moves into the Chuckchi Sea in late summer and autumn, occasionally wandering N to Point Barrow, Alaska, and Cape Serdse Kamen' in northern Siberia. Vagrants have reached coastal waters of British Columbia, Washington, Oregon, and California,

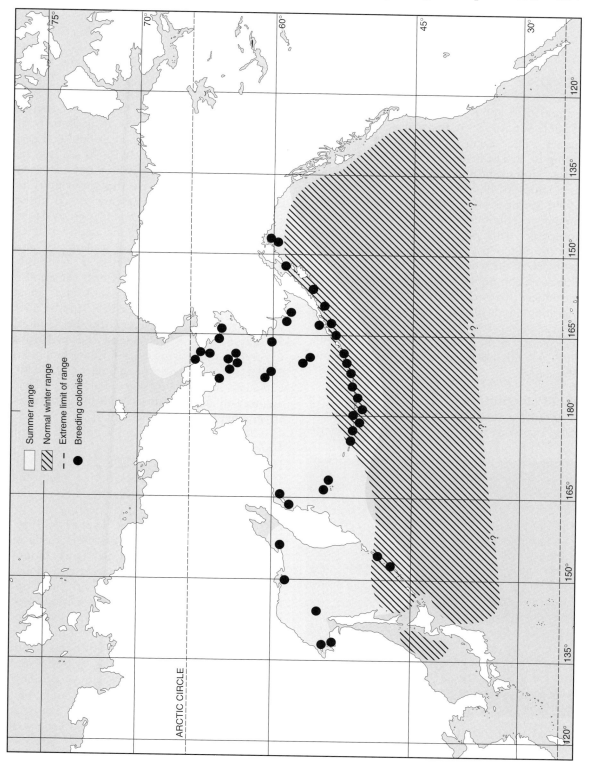

and Kure and Midway Atolls in the Hawaiian Islands, most records from the winter months. Accidental in Sweden on a freshwater lake.

STATUS
The total Alaskan population is officially estimated as about 1 million (U. S. Fish and Wildlife Service 1993), but may be considerably higher. The Asian population is unknown owing to lack of inventory of most of breeding range, but may be about 400 000, with 300 000 at colonies in the Sea of Okhotsk (Konyukhov 1989; Springer *et al.* 1993). Population trends unknown. Numbers probably low in Aleutians compared with populations prior to Arctic Fox introductions, although not noted to have been one of the worst hit species (Bailey 1993). Effects of chronic oil spills, gill-net fishing, and introduced foxes need to be studied. Prone to ingesting small plastic particles when feeding at sea; 94% of birds shot in a recent study had plastic in their crop (Robards *et al.* 1995; Fig. 1), with unknown effects on health and physiology.

Field characters
A chunky-bodied alcid with dark grey upperparts and white underparts, superficially resembling other small Beringian auklets and also Rhinoceros Auklet. Unmistakable when observed ashore at colonies and the facial ornaments and bill-shape are easily visible. In this situation resembles a giant Least Auklet, looking at least twice as large, and often seen standing among groups of the smaller species at mixed colonies. Hence it is known as 'choochki-mama' by Aleuts at St Paul I. (Least Auklet is called the 'choochki'). At sea may be more difficult to distinguish, but fortunately only a few observers have the opportunity to face this challenge while being tossed about aboard ship in the Bering Sea. Closest in size to Crested Auklet, it is easily distinguished from that species by its white underparts and rounded wings (Crested Auklets are uniformly dark and have more pointed wings). At sea also resembles Least and Cassin's Auklets most closely, but size, slower wing-beat, and broader wings are useful characters. In the Gulf of Alaska and off SE Alaska and coastal British Columbia could be mistaken for the Rhinoceros Auklet, but that species has a longer pointed bill and a less chunky body shape. Parakeet Auklets are unlikely to be confused with the even more streamlined murrelet species.

1 Crop of a Parakeet Auklet containing plastic particles, Buldir I., Alaska (ILJ photo).

Voice
Based on observations of ILJ at Pribilof Is. and Buldir I., A highly vocal species, sounding quite similar to Cassin's Auklet. Lone males and pairs vocalize near their nesting sites in cliff crevices or boulder piles. Typical vocalization is a 'whinny', a harsh grating screech or whinny repeated rhythmically and monotonously for long periods, apparently by males. Vocalizing birds adopt posture with head uptilted at 45° angle. A similar vocalization 'duet whinny' is given by pairs during courtship. The 'squeal' is an alarm call, given mostly as individuals take flight from disturbance, both on the colony and at sea. The 'whinny' and 'duet whinny' are similar and almost certainly homologous to the 'display kreek' and 'duet kreek' calls of Cassin's Auklet. Parakeet Auklet vocalizations have not been well described and need further study.

Habitat, food, and feeding behaviour

MARINE HABITAT

Summer and winter ranges defined by ocean temperatures of 2–15 and 2–8 °C. Disperses more widely over deep offshore waters of open ocean than other auklet species. Feed over stratified waters, avoiding, or at least not concentrating at, areas of turbulence and upwelling (Hunt *et al.* 1993*b*). In winter occurs further offshore and further S than other Beringian auklet species.

FEEDING BEHAVIOUR

The unusual bill structure is probably an adaptation for handling their slimy prey, gelatinous zooplankton, although foraging has not been observed underwater and how exactly the bill is used is unclear. Several at-sea studies have found Parakeet Auklets at low densities, scattered over wide areas rather than clumped near oceanographic features like Least and Crested Auklets. At sea normally seen singly or in small groups of two or three individuals, never in dense flocks. It appears that rather than sticking to powerful upwelling currents that drive zooplankton to the surface and concentrate it, Parakeet Auklets forage on jellyfish which are a reliable but thinly distributed source of food (Hunt *et al.* 1993), although like other auklets they are opportunistic and will take other prey when it becomes available. There have been no detailed studies of dive depths or behaviour, but this species may have rather poor diving ability and low underwater speed and manoeuvrability related to the fact that their usual prey are found near the surface and are probably unable to take evasive action.

DIET

Relatively well studied, with data from several locations in the Bering Sea emphasizing the wide diversity of prey taken. Feeds on a wider variety of small planktonic animals than other auklet species, and more often preys on gelatinous zooplankton (jellyfish and ctenophores) and plankton associated with jellyfish such as hyperiid amphipods and certain fish larvae

(Harrison 1990). Differences between studies probably reflect stochastic effects rather than real geographical variation in preferred prey. At Buldir I., Day and Byrd (1989) found the summer diet to be copepods (mostly *Neocalanus cristatus*, 58.2% of index of total importance), amphipods (mostly the hyperiid amphipod *Parathemisto pacifica* 23.2%), decapods (16.2%), Brachyura (*Erimacrus isenbeckii*, 13.4%), euphausiids (0.9%), and chaetognaths (0.1%). At the Pribilof Is., Hunt *et al.* (1981) found the diet to be euphausiids (mainly *Thysanoessa inermis* and *T. raschii*, 32.5%), copepods (*Calanus* spp. 21.6% by number), amphipods (mainly the hyperiid amphipod *Parathemisto libellula*, 15.3%) and fish larvae, including walleye pollock (*Theragra chalcogramma*, 8.3%). At St Matthew I., Harrison (1990) found a euphausiid *Thysanoessa raschii* (dominant in 23% of samples), gelatinous Scyphomedusae (13%), walleye pollock (13%) and *Parathemisto libellula* (3%); at the Chirikov Basin, northern Bering sea, diet included the gelatinous zooplankton Ctenophora (dominant in 38% of samples) and Scyphomedusae (15%), and the copepod *Neocalanus cristatus* (8%). At St Lawrence I., Bédard (1969*c*) found *Parathemisto libellula* (60.8% of prey individuals), *Limacina* (35.9%), and polychaete worms (2.2%). Winter diets have not been recorded.

Displays and breeding behaviour

COLONY ATTENDANCE AND FLIGHT DISPLAYS: colony attendance is much less synchronized than other *Aethia* species, and nesting densities much lower, paralleling behaviour at sea. Consequently Parakeet Auklets do not have conspicuous flight displays, although pairs sometimes circle over breeding habitat in peculiar 'butterfly-flight' resembling flight displays of other auk species. Where they nest in association with *Aethia* auklets, they rarely join massive flocks circling over breeding habitat. Normally, birds arrive and stage in small rafts of 10–40 birds just offshore from colony early in the morning and fly to nest-sites in ones and twos. Frequently, flocks on the sea drift in very close to shore and birds climb out on wet or wave-washed rocks to socialize, which *Aethia* auklets never do. Daily activity pattern is relatively unsynchronized. Activity near nesting crevices lasted all day in the Pribilofs, where Least and Crested Auklets restricted themselves to well-defined morning and evening activity periods. Activity at night as well as during daylight was observed at Buldir I., where birds were caught after dark in mistnets set for Whiskered Auklets, and calling birds were active after dark (ILJ and F. M. Hunter, personal observations).

AGONISTIC AND SEXUAL BEHAVIOUR: socially monogamous, extra-pair copulation attempts rarely occur. Social behaviour little studied (but see Manuwal and Manuwal 1979). Compared with Least and particularly Crested Auklets, Parakeet Auklets are a mild-mannered and non-aggressive species. Fights are rare, but birds are occasionally seen grappling with bills and kicking each other with their feet in combats on the sea near breeding sites. Courtship takes place near nesting crevices, on cliff tops, on flat tops of large boulders, and at ocean gathering grounds nearby. Pairing process appears to begin when ♀ approaches and lands near a displaying ♂ (see 'whinny', Voice above). Lone birds (presumably mostly ♂♂) display from tops of boulders and from perches on cliffs, and also from the mouths of nesting crevices. After landing nearby, ♀♀ cautiously approach and exchange long looks with the calling ♂. Courtship begins when ♀ adopts submissive posture beside or in front of ♂ and begins to vocalize along with ♂ (duet whinny). This is sometimes accompanied by wing-raising, and usually followed by billing and further mutual vocalizing. Copulations take place only on the sea, preceded by lengthy billing and mutual vocalization. Pairs averaged about 1.5 copulations per h at the staging area, less than a third the rate in Least, Crested, and Whiskered Auklets. ♂♂ occasionally attempt extra-pair copulations but these are rarely successful (F. M. Hunter and ILJ).

Breeding and life cycle

Biology is similar to other auklet species, but they are much less dependent on mass aggregations for feeding and breeding. It is unclear how they are able to escape predation with their diurnal colony attendance and low nesting densities. Timing of breeding slightly later than *Aethia* species, both at St Lawrence I., where snow melted from nesting habitat later than for other auklets (Sealy and Bédard 1973), and at Buldir where snow-cover was not a factor (Hipfner and Byrd 1993).

BREEDING HABITAT AND NEST DISPERSION: greater diversity of nesting habitat than other auklets. On islands where there are no mammalian predators, select sites with rock crevices such as among beach boulders, recent talus slopes, scree slopes, and cliffs, or where soft soil for burrowing is present, such as on steep grassy slopes with scattered boulders, and old talus slopes densely vegetated with grasses and umbels (Hipfner and Byrd 1993). Restricted to cliff habitat where foxes are present, such as in the Pribilof Is., where the biggest concentrations occur on fragmented volcanic cliffs with numerous small caves, rock cracks, and crevices (Fig. 2). A loosely colonial species, usually nesting in groups of a dozen to a hundred pairs. Nest-sites are usually several metres apart. Less gregarious than Least or Crested Auklets and rarely congregates in dense flocks. Coexists with Least and Crested Auklets at many sites but usually avoids nesting among dense concentrations of these other species.

NEST: in bare rock crevices, either on cliffs or in talus, in dirt floored chambers under boulders, or in earth burrows excavated by the birds themselves (burrowing habit possibly explains why this species has unusually large feet among the Beringian auklet species). Nest depths usually shallow, the incubating bird occasionally visible from the surface. Egg normally laid in slight depression on floor of crevice. Crevice entrance dimensions: height 81.0 ± 1.8 mm, width 111.9 ± 2.9 mm

2 Typical breeding habitat, St Paul I., Pribilof Is., Alaska (ILJ photo).

($n = 76$) at Buldir I., Alaska (Hipfner and Byrd 1993).

EGG-LAYING: peak of laying usually late May–mid June (western Aleutians *c.* 52° N), early–mid June (Pribilofs *c.* 57° N) or early July (St Lawrence I. *c.* 63° N). At northern colonies, timing of breeding is affected by the time of snow melt on the nesting habitat (Sealy 1975*c*).

EGGS: one egg, relaying of lost first eggs rarely or never occurs (Piatt *et al.* 1990*c*). Egg shape: elliptical, smooth textured, white sometimes with a slight bluish or greenish tinge, often becoming stained brownish from dirt in nesting crevice. Egg size and mass: from Russia (Konyukhov 1990*c*), 53.8 (48.9–58.7) × 37.4 (33.0–40.0, $n = 88$) mm; mass of fresh egg 37.6 g (35.4–43.5, $n = 3$) or about 14% of average adult ♀ mass. From Buldir I., Alaska (M. Hipfner, unpublished data), 54.0 ± 1.6 × 37.4 ± 1.1 ($n = 34$) mm.

INCUBATION: breeding adults have two lateral brood patches which become vascularized and thickened by onset of incubation, and rapidly shrink and become refeathered after hatching. Incubation period days 35.2 (35–36, $n = 4$) at St Lawrence I. (Sealy 1984). Incubation duties are by ♂ and ♀ with incubation shifts usually 24 hrs although longer shifts were recorded by Hipfner and Byrd (1993).

HATCHING AND CHICK-REARING: duration of pipping unrecorded. Chicks semi-precocial, brooded continuously for first day after hatching, but diurnal brooding decreases steadily until chick is left by itself during the day after 1 week of age. Adults carry food in sublingual pouch and make several deliveries per day; food load mass: 20.1 ± 5.3 g (14.5–25.3, $n = 13$; Bédard 1969*c*).

CHICK GROWTH AND DEVELOPMENT: chick mass at hatching: 26.6 ± 2.1 g (23.5–29.5, $n = 12$, Buldir I.; Hipfner and Byrd 1993), 28.1 g (25.6–34.4, $n = 4$, St Lawrence I., Sealy and Bédard 1973); at asymptote (day 28) 253.1 ± 21.7 g (234–296 $n = 7$; Buldir I.; Hipfner and Byrd 1993); at fledging 207.6 ± 19.9 g (189–244, $n = 10$, Buldir I.; Hipfner and Byrd 1993), 222.6 g (194–259, $n = 6$, St Lawrence I.; Sealy and Bédard 1973). Thus chicks fledge at about 80% of average adult mass. Chick growth rate (g/day during linear phase): 8.6 ± 3.2 ($n = 66$, Buldir I.; Hipfner and Byrd 1993). Chick wing chord about 22 mm at hatching and increased at 3.6 mm/day to reach about 120 mm at fledging (80% of adult length, Buldir I., Hipfner and Byrd 1993). Age at fledging was 36.1 ± 1.5 days (34–38, $n = 9$) at Buldir (Hipfner and Byrd 1993), and 35.3 ± 0.9 days (34–37, $n = 6$) at St Lawrence I. (Sealy and Bédard 1973). Fledglings fly to sea from their nesting crevice, probably during the night. There is no post-fledging parental care.

BREEDING SUCCESS: productivity is probably about 0.5 chicks fledged per breeding pair per year. At Buldir I., hatching success was 77%, fledging success 66%, and overall reproductive success 52%. Productivity at frequently checked sites was compared with productivity at control (less disturbed) sites and there was slightly higher productivity (but not statistically significantly different) at the disturbed sites, suggesting that the protocol monitored natural reproductive success (Hipfner and Byrd 1993). At St Lawrence I., hatching success was 67.7%, fledging success 76.2, and overall reproductive success 51.6% (Sealy and Bédard 1973).

PREDATION: mammalian predators include Arctic and Red Foxes which take adults at colonies, and voles (*Microtus* and *Clethrionomys*) which prey on eggs and nestlings in crevices (Sealy 1982). Avian predators include Slaty-backed, Glaucous and Glaucous-winged Gulls, Bald and Steller's Sea Eagles, Gyrfalcons, and Peregrine Falcons (ILJ; Konyukhov 1990c).

MOULT: from Bédard and Sealy (1984). Juveniles acquire their full plumage by about 30 days of age and retain this through their first winter, when they closely resemble winter-plumage adults. First-summer (Alternate I) birds apparently do not attend breeding colonies and their appearance has not been described, although they probably resemble winter adults. Subadults (2-year-olds) attend colonies with worn brownish upperwing coverts and primaries in moult, and whitish spotting on throat and upper breast. They undergo a complete moult of body feathering late in the breeding season (Definitive Pre-basic moult). Adult breeding plumage is attained in the third summer. Adults undergo moult of underparts, head, and neck feathering (but not flight feathers, greater coverts, or thighs) during late winter and spring when facial plumes develop (Definitive Pre-alternate moult, Mar–May). Gradual primary moult begins late in or just following breeding season; the primaries are replaced sequentially starting with the innermost and working outward, followed by complete moult of body feathers (Definitive Pre-basic moult, July–Oct). Facial plumes look worn by end of breeding season, with most adults experiencing partial loss of white plumes and peeling of outer covering of their bills late in chick-rearing.

Population dynamics

Age at first breeding is uncertain, probably 3 or 4 years. Small numbers of immatures which may be 2-year-old sub-adults attend colonies, but these individuals remain unpaired. Survival has not been measured. At Talan I., five of six pairs monitored stayed together to a following year, the split being due to the death of one partner, and one pair remained together for at least 4 years (Zubakin and Zubakina 1993). Re-mating pairs often return to the same crevice used in a previous year (Sealy 1968; M. Hipfner, unpublished data), and 93% of marked birds returned to the vicinity of the previous year's nest-site (Zubakin and Zubakina 1993).

Crested Auklet *Aethia cristatella*

Alca cristatella (Pallas, 1769, Spicil. Zool. I, V: 18, plates iii and v—between Hokkaido and Kamchatka).

PLATE 6

Description

A small alcid with uniform grey plumage and extraordinary ornaments during the breeding season that give this bird either a bizarre or comical appearance, depending on your perspective. A medium-sized auklet, similar in body size to Parakeet Auklet and Dovekie.

They have an upright posture, and adults have little difficulty climbing and leaping about through their rocky breeding habitat or in taking off from the ground. In alert posture, often adopt a reptilian long and thin-necked shape not found in most other small alcids. Flight fast and direct on relatively narrow pointed wings compared with other auklets; feet strong and well adapted for agility on land.

ADULT SUMMER: dark grey from crest to tail, except for narrow silvery-white tracts of plumes (auricular plumes) that extend from behind eye to nape; upper- and underwing coverts dark grey or fading to grey-brown when worn; flight and tail feathers blackish grey or blackish brown; ornaments consist of a spectacular forehead crest consisting of 12–20 narrow elongated (2–3 mm width by 15–50 mm length) blackish feathers, the white auricular plumes, and a brilliant orange bill with accessory plates; crest flattens against crown when bird is in flight; plumage has a strong distinctive citrus odour which may be considered to be an additional form of adornment; irides white; legs and feet normally pale bluish grey (rarely brown or dark pinkish) with black webs. In early summer adults' bills are chrome orange with a buffy tip, enlarged by four or five brightly pigmented horny plates, including: the nasal—attached saddle-like over base of mandible; the subnasal—along cutting edge of mandible, extending from below nostril to rictal plate; the rictal—a peculiar semi-circular structure at gape; and the maxillary—a crescent-shaped structure attached to base of maxilla, curving upward to merge evenly with rictal plate when beak closed. Bill of ♂ distinctly deeper than ♀'s with culmen more sharply decurved, and with stronger hooked tip to upper mandible (Jones 1993*b*; see Fig. 1). These brightly coloured plates begin

to shed during chick-rearing (adult breeders) or at end of breeding season (adult non-breeders) leaving birds with a smaller dull bill.

ADULT WINTER: overall plumage similar in summer and winter, but in non-breeding season white head plumes much reduced or absent, and brightly coloured bill plates shed after breeding as in puffins. Some birds (presumably immatures and subadults, 1- or 2-year-olds) remain in winter-like plumage during summer.

SUBADULT: overall plumage similar to summer adults but wing coverts and remiges brown or tan coloured from wear, with small crests (<20 mm length), and small dull carrot-coloured bills; these birds are probably 2-year-olds (ILJ).

JUVENILE AND FIRST WINTER: resembles winter plumage adults but has smaller blackish bills, greyish eyes, lacks crest and auricular plumes, and has dull blackish grey feet.

CHICKS: dense woolly down uniformly blackish grey except for a contrasting pale grey spot on lower back near rump; feet and eyes blackish brown.

VARIATION
Monotypic; no notable patterns of geographical variation have been described.

MEASUREMENTS
♂♂ are slightly larger than ♀♀ in most measurements, especially bill depth (adults can be

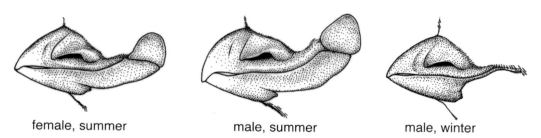

female, summer   male, summer   male, winter

1 Bill shapes of adult male and female in summer and male in winter.

sexed by bill shape and depth, Jones 1993*c*), but ♂ and ♀ ornament size is virtually identical—see Table, (1) Buldir I., Alaska (Jones 1993*ab*; I. L. Jones and F. M. Hunter, unpublished data); (2) St Lawrence I., Alaska (Bédard and Sealy 1984).

WEIGHTS

Body mass varies with time of year, time of day, breeding status, age and sex. ♂♂ are slightly larger than ♀♀—see Table, (1) Buldir I., Alaska (Jones 1993*a,b*; I. L. Jones and F. M. Hunter,

unpublished data); (2) St. Lawrance I., Alaska (Piatt *et al.* 1990); (3) Chukotka, Siberia (Konyukhov 1990*a*); (4) Alaska Peninsula (J. F. Piatt, unpublished data).

Range and status
RANGE

Restricted to the Bering and Okhotsk Seas and adjacent areas of the North Pacific.

BREEDING: In Asia breeds in Kuril Is., including Chirinkotan, Lovushki, Raykoke, Matua,

Table: Crested Auklet measurements (mm)

ADULT ♂♂	ref	mean	s.d.	range	n
wing length	(1)	143.0	2.5	140–148	9
	(2)	137.5	0.9	—	30
tarsus	(1)	29.0	1.0	26.9–31.3	153
culmen	(1)	12.6	0.6	11.7–13.4	15
	(2)	13.1	0.6	—	30
bill depth	(1)	12.3	0.5	11.1–13.0	15
	(2)	13.0	0.4	—	30
rictal plate height	(1)	5.1	0.9	2.1–7.1	123
	(2)	6.0	0.9	—	30
crest length	(1)	38.4	5.5	16.2–58.5	90
	(2)	43.3	5.8	—	30
auricular plume length	(1)	29.3	4.2	17.3–41.3	164

ADULT ♀♀	ref	mean	s.d.	range	n
wing length	(1)	140.3	2.3	136–143	15
	(2)	134.1	0.2	—	30
tarsus	(1)	28.4	0.9	25.8–30.2	184
culmen	(1)	11.8	0.5	11.1–12.8	21
	(2)	12.9	0.6	—	30
bill depth	(1)	10.6	0.5	9.9–11.9	21
	(2)	11.3	0.4	—	30
rictal plate height	(1)	4.4	1.2	0.1–6.8	113
	(2)	5.7	0.6	—	30
crest length	(1)	35.0	6.8	8.1–45.5	105
	(2)	39.3	6.4	—	30
auricular plume length	(1)	27.3	4.1	15.4–36.7	202

UNSEXED BIRDS (SUBADULTS)	ref	mean	s.d.	range	n
bill depth	(1)	10.5	0.9	8.6–12.8	16
rictal plate height	(1)	3.9	1.0	2.4–6.3	13
crest length	(1)	29.5	9.7	16.0–38.5	4
auricular plume length	(1)	25.2	3.6	17.2–32.2	16

Table: Crested Auklet weights (g)

ADULT ♂♂	ref	mean	s.d.	range	n
at colony, during incubation	(1)	267.7	17.1	227–322	179
ADULT ♀♀	ref	mean	s.d.	range	n
at colony, during incubation	(1)	251.4	17.6	211–300	229
UNSEXED BIRDS (ADULTS)	ref	mean	s.d.	range	n
at colony	(2)	260.0	14.0	227–285	34
at colony, laying period	(3)	300.3	—	—	84
at colony, fledging period	(3)	265.1	—	—	77
shot at sea	(4)	260.3	21.7	—	89
UNSEXED BIRDS (SUBADULTS)	ref	mean	s.d.	range	n
at colony, July	(1)	239.6	12.3	215–263	16

Torporkov, Rassua, Ushishir, Simushir, and Chernye Brat'ya, and islands in the Sea of Okhotsk including Iona, Sakhalin, Talan, Matykil', the Yama, and the Shantar Is. In the western Bering Sea also breeds in the Commander Is., the SE coast of the Chukotka Peninsula, and Ratmanov (Big Diomede) I. in the Bering Strait (Konyukhov 1990a). In Alaska, breeds on remote islands in Aleutians, from Krenitzen Is. W to Buldir I., and in the central and northern Bering Sea in the Pribilofs, St Matthew I., St Lawrence I., King I., at Little Diomede Is. in the Bering Strait, and also in the Shumagin Is. (Big Koniuji) in the Gulf of Alaska S of the Alaska Peninsula (Sowls *et al.* 1978).

WINTER: most birds move S of the pack-ice, although some birds winter in ice-choked parts of the Sea of Okhotsk (Konyukhov 1990a). In Asia, winters in large numbers around the central Kuril Is., Sakhalin Is. (Konyukhov 1989a) and around northern Hokkaido (Brazil 1991). In the east, winters in Aleutian I. passes and eastwards to Kodiak I. Vagrants have occurred at Wrangel I. (Konyukhov 1990a), in interior Alaska 600 km up the Yukon River, as far SE as British Columbia (Godfrey 1986) and Baja California (a July record off Isla Cedros, R. Pitman), and at sea off Iceland in the North Atlantic.

STATUS
Large breeding colonies (>100 000 individuals) are located at Buldir, Kiska, and Gareloi Is. in the central Aleutians, St Lawrence, St Matthew, Ratmanov (Big Diomede) I., and Cape Yagnochymlo in the northern Bering Sea, Chirinkotan and Ushishir I. in the Kurils, and Talan and Matykil I. in the Sea of Okhotsk. The sizes of the largest colonies are difficult to estimate but probably exceed 500 000 breeding paris (Springer *et al.* 1993). There are estimated to be at least 2 million in Russia (Konyukhov 1990a) and 3 million in the U. S. A. (U. S. Fish and Wildlife Service 1993), but populations are likely to be much higher. No evidence of any recent large-scale population changes, but the difficulties of monitoring populations precludes scientific assessment of this at present. Introduction of foxes certainly caused population declines on some Aleutian Is. (Bailey 1993), but currently several large auklet colonies in the Aleutians

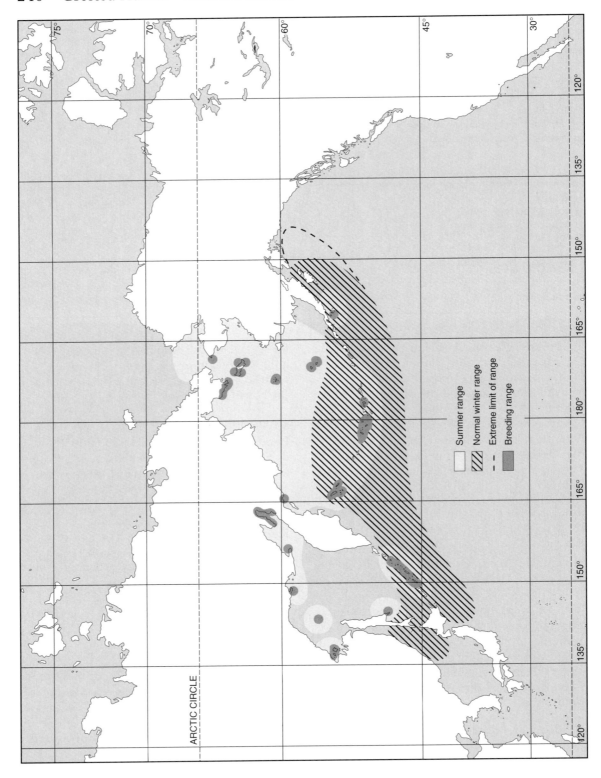

ARCTIC CIRCLE

Summer range
Normal winter range
Extreme limit of range
Breeding range

apparently coexist with introduced Arctic Foxes. Nevertheless, removal of introduced foxes from the Aleutians will probably restore populations to historical levels on some islands. Oiled corpses have been found at breeding islands near the Pacific great circle shipping route (Buldir, Kiska, Segula), suggesting oil pollution could be a significant source of mortality. Ainley *et al.* (1981) documented mortality in high seas drift gill-nets. Artificial lights on vessels attract and kill large numbers of birds on foggy nights (e.g. Dick and Donaldson 1978). Introduced rats may threaten populations at Kiska I. in the Aleutians and St Paul I., Pribilofs.

Field characters

Similar to several small North Pacific alcid species, but the remoteness of their breeding and winter ranges makes them unlikely to be seen by most birdwatchers. Breeding adults have conspicuous forehead crest, bright orange bill with accessory plates, and small white auricular plume, distinctive ornaments which are likely to render them easily identifiable at colonies. The only small alcid that is uniformly dark in overall coloration. Similar in body size to Parakeet Auklet and Dovekie, but their wings are somewhat longer and more pointed; these other species also have white underparts. They also superficially resemble Whiskered Auklet, but Crested Auklets are more than twice as large, they lack whitish lower belly and vent, their forehead crest is bushier, they lack white V-shaped loral plumes, and their bill colour in summer is orange rather than red. At sea usually seen in large flocks or not at all.

Voice

A highly vocal species, especially at breeding colonies, with a variety of barking and hooting calls sounding unlike any other alcid. The most frequently heard vocalization is a short hoarse call like the yap of a small dog—'bark'. This is normally performed singly or repeated several times in quick succession, and is associated with locomotion and contact calling by

pair members. It may also serve as an alarm call. The 'bark' call is heard mostly on the colony surface, but also from nesting crevices, and on the sea at the staging area. The most spectacular vocalization, 'trumpeting', is a vocal advertising display performed by ♂♂ (see Sexual behaviour below). Pairs perform the 'cackling' display, a duet associated with courtship on the colony or offshore at the staging area. Solitary birds incubating or brooding in crevices sometimes make a soft whimpering call. (Based on observations and recordings by ILJ, Alaska; Jones 1993*c*).

Habitat, food, and feeding behaviour
MARINE HABITAT

Summer and winter ranges have mean ocean surface temperatures of 2–10 and 2–5 °C. Feeds near or far from shore, often more than 100 km from breeding colonies, where water column is stratified and dense layers of zooplankton are present, especially where strong currents, tide rips, and upwellings transport and concentrate prey (Hunt *et al.* 1993). Tends to occur upstream of surface features (R. W. Russell & G. L. Hunt, personal communication), diving deep to enter zone of maximum vertical water motion where they use their powerful underwater flying abilities to intercept zooplankton. Comes ashore only during summer breeding season.

FEEDING BEHAVIOUR

Often congregates in huge dense feeding flocks near violent tide rips.

DIET

Little published data. Winter foods unknown. In summer feeds on marine invertebrates, particularly *Thysanoessa* euphausiids, also mysids, hyperiids, gammarids, calanoid copepods, larval fish, and squid (Bédard 1969*c*; Piatt *et al.* 1990; A. Kitaysky, unpublished data; see recent review by Hunt *et al.* 1993). Bédard (1969*c*) found that adult diet at St Lawrence Island was composed of a wide variety of crustacea before hatching period (including euphausiids, mysids, hyperiids, and gammarids, based on gut contents), but consisted mainly of *Thysanoessa* spp. during chick-rearing (based on undigested food loads brought to chicks).

ADULTS: Gut contents at St Lawrence I., in order of abundance, before hatching (% by number): Gammaridea 48.0, Mysidacea 24.5, *Parathemisto libellula* 19.5, *Thysanoessa* spp. 5.7, *Calanus finmarchicus* 1.3, *Calanus cristatus* 1.1, Caridea 0.6, Decapoda 0.4, Cumacea 1.0 and *Limacina* 0.1 ($n = 107$ adults collected); after hatching: *Thysanoessa* spp. 56.0, *Calanus finmarchicus* 28.7, *Calanus cristatus* 7.0, Caridea 4.4, and Mysidacea 0.7 ($n = 135$ chick meals; Bédard 1969c). At St Matthew I., Harrison (1990) found the euphausiid *Thysanoessa raschii* (83% of samples) and walleye pollock larvae (17%) to be significant diet components. In the northern Sea of Okhotsk in summer, *Thysanoessa* euphausiids composed 69.8% (by mass) and 88.2% (by number) of adult diet (A. Kitaysky, unpublished data).

CHICKS: Piatt *et al.* (1990c) analysed food loads ($n = 54$) brought to chicks at St Lawrence I.; in order of abundance (% composition by number/ mass): *Thysanoessa* spp. adults 80.3/97.8, *Neocalanus plumchrus* 10.4/1.2, *Neocalanus cristatus* 3.2/0.6, *Calanus marshallae* 3.3/0.2, Calanoid copepod 0.2/0.1, *Parathemisto libellula* 0.1/0.1, *Parathemisto pacifica* 0.3/0.1, *Parathemisto* spp. 0.3/0.1, *Thysanoessa* spp. furcilia 0.1, *Pandulus* spp. zoea 0.1.

Displays and breeding behaviour

COLONY ATTENDANCE AND FLIGHT DISPLAYS: spectacular flight displays at breeding colonies involve huge flocks circling between sea and nesting slope, often rising more than 500 m above sea/colony. Extremely gregarious, Crested Auklets stick together closely in dense flocks at all times. 'It takes nine Crested Auklets to make a decision.' (Douglas J. Forsell, personal communication). Mass manoeuvring at nesting colonies resembles group behaviour of wader (shorebird) flocks. Individuals seem nervous about losing close proximity to their flock mates, to the point where birds at the front of a flock slow down, while birds in the rear speed up, resulting in chaotic situations. Typical massed flight behaviour involves an elongate flock formation slowly congealing into a dense spherical shape, which may abruptly break into separate flocks or rapidly change direction. Sometimes a flock forms a narrow undulating snake-like shape a kilometre or more in length and about 10 m diameter stretching from ocean gathering area to the colony site. Flocks diving off colony or rapidly changing direction produce a loud roaring sound. Colony attendance behaviour differs strikingly between colonies: where predatory gulls are present, surface activity is restricted and most socializing occurs underground; where gulls absent (usually related to presence of Arctic Foxes) activity takes place on surface and lasts longer. Number of birds attending colony and remaining on surface varies greatly from day to day, within the season, and among years. Daily pattern of attendance typically includes two activity periods when numbers are present on colony surface. At other times colony appears almost deserted, with visible activity limited to a few birds departing from nesting crevices, or (during chick-rearing) delivering food loads to chicks.

AGONISTIC BEHAVIOUR: violence frequent at colony, apparently relating to competition for access to potential nest-sites or mates. Adult ♂♂ intolerant of close approach by other ♂♂ and subadults, normally reacting with aggression (see discussion of 'scrum' behaviour, below). Fights between ♂♂ involve lunging and pecking with bill (often aimed at eyes), seizing gape, nape, or crest of opponent with bill, or grappling with bills locked and beating with wings. During breeding season ♂♂ have a more strongly hooked bill than ♀♀ (Jones 1993b), which is used for fighting. Some males become quite battered looking, with crests ripped out or twisted askew. Crest size is correlated with dominance within both sexes, with winner of aggressive interaction usually having larger crest (ILJ). Inter-male aggression is also frequent at ocean gathering ground, particularly when pair ♂♂ defend their mate from approaches by other ♂♂ and extra-pair copulation attempts. Subadults invariably lose agonistic interactions with adults of either sex. ♀♀ are usually subordinate to adult ♂♂ and engage in agonistic behaviour less frequently, but lunge or peck at one another in competition for access to displaying ♂♂.

SEXUAL BEHAVIOUR: pairing can follow: (1) adults in breeding condition engaging in intense courtship activity on colony surface near nest-site during pre-laying period, resulting in breeding in same year, (2) non-breeding adults attending colony and seeking mates through entire breeding season, normally engaging in courtship behaviour with several individuals, some eventually forming lasting pair-bonds for breeding in a following year, or (3) off-duty breeding birds of both sexes engaging in courtship encounters with extra-pair individuals during incubation and chick-rearing periods, for possible pairing in following year. Courtship normally follows approach by ♀ to displaying ♂ or (less commonly) follows ♂ approach to ♀. Both paired and unattached ♂♂ advertise vocally ('Trumpeting' display) while in crouched posture with nape feathers erected. Pupils of ♂♂ contract during intense displays, creating peculiar glassy-eyed

2   Courtship behaviour of a pair of Crested Auklets (ILJ photo).

expression. ♀♀ initially extend their neck as they stare intently at ♂ then may show increased interest and attentiveness by crouching with breast low and head at upward angle. If ♂ continues to accept ♀'s advances, courtship proceeds with mutual 'Cackling' vocalizations and intense billing, nibbling of ♂'s bill by ♀ and mutual burying of bills in nape and neck feathers ('Ruff-sniff' display), apparently to smell their distinctive plumage odour. ♂♂ also display to ♀♀ with whom they are courting, but at a lower frequency than in ♀♀ (Jones and Hunter 1993). These courtship displays are rarely performed to same sex individuals. Similar displays are given by courting pairs at staging area on sea where large numbers gather daily in rafts after meeting on the colony surface. Here, homologues to terrestrial courtship displays are followed by copulation, which apparently takes place only at sea near colony. Pairs averaged just under six copulations per h at the staging area, extra-pair copulations have been observed (ILJ and F. M. Hunter, unpublished data). Mate choice is mutual and field experiments have shown that both sexes have an active preference for partners with large crests (Jones and Hunter 1993). Both sexes are attracted by lone displaying ♂♂ and actively courting couples, resulting in 'scrums', when as many as 10 or more auklets rush to join and interact with the ♂ or couple. Scrum participants include subadults of unknown sex, ♀♀ and smaller numbers of adult ♂♂. ♀♀ jostle one another in attempts to make physical contact with central displaying ♂, as do ♂♂ to reach central ♀, or attempt to place their necks along neck of opposite sex central birds, or briefly touch other same sex scrum participants. Meanwhile the central adult ♂ gives Trumpeting display and has his chin nibbled by ♀ partner. This ♂ appears to watch constantly for other adult ♂♂ among scrum participants, and to drive them off with bites and stabs with bill. ♀♀ other than central ♂'s partner are also frequently driven off by central ♂ or ♀. If another adult ♂ attempts to display nearby, central ♂ will often break out of scrum to drive it off (see Jones 1993*c*, for further details).

Breeding and life cycle

Not well known, but studies by Bédard (1969*d*), Knudtson and Byrd (1982), Piatt *et al.* (1990*c*), Sealy (1968), and Zubakin (1990) have outlined basic biology (see also reviews by Konyukhov 1990*c*; Jones 1993*c*). Birds remain at sea from Aug to early May (Sept–early June at northernmost colonies), beginning to come ashore at colonies in early May–early June. Highly colonial, nesting in talus slopes at high density, with most colonies having tens to hundreds of thousands of breeding pairs.

BREEDING HABITAT AND NEST DISPERSION: breeds on remote islands and headlands where sea-facing talus slopes, boulder fields, lava flows, cliffs, pinnacles, and offshore stacks provide abundant rock crevices. In colonies with deep talus, dozens, or even hundreds of pairs may use single large, cave-like openings on talus slopes with nests located 10 m or more beneath surface. Most colonies are shared with larger numbers of Least Auklets (coexists with smaller species at 32/42 or 76% of Alaskan colonies). Occasionally nests on slopes where talus is partly covered with vegetation (mosses, grasses, umbels), or in rock crevices on vegetated cliffs, in Aleutians (Knudtson and Byrd 1982), but densest nesting concentrations normally occur on un-vegetated talus. Nests from sea level up to more than 500 m elevation. Compared with Least Auklet, prefers areas with larger boulders and crevices, usually occupied central area of colony site, and nests deeper beneath colony surface where light levels are lower. At St Lawrence I., nests in areas with rock diameters of 0.2–1.1 m, greatest densities occurring among 0.75 m rocks (Bédard 1969*d*). Nests at very high densities in suitable habitat. One high density 100 m² plot at Buldir I., Alaska has 1500–2000 breeding pairs in talus of about 10 m depth, or about 1.5–2.0 pairs per cubic metre of occupied talus volume (ILJ

and F. M. Hunter unpublished data); overall average nesting densities may be lower.

NEST: in rock crevices; incubating bird is seldom visible from surface, some birds incubate in complete darkness in deep talus. Nest-sites are usually on soft damp soil on floor of rock crevice, or on collections of small pebbles and detritus in a rock crack, perhaps occasionally on bare rock, sometimes in a shallow depression in dirt on the floor of crevice, apparently excavated by bird where possible (Knudtson and Byrd 1982). Crevice volume: 1568 ± 342 cm³ ($n = 55$) at Buldir (Knudtson and Byrd 1982).

EGG-LAYING: peak of laying usually late May (Aleutians *c.* 52° N), early June (Pribilofs *c.* 57° N), mid–late June (St Matthew I. *c.* 60° N) or early July (St Lawrence I. *c.* 63° N). At northern colonies, timing of breeding is affected by snow melt on nesting habitat.

EGGS: clutch size is one egg, relaying of lost first eggs at same site rarely or never occurs (Piatt *et al.* 1990). Occasionally more than one female lays in a crevice, usually leading to complete abandonment. Egg shape; oval to short oval, smooth textured, white becoming stained brownish from dirt in nesting crevice. Egg size and mass: from various collections (Bent 1919), 54.2 (50.0–60.0) × 37.9 mm (32.5– 42.5, $n = 30$); from Buldir Is., Alaska (Jones 1993*a*), 53.7 ± 2.3 (48.5–57.5) × 37.1 ± 1.1 mm (35.3–39.5 $n = 19$), mass of fresh egg 36.3 ± 3.2 g (30–40, $n = 8$), averaging about 14 per cent of mass of adult ♀.

INCUBATION: period is about 34 days: at St Lawrence I., 33.8 ± 0.6 (29–40, $n = 20$; Piatt *et al.* 1990*c*); 35.6 ± 1.4 ($n = 6$; Sealy 1984), shared by ♂ and ♀, but relative time incubating has not been measured. Duration of incubation shifts not measured, may be 24 hrs.

HATCHING AND CHICK-REARING: duration of pipping unrecorded. Chicks semi-precocial, brooded almost continuously for first day after hatching, but diurnal brooding (mostly by ♂ parent) decreases steadily to about 25% of the time by day 6, and to zero at about 20 days following hatching (Piatt *et al.* 1990*c*; ILJ and F. M. Hunter, unpublished data). One or both adults usually spend night with chick in nesting crevice during first 10 days after hatching. Adults carry food in sublingual pouch and make deliveries 2–4 times per day; food load mass 20.5 ± 7.5 g (4.7–36.3, *n* = 59; Jones 1993*c*); 21.0 ± 6.2 g (14.5– 27.0, *n* = 40; Bédard 1969*d*). ♀ appears to contribute more to chick feeding than ♂, particularly during first week. After arriving with food for small chicks, parent remains in crevice and feeds it at intervals. Adults delivering food to large chicks normally remain in crevice for only 5–10 min before departing, excepting deliveries in late evening.

CHICK GROWTH AND DEVELOPMENT: chicks remain in crevice for about 33 days before fledging: at St Lawrence I. (Piatt *et al.* 1990*c*), mean = 33.2 ± 0.1 days (27–36, *n* = 6). Maximum growth rate 11.1–12.8 g/day (Searing 1977; Piatt *et al.* 1990*c*). Chick mass asymptote 269 ± 6.1 g, (258–279, *n* = 3) at St Lawrence I. Before departure chicks begin leaving nest-sites regularly to exercise flight muscles by flapping their wings for 15–30 sec at a time, and occasionally come close to the surface to peer out of talus. Average chick mass at fledging probably about 250 g (94% of adult mass), but this is difficult to estimate because fledglings caught and measured on the surface are mostly weak, underweight ones soon to perish. Colony departures take place at all times of day, chicks deliberately emerge from crevice, climb to top of nearby boulder, and take flight. Many fly directly out to sea, others crash-land on the talus and injure themselves or are caught by gulls, falcons, and foxes. Chicks fledge unaccompanied by adults and there is no post-fledging parental care.

BREEDING SUCCESS: natural reproductive success difficult to measure because of sensitivity to investigator disturbance of nest-sites, probably about 0.50–0.55 chicks fledged per breeding pair per year. Hatching success averaged 74.2% (73.3–76.3), fledging success 73.9% (66.7–76.2), and breeding success 54.8% (51.0–56.7) at Buldir I. Alaska (5 years of monitoring, Knudson and Byrd 1982; G. V. Byrd unpublished data). At St Lawrence I. in 1987, hatching success at regularly checked crevices was 75.7%, fledging success 64.1%, and overall breeding success 48.5% (Piatt *et al.* 1990*c*).

PREDATION: mammalian predators include Arctic and Red Foxes which take adults at colonies (Bailey 1993), brown bears (*Ursus arctos*) which excavate talus to predate adults, nestlings, and eggs (N. Konyukhov, personal communication), and voles (*Microtus* and *Clethrionomys*) which prey on eggs and nestlings in crevices (Sealy 1982). Rats kill adults in crevices and take eggs and chicks (ILJ). Avian predators include Slaty-backed, Glaucous, and Glaucous-winged Gulls, Snowy Owls, Bald and Steller's Sea Eagles, Gyrfalcons, and Peregrine Falcons (ILJ; Konyukhov 1990*a*).

MOULT: from Bédard and Sealy (1984). Adults undergo moult of underparts, head, and neck feathering (but not flight feathers, greater coverts, or thighs) during late winter and spring when crest and facial plumes develop (Definitive Pre-alternate moult, Mar–May). Gradual primary moult begins late in breeding season; primaries are replaced sequentially starting with the innermost and working outward, followed by complete moult of body feathers (Definitive Pre-basic moult, July–Oct). Crest feathers begin to drop out and facial plumes look worn by end of breeding season, with most adults experiencing partial loss of white plumes and loss of bill plates during chick-rearing. Subadults (2-year-olds) attend colonies with very worn brownish or tan coloured upperwing coverts and primaries which sequentially moult starting late in breeding season, and they undergo a complete moult of body feathering (Definitive Pre-basic moult, July–Oct).

Population dynamics

SURVIVAL: adult annual survival rates at a colony at Buldir I., Alaska in 1991–5 were (with 95% confidence limits): 1991, 93% (87–96), 1992, 80% (83–92), 1993, 91% (85–94), 1994, 84% (77–88), 1995, 89 (80–94), giving an average survival rate of about 89% and an estimated lifespan of about 9.6 years (ILJ).

AGE AT FIRST BREEDING: uncertain; 1-year-olds rarely observed at colonies, 2-year-old subadults attend colonies for about 3–4 weeks, but do not breed or form pair-bonds. Possibly breeds at 3 years of age.

PAIR- AND NEST-SITE FIDELITY: limited information on mate-retention, but of 44 pairs, 20 (46%) re-mated in following year, 11 divorced, and 13 split after disappearance of one partner (at Talan., I. Sea of Okhotsk; Zubakin and Zubakina 1994). This suggests that Crested Auklets have a similar high divorce rate to Least Auklets, with divorces taking place in about one third of pairs with both members surviving to a following year. Breeders normally use same crevice from year to year: 75% of males and 62% of females returned to use same crevice, and in cases of divorce, burrow was usually retained by ♂ (at Talan I., Russia; Zubakin and Zubakina 1994). Occasionally evicted from crevices by Horned Puffins.

Least Auklet   *Aethia pusilla*

Uria pusilla (Pallas, 1811, Zoogr. Russo-Asiatica 2: 343—Kamchatka).   PLATE 6

Description

A tiny alcid barely larger than a sparrow; the smallest auk species. On land, posture is erect, usually stand on their toes with no other part of the body touching the ground. Very agile when climbing about their rocky breeding habitat. Rather short and rounded wings, and although these birds are capable of rapidly taking off from the sea, their flight is rather weak and they characteristically weave from side to side. They are incapable of forward progress against strong winds, and during gales are often seen flying backwards at high speed!

ADULT SUMMER: typically dark grey or blackish upperparts with white patches on scapulars and tertials, and whitish underparts variably marked with irregular dark spotting and barring; however, unlike all other alcid species, adult population varies continuously in summer underpart coloration from almost black to unmarked white, with a similarly variable amount of white in the scapulars, but almost always with an unmarked white throat (Fig. 1). Adults with pure white or black underparts are rare and amount to 5% of most populations. Irides white, giving this and other auklet species an odd doll-eyed look. Like other *Aethia* species, tarsi and toes are bluish grey with black webbing. Breeding adults have white facial

1   Two adult Least Auklets at St Paul Island, Alaska, showing variation in underpart coloration (ILJ photo).

plumes both on foreheads and behind eye, a small horny knob-shaped bill ornament at base of culmen, and reddish bill (Jones and Montgomerie 1992). Forehead plumes are short (2–4 mm long) and scattered evenly over front of head, extending onto crown and lores. White auricular plumes (1 mm wide by 5–15 mm long) extend from behind and below eye onto sides of nape; these are similar in shape and structure to those of Parakeet, Crested, and Whiskered Auklets. Bill knob black and varies in size from a small bump on the culmen to a conspicuous forward- curving appendage, otherwise bill rather small, symmetrical, and lacks other ornamental plates. Knob is shed late in breeding season. Bill colour also quite variable in adults from nearly black to bright red with black base, usually with light straw-coloured tip.

ADULT WINTER: similar to summer adults, but all birds have uniformly white underparts, the ornaments disappear (bill black, no bill knob, only a trace of white facial plumes), and the white scapular patches are more prominent.

SUBADULT: (2-year-olds) in summer have brown foreheads with sparse or no plumes present, dull coloured bills, and smaller bill knobs than breeding adults (Jones and Montgomerie 1992). Also vary in underpart coloration but average darker than adults and throats often entirely dark grey, or white heavily spotted with dark grey flight and tail feathers blackish grey (fresh) or brownish (worn).

JUVENILE AND FIRST WINTER: closely resemble winter-plumage adults, with plain black-and-white plumage, but irides pale greyish rather than white.

CHICK: blackish grey down that is retained until about a week before colony departure.

VARIATION
Monotypic. There is some evidence for differences in coloration between colonies: birds in the Pribilofs have lighter underparts than birds at Buldir I. in the Aleutians (Jones 1993a). The geographical pattern and significance of plumage variation in Least Auklets require further study.

MEASUREMENTS
Minimal sexual dimorphism, although ♂♂ are slightly deeper-billed than ♀♀—see Table, (1) St Lawrence I., Alaska (Bédard and Sealy 1984); (2) St Paul I., Alaska (Jones 1994); (3) St Lawrence I., Alaska (Piatt *et al.* 1990c); (4) Buldir I., Alaska (Jones 1993a; Jones and Montgomerie 1992).

WEIGHTS
Weights vary with time of day, time of year, breeding status, and age (Jones 1994)—see Table, (1) Buldir I., Alaska (Jones 1993a); (2) St Paul I., Alaska (Jones 1994); (3) St Lawrence I., Alaska (Piatt *et al.* 1990c); (4) St George I. (Roby and Brink 1986a).

Range and status
RANGE
BREEDING: in North America, on Aleutian Is. from Koniuji I. W to Buldir I., on the Semidi and Shumagin Is. S of the Alaska Peninsula, and in the Bering Sea on the Pribilof Is., St Matthew I., St Lawrence I., King I., and the Diomedes (Sowls *et al.* 1978). In Asia, breeds locally in the Kuril Is. (Ushishir), on Iona, Talan, and Yama Is. in the Sea of Okhotsk, on the Commander Is., and northeastwards along the coast of the Kamchatka Peninsula to the southern coast of the Chukotka Peninsula (Ratmanov [Big Diomede] I., Cape Ulyamkhpen, Cape Lysaya Golova; Konyukhov 1990b). The largest breeding colonies are located at Kiska I. in the central Aleutians, St Lawrence I., and at Peter the Great I. (each with more than 1 million breeding individuals). Wanders occasionally as far N as Point Barrow, Alaska and Wrangel I., Russia in late summer.

WINTER: Remain near breeding areas year round where the sea remains free of ice. Some southward movement in winter: in North

Table: Least Auklet measurements (mm)

ADULT ♂♂	ref	mean	s.d.	range	n
wing length	(1)	89.9	2.4	—	20
culmen	(1)	9.3	0.4	—	20
bill depth, including knob	(1)	8.5	0.4	—	20

ADULT ♀♀	ref	mean	s.d.	range	n
wing length	(1)	90.8	1.3	—	20
culmen	(1)	9.3	0.4	—	20
bill depth, including knob	(1)	8.1	0.6	—	20

UNSEXED BIRDS (ADULTS)	ref	mean	s.d.	range	n
wing length	(2)	97.1	2.5	91–107	233
	(3)	97.6	2.1	95–102	18
tarsus	(4)	19.7	0.8	17.7–22.6	135
	(2)	19.4	0.7	16.5–21.1	233
culmen	(4)	9.2	0.5	8.2–10.2	63
	(2)	9.3	0.5	7.8–10.6	229
bill depth	(2)	6.7	0.4	5.9–7.6	227
bill knob height	(2)	1.7	0.4	0.9–3.2	217
auricular plume length	(4)	14.5	3.3	4.7–23.4	129
	(2)	16.7	3.0	6.8–27.0	235

UNSEXED BIRDS (SUBADULTS)	ref	mean	s.d.	range	n
wing length	(2)	95.4	2.2	90–101	62
culmen	(2)	9.2	0.5	7.9–10.1	63
bill depth	(2)	6.2	0.3	5.7–6.9	63
bill knob height	(2)	1.5	0.3	0.7–2.2	54
auricular plume length	(2)	13.8	2.6	6.8–20.0	62

Table: Least Auklet weights (g)

UNSEXED BIRDS (ADULTS)	ref	mean	s.d.	range	n
incubation	(1)	81.4	5.1	70–98	143
	(2)	85.7	5.1	70–101	252
pre-laying	(3)	86.6	6.4	77–101	18
fledging	(4)	84.5	5.2	—	44

UNSEXED BIRDS (SUBADULTS)	ref	mean	s.d.	range	n
	(1)	78.4	4.7	70–91	72
	(2)	79.4	5.9	64–95	797

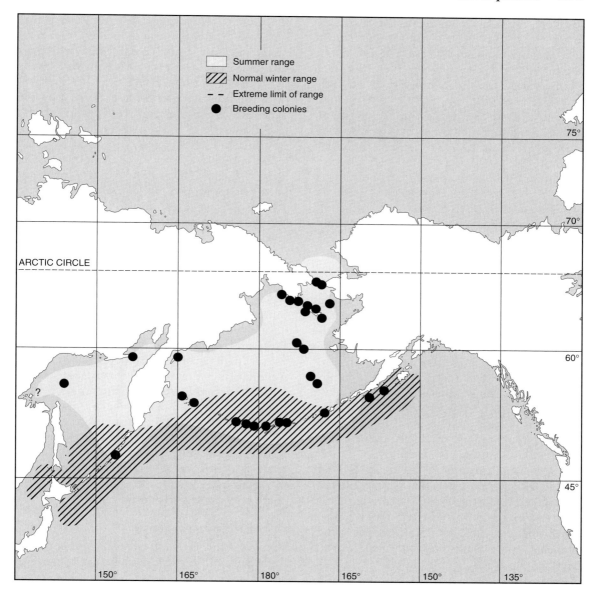

Summer range
Normal winter range
Extreme limit of range
Breeding colonies

America, birds disperse eastwards into the Gulf of Alaska, and in Asia, commonly winter S to northern Japan (Brazil 1991). One record of a bird found inland on a frozen lake in NW Mackenzie District, Northwest Territories, Canada (Godfrey 1986).

STATUS

The Russian population is probably several million (Konyukhov 1990b), that in Alaska is officially estimated at 9 million (U. S. Fish and Wildlife Service 1993), but these estimates are probably very low, actual population could be 20 million or more. There is some evidence of a population increase at St Lawrence I. (Piatt *et al.* 1990b), but population trends difficult to assess because of extreme day to day and inter-year variation in colony attendance (Piatt *et al.* 1990b; Jones 1992a). A large colony at St George I.,

Pribilofs, seems to have decreased drastically in size in the last 100 years owing to plant succession (Roby and Brink 1986*b*). Some Aleutian I. colonies were reduced or even eliminated earlier this century by introduced Arctic Foxes (Bailey 1993), but auklets coexist with native foxes and other mammalian predators including bears and weasels in other areas. At Kiska, as many as a million or more Least Auklets colonized a recent (1969) lava flow in the presence of both introduced foxes and rats. Nevertheless, removal of introduced foxes from the Aleutian Is. will be crucial for restoration of former numbers.

Field characters

The smallest seabird in most parts of their range. Their tiny size and dumpy body shape easily eliminates all other alcids except Crested, Whiskered, Parakeet, and Cassin's Auklets. Easily identified at colonies by the distinctive ornaments. At sea identification can be tricky because they can so easily disappear as they dart and dive among foaming waves. Superficially most similar to Parakeet Auklets, which are larger and have a slower wing-beat. Least Auklets also almost always show white scapulars, which are lacking in all other auklets. Can be identified at sea by voice, as small groups at sea often give the distinctive chattering call that is very different from other auklet species. They are closest in size to Whiskered Auklets, which are slightly larger, blacker, have somewhat longer wings, and show pale feathering only on the vent region.

Voice

Based on observations of I. L. Jones at Pribilofs and Buldir I., Alaska—Jones 1993*a*. A highly vocal species. Has a variety of vocalizations that sound like high-pitched, high-frequency chattering. The most characteristic sound is 'display chatter,' which is performed mostly by ♂♂. This is a sexual advertising display that is performed on the colony surface near the nesting crevice, it attracts ♀♀ and curious subadults (probably of both

sexes) and incites aggression from competing ♂♂ (see Sexual behaviour, below). Courting pairs perform 'duet chattering' which is similar to the display chatter display but is performed as a duet. The 'chirr' call is a short rasping note given by birds disturbed by the approach of a human or predator and seems to function as a low intensity alarm signal. Another call, the 'squeak', is given by birds taking flight and serves as a high intensity alarm signal. The Least Auklet vocal repertoire apparently lacks homologies with other auklet species, perhaps indicating this species has diverged considerably from other *Aethia* species, and from *Ptychoramphus* and *Cyclorhynchus*.

Habitat food and feeding behaviour

MARINE HABITAT

Summer and winter ranges are delimited by ocean temperatures of 2–11 and 1–5 °C. Preferred foraging habitat occurs in stratified waters with strong thermoclines where upwelling brings thermocline to surface, such as at oceanographic features including fronts, tide rips over shallow sills, and tidal pumps (Harrison *et al.* 1990; Hunt and Harrison 1990; Hunt *et al.* 1990; Piatt *et al.* 1992). Tend to congregate in areas of surface turbulence downstream of features, avoiding areas of maximum vertical water motion. Roost on shore only during breeding season.

FEEDING BEHAVIOUR

Forages while in wing-propelled underwater flight, but may be a weak swimmer because of small wing size. Haney (1991) suggested diving ability poor owing to low body mass and high buoyancy, and estimated maximum dive depth to be about 15 m, but Hunt *et al.* (1993*b*) suggested maximum diving depths of 25 m or more. No measurements of diving depths have been obtained. Food items probably taken individually. It is not known whether prey are capable of taking evasive action, possibly Least Auklets approach and at their leisure graze defenceless plankton swarms.

DIET

Calanoid copepods, particularly *N. plumchrus*, are main adult food during breeding season, and main food delivered to chicks. For a review of published and unpublished data see Hunt *et al.* (1993*b*). Some data on prey utilized come from analysis of the contents of undigested food loads delivered to chicks (chick meals), which may differ in composition from adult diet (Bédard 1969*c*). In the Pribilofs, Hunt *et al.* (1981) found calanoid copepods to comprise 93.4% of prey (% of number), *Calanus marshallae* (18.6%) and *Neocalanus cristatus* (4.7%) predominating (*n* = 258 adults collected), Bradstreet (1985) found *Neocalanus plumchrus* (38.5%), *C. marshallae* (27.3%), and other calanoid copepods (18.3%) predominating, while Roby and Brink (1986*a*) noted that *N. plumchrus* and *N. cristatus* were most abundant in samples (*n* = 66 chick meals). At St Lawrence I., Bédard (1969*c*) found *N. plumchrus* (39.2%), Caridea (33.3%), Gammaridea (9.7%), and the pelagic hyperiid amphipod *Parathemisto libellula* (6.6%) most common in adult gut contents prior to hatching (*n* = 145), and *N. plumchrus* (88.9%) predominated during chick-rearing (*n* = 124 chick meals). Nearly 20 years later Piatt *et al.* (1990*c*) found *N. plumchrus* (82.9%) and *C. marshallae* (10.5%) predominating at St Lawrence I. during chick-rearing (*n* = 74 chick meals). Also working near St Lawrence, Hunt *et al.* (1990) found, in order of decreasing abundance: *N. plumchrus*, *N. cristatus*, shrimp (Pandalidae) larvae, hyperiids, and euphausiids (*n* = 87 stomachs and chick meals). Near King I., Hunt and Harrison (1990) found *Neocalanus* copepods composed about 85% of summer adult diet, in order of decreasing abundance: *N. plumchrus*, *N. cristatus*, *Eucalanus bungii*, Pandalidae larvae, Brachyura larvae, and *Thysanoessa* (n = 21 stomachs and sublingual pouches). Preliminary work suggests Least and Whiskered Auklets have very similar diet where they both occur in the western Aleutians (Day and Byrd 1989); this requires further study. Winter foods have not been studied.

Displays and breeding behaviour

COLONY ATTENDANCE AND FLIGHT DISPLAYS: immense numbers normally circle between sea and nesting slope in disorganized flocks, so large breeding colonies resemble huge Starling (*Sturnus vulgaris*) roosts. Least Auklet flocks are less ordered and less tightly packed than Crested Auklet flocks, and their flight speed is lower, so during gales may have difficulty reaching higher nesting slopes against strong winds. The simultaneous chattering of many thousands of birds can be nearly deafening at dense colonies. Daily pattern of attendance typically includes two activity periods when numbers are present on colony surface. In S (Aleutians and Pribilofs) the daytime activity peaks occur in the morning to early afternoon (09:00–14:00 hrs) and the evening period at dusk (23:00–24:00 hrs), with daytime period gradually shifting to later in day as breeding season progresses (moving to 11:00–16:00 during chick-rearing; Byrd *et al.* 1983; Piatt *et al.* 1990*b*; Jones 1992*a*). In far N (e.g. Little Diomede I.), where daylight is continuous or nearly continuous during breeding season, morning and evening activity periods are less well defined. At other times colony appears almost deserted, with visible activity limited to a few birds departing from nesting crevices, or (during chick-rearing) delivering food loads to chicks. Number of adults socializing on colony surface varies greatly from day to day, often changes unpredictably within the breeding season, and among different years (Piatt *et al.* 1990*b*; Jones 1992*a*). Least Auklets congregate in dense flocks year round, but are less gregarious than Crested Auklets. Least Auklets returning to colony usually land before Crested Auklets, the presence of the former species attracting the latter to alight on the colony surface.

AGONISTIC BEHAVIOUR: agonistic interactions frequent at colony, apparently relating to competition for access to potential nest-sites or mates. Adult ♂♂ intolerant of close approach by other ♂♂ and subadults, normally reacting with aggression. Fights between ♂♂ involve lunging and pecking (often aimed at eyes), seizing gape, nape, or facial feathers, and grappling with bills locked and wings beating. During breeding season, breast plumage colour is variable (see Description) and correlated with dominance; winner of aggressive interaction usually has lighter plumage (Jones 1990). Experimental manipulations of plumage colour have confirmed it functions as a status signal (Jones 1990). Inter-male aggression is also frequent at gathering ground, particularly when pair ♂♂ defend their mate from extra-pair copulation attempts. Subadults invariably lose agonistic interactions with adults of either sex. Highly gregarious, particularly at staging areas on sea near breeding colonies. Little evidence of territoriality outside defence of a nesting crevice; intense fighting for access to disputed crevices by rival pairs often occurs. Individual ♂♂ usually perform vocal advertising displays from an area encompassing a few square metres of colony surface, but because of high nesting densities and intermittent visits to this area, it is not possible for individuals to defend surface display areas exclusively from other ♂♂. Subordinate to other *Aethia* species, particularly Crested Auklet, by which displaced from both nesting crevices and display areas on surface of colonies (Knudtson and Byrd 1982). Dead Least Auklets found in crevices may be victims of Crested Auklet aggression.

SEXUAL BEHAVIOUR: socially monogamous, but extra-pair copulations occur more frequently than among other *Aethia* species, two extra-pair fertilizations recorded among 20 families DNA fingerprinted (F. M. Hunter and ILJ). Courtship takes place on the colony surface and at ocean gathering ground nearby. Pairing follows three patterns: (1) adults in breeding condition exhibit brief intense courtship activity on ocean gathering ground and on colony surface near nest-site during pre-laying period, resulting in re-mating of pairs from preceding years and formation of new pairs for breeding attempts in same year; (2) adult birds not attempting breeding attend colony and engage in courtship behaviour with several individuals through entire breeding season, occasionally forming lasting pair-bonds for breeding in a following year; and (3) off-duty breeders of both sexes seek out courtship encounters with extra-pair individuals during incubation and chick-rearing periods (Jones 1989; Jones and Montgomerie 1991, 1992; observations of colour-marked individuals by ILJ and F. M. Hunter at Buldir Is., Alaska). As in Crested Auklet, courtship normally follows approach by ♀ to a displaying ♂ (see vocalizations) on colony surface, or by ♂ approach to a ♀. Both sexes perform stereotyped postural displays that increase in intensity as courtship proceeds. ♀♀ initially extend their neck as they stare intently at a displaying ♂, then may show increased interest by crouching with breast low and head at upward angle with nape feathers erected. If ♂ shows interest with directed 'Chattering' displays, she may enter an exaggerated neck-extended posture with head tilted to one side and cautiously touch ♂'s neck, face, or bill. Both paired and unmated ♂♂ advertise vocally while in crouched posture. If couple continue to accept one another's advances, courtship proceeds with mutual 'Duet-chatter' vocalizations and intense billing (Fig 2). Courtship displays are often initially performed to same sex individuals, because the sexes are indistinguishable in

2   Mutual Duet-chatter display (ILJ).

the absence of behavioural cues (Jones and Montgomerie 1992). Individuals that have exchanged courtship displays may then meet regularly on colony surface and eventually form a lasting pair-bond. After intense courtship displays, ♂♂ perform rodent run display, in which they adopt an extreme hunched and flattened posture and with continuous 'Chattering' vocalization, rapidly run in a circle around their mate then turn towards a nesting crevice. ♀♀ often follow their partner into the crevice after this display. Similar displays are given by courting pairs at gathering ground on sea, where homologies to terrestrial displays are followed by copulation, which has been observed only at sea near colony. Pairs averaged about five copulations per hr at the staging area. ♀♀ invariably resist ♂ attempts to mount on land, but on the sea engage in unforced extra-pair copulations with roving ♂♂ (F. M. Hunter and ILJ). Mate choice appears to be mutual and field experiments have shown that both sexes have an active preference for partners with elaborate facial ornaments (Jones and Montgomerie 1992). Although highly variable in expression and favoured by inter-sexual selection, the ornaments reveal little about individual mate quality (Jones and Montgomerie 1992).

Breeding and life cycle

Other than Cassin's Auklet, perhaps the best-studied auklet species. Most colonies are shared with larger numbers of Crested Auklets (co-exists with smaller species at 33/38 or 87% of Alaskan colonies).

BREEDING HABITAT AND NEST DISPERSION: colonies are situated in habitat where suitable crevices are present, including beach boulders, talus and scree slopes and slides, porous lava fields, and in cliffs with cracks and crevices. Most colonies are on seaward facing slopes adjacent to the sea, but where the most suitable talus habitat is situated inland, colonies may occur on mountainsides and cirques as much as 1.5 km from the sea. Most colonies are in areas with little or no vegetation, but in

Aleutians some breeding areas are on older parts of lava flows that are heavily vegetated. The four biggest Aleutian colonies, at Gareloi, Semisopochnoi, Segula, and Kiska are all situated on recent lava flows on the sides of large volcanoes. Plant succession inevitably leads to nesting areas becoming uninhabitable owing to clogging of crevices with humus and vegetation (Roby and Brink 1986b), so in the long run this species depends on erosion and volcanic activity. A highly gregarious species, nesting in colonies varying in size from a few hundred to over 1 million pairs.

NEST: in rock crevices; the incubating bird occasionally visible from the surface. In colonies with deep talus, nests may be located several metres beneath surface, but Least Auklet nest-sites are normally shallower than Crested Auklets. Egg normally laid directly on bare rock, on collections of small pebbles and detritus in a rock crack, or on unmodified soil substrate; normally laid on flat surfaces and in slight depressions, but occasionally on quite precarious sites (Knudtson and Byrd 1982). Crevice volume 117.6 ± 65.5 cm^3: ($n = 40$) at Buldir (Knudtson and Byrd 1982).

EGG-LAYING: peak of laying usually late May (western Aleutians c. 52° N), early June (Pribilofs c. 57° N), mid-late June (St Matthew I. c. 60 ° N), or early July (St Lawrence I. c. 63° N). At northern colonies, timing of breeding is affected by the time of snow melt on the nesting habitat. In late years, when snow cover is extensive, laying may be delayed by 2 weeks or more and eggs may be laid on snow and abandoned (Sealy 1975c). Bédard (1969d) recorded laying dates from 12 June to 5 July at St Lawrence Is. In mixed colonies, Least Auklet chicks hatch and fledge earlier than Crested Auklet chicks.

EGGS: clutch size is one egg, relaying of lost first eggs in the same crevice rarely or never occurs (Piatt et al. 1990c). Egg shape oval to short oval, smooth textured, white becoming stained brownish from dirt in nesting crevice.

Egg size and mass: from St George I., Pribilofs (Roby and Brink 1986*a*), 39.4 ± 1.4 × 28.4 ± 0.8 mm (*n* = 65); from St Lawrence I. (Sealy 1968), 41.2 ± 1.9 × 29.0 ± 1.2 mm (*n* = 10), mass of fresh egg 17.4 ± 1.2 g (15.0–19.6, *n* = 20) at St George I. (Roby and Brink 1986*a*), 18.7 ± 1.5 g (17–22, *n* = 10) at Buldir (Jones 1993*a*), averaging about 22% of mass of adult ♀.

INCUBATION: breeding adults have two lateral brood patches with maximum dimensions of about 25 by 15 mm which defeather a few days prior to laying, become vascularized and thickened by onset of incubation, and rapidly shrink and become refeathered after hatching. Non-breeding adults and subadults occasionally show some brood patch development (ILJ). Incubation period is about 30 days: at St Lawrence I., 30.1 ± 0.5 days (25–39, *n* = 31; Piatt *et al.* 1990*c*); 31.2 ± 2.7 days (*n* = 15; Sealy 1984). Incubation duties are shared equally between ♂ and ♀, with incubation shifts of 24 hrs (Roby and Brink 1986*a*). Interruptions due to human disturbance, food shortages, or unknown causes can lead to egg neglect and extension of the incubation period. Eggs survive near freezing conditions in absence of incubating parent.

HATCHING AND CHICK-REARING: duration of pipping unrecorded. Chicks semi-precocial, brooded almost continuously for first day after hatching, but diurnal brooding (by both parents alternating) decreases steadily to about 25% of the time by day 6, and to zero at about 20 days following hatching (Piatt *et al.* 1990*c*; ILJ and F. M. Hunter, unpublished data). One or both adults usually spend night with chick in nesting crevice during first 10 days after hatching. Adults carry food in sublingual pouch and make deliveries 2–4 times per day; meal mass was estimated to be 5.4 ± 2.3 g (0.9–12.3, *n* = 55) at St George Is. (Roby and Brink 1986*a*); 8.2 ± 1.5 (6–10, *n* = 23) at St Paul Is. (Jones and Montgomerie 1992); 5.0 ± 1.5 (4.5–7.8, *n* = 11) at St Lawrence I. (Bédard 1969*c*). After arriving with food for small chicks, parent remains in crevice and feeds it at intervals. Adults delivering food to large chicks normally remain in crevice for only 5–10 min before departing, except for deliveries in late evening. Pair members make almost equal numbers of food deliveries to the chick, with a slight bias in favour of ♂♂ (ILJ). Chicks received 2.6 ± 0.7 (1–4, *n* = 64) meals/day from each parent at St George I., Alaska (Roby and Brink 1986*a*). Chicks' daily food intake as a percentage of adult mass was about 42% (St George; Roby and Brink 1986*a*) to 57 per cent (St Paul; ILJ).

CHICK GROWTH AND DEVELOPMENT: chick mass at hatching was: 11.9 ± 0.9 g (*n* = 10; St George I., Roby and Brink 1986*a*); 12.6 (*n* = 22; St Lawrence I.; Piatt *et al.* 1990*b*). Maximum growth rate was about 6 g/day in Pribilofs (Roby and Brink 1986*a*), 4.9 g/day at St Lawrence I. (Piatt *et al.* 1990*c*). Mean chick mass asymptote was 90.8 ± 3.1 g (81–100, *n* = 6) at St Lawrence I. (Piatt *et al.* 1990*c*). At St George, fledging mass was 91.5 ± 5.2 g (*n* = 11) or 108 per cent of mean adult body mass (Roby and Brink 1986*a*); at St Lawrence fledging mass was 82.2 ± 2.8 (72–99 *n* = 12) or 100% of mean adult body mass (Piatt *et al.* 1990*c*). Age at fledging: St George Is. (Roby and Brink 1986*a*), 28.6 ± 1.5 days (26–31, *n* = 20); St Lawrence I. (Piatt *et al.* 1990*c*), 29.3 ± 0.4 (25–33). Body mass increases rapidly during first 20 days, changes little during last 6–8 days of nestling period. Wing length increases continuously to fledging age, and fledgling wing length similar to adults' (Roby and Brink 1986*a*). Several days to 1 week prior to departure, chick occasionally leaves nest-site and emerges briefly to look out. Older chicks regularly leave nest-site and move to nearby unobstructed space (within talus, up to 2 m from nesting crevice) to exercise flight muscles by flapping wings rapidly for 15–30 sec. To depart colony, chicks deliberately emerge from crevice, climb to top of nearby boulder, and take flight. Many chicks fly directly out to sea (at low velocity compared with adults), fledging successfully. Underweight fledglings either try to walk to the sea or fly weakly and crash on talus slope and are killed by predators. During departure, chicks particularly vulnerable to predation by gulls and foxes. Chick mortality also results from crashes during abortive first flights, becoming trapped among rocks, and being crushed and drowned in surf. Colony departures take place at all times of day, perhaps peaking in late evening and predawn (Roby and Brink 1986a; ILJ). Careful observation has revealed no evidence of post-fledging parental care.

BREEDING SUCCESS: natural reproductive success estimates need to be treated with caution because of sensitivity to investigator disturbance of nest-sites. However, productivity is probably 0.5–0.7 chicks fledged per breeding pair per year. At Buldir I. (5 years of monitoring, Knudtson and Byrd 1982; G.V. Byrd and J.C. Williams, unpublished data) hatching success averaged 77.7% (67.9–94.3), fledging success averaged 71.6% (53.2–90.9), and breeding success averaged 56.5% (37.5–85.7). At St George I. (2 years of monitoring; Roby and Brink 1986a) hatching success averaged 88.5% (possibly inflated due to undetected nest failures early in incubation), fledging success 78.8%, and breeding success 69.7%. At St Lawrence I. in 1987, hatching success at regularly checked crevices was 69.7%, fledging success 53.9%, and overall breeding success 37.6%; at less frequently checked crevices (less investigator disturbance) hatching success was 75.3%, fledging success 71.1%, and overall breeding success 49.6% (Piatt *et al.* 1990c).

PREDATION: mammalian predators include Arctic and Red Foxes which take adults at colonies, Brown Bears which excavate talus to take adults, nestlings, and eggs (N. Konyukhov, personal communication), and voles (*Microtus* and *Clethrionomys*) which prey on eggs and nestlings in crevices (Sealy 1982). Ermine, and introduced rats kill adults in crevices and take eggs and chicks (Konyukhov 1990b). Avian predators: Slaty-backed

Glaucous, and Glaucous-winged Gulls, Snowy and Short-eared Owls (*Asio flammeus*), Bald and Steller's Sea Eagles, Gyrfalcons, and Peregrine Falcons (ILJ; Konyukhov 1990a).

MOULT: from Bédard and Sealy (1984). Adults undergo moult of underparts, head, and neck feathering (but not flight feathers, greater coverts, or thighs) during late winter and spring when facial plumes develop and underparts become spotted (Definitive Pre-alternate moult, Mar–May). Gradual primary moult begins during breeding season; primaries are replaced sequentially starting with innermost and working outward, followed by complete moult of body feathers (Definitive Pre-basic moult, July–Oct). Facial plumes look worn by end of breeding season, with most adults experiencing partial loss of white plumes and loss of bill knob ornament. Subadults (2-year-olds) attend colonies with very worn brownish or tan coloured primaries which sequentially moult starting late in breeding season, and they undergo a complete moult of body feathering (Definitive Pre-basic moult, July–Oct). These subadults also show extreme feather wear on face and head in May. By Sept the Definitive Pre-basic moult is complete and these birds have fresh primaries, contour feathering, and rectrices, and they resemble adults.

Population dynamics

SURVIVAL: mean adult survivorship was 79% (95% confidence limits 0.73–0.84, $n = 3$ years), based on colour band resightings (Jones 1992b; ILJ) at St Paul I., Alaska, and 81% varying from 0.71 to 0.95, $n = 5$ years; colour band resightings; ILJ and F. M. Hunter) at Buldir I., Alaska, giving a predicted mean life expectancy of about 5.5 years (note, there was no evidence that these populations were stable at the time of the estimates).

AGE AT FIRST BREEDING: uncertain; 1-year-olds rarely observed at colonies, 2-year-old subadults attend colonies for about 3–4 weeks, but do not breed and rarely form pair-bonds

(Jones 1992*a*; but see Bédard and Sealy 1984). probably 3 years, although the proportion of 3-year-olds breeding is probably lower than older birds (Jones 1992*b*).

PAIR- AND NEST-SITE FIDELITY: relatively low pair-fidelity; among 81 pairs (in two breeding seasons), both members of 55 pairs survived to following year, 35 (64%) remated in following year and 20 (36%) divorced (at St Paul Is.; Jones and Montgomerie 1991). In addition, 21 (26%) of the 81 pairs split owing to disappearance of one partner. Thus, over 50% of breeding individuals must find a new mate each year, owing to death of their mate or divorce. Re-mating pairs usually return to the same crevice used in a previous year (ILJ).

Whiskered Auklet   *Aethia pygmaea*

Alca pygmaea (Gmelin, 1789, Syst. Nat. I: 555— Bird Island, between Asia and America).

PLATE 6

Description

Among alcids, only Least Auklets are smaller, but Whiskered Auklets are certainly the most ornamented alcid species and perhaps the most decorated of all seabirds. Adults and subadults have quite similar plumage at all times of year, except for their distinctive ornaments and bill colour which are lost after the breeding season and regrown in late winter. This species lacks the elaborate plate-like bill appendages possessed by Crested Auklets. Bill short and vertically symmetrical except that ♂♂ have a slightly deeper bill, usually with more hook on the end of the upper mandible. Like Crested Auklets, Whiskereds have a faint citrus odour to their plumage. Less agile on land than other *Aethia* species; prefers to cling to steep cliffs and near vertical sides of boulders. On flat surfaces body tilts forward at a steep angle, rests weight on tarsi and toes.

ADULT SUMMER: upperparts and underparts nearly uniform blackish-grey, darker and more bluish-tinted than similar Crested Auklet, but with undertail coverts, vent, and lower belly light grey to whitish; slightly darker on head and neck; light belly and vent colour graduates evenly into dark breast, and can be difficult to see in normal poses on land; flight feathers, upperwing, and underwing coverts blackish grey or blackish brown; adults in summer normally have mix of worn brownish flight feathers and newer, recently moulted, clean blackish primaries and secondaries; short blackish-grey tail; irides white and eyes relatively slightly larger than other *Aethia*, probably owing to this species' nocturnal colony attendance or night foraging; foot and leg colour virtually identical to other *Aethia* species: tarsi and toes bluish grey with black webbing. The spectacular ornaments consist of a long slender forehead crest, white facial plumes, and a brightly coloured bill. Crest consists of 6–18 long slender black feathers (9–57 mm length by 1–2 mm width) that originate from middle of forehead and curve forward and downward. Crest is stiff enough to remain erect when bird is resting on land or on water, but sufficiently flexible to be blown back flat against crown when flying or swimming underwater. When wet, crest resembles a stiff curved wire sprouting from forehead. The white facial plumes are more complex than those of other *Aethia* species. As in Least, Crested, and Parakeet Auklets there is a narrow sheaf of silvery white plumes extending backwards from just behind eye (auricular plumes), but Whiskered Auklets also have two other tracts of plumes that meet at a 90° angle at the lores forming a sideways V-shaped ornament. The longer of these two plumes (super-orbital plumes) extend upwards and backward over eye and out to side, resembling antennae; the shorter (sub-orbital plumes) project downwards along sides of neck. During breeding season adults' bill colour is bright blood red with a whitish tip to the upper mandible.

ADULT WINTER: resembles summer adults but either no crest or a short crest, reduced white facial plumes, and a dull red bill. Adults regrow their nuptial plumes earlier than other auklets, with some birds in breeding plumage by Feb (G. V. Byrd, personal communication).

SUBADULT: (1- or 2-years old) those attending colonies resemble summer adults but have either no crest or a short crest; facial plumes shorter and sometimes brownish-tinted; brown forehead and faces produced by worn contour feathers that contrast with the crisp dark coloration of rest of upperparts bill dull red.

JUVENILE AND FIRST WINTER: resembles winter-plumage adults but irides grey only a faint trace of facial plumes, and bill black.

CHICK: dense black down retained until a few days before colony departure; feet and bill black.

VARIATION
A. p. pygmaea Aleutian Is., *A. p. camtschatica* Kuril Is., but variation may be continuous between these two forms (Williams and Byrd 1994). Clinal size variation with birds from the eastern Aleutians smallest, and the Kuril birds largest.

MEASUREMENTS
Sexual dimorphism minimal, clinal increase in body size from eastern to western Aleutians (R. H. Day and G. V. Byrd, unpublished data)—see Table, (1) *A. p. pygmaea* from Buldir I., Alaska (R. H. Day and G. V. Byrd, unpublished data); (2) *A. p. pygmaea* (ssp.?) from Commander Islands (R. H. Day and G. V.

Byrd, unpublished data); (3) *A. p. camtschatica* from Kuril Is. (R. H. Day and G. V. Byrd, unpublished data); (4) *A. p. pygmaea* from eastern Aleutians (J. F. Piatt, unpublished data); (5) from Buldir I., Alaska (I. L. Jones and F. M. Hunter, unpublished data); (6) from the western Aleutians (J. F. Piatt, unpublished data).

WEIGHTS
Non-breeding subadults lighter than adults—see Table, (1) *A. p. pygmaea* from Buldir I., Alaska (R. H. Day and G. V. Byrd, unpublished data); (2) from eastern Aleutians (J. F. Piatt, unpublished data); (3) from Buldir I., Alaska (I. L. Jones and F. M. Hunter, unpublished data); (4) from the western Aleutians (J. F. Piatt, unpublished data).

Range and status
RANGE
A locally common but cryptic species occurring for the most part in the Kuril and Aleutian Is.

BREEDING: in Asia, known to nest very locally in the northern Sea of Okhotsk (Matykil' and Iona Is., recorded from, and possibly breeds at, Moneran I. and islands in Penzhina Gulf), in the central Kuril I. (including Iturup, Urup, Simushir, Srednego, Rasshua, Matua, Raykoke, and Ekarma) and Commander Is. (Golovkin 1989). In Aleutians nests in Krenitzen Is. in eastern Aleutians, Islands of Four Mountains (Chagulak and Carlisle Is.), locally in Andreanof Is. (Little Tanaga and Kanaga), and in the Rat Is. at Segula and Buldir (Sowls *et al.* 1978; Byrd and Gibson 1980) Actual distribution may be more extensive because of the difficulty of detecting birds and locating breeding sites. For example, a new breeding site at Kanaga Volcano was discovered in 1994 only because a research ship approached the island to observe a massive lava flow at night; a large colony of Whiskered Auklets was detected by their chorus of vocalization on a steep headland between two active flows! Largest known breeding colony (>20 000 pairs, estimated by ILJ) is at Buldir

Table: Whiskered Auklet measurements (mm)

ADULT ♂♂	ref	mean	s.d.	range	n
wing length	(1)	107.5	2.5	—	28
	(2)	111.1	0.7	—	21
	(3)	114.8	0.5	—	24
tarsus	(1)	21.0	0.8	—	27
	(2)	21.6	0.4	—	17
	(3)	22.6	0.2	—	25
culmen	(1)	9.3	0.3	—	28
	(2)	9.5	0.1	—	20
	(3)	9.9	0.1	—	25

ADULT ♀♀	ref	mean	s.d.	range	n
wing length	(1)	105.6	3.3	—	9
	(2)	110.9	0.2	—	10
	(3)	110.6	1.3	—	7
tarsus	(1)	20.6	0.7	—	9
	(2)	21.5	0.2	—	8
	(3)	21.7	0.2	—	9
culmen	(1)	8.9	0.4	—	9
	(2)	9.5	0.1	—	9
	(3)	9.7	0.2	—	9

SEXES COMBINED (ADULTS)	ref	mean	s.d.	range	n
wing length	(4)	109.1	0.7	—	48
	(6)	109.2	2.6	—	10
tarsus	(4)	21.6	1.1	—	10
	(5)	21.9	0.8	20.3–24.2	169
	(6)	21.3	1.1	—	9
culmen	(4)	10.1	0.5	—	10
	(5)	9.5	0.5	—	129
	(6)	10.0	0.5	—	10
bill depth	(5)	7.4	0.4	6.4–8.4	168
	(5)	7.3	0.4	6.7–8.4	55
crest length	(5)	36.8	6.7	8.5–57.0	225
superorbital plume length	(5)	29.2	8.5	0–46.6	158
auricular plume length	(5)	32.6	4.1	20.5–43.9	124
sub-orbital plume length	(5)	29.8	3.6	16.1–37.8	123

SEXED COMBINED (SUBADULTS)	ref	mean	s.d.	range	n
crest length	(5)	19.8	8.8	0–38.0	55
superorbital plume length	(5)	22.0	7.6	0–38.0	32
auricular plume length	(5)	28.2	3.1	21.6–34.8	32
sub-orbital plume length	(5)	26.9	3.2	19.1–32.9	32

(52° 21'N 175° 56'E), but even larger concentrations probably occur in the Krenitzen Is. near Unimak Pass (J. F. Piatt, personal communication).

WINTER: apparently winters within nesting range without long-distance movements. Some eastward movement may occur as enormous concentrations have been observed in passes

Table: Whiskered Auklet weights (g)

ADULT ♂♂	ref	mean	s.d.	range	n
	(1)	127.0	9.8	—	27
ADULT ♀♀	ref	mean	s.d.	range	n
	(1)	118.1	7.0	—	8
SEXES COMBINED (ADULTS)	ref	mean	s.d.	range	n
	(2)	116.7	10.8	—	43
	(3)	116.6	8.4	99–141	228
	(4)	123.8	14.0	—	11
SEXES COMBINED (SUBADULTS)	ref	mean	s.d.	range	n
	(3)	107.0	6.3	94–120	56

among the Krenitzen Is. and at Unimak Pass, both in the eastern Aleutians, in winter.

STATUS

The total population in Alaska was officially estimated to be 30 000 (U.S. Fish and Wildlife Service 1993) but is certainly much higher, perhaps 200 000–300 000. The total Kuril and Commander Is. population may be similar but no censuses or inventories have been made for these areas (Golovkin 1990*a*). No quantitative censusing or monitoring methods have been developed for this nocturnal species, although listening for vocalizations at night is the best way to detect breeding birds. Further inventories of populations are urgently required for this poorly known seabird, particularly in the Kurils. Like other nocturnal seabirds, Whiskered Auklets are apparently extremely vulnerable to introduced mammalian predators, and thus may have been easy pickings for Arctic Foxes put on Aleutians during early twentieth century (Bailey 1993). This may explain why the largest known population is at Buldir I., one of the few Aleutian Islands that escaped fox introductions. Most significant current threats are accidental introduction of rats to breeding islands and mortality due to light attraction. Like other nocturnal

birds, this species is strongly attracted to artificial lights such as those used on fishing vessels, particularly on foggy nights, and often fatally (see account of Dick and Donaldson 1978, for Crested Auklets). In the near future it may be necessary to restrict use of bright lights at sea near breeding colonies. Oiled birds have been found at Buldir, which is near to the Pacific great circle shipping lanes, raising concerns about the effects of chronic dumping of oil at sea near breeding colonies.

Field characters

Adults similar to Crested Auklet, but are much smaller (less than half their body size), and differ in having a small red bill (Crested Auklet has a relatively larger orange bill with conspicuous plates curving backward onto face), more extensive white plumes on face (Crested Auklets have a single, small, white auricular plume, while Whiskered Auklets have three large white plumes), slaty blue-black rather than grey plumage, and white belly and vent (Crested Auklets have entirely grey plumage). On land, may occasionally be seen by day where it occurs in mixed colonies with other *Aethia* species. In this situation it is easily picked out by its distinctive incessant 'mew' call. Most of the population is active at

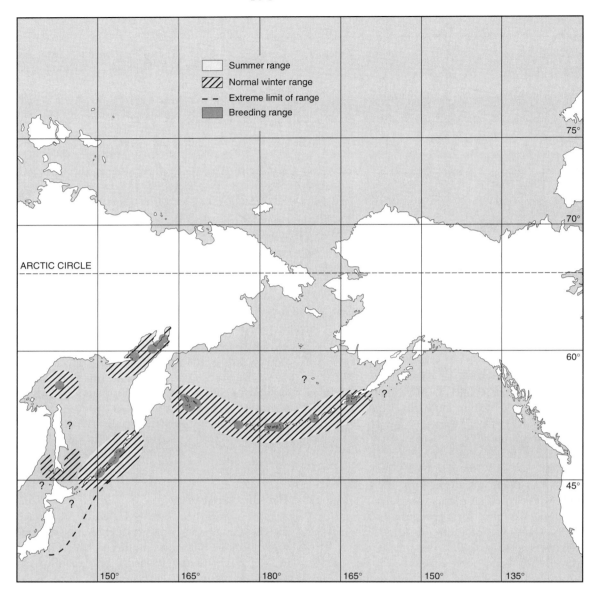

colonies only at night, and nesting concentrations are best detected at night by the distinctive vocalizations. At sea, their small size and nearly all-dark plumage (appearing black except for the lighter vent region) eliminates all other auk species except Crested Auklet. Whiskered Auklets are usually seen in groups, solitary individuals are rarely encountered. Adults' white head plumes are visible from quite far off, even in flight. Juveniles most easily distinguished by their small size, pale vent, and (at close range) traces of the V-shaped facial ornament.

Voice

Highly vocal, although noisy mainly at night when attending colonies, and early in the morning at sea near the colony. Individuals on colony surface incessantly give a short, plaintive, high-pitched sound like the mewing of a

1 Facial ornamentation of an adult breeding plumage Whiskered Auklet (ILJ photo).

kitten, every few seconds ('mew'). This call is also given at night in crevices (individuals respond to imitations and tape recordings by day or night) and at the gathering ground at sea. The loudest and most distinctive call is a vocal advertising display performed mostly by males: 'bee-deer bee-deer bee-deer bee-deer bideer bideer bideer bidi bidi bidi bidi bidi bidee' ('staccato beedoo', a rapid series of sharp two-syllable notes in varying tempo. It is normally performed at night on the colony surface by individuals in the presence of other actively socializing and vocalizing auklets. It is also heard from crevices and at the staging area on the sea by day. During courtship, pairs performed 'duet beedoo' which resembles the 'staccato beedoo' but is less musical and is performed only simultaneously by mated pairs. Another call, the 'squeak' an alarm call given by birds taking flight from disturbance. The 'mew', 'staccato beedoo, and duet beedoo are perhaps homologous to the lower pitched 'bark', 'trumpeting' and 'cackling' vocal displays of the closely related Crested Auklet (observations by ILJ at Buldir I., Alaska, recordings by J. C. Williams).

Habitat food and feeding behaviour

MARINE HABITAT

Summer and winter ranges defined by average ocean surface temperatures of 9–12 and 2–4 °C, actual temperatures preferred in summer may be colder because this species spends much of its time in or near powerful tide rips where cool water is forced to the surface. Normally found in large flocks at tidal convergences, tidal pumps, areas of standing waves up to 10 m amplitude, and other rough water areas close to islands and offshore reefs, usually within 10 km of breeding colonies in summer, within 16 km of shore at all times of year (Byrd and Gibson 1980). Occasionally seen on freshwater lagoons in Aleutians.

FEEDING BEHAVIOUR

Feeding flocks congregate at convergent tidal fronts where there are violent tide rips, often in water less than 100 m deep (Byrd and Gibson 1980). Actively feeding birds observed closely near Buldir surfaced with throats bulging with excess water, which was expelled in a choking motion lasting several seconds, with head pointed nearly straight up, after which the birds dived to resume feeding (ILJ). No information on dive depths or other aspects of feeding behaviour.

DIET

Diet apparently very similar to Least Auklet (Day and Byrd 1989). Calanoid copepods, particularly *Neocalanus plumchrus* are apparently the preferred prey in the western Aleutians, and euphausiids are preferred in the eastern Aleutians, but sampling of diet has not been very intensive or geographically diverse and this generalization needs to be interpreted with caution. Based on 25 birds' gut contents examined in 1 year at Buldir I., 76% contained *N. plumchrus*, and this copepod comprised 96% of all food items identified; 32% of the stomachs contained chaetognaths (probably *Sagitta elegans*) but these amounted to 1% of all items (Day and Byrd 1989). In a different year, a smaller sample ($n = 8$) of stomachs contained a mixture of copepods and euphausiids (Day and Byrd 1989). In the eastern Aleutians in autumn, prey consisted of 93.3% euphausiids, 2.6% unidentified crustacea, 1.5% *N. plumchrus*, and lesser quantities of four other invertebrates by wet weight, in winter 99.8% euphausiids, in early spring 99.3% euphausiids (mostly *Thysanoessa inermis*; Troy and Bradstreet 1991). Stejneger's (1885) samples from the Commander Is. contained gammarids, amphipods, gastropods, and a decapod.

Displays and breeding behaviour

COLONY ATTENDANCE AND FLIGHT DISPLAYS: unlike other *Aethia* species, has mostly nocturnal colony attendance, but individuals or pairs occasionally appear on the surface by day where the species nests in colonies with other *Aethia* species, particularly during pre-laying period. Most birds arrive and gather in dense rafts offshore from colony an hour or more before dark. Birds go ashore after darkness falls and the colony remains active until just before first light. There are no flight displays; birds arrive individually and fly straight to nesting crevice at high speed, often flying directly into crevice without landing on surface. Unlike Least and Crested Auklets, most social activity takes place below ground in cavities in talus. There are two nightly peaks of activity when most arrivals, departures, and vocal activity occur, the first starting when complete darkness has fallen and lasting 1–2 hrs the second in the early morning in the last 2 hrs before first light (V. Zubakin, unpublished data). Like other nocturnal seabirds, this species exhibits lunar phobia, with activity and vocalizations drastically reduced on clear moonlit nights, presumably relating to predation risk. When underground spaces are not available, groups of 4–10 birds may cluster around crevice openings, but surface activity is generally limited. Nocturnal habits may permit dispersed nesting at low density because birds arriving are immune to diurnal predators such as gulls and falcons, reducing competition for nest-sites, and probably allowing this species to breed at more dispersed locations than other *Aethia* species. Rafts of birds remain near shore at gathering ground in early morning. Colony attendance by juveniles and non-breeding adults continues after end of breeding season; may occur nearly year round as in some Cassin's Auklet populations, as this species remains close to its breeding colonies all year.

AGONISTIC AND SEXUAL BEHAVIOUR: information on social behaviour is limited owing to nocturnal timing and subterranean location of activity ashore, and to difficulty observing behaviour in flocks at sea at dawn and dusk. Fights and aggressive displays occur on colony site, probably relating to competition for nesting sites. Courtship and pairing apparently take place at the breeding site, with mated pairs leaving colony together early in the morning during the pre-laying period to court and copulate offshore. Pairs on surface stand facing each other and perform 'duet beedoo' mutual vocalization while billing. Other individuals, presumably unpaired ♂♂, perform staccato beedoo advertising calls from atop boulders, perches on cliffs near crevice entrances, and from within crevices. Vocalizations are probably more important in Whiskered Auklets than in other species, owing to restriction on visual communication in nocturnal situations. The showy white facial plumes are likely to be visible at night, but the facial ornaments probably more important at the gathering area on the sea. Courting pairs at sea follow each other closely, circle, and engage in unique side-to-side head waving display while vocalizing (V. Zubakin, unpublished data). Copulation occurs on the sea in early morning, and has not been observed on land. Whiskered Auklets had the highest copulation rate observed among *Aethia* species at Buldir I., pairs averaging more than six copulations per hr at the gathering ground (F. M. Hunter and I. L. Jones, unpublished data).

Breeding and life cycle

BREEDING HABITAT AND NEST DISPERSION: nests in rock crevices among beach boulders, in talus slopes and cliffs, also uses crevices and earthen cavities on steep grassy slopes; nest elevations 3–250 m above sea level. Nests in a wider variety of habitats and in a more dispersed manner than other *Aethia* species, but sometimes nests within dense colonies of Least and Crested Auklets, or in areas used by Parakeet Auklets. Some pairs' nesting crevices may be located tens of metres from nearest nesting conspecifics. Not as colonial as Least and Crested Auklets, nesting crevices relatively dispersed along shorelines with suitable rock crevices. At Buldir, nesting density averages about 1–3 breeding pairs per 100 m^2 in best areas, may be an order of magnitude less in low density nesting areas.

NEST: usually in rock crevices, also in cavities beside isolated boulders on densely vegetated grassy slopes; the incubating bird rarely visible from the surface. Nests may be located up to 1 m beneath surface in deep talus, but are usually close to the surface. Egg normally laid directly on dirt, or on collections of small pebbles and detritus, usually in a slight depression (Knudtson and Byrd 1982). Crevice volume (cm^3): 162.5.6 ± 70.3 ($n = 12$) at Buldir (Knudtson and Byrd 1982) intermediate in size between Least and Crested Auklet crevices, but Whiskered Auklet crevices often give the impression of being constricted relative to the other species, often with a very narrow entrance.

EGG-LAYING: peak of laying early to mid May (Buldir I.), 1–3 weeks earlier than Least and Crested Auklet species breeding at the same island.

EGGS: one egg, relaying of lost first eggs has not been recorded. Egg shape: pointed ovate to elongate ovate, smooth textured, white, sometimes becoming stained brown with dirt from nesting crevice during incubation. Egg size: from Buldir Is. Alaska (M. Hipfner, unpublished data), 44.0 ± 1.5 (41.0–46.3) × 31.2 ± 1.0 mm (29.2–32.8, $n = 26$); from Russia, 45.2–48.5 × 31.5–33.5 mm (sample size not provided; Golovkin 1990a).

INCUBATION: breeding adults have two lateral brood patches with maximum dimensions of about 33 by 24 mm which defeather a few days prior to laying, become vascularized and thickened by onset of incubation, and rapidly shrink and become refeathered after hatching. Non-breeding adults and subadults occasionally show some brood patch development (ILJ). Incubation period is about 35–36 days (Knudtson and Byrd 1982). Incubation duties are shared by ♂ and ♀, with incubation shifts of 24 hrs; change-overs take place at night (Williams and Byrd 1994).

HATCHING AND CHICK-REARING: duration of pipping unrecorded. Chicks semi-precocial, brooded almost continuously for first few days after hatching, but daytime brooding intermittent after fourth or fifth day (ILJ and F.M. Hunter, unpublished data). One or both adults usually spend night with chick in nesting crevice during first 10 days after hatching. Individual adults carry food in sublingual pouch and make deliveries 1–2 times per night, the first delivery soon after dark, the second 1 hr or more before first light (indicating that this species can successfully forage at night); no measures of food load mass but this may be large relative to other *Aethia* species (ILJ). Food deliveries also occur during the day where this species nests in dense colonies of Least and Crested Auklets, such as at Buldir I., Alaska.

CHICK GROWTH AND DEVELOPMENT: relative to other *Aethia* species, chick development is slower and time in crevice before fledging longer, undoubtedly related to the constraints on adult provisioning imposed by nocturnality. From Buldir (Williams and Byrd 1994): newly hatched chick mass, 12.5–14.0 g ($n = 2$), maximum growth rate about 3.5 g/day, mean chick mass asymptote 113.4 ± 7.0 g (100.4–122.6, $n = 8$), fledging mass 106.0 ± 6.8 g ($n = 8$) or 92% of mean adult body mass, age at fledging 35–45 days. Body mass increases steadily during first 20 days, changes little during last 7–10 days of nestling period. Chick fully feathered by day 33, wing length increases continuously to fledging age, and fledgling wing length about 94% of adults'. Fledging has not been observed, probably takes place at night. Uniquely among alcids, Whiskered Auklet fledglings return to colony and roost on shore at night in breeding habitat after initial departure, regularly returning for up to 6 weeks after most of population has departed (early Sept at Buldir I., V. Zubakin and N. Konyukhov, unpublished data). Some fledglings get lost and have been found grounded more than 1 km inland and 500 m elevation (ILJ).

BREEDING SUCCESS: natural reproductive success estimates need to be taken with caution because of sensitivity to investigator disturbance of nest-sites. However, productivity is probably about 0.6–0.8 chicks fledged per breeding pair per year. At Buldir I. (5 years of monitoring, Knudtson and Byrd 1982; G. V. Byrd and J. C. Williams, unpublished data) hatching success averaged 82% fledging success 85%, and breeding success 73%.

PREDATION: introduced mammalian predators include Arctic and Red Foxes and rats which took adults at colonies and probably greatly reduced or extirpated this species from several Aleutian and Kuril Is. (Voronov 1982; Bailey 1993; Williams and Byrd 1994). Avian predators include Slaty-backed and Glaucous-winged Gulls which take adults and fledglings near colonies and kill many birds that become dazzled at night at brightly lit fishing vessels (ILJ) and Peregrine Falcons which take birds at sea.

MOULT: little information (but see Williams and Byrd 1994). Adults undergo moult of head and neck feathering during Jan–Mar, during which the crest and plumes develop. Birds have full set of breeding ornaments by mid-to late-winter, earlier than other *Aethia* species. Gradual primary moult begins before arrival at breeding colonies in May. Primaries are replaced sequentially during breeding season, one at a time starting with the innermost and working outward. Most birds probably have a complete set of fresh primaries by Sept. Ornamental facial plumes look worn by end of breeding season, with most adults experiencing partial loss of crest and plume feathers. Subadults (both 1- and 2-years-olds) attend colonies with very worn brownish or tan coloured primaries which sequentially moult starting in May beginning with the innermost. These subadults also show extreme feather wear on the face and head in May. By Sept the Definitive Pre-basic moult is complete and these birds have fresh primaries, contour feathering, and rectrices.

Population dynamics

SURVIVAL, PAIR- AND NEST-SITE FIDELITY: adult annual survival was 82% (based on analysis of mistnet recaptures in 1993 and 1994; ILJ and F. M. Hunter, unpublished data) at Buldir I. Alaska, giving an overall mean life expectancy of about 5 years. No information on pair-fidelity. Re-mating pairs sometimes return to the same crevice used in a previous year (ILJ).

AGE AT FIRST BREEDING: uncertain probably 3 years; 1-year-olds regularly come ashore at colonies 2-year-old subadults enter crevices, but apparently neither breed nor form pair-bonds (ILJ).

Rhinoceros Auklet   *Cerorhinca monocerata*

Alca monocerata (Pallas, 1811, *Zoographia Russo-Asiatica* 2:362—Alaska).   PLATE 3

Description

A medium sized, dark brownish-grey auk, with a thick neck, and stubby bill surmounted, in summer, by a vertical 'horn' up to 25 mm high at the base of the upper mandible. Somewhat reminiscent of puffins in shape, but lacking the rounded head and massive bill. On land their posture is rather horizontal, resting on the tarsus, sometimes lying forward on the sternum. Runs, rather than walks, sometimes using the wings for assistance, especially when going uphill. Markedly more clumsy in walking than puffins *Fratercula*. A powerful, but rather heavy flier, compared with smaller auklets and murrelets, keeping just above the water except when approaching the colony. As it approaches land, flies steeply upwards below forest canopy, rather than crashing in from above, as is typical of Ancient Murrelets arriving at the same islands.

ADULT SUMMER: above, dark sooty brownish-grey, medium grey-brown on cheeks, throat, and breast, becoming dirty white on belly; long, narrow tracts of white plumes form two parallel lines, one running back down neck from eye, another running backwards and downwards from base of gape; bill orange, with black stripe running the length of culmen, another along gonys, and a thin black line on cutting edge; the vertical, cream coloured horn forms a saddle above nostrils, standing 10–20 mm above top of bill; irides light yellow; narrow, dark-grey, fleshy eye-ring; legs and toes yellow with brownish webs and black nails; inner toenail is not turned inwards, unlike puffins.

ADULT WINTER: similar to breeding, but without the horn or plumes, with bill dull brown, and with darker belly.

JUVENILE AND FIRST WINTER: similar to winter-plumage adult, but smaller, with a shorter, more slender, dark grey or brownish bill; upperparts dark blackish brown, sides to head dark grey, becoming slightly paler on throat and greyish-brown on breast; usually, a sharp line divides the upper breast from white lower breast and belly, which are variably sullied with grey feathers forming faint bars.

CHICK: down dark brown above and paler, greyish brown on breast and belly.

VARIATION

No subspecies recognized; geographical variation poorly known. Measurements given by Shibaev (1990e) suggest that birds from the Russian far east may be slightly larger than those from North America. The genus is known from the late Miocene of California and Pliocene of the Sea of Cortez (Howard 1971; Warheit 1992). Pliocene specimens from Sea of Cortez (*C. minor*) were approximately 25% smaller than recent specimens and may have represented a smaller, warm-water, form, rather than an ancestor of modern *C. monocerata*.

MEASUREMENTS

Little sexual dimorphism, with means for ♂♂ slightly larger in all dimensions (except horn height), but broad overlap. Plumes on face become shorter during breeding season, perhaps through wear. Likewise, wing length is affected by abrasion, with primaries that show signs of fading in July shorter than those appearing fresh in the same month. All measurements (mm) of breeders: (1) Royal British Columbia Museum specimens (Gaston and Dechesne 1996); (2) Triangle I., BC (ILJ).

WEIGHTS

♂♂ generally heavier than ♀♀, but there is extensive overlap. Birds collected while moulting in July and Aug averaged heavier than breeders at any breeding stage, but wide variation occurs. Breeding adults in July (rearing chicks) in British Columbia (Royal British Columbia Museum specimens): ♂ 510 ± 11 g (441–569, $n = 20$); ♀ 456 ± 9 g (389–616, $n = 25$). At Triangle I., unsexed birds trapped over the whole season and including some non-breeders, averaged 491.9 ± 35.5 g (353–606, $n = 657$; ILJ). Wing loading is 1.46 g cm^{-2} (Vermeer 1979).

Range and status
RANGE
Pacific, boreal, and sub-Arctic, nesting on offshore islands from the Aleutians S to Korea, northern Honshu, Japan, and central

Table: Rhinoceros Auklet measurements (mm)

♂♂	ref	mean	s.d.	range	n
wing	(1)	182.9	3.8	177–191	19
tarsus	(1)	29.9	0.8	28.6–31.3	19
culmen	(1)	33.3	1.3	30.9–34.5	18
bill depth in front of horn	(1)	15.7	1.0	14.3–18.2	12
plume above eye (May)	(1)	54.7	6.8	45–60	3
(July)	(1)	40.8	5.3	29–49	22
♀♀	ref	mean	s.d.	range	n
wing	(1)	177.7	4.3	170–186	24
tarsus	(1)	28.8	1.4	24.0–31.3	24
culmen	(1)	31.9	1.2	29.7–34.3	21
bill depth in front of horn	(1)	15.5	1.2	14.0–17.9	15
plume above eye (May)	(1)	46.2	8.1	32–54	4
(July)	(1)	36.7	5.4	29–49	25
UNSEXED	ref	mean	s.d.	range	n
tarsus	(2)	31.5	1.3	28.2–36.8	436
culmen	(2)	32.9	1.6	25.8–38.8	444
horn height	(1)	9.5	1.9	4.8–11.7	19
	(2)	10.1	1.6	1.2–15.0	441
plume above eye	(2)	52.6	5.8	41.8–65.2	29
plume behind gape	(2)	47.2	5.7	32.0–64.0	132

California. Occurs on both sides of the Pacific, but uncommon between the Alaska Peninsula and N Japan.

BREEDING: in Asia, concentrated mainly on Teuri I., off W coast of Hokkaido, where 350 000 breed; otherwise there are colonies of up to a few thousand pairs on other Japanese islands S to Honshu, on islands off Korean peninsula, the coast of southern Primoriye, N to central Sakhalin, and in the southern Kuril Is. In North America, a few breed as far S as Farallon Is. and Ano Nuevo I., California, but main concentration is in Washington State, BC and SE Alaska, where the seven largest colonies support about 400 000 breeding pairs: Destruction and Protection Is., Washington (24 000 and 34 000, Speich and Wahl 1989), in BC, Triangle (42 000), Pine and Storm (160 000), and Lucy (25 000) Is. and the Moore Group (92 000, Rodway 1991), and Forrester I., SE Alaska (55 000). Smaller numbers, amounting to a few tens of thousands, breed on islands throughout the Gulf of Alaska (Alaska seabird catalogue).

WINTER: there is a general post-breeding dispersal, followed by a southward movement, especially in the eastern Pacific, where the bulk of the population appears to winter off California. Off the W coast of Vancouver I., where few breed, the species is common in Aug–Sept, but rare in Oct–Dec, suggesting a passage through the area (Morgan *et al.* 1991). In W Pacific, Rhinoceros Auklets do not move far outside the breeding range, although common in winter off the main islands of Japan as far S as Tokyo, occasionally Kyushu and NE China (Meyer de Schauensee 1984; Brazil 1991). In E Pacific, they are common

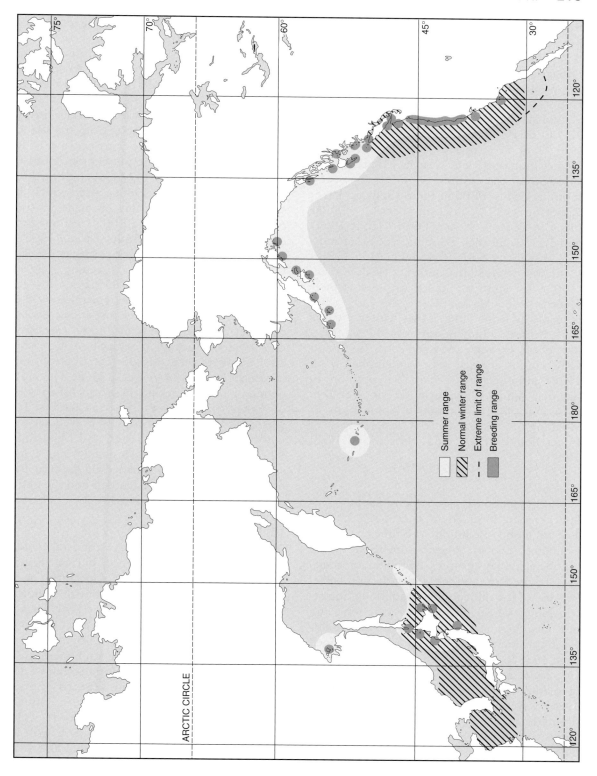

Summer range
Normal winter range
Extreme limit of range
Breeding range

ARCTIC CIRCLE

from Washington State to S California. Peak numbers occur off California in Feb, when 2–300 000 were estimated to be present (Briggs *et al.* 1987); far in excess of the breeding population of the U.S. Christmas Bird Counts over the period 1961–92 at Monterey, California averaged over 1000 (maximum 12 750 in 1975) and several other California sites N of Los Angeles averaged over 100. 'Many thousands' are seen in Monterey Bay throughout the winter, with regular early morning flights off Point Pinos (Stallcup 1976; Roberson 1980). Small numbers are regular offshore as far S as San Diego (Unitt 1984) and Baja California. Banding recoveries indicate that many birds from colonies in British Columbia winter off California (Kaiser *et al.* 1984). In contrast, there are few winter records from Alaska, where the only CBC record was from Sitka. Most winter records from British Columbia are from the southern part of the province (Morgan *et al.* 1990) and CBC counts average only 4 (Campbell *et al.* 1990) although birds may be common in Saanich Inlet, Vancouver I. in Oct–Dec (Morgan 1989). Common in some parts of Washington, especially the inner parts of Puget Sound (CBC data, Speich and Wahl 1989).

STATUS

World population is estimated at about 1 million breeding birds (Byrd *et al.* 1993: implies 1–2 million, including pre-breeders); our estimates (above) suggest somewhat higher numbers, perhaps nearer 1.25 million breeders. Population estimates are generally unreliable, because of the difficulty of establishing burrow occupancy where burrows are very long and nest chambers inaccessible. Population trends are poorly known. In Russia, numbers declined from 1938 to 1949, with no subsequent recovery, while at Teuri I., Japan, they probably increased between 1951 and 1970 (Fujimaki 1986). Several small Japanese colonies showed dramatic increases between the 1970s and 1980s (Brazil 1991). In North America, populations in California, Oregon, and British Columbia have been increasing at least since the 1970s (Rodway 1991; Byrd *et al.* 1993). After removal of sheep from Destruction I. in 1969, the population increased from 3500 burrows (Richardson 1961) to 28 000 pairs in 1974 (Wilson 1986). Likewise, on SE Farallon I., recolonized by Rhinoceros Auklets in the 1970s, the removal of rabbits from the island may have made suitable burrows available (Ainley *et al.* 1990*b*): the population doubled annually from 1973 to 1977 and continued to increase through the 1980s, after which it stabilized. The population on nearby Ano Nuevo I., colonized in the 1980s continued to increase up to 1994 (Ainley *et al.* 1996). Populations appear to be increasing in British Columbia: at Cleland I., the breeding population rose from a few to 2700 pairs between 1967 and 1982 and the area occupied by the main colony on Triangle I. expanded by 14 per cent between 1984 and 1989 (Rodway *et al.* 1992*b*). Other colonies may have increased also; numbers seen offshore in Hecate Strait increased from 1984 to 1994 (AJG). The only definite record of decline in North America is of several colonies in the Queen Charlotte Is. where birds have been heavily preyed on by introduced Raccoons (Gaston and Masselink 1997) and (at Langara I., probably) rats. The species was adversely affected in the past by introduced foxes in the Aleutians and the Gulf of Alaska (Bailey and Kaiser 1993).

Field characters

Adults in breeding plumage unmistakeable, owing to pale 'horn' on top of bill and white plumes on face. The general appearance at a distance is of a drab, brown auk with a short, stout, orange bill. Immature birds have smaller, dark bills, and no ornaments, but can be distinguished from other auks by the combination of pale, but not white, underparts and size. The lack of clean white underwings and belly distinguishes it from Horned Puffin and murres in flight. The immature Tufted Puffin is darker below, more sooty above, with a deeper bill and more rounded wings. Beached birds are distinguished from all other auks of similar size by the colour of the under-parts. A medium-sized auk with pale yellowish legs is likely to be this species; the bill is unmistakeable, if present.

Voice

ADULTS: several adult call types have been described (Gaston and Dechesne 1996). 'Mooing' is frequently heard on the breeding colony at night: a series of about 7–12 mellow notes, slower and longer in the middle, faint at the end (Fig. 1); 'barks' are a nasal series of four short, rapid notes (Heath 1915); 'groan', a single call note given on the water; 'braying', single, low-pitched notes heard occasionally from burrows (Richardson 1961); considered by Wehle (1980) a threat call; 'rasping squeak', given occasionally by birds on the sea; 'tremolo' is a vibrato call given by birds in flight while leaving the colony and created by the strong pumping of the wings (as noted in other alcid calls); 'huffs', loud exhalation given during certain posture displays (see Agonistic Behaviour); occasional grunts are given on capture. Calling occurs nightly throughout the period of colony attendance, with the greatest activity during hatching and fledging (Wilson and Manuwal 1986). It begins soon after the first arrivals at the colony and ceases 1–2 hrs before sunrise, with a peak after midnight. Most calling occurs among loose aggregations or displaying groups of birds on the colony surface. Some occurs inside burrows and may be linked to activities such as nest relief, pair formation, or territoriality. No differences in calls between sexes known, but individually unique vocalizations probably occur, facilitating individual recognition (Richardson 1961; Welham and Bertram 1993).

CHICKS: calls are exchanged between hatchling and parents while egg is pipping (Wilson and Manuwal 1986). After hatching, the 'chirp' is a monotonous note repeated 20–60 times/min, occasionally interrupted by 'peeps'; a series of high-pitched calls, repeated rapidly. At night chicks may 'chirp' continuously from burrow. Variation in chick calling rates not correlated with provisioning rate or nutritional state of chick (Welham and Bertram 1993).

Habitat, food, and feeding behaviour
MARINE HABITAT
Found mainly in continental shelf waters, and along the shelf break (Morgan *et al.* 1991). During the breeding season, found in waters with surface temperatures between 5 and 16 °C. Generally an offshore feeder, but found close to shore during the breeding season and in inshore waters in Puget Sound and Monterey Bay outside the breeding season. Prefers water >15 m deep (Burger *et al.* 1993). In winter, off California, found concentrated over the edge of the continental shelf. In British Columbia, in summer, occurs in sheltered bays and throughout continental shelf waters, often close to shore, but not usually in small inlets, estuaries, or other waters with reduced salinity. Makes use of tide-rips where prey is forced to the surface.

FEEDING BEHAVIOUR
Prey mostly taken in mid-water, rather than at the bottom. Dives to a mean maximum depth of 30 ± 3.7 m ($n = 16$, range 12–60, Burger *et al.* 1993), but most time is spent at <10 m. Mean dive times recorded from shore were 45 ± 2.5 sec, with a mean inter-dive interval of 10.8 ± 2.0 sec during diving bouts, although this was in shallow water. Underwater swimming speed while foraging probably exceeds 1 m/sec, with bursts of 1.5–2.0 m/sec (Burger *et al.* 1993). Underwater wing-beat rate 82 per min (Fink and Johnsgaard in Johnsgaard 1987). Heart rate and arterial blood pressure unchanged during feeding and escape dives, compared with pre-dive levels (Stephenson *et al.* 1992). Feeds during day and at dusk (Vermeer *et al.* 1987). Analysis of bite marks on fish delivered to chicks showed that 73 per cent were seized from below and behind (Burger *et al.* 1993). Frequently occurs in mixed species flocks with Black-legged Kittiwakes, Pacific Loons (*Gavia pacifica*), Common Murres, Ancient Murrelets, gulls, cormorants and other species, usually concen-

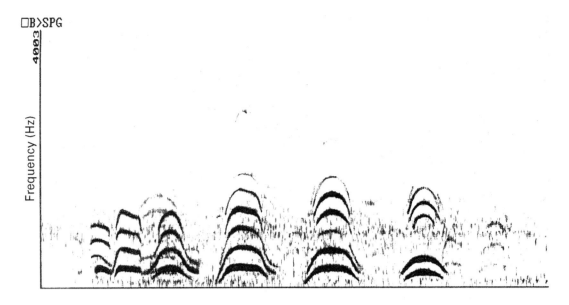

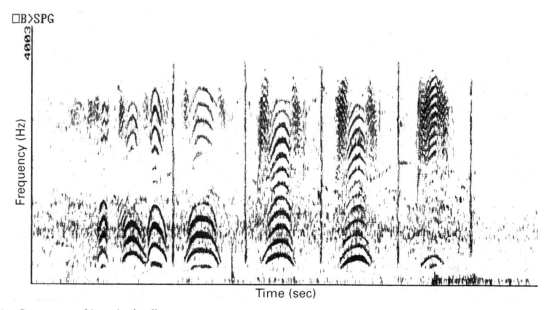

1   Sonogram of 'mooing' call.

trating on fish schools, or swarms of euphausiids (Porter and Sealy 1981; Gaston 1992*a*). Apparently cue on feeding flocks of kittiwakes, or the sight of conspecifics on the move; rarely initiate flocks (Hoffman *et al.* 1981). May use bubbles, emitted from the mouth, to herd sandlance and herrings close to the sea surface and prevent the schools from breaking up (Sharpe 1995). In June and July, off Protection I., Rhinoceros Auklets were present in 100 per cent of mixed feeding flocks and frequently seen to surface around sandlance balls, as though corralling them. Other common participants were Glaucous-winged and Heerman's Gulls (*Larus glaucescens* and *L. heermani*), sometimes Common Murres. Average of 70 in mixed feeding flocks off Washington and 44 off Langara I., BC (Sealy 1973a; Hoffman *et al.* 1981). Little evidence of intraspecific aggression in feeding flocks. Although foraging groups of up to several hundred may occur, there is no evidence of social foraging (i.e. co-operative herding of fish schools). Non-feeding Rhinoceros Auklets concentrated in 'loafing rafts', usually of 20–40, but sometimes up to 400 birds (Grover and Olla 1983). Birds carrying fish in areas far from breeding colonies suggest that foraging range may be as much as 100 km (AJG, ILJ).

DIET

ADULTS: relatively little information available. Flocks feeding off Washington, Alaska and British Columbia in summer appeared to be feeding on sandlance, capelin, and herring (Sealy 1975b; Hoffman *et al.* 1981). In Gulf of Alaska, 99% of stomachs contained fish remains and 1% cephalopods; 61% of identifiable fish were capelin and 21% sandlance (Sanger 1986, 1987a). Stomach analysis of birds collected in Monterey Bay in winter showed 85% contained market squid, 38% northern anchovy (Baltz and Morejohn 1977).

CHICKS: food delivered to nestlings can be easily intercepted by researchers and hence good information is available on nestling diets. Several fish are usually delivered in a 'bill-load', in British Columbia averaging 5.7 ± 0.2 fish (*n* = 158, range = 1–19). Bill-loads comprising more than one prey species constituted 40% of those delivered, with a third of these combining sandlance and rockfish (Burger *et al.* 1993). At Middleton I., Alaska diet comprised 59% first year sandlance, 36% second year sandlance (*n* = 68); at Semidi Is. 75% sandlance, 22% capelin (*n* = 82, Hatch 1984). At Protection I., sandlance were 71% and 64% of diet by weight in two years, and herrings 26% and 26%. Average mass of individual fish was 5.6 g (Wilson 1977). At Destruction I., in another 2-year study, sandlance were 6% and 32%, northern anchovy 56% and 27%, night smelt (*Spirinchus starksi*) 0% and 32%, herring 21% and 4%, and surf smelt 15% and 2% (Leschner 1976). At Teuri I., Japanese greenling (*Pleurogrammus azonus*) was the commonest food during the early part of the chick-rearing period, with sardine, Pacific saury and sandlance becoming important later (Watanuki 1987). At Daikoku I., Japan, anchovy, salmon (*Onchorhynchus*), and greenling were all common (Watanuki 1990). At Cleland I., BC two thirds of food items delivered were sandlance, and the rest juvenile rockfishes (Summers and Drent 1979). Most sandlance were 70–110 mm. Pacific saury, fed to chicks at Triangle I., especially in 1976, were often found uneaten in burrows and may have been unsuitable for chick manipulation (Vermeer 1980). See Table: (1) Protection I., WA (Wilson 1977); (2) Destruction I., WA (Leschner 1976); (3) Cleland I., BC (Summers and Drent 1979); (4) Seabird Rocks, BC (Burger *et al.* 1993); (5) Triangle I., BC (ILJ and H. Knechtel, unpublished data); (6) Pine I., BC (Vermeer and Westrheim 1984); (7) Lucy I., BC (Vermeer and Westrheim 1984, Bertram 1988); (8) Helgesen I., BC (AJG); (9) Middleton I., AK (Hatch 1984); (10) Semidi Is., AK (Hatch 1984); (11) Teuri I., Japan (Watanuki 1987); (12) Daikoku I., Japan (Watanuki 1990).

Displays and breeding behaviour

COLONY ATTENDANCE AND FLIGHT DISPLAYS: usually nocturnal, especially at major colonies, but crepuscular in some areas, and sometimes diurnal (Scott *et al.* 1974; Summers and Drent 1979; Thoresen 1983). Birds gather at sea

Table: Rhinoceros Auklet chick diet

ref	Primary prey	Secondary prey
(1)	sandlance	herring
(2)	northern anchovy	night smelt, sandlance, herring
(3)	sandlance	juv. rockfish
(4)	sandlance	surf smelt, herring, juv. salmon
(5)	sandlance	rockfish, greenling, Pacific saury
(6)	herring	sandlance, rockfish, salmon
(7)	sandlance	herring
(8)	sandlance	—
(9)	sandlance	—
(10)	sandlance	capelin
(11)	greenling, sandlance	salmon
(12)	greenling, anchovy	salmon, sardine, Pacific saury, sandlance

close by, up to 4 hrs before they begin to land on their colony, or even earlier (14:00–15:00) if food is very abundant (A. Harfenist, personal communication). In Washington, rafting occurred on water close to the colony island up to 3 hrs before sunset; with birds circling over colony from sunset onwards and arrivals 60–90 min after sunset, earlier if overcast. Circling flights not seen at all colonies: may reflect the abundance of pre-breeding recruits. Departures begin 2 hrs before sunrise and continue to sunrise, with most birds departing in the last hour (Wilson 1977). Pairs depart together in pre-laying period (Richardson 1961). Peak circling activity occurred about the beginning of hatching. Activity and calling are reduced on moonlit nights (Leschner 1976; Wilson 1977; ILJ). At Forrester I., on foggy days, some birds remained on the surface of the colony after dawn (Heath 1915). Young were fed throughout the day at cave nest-sites in Oregon (Scott *et al.* 1974) and California (E. McLaren, personal communication). At Teuri I., Japan, birds gathered on sea 2 hrs before sunset, and arrived on land from 1 hr before sunset in June (chick-rearing), with some arriving as late as mid-morning on foggy days (Thoresen 1983). Watanuki (1990) gives different timings for same colony, with gathering 300–500 m offshore from 30 min before sunset and peak landings 1 hr after sunset during incubation, 45–105 min during chick-rearing. At dawn, birds follow well-used pathways to take-off points. Last birds have left 15 min before sunrise (Summers and Drent 1979). Some birds spend the day in the burrow during the pre-laying period (Leschner 1976) and occasionally during chick-rearing (A. Harfenist, personal communication). On leaving colony, may dive steeply, with back-swept wings (making a roaring sound), use normal flapping flight, or 'butterfly flight', using shallow, gentle wing-beats (Thoresen 1983). Landing on sea may dive in head first, or land parallel to water, after braking.

AGONISTIC BEHAVIOUR: defence of burrows has been observed and territory may include the runway to the burrow and a raised area nearby, used for take-off, landing, and loafing. Area defended may vary with burrow density; intensity of defence highest in pre-laying and laying periods (Wehle 1980). Fighting at night, on the colony, is not uncommon, and often occurs among groups of socializing birds. Birds indulge in aggressive chases, with raised wings. At high intensity, they will lock bills and claw at one another's belly (Thoresen 1983). Chicks are sometimes killed by conspecifics, possibly when they wander into the wrong burrow (Wilson and Manuwal 1986). May share nesting habitat with Tufted Puffins and Cassin's Auklets, and competition for burrows may occur; Rhinoceros Auklets may occasionally break eggs or kill chicks of Cassin's Auklet (Vermeer and Cullen 1979; Wallace *et al.* 1992). At Farallon Is., often driven off by Tufted Puffins, which may exclude auklets from some parts of their range (Ainley *et al.* 1990*a*). Seen to displace aggressively Black-tailed Gulls (*Larus crassirostris*) on Teuri I., Japan (Thoresen 1983).

SEXUAL BEHAVIOUR: presumably monogamous (Richardson 1961; Leschner 1976). No information on the role or function of ornaments (horn, plumes Fig. 2) available. Posture displays poorly known because of nocturnal habits. Best information from diurnal behaviour at Teuri I., Japan (Figure 3 in Thoreson 1983): 'hunch walk' or 'low neck-forward' posture involves walking slowly and deliberately, with high steps, used by birds close to their burrow in the presence of intruders (conspecifics and other species); 'gaped bill with raised wings' (degree to which wings are raised relates to intensity) forms part of a threat display; 'upright-huff stance', involves standing erect with body almost vertical, often with wings partly spread and bill open, pointing upwards and accompanying 'huffs' of exhaled air; apparently used in burrow defence. On the arrival of one pair member on the colony, the pair 'hunch-walk' towards one another and the arriving bird assumes the 'upright-huff' position. 'Billing' is important for pair maintainance and is seen on water and land; pairs face in 'semi-hunched position and with slow, deliberate movements passed and repassed each other's bill. The bills did not appear to actually touch' (Thoresen 1983). 'Bowed head' display, similar to that of puffins *Fratercula*, for which it serves to invite billing, is seen in Rhinoceros Auklets, but the function for this species is not known. Pair formation presumably occurs on land at the colony. The strong vocal chorus during the chick-rearing period suggests that courtship of pre-breeders peaks at that season and pairs thus formed may reconvene the following season.

Breeding and life cycle

Detailed studies of colony attendance and breeding have been carried out in Washington (Richards 1961; Leschner 1976; Wilson 1977) and British Columbia (Summers and Drent 1979; Bertram 1988), but some aspects, especially behaviour at the colony, are not well known, because of nocturnal habits. Birds begin to visit their colonies as early as Feb in California and Japan (Thoreson 1983; Ainley *et al.* 1990*a*), and Mar in Washington and BC (Wilson 1977; ILJ), but are not usually seen until Apr in Alaska. Peak of arrivals in Washington about 23–24 Apr, when much burrow maintainance occurs (Leschner 1976), followed by 'honeymoon' trough in early May and a gradual build up in burrow visits until late June. Burrow digging peaked in late Mar at Protection I. (Richardson 1961). A rapid decline in colony visits occurs after mid-Aug, as young fledge (Wilson 1977), although some birds continue to visit burrows for several weeks after the last chicks depart (Richardson 1961).

BREEDING HABITAT AND NEST DISPERSION: breeds colonially on islands up to several thousand hectares, usually forested (SE Alaska, BC, apart from Triangle and Cleland Is., and

2 Facial ornamentation of the Rhinoceros Auklet (ILJ photo).

Japan), or covered in grass or dense forbs (Washington, Aleutians, and Gulf of Alaska). Burrow entrances typically more than 1 m apart, but may be within 25 cm in some areas. Burrow occupancy at Protection I., 62–65% (Wilson 1977), at Triangle I., 68% (Rodway *et al.* 1990a).

NEST: normally a burrow 1–5 m long in deep soil, either among grass, low shrubs, or beneath forest. On Protection I., burrows were 0.3–6.1 m long, mainly on grassy slopes, more than 10 m above sea level, with mean density of *c.* 0.5 per m² (Wilson 1977). At Cleland I., burrows 0.3–2.3 m, usually simple, mainly among salmonberry, or rose bushes, interspersed with rocky knolls from which birds took off (Summers and Drent 1979). In SE Alaska and in British Columbia, except Triangle and Cleland Is., burrows occur up to 200 m from sea, typically under forest canopy, with sparse ground vegetation, often among tree roots, under logs or stumps (Rodway 1991; AJG). On forested islands burrows tend to be long (>1 m) and sometimes branched (Heath 1915; A. Harfenist, personal communication). At Farallon Is., nests mainly on the floor of a cave. Burrow digging occurs during the early pre-laying period (Leschner 1976; Wilson 1977). Bill and claws are used and both sexes take part; rate of excavation may be as much as 18 cm per night (Richardson 1961). As in puffins, claws long and sharp, well suited for digging. Nest chamber generally lower than burrow, saucer-shaped; lined with local vegetation and litter, e.g. grass, leaves, and twigs (Heath 1915).

EGG-LAYING: median date of laying varies from mid-Apr (Farallon Is., California, Ainley *et al.* 1990a) to early June (Semidi Is., Alaska, Hatch 1984). At Protection I., laying 30 Apr–10 June, with peak 12 May–2 June (Wilson 1977) and at Destruction I., 2 May–17 June, with peak 5–20 May (Leschner 1976). At Cleland I., BC, laying peaked 28 Apr–10 May (Summers and Drent 1979), while at Pine I., median hatching occurred on 16 June in 2 years (range 6–29 June, *n* = 142 and 145). In Alaska, at Forrester

I. laying peaked in late May (Heath 1915), at Middleton I. laying was similar; early May–10 June, median about 20 May and at Semidi Is., about 25 May (Hatch 1984). Mean laying in 2 years at Teuri I., Japan, was 8 June and 25 May, the later date caused by persistent snow cover (Watanuki 1987).

EGGS: clutch of one, elliptical or subelliptical, off-white with faint lilac or purple blotches. Measurements: Protection I. (Wilson 1977, *n* = 15); 68.9 (63.2–73.3) × 45.7 (42.8–48.4) mm; fresh mass 77.7 ± 6.4 g (about 16% of adult body weight), shell thickness 0.4 mm. Composition: shell 8%, yolk 32%, albumen 60% (Kuroda 1963).

INCUBATION: starts immediately after egg is laid and may be irregular for first 8 days after laying, with gaps of up to 4 days, which do not affect hatchability. Incubation period estimated as 44.9 days (39–52, *n* = 28; Wilson 1977) at Protection I. and 45.6 days at Destruction I. (42–49, *n* = 10; Leschner 1976). There are two incubation patches, laterally on the belly. Refeathering occurs during chick-rearing. Most incubation shifts 24 hrs, but occasionally up to 4 days, with longer shifts early in incubation (Wilson 1977). Replacement periods after egg loss, 9–22 days (Leschner 1976).

HATCHING AND CHICK-REARING: eggs neglected for 4–5 days during pipping nevertheless hatched successfully (Summers and Drent 1979). Chicks semi-precocial and nidicolous, covered in dense down; capable of walking as soon as hatched and thermoregulating within 2 days (Richardson 1961). Hatchling mass 54.5 ± 1.1 g (46–64, *n* = 24; Bertram 1988); 53.3 g (*n* = 8; Summers and Drent 1979). Signs of an incipient horn are already present on the bill. Duration of brooding averaged 3.9 ± 2.5 days (*n* = 23, Wilson 1977): presumably both parents participate. The role of the sexes in chick-rearing is unknown, but most chicks are fed nightly by both parents (occasionally more than once by the same bird), so approximately equal contributions are indicated. Bill loads vary in size with chick age, averaging 36.3 g ($n = 427$) in British Columbia, with annual means ranging from 23.8 to 33.9 for 9 colony-years, while the mean mass of constituent fish ranged from 1.7 to 5.5 g. Bill loads comprising rockfish were lighter than those comprising sandlance in 2 years (Bertram and Kaiser 1993). At Lucy I., BC, most chicks receive 20–70 g/night. Provisioning peaked at 30 days, falling to less than 50% of peak by 50 days in 1985. In 1986, the peak was at 20 days, falling by 40% at 50 days. In 1987 there was slight decline in burrow loads (total amount delivered per burrow/night) from peak at 30 days, but rate at 50 days was 80% of peak. Date had no effect on provisioning rate when age was taken into account (Bertram 1988). Mean bill loads averaged 23.8–33.9 g (8 colony-years), with the size of loads varying among years, but not among colonies, when Lucy, Pine, and Triangle Is. were compared. There was some evidence of correlation between sandlance as proportion of food fed to chicks and their availability at sea (Bertram and Kaiser 1993). Bill loads at Triangle I. in July averaged 25–35 g and in Aug 32–37 g (3 years, Vermeer and Cullen 1979). Mean amounts delivered to burrows at Semidi Is., Alaska were 39.6 ± 16.5 g ($n = 68$; Hatch 1984), suggesting smaller bill-loads, or less frequent feedings, than recorded in British Columbia and Washington. Chicks approach burrow entrance at night, but usually remain in nest chamber during day. Older chicks may emerge from burrow at night to exercise, sometimes wandering about (Harfenist 1991). No parental care occurs after fledging.

CHICK GROWTH AND DEVELOPMENT: remiges and greater coverts burst from their sheaths in fourth week; contour feathers appear on head and elsewhere soon afterwards (Richardson 1961). Rate of increase in mass between 10 and 40 days at Lucy I. averaged 5.2–7.1 g/day (3 years, Bertram 1988). At Pine I., in two seasons, mass at 10 days old declined as the season progressed, but growth rate did not.

Chicks grew faster and fledged at a younger age and heavier when fed supplementary food (Harfenist 1991). Peak mass is attained at about 40 days, means 299–416 g (15 colony-years, Gaston and Dechesne 1996). A small mass recession occurs before departure which may reduce wing loading for the first flight (Wilson 1977). Mass at departure is 266–384 g (15 colony-years); the plumage is fully grown but down sometimes remains on nape and rump. The wings are only 83–86% of adult length (Vermeer 1980). An incipient horn, less than 0.5 cm high, is present on bill. Fledging period is estimated as 50 days (38–56, $n = 12$, Cleland I., Summers and Drent 1979)), 49.3 ± 3.6 days ($n = 27$, 1975) and 50.6 ± 2.7 days (42–58, $n = 37$, 1976, Protection I., Wilson 1977). Age at departure is affected by rate of growth and threat of predation to adults (Bertram 1988; Harfenist 1991). Fledglings walk or flutter to the water when they leave (Leschner 1976).

BREEDING SUCCESS: much affected by observer disturbance, with 28–30% of burrows at Protection I. deserted because of observer activity during incubation (Wilson 1977). Reproductive success at undisturbed sites on Protection I. was estimated at 81% and 91% in 2 years. Hatching success at Cleland I. was 90% ($n = 49$, Summers and Drent 1979) and at Pine I. 98% ($n = 182$, Harfenist 1991). These figures are substantially higher than those estimated at Triangle I. over 4 years (32–62%, Vermeer and Cullen 1979).

PREDATION: several avian predators take adults at the breeding colony, including Bald Eagles (DeGange and Nelson 1982; Kaiser 1989), Great Horned Owls (Wilson and Manuwal 1986; Hayward et al. 1993), and Peregrine Falcon (Paine et al. 1990; Shibaev 1990e). At Lucy I., Bald Eagles converged on the colony at sunset to hunt auklets (Kaiser 1989), while at Helgesen I., eagles perched close to take off areas at dawn, possibly to intercept departing auklets (AJG). At Pine I. nestling auklets delayed fledging in areas with heavy eagle predation (Harfenist 1991). Peregrine predation can be a significant source of adult mortality: at Tatoosh I., Washington, they took at least 69 birds out of an estimated breeding population of 300 (Paine *et al.* 1990). Kleptoparasitism by gulls is common at some colonies: at Teuri I., Japan, Black-tailed and Slaty-backed Gulls are not active after dark during the incubation period of the auklets, but there was a peak of activity at 1 hr after sunset during chick-rearing, with gulls chasing auklets mainly on the ground. The risk of kleptoparasitism declined as it got darker, but more auklets hit artificial obstacles later in night, suggesting that risk of injury may penalize late arrivals (Watanuki 1990). Despite suggestions to the contrary, avoidance of predation and kleptoparasitism seems the most likely cause of nocturnal colony activity in Rhinoceros Auklets.

MOULT: timing and duration are very poorly known. Non-breeding plumage appears to be somewhat variable, especially the extent and tone of dark feathers on the underparts, but this variation may be partly because of overlap between body moults. Juvenile plumage fully developed at departure from nest and similar to first-winter plumage. Extent and timing of moult in first winter not known. A partial or complete body moult presumably occurs in spring of first year, but little evidence to suggest any substantial change in plumage. Few birds collected during May–Aug lack facial plumes, but practically all birds collected at that season were taken at colonies, which are presumably not visited by birds in their first summer. Birds without plumes collected in July, presumably first years, have pale bellies, characteristic of first-winter plumage. Some pre-breeders (without facial plumes) collected (net-drowned) in British Columbia in July and Aug were in heavy wing moult; these were probably undergoing first complete moult (Definitive Pre-basic). For breeders, wing moult is presumably initiated soon after chicks have fledged, in Sept for most populations. By Oct, practically all specimens are in full winter plumage; similar to first winter, but lacking barred appearance on belly. Wing moult is presumably rapid, lasting less than 6 weeks. Primary feathers are shed simultaneously; body moult may be extended, as suggested by great variation in belly plumage throughout year. The pre-breeding moult (Definitive Pre-alternate) is partial and occurs in Mar and Apr. In May, all appear to be in full breeding plumage.

Population dynamics

Probably do not breed before 3 years old. Retrapping of breeding individuals suggests that practically all breed annually. There is no information on survival rates or lifespan, although birds have been retrapped at more than 15 years old. Pairs often breed together in successive years, either in same burrow or nearby (Richardson 1961).

Atlantic Puffin or Puffin   *Fratercula arctica*

Common Puffin, Sea Parrot (Newfoundland)

PLATE 7

Fratercula arctica (Linnaeus, 1759, Syst. Nat.)

Description

A stout, medium-sized auk, with very deep, laterally compressed, and brightly coloured bill. Head large, rounded, neck short, and body plump, giving a rather satisfied appearance. Legs are stout, with relatively little flattening of tarsus, and stance is erect, standing on toes, rather than tarsus; runs well, with rather hunched posture. Flight is strong and direct and birds can land and take off from flat ground.

ADULT SUMMER: glossy black above, including wings and tail, but paler, grey-brown on crown, which is almost separated from mantle by a narrow grey collar extending round from rear of cheek; upperparts become dull and worn during course of breeding season; underparts white, except for black collar and greyish chin; face silvery white; bill very gaudy, blue at base and red at tip, these areas separated by a crescentic yellow band, upper mandible with 1–3 vertical grooves, divided from feathers of head by a fleshy cere running between rictal rosettes at the corners of the mouth; eye set in a triangular dark area of bare skin and surrounded by a red ring; irides black or brown; legs and feet bright orange. Parts of the bill decorations (rictal rosettes, etc.) lost as breeding is completed. Albinos or partial albinos are not infrequent.

ADULT WINTER: as for breeding, but facial disk dusky, darker in front of eye; bill smaller and dull brown, with a noticeable constriction at base, caused by loss of horny sheath from upper mandible; legs are orange-yellow.

JUVENILE AND FIRST WINTER: similar to winter adult, but bill black, smaller and more triangular in shape. Some first summer birds retain dark faces and do not develop red eye-rings. Second year birds in breeding season have markedly more triangular bills than older birds.

CHICK: covered in black down with a brownish tinge, but with a white patch on the belly; bill and legs dark grey.

VARIATION

Considerable clinal size variation, with smallest from British Isles and largest from Arctic. Three races generally recognized, based entirely on measurements, but clinal variation appears to be continuous between the southern *F. a. grabae* and intermediate *F. a. arctica*. The high Arctic *F. a. naumanni* has a highly disjunct range, with a small population in NW Greenland separated by several thousand kilometres (by sea) from other representatives in Jan Mayen (possibly, on the basis of measurements given by Camphuysen 1989) and Spitzbergen. Much of the variation in size seems to be related to temperature, although birds from Norway appear a little bigger than might be expected on this basis and those from Bear I. a little smaller (Fig. 3.3). There is a parallel cline in egg measurements. Moen (1991), using allozyme analysis, found little genetic variation among populations differing considerably in size (as found also among Thick-billed Murres). She suggested that the clinal size variation found in the species is related to diet during the chick growth period. Salomonsen (1944) commented on the occurrence of dwarf individuals, with measurements 10 per cent smaller than normal, among Arctic populations.

MEASUREMENTS

♂♂ slightly larger than females, especially in bill measurements. Wing length and bill depth, but not weight, increase with age: first years, wing 156.3, culmen 41.2, bill depth 30.9, weight 390 g; third years, wing 157.6, culmen 43.3, bill depth 34.5, weight 369 g (*F. a. grabae*, Isle of May, Harris 1981, compare with Table for breeders). Number of grooves on bill provides a rough guide to age. First summer birds have only a trace of a groove, second summer usually have 1, third summer mostly 1.5 or 2, fourth summers 1.5–3 (mostly 2 or 2+), and fifth summer and breeding birds 1.5–3, mostly 2–3 grooves (Harris 1981). For a given population, the sex of most breeders can be distinguished by a combination of bill length and depth (Corkhill 1972); see Tables, all breeding adults (1) *F. a. grabae* Isle of May, Scotland (Harris 1979); (2) *F. a. arctica* Røst, Norway, (Barrett *et al.* 1985); (3) Westmann Is., Iceland (Petersen 1976); (4) *F. a. naumanni* Spitzbergen (Vaurie 1965); (5) Spitzbergen (Lovenskiold 1954).

WEIGHTS

Weight of breeding birds declines during breeding by about 10% (Barrett *et al.* 1985).

Table: Atlantic Puffin measurements (mm)

ADULT ♂♂	ref	mean	s.d.	range	n
wing	(1)	162	3.2	—	29
	(2)	173	7.4	—	55
	(3)	164	5.7	—	9
	(4)	185	—	177–195	24
	(5)	184	3.4	178–188	8
culmen	(1)	46.5	1.4	—	30
	(2)	48.1	3.0	—	55
	(3)	45.7	2.4	—	9
	(4)	55.4	—	51–59	24
bill depth	(1)	37.1	1.1	—	30
	(2)	37.5	2.2	—	55
	(3)	37.7	1.2	—	9
	(4)	47.1	—	42–51	24
tarsus	(3)	26.5	0.9	—	9
	(5)	31.0	1.1	30–33	8

ADULT ♀♀	ref	mean	s.d.	range	n
wing	(1)	162	2.6	—	19
	(2)	171	x.x	—	xx
	(3)	162	3.5	—	12
	(4)	183	—	175–187	24
culmen	(1)	44.2	1.4	—	22
	(2)	45.9	x.x	—	xx
	(3)	43.1	2.1	—	12
	(4)	53.5	—	50–55	24
bill depth	(1)	34.6	0.9	—	22
	(2)	36.5	x.x	—	xx
	(3)	35.3	1.4	—	12
	(4)	44.0	—	40–47	24
tarsus	(3)	25.6	1.0	—	12
	(5)	29.7	1.5	28–31	3

This effect occurs irrespective of whether chicks are being provisioned at a high or low rate, suggesting that, in years of poor food supply, Puffins adjust the amount of effort expended on feeding their chicks, rather than altering their own body condition (Barrett and Rikardsen 1992). Isle of May, sexes combined, mean 387 g (Harris 1979); Skomer Island, Wales, ♂ 391 ± 22 g (323–450, n = 164); ♀ 361 ± 23 g (300–433, n = 246; Ashcroft 1976); Spitzbergen, ♂ 657 ± 34 g (620–710, n = 7; Lovenskiold 1954).

Range and status

Atlantic, boreal, low Arctic and high Arctic.

RANGE

BREEDING: on offshore islands from 43° (Gulf of Maine) to 79° N (Coburg Island) in the W Atlantic and from 50° (Brittany) to 80° N (Spitzbergen) in Europe. Occurs in the English Channel, around the W coast of Britain and Ireland, in the North Sea, Barents Sea, Norwegian and Greenland seas, Davis and Hudson Straits and Baffin Bay, Labrador Sea, Gulf of St Lawrence, waters off Newfoundland, and Nova Scotia, and the Gulf of Maine. Not usually seen close to shore except in the vicinity of breeding colonies. Abundant, except at extremes of its range.

WINTER: rarely seen, because majority of population is far offshore. Movements are poorly known, but the W Atlantic population remains in its summer range throughout the year, with a large proportion probably wintering on the Newfoundland Banks and smaller numbers as far S as George's Bank, S of Nova Scotia (Powers 1983). Birds from northern Norway, Iceland, and northern Britain also appear off Newfoundland in winter, especially in the first year. Most other European Puffins winter offshore, presumably scattered over large areas, including those beyond the continental shelf (Brown 1985). British Puffins are fairly well segregated in their wintering areas, with those from NE Britain wintering mainly in the North Sea and those from western Britain wintering in the Bay of Biscay and S to the Mediterranean (Harris 1984).

STATUS

The largest breeding populations are found in Europe in the British Isles, Iceland, and Norway, in E North America in Newfoundland and Labrador, and Gulf of St Lawrence. In the 1980s the world population was estimated as 7.6–16.4 million breeders, with a best estimate of 11.8 million (Nettleship and Evans 1985), the majority being distributed as follows: Iceland 51%, Norway 21%, British Isles

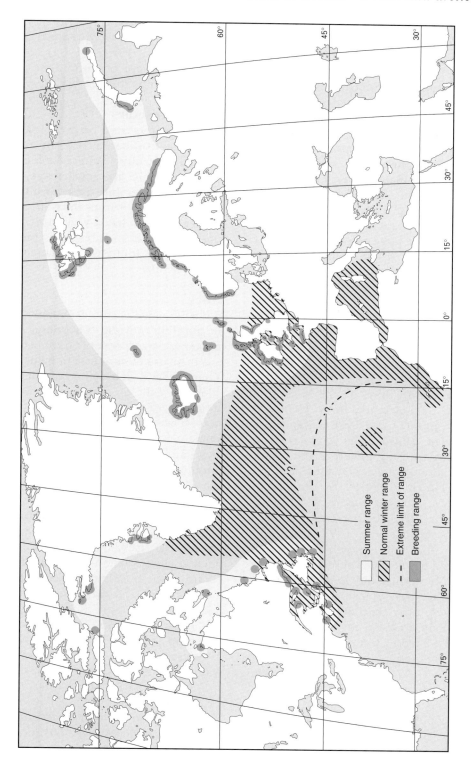

Summer range

Normal winter range

Extreme limit of range

Breeding range

12 %, Faeroes 9%, eastern Canada and New England 6 %. The wide range of uncertainty regarding the size of the population is due to lack of precision in estimates of the very large Icelandic population. Recent estimates for Iceland (A. Petersen, personal communication) suggest that the population there is 1–2 million birds; at the low end of previous estimates. The true figure for the total world population is probably between 5 and 10 million breeding birds. Populations have been severely affected during the last few centuries by human harvesting of birds and eggs, and by the introduction of rats to some important colony islands. The large colonies at Lundy I. and Ailsa Crag in Britain were reduced to remnants following the introduction of rats, while formerly very large colonies in Brittany, the Gulf of St Lawrence, and Gulf of Maine were decimated by over-harvesting. Over the past few decades, the Atlantic Puffin has been expanding in some areas, especially E Scotland, where numbers at Isle of May increased from 5 pairs in 1959 to 10 000 in 1982 (Harris 1984); those on the N shore of the Gulf of St Lawrence from 15 000 to 35 000 between 1977 and 1988 (Chapdelaine and Brousseau 1991). The population of St Kilda decreased substantially in the twentieth century, although the scale of the decline is uncertain (Harris 1984); the population appears to have declined somewhat during the period prior to 1970, before experiencing a modest increase between 1977 and 1987. Suggestions that the population of SE Newfoundland has declined substantially since the 1960s (Nettleship and Evans 1985) are based on rather slender evidence. More recently, the very poor reproductive success at the huge colony on Røst, in the Lofoten Is., during 1969–82 led to a steady decline in the Norwegian population, by 14% per annum during 1983–7 (Anker-Nilssen and Rostad 1993) and breeding failures were also recorded in the Shetland Is. in the late 1980s. Overall, the British population remained virtually unchanged from 1969–70 to 1985–7 (Lloyd *et al.* 1991). There is no information concerning

trends in the Icelandic population, hence the overall state of the world population is impossible to assess. Harris (1984) concluded '... the general state of Puffindom is far better than at any time this century.'

Field characters
In summer the large brightly coloured bill is unmistakable. There is no range overlap with the superficially similar-looking Horned Puffin. In winter the smaller size, dusky face, short, triangular bill, large head, and short neck distinguish it from the murres; the much smaller Dovekie has a very short, stubby bill. Juvenile and first winter birds can be distinguished from adults with practice by shorter, shallower bills.

Voice
Not a big range of sounds; mainly low guttural or gurgling noises. 'Growling', 'grunting', 'moaning', and 'creaking' have been described and their use in various situations categorized (Cramp 1985). It may be more useful to consider the calls as constituting a continuum of noises of varying intensity, without any sterotyped relationship to particular behaviours. Adults make a soft clicking sound as they enter burrows; a low groaning sound is commonly given inside the burrow. Chick call, a low 'peep-peep'.

Habitat, food, and feeding behaviour
MARINE HABITAT
During the breeding season occurs in continental shelf waters and mainly feeds within 10 km of the breeding colony, although some in Newfoundland have been seen carrying fish from 70 km away (Schneider *et al.* 1990a) and under conditions of extreme food scarcity, from up to 137 km (Anker-Nilssen and Lorentsen 1990). In winter, rarely occurs within sight of land, except when storm driven.

FEEDING BEHAVIOUR
Will dive down to 60 m (Piatt and Nettleship 1985), but perhaps more commonly to shallower depths. Ten birds in Newfoundland,

fitted with MDRs, reached maxima of 40–68 m over periods of up to 17 days; presumably their average dives were much shallower (Burger and Simpson 1986), while six birds in Norway reached 10–45 m (Barrett and Furness 1990). Radio-tracked Puffins feeding near Isle of May, Scotland, submerged for less than 39 sec in 80 per cent of dives (suggests a maximum depth of less than 30 m), with a mean dive duration of 28 sec and a maximum of 115 sec. One bird made 194 dives in 84 min, with an average of only 3 sec between each dive (Wanless *et al.* 1988). During dives, several fish may be taken and either swallowed (Boag and Alexander 1986), or held in the bill, the first fish being clamped against the upper mandible with the tongue while the second is captured.

DIET

Chick diet well known from food collected at the colony, but only scanty data are available on adult diet.

ADULTS: probably contains a substantial proportion of zooplankton, especially in winter. Adults wintering off the Faero Is. contained 85 % by volume euphausiids, while a sample taken in the Norwegian Sea at 66° N contained more or less equal proportions of squid and lanternfish (Falk *et al.* 1992). Zooplankton in the form of euphausiids, mysids, and copepods, is an important component of the adult diet at Jan Mayen and Spitzbergen in summer (Bird and Bird 1935; Hartley and Fisher 1936), although all of 13 birds collected by Lydersen *et al.* (1985) at Hornsund contained mainly juvenile Arctic cod.

CHICKS: in waters around Britain and S Norway, young are fed mainly on sprats, herrings, and sandlance, with juvenile whiting and haddock (*Melanogrammus*) occasionally important (Harris and Hislop 1978). In sub-Arctic waters, around Newfoundland, Labrador, and the Barents Sea, capelin is the main prey delivered, with some sandlance, juvenile cod, and

occasionally squid (Brown and Nettleship 1984; Barrett *et al.* 1985; Barrett and Furness 1990). All of the fish form dense schools in mid-water. Off SE Newfoundland, puffins feed in similar areas to Common Murres, but usually at lower densities, apparently making use of smaller, or less dense, capelin schools than the murres, perhaps because of competition from the murres at denser schools (Piatt 1990). Diet of chicks at high Arctic colonies is not known, but the only likely schooling fish is Arctic cod, which was found in 7 out of 10 adult stomachs examined at Spitzbergen (Hartley and Fisher 1936).

Displays and breeding behaviour

Account mainly from Myrberget (1962), Nettleship (1972), Ashcroft (1976, 1979), Harris (1984), and Taylor (1993).

COLONY ATTENDANCE: birds begin to visit their breeding colonies in Mar–May, occasionally in Feb in E Scotland, and laying commences in Apr–June, depending on latitude. In E Scotland, arrival has become earlier since 1970 than in previous decades (Harris 1984). Attendance during pre-laying period varies greatly from day to day, with peaks of numbers alternating with days when few or no birds are present. This variation is affected by weather, with smaller numbers present on stormy days, but is also influenced by a cyclical pattern independent of weather. Numbers at Great I., Newfoundland fluctuated over 5–6 days (Nettleship 1972), at Skomer and Isle of May 4–7 days (Ashcroft 1976; Harris 1984), and at Lovunden, Norway 4–10 days (Myrberget 1959). Attendance at Skomer and Skokholm Is., 5 km apart, was highly synchronized. Highest numbers of adults are present at the colony during the afternoon and evening. Birds waiting to come ashore assemble in rafts just offshore from mid-day onwards. Second year birds come ashore only during chick-rearing period, and third years from about middle of incubation; older birds are present from start of laying (Harris 1983*a*).

AGONISTIC AND SEXUAL BEHAVIOUR: breeders in pre-laying period and non-breeders throughout season spend a substantial proportion of the day on the ground near their burrows and social interactions are very frequent. The most obvious courtship activity is 'billing' (Taylor, in Harris 1984). The initiating bird approaches another with the body held low and horizontal ('low profile walk'), swinging its bill from side to side. The birds then knock their bills together sideways for up to a minute, making a noise that may attract other birds. Birds arriving without fish in the vicinity of others give a landing display, in which the wings are raised and the body bent forward, but with the head held high. Some landing displays are followed by 'skypointing', holding the body erect and pointing the bill vertically upwards (Danchin 1983). A common response by burrow-owners nearby is the 'pelican walk', in which the bird pulls itself erect with the beak pointed down at the breast and struts with the feet raised high at each step. This appears to signal burrow ownership (Taylor 1984). 'Head flicking' or 'jerking' is a characteristic behaviour seen both on the sea, among rafting birds close to the colony, and on land, especially among loafing birds late in the season (Hudson 1979b). ♂♂ perform head flicking persistently at a rate of 1–2 flicks/sec for several minutes at a time, while swimming behind ♀♀. These bouts are sometimes followed by copulation attempts, although most mating occurs at sea. Copulations on land are usually unsuccessful. Consequently, we must assume that pairs recognize one another and associate together on the sea. Head flicking on land is most common just before dusk and may signal readiness to depart for the night (Taylor 1984).

BEHAVIOUR OF PRE-BREEDERS: pre-breeding birds visit colonies from the first year (rarely seen on land, but more frequent in rafts off the colony) onwards, generally arriving later than breeders. Wheeling flights over colony are a common feature during late incubation and chick-rearing periods and many of those involved are second and third year birds. Birds fly against the wind over the colony, then return downwind over the sea (this is also true for murres). On large colonies there may be separate wheels for different sub-areas; Taylor (1993) recorded 21 on Dun, St Kilda, which were in the same place in successive years. Wheels may be 'stacked' above each other and may rotate in opposite directions. Incoming breeders may join the wheel for a few circuits before landing, adopting a trajectory over land that takes them over their burrow. This behaviour may function as a reconnaissance (Taylor 1984). Pre-breeding birds gradually acquire confidence as they visit the colony. Two-year-olds rarely come ashore and those that do so do not usually enter burrows. Three-year-olds inspect burrows, but do not occupy them for long; they also form 'clubs' (social aggregations of non-breeders), usually on boulders or other raised areas. Many fourth years occupy burrows for prolonged periods, and a few breed. Second and third years are more likely than older birds to visit different colonies, especially in July (Scotland). In Maine, second year birds rarely attended the colony more than 2 days consecutively, but third years frequently came ashore on several days in succession (Kress and Nettleship 1988).

Breeding and life cycle

Information mainly from Lockley (1953), Myrberget (1962), Nettleship (1972), Ashcroft (1979), and Harris (1984).

BREEDING HABITAT AND NEST DISPERSION: breeds colonially on turf-covered offshore islands, at densities of up to 3 burrows/m², normally in open grassland on slopes facing the sea. At Great I., Newfoundland, burrowing was densest on slopes steeper than 15° (2.1 burrows/m²), and less than 1 burrow/m² on slopes of <10°.

NEST: in burrows, usually 70–110 cm long, occasionally up to 15 m. In dense areas, burrows may intersect, or be multi-storied. In Iceland burrows are placed close to large boulders, apparently to provide landmarks to arriving birds (Grant and Nettleship 1971). Some birds nest in rock crevices: this may be the typical site at Arctic colonies, where soil does not develop on offshore islands. The sexes play an equal role in most breeding activities. Digging is performed with bill and feet and excavated soil is expelled from the burrow, forming a fresh heap at the entrance. A substantial amount of nest material (grass, leaves, plant stalks, and feathers) is placed in the nest chamber. Snow and soil moisture may affect access to burrows early in the season and hence alter timing of breeding (Skokova 1967; Hornung and Harris 1976). At Skomer I., Wales, rarely, if ever, dig own burrows, relying on Rabbits (*Oryctolagus cuniculus*) to perform this function for them. This is despite an apparent shortage of burrows at that colony. Rabbits have been present in Britain only since the twelfth century, and on Skomer only since the seventeenth century; hence the loss of the burrow-digging habit must be fairly recent. Competition for burrows with Manx Shearwaters (*Puffinus puffinus*) was demonstrated at Skomer I. by Ashcroft (1976). Shearwaters generally began to occupy burrows earlier in the season than Puffins and were able to exclude them. Some segregation in breeding areas helped to reduce competition, with most Puffins nesting on the periphery of the island, and most shearwaters in the interior. However, the presence of shear-waters may have limited the availability of burrows for Puffins.

EGG-LAYING: in Britain, first eggs usually in Apr; in Newfoundland and Norway, May. Median laying at Great I., Newfoundland was 17–20 May (3 seasons). At St Kilda, mean laying occurred between 10 and 25 June (4 years, Harris 1980). A replacement egg may be laid about 2 weeks after the loss of a first egg (Ashcroft 1979).

EGGS: one, ovoid-elliptical, dirty white, sometimes with faint bluish or purplish spots or smudges, occasionally with darker marks, often stained by peat or soil. Occasionally two eggs are found in the same burrow, probably because of desertion of the first egg, followed by relaying. Only one egg ever hatches. Composition of fresh egg: shell 7%, yolk 33%, albumen 57%. Pipping weight 87% of fresh weight, of which the neonate chick constitutes 60% and the remaining yolk 19% (Birkhead and Nettleship 1984*b*). Egg weight is approximately 15% of adult body weight. For measurements, see Table: (1) *F. a. naumanni*, Greenland (BMNH); (2) *F. a. arctica*, North America (Nettleship 1972); (3) North America (Bent 1919); (4) Iceland (BMNH); (5) Lovunden, Norway (Myrberget 1962), (6) N Norway (BMNH) (7) *F. a. grabae* Irish Sea (BMNH); (8) British Isles (Witherby *et al.* 1941).

Table: Atlantic Puffin egg dimensions

ref	length (mm)			breadth (mm)			fresh weight (g)	n
	mean	s.d.	range	mean	s.d.	range		
(1)	65.8	±2.1	63.1–68.5	44.2	±1.1	43.6–45.5		5
(2)	62.9	±2.1		44.5	±1.1			150
(3)	63.0		58–67	44.2		41.5–47.0	65.0	41
(4)	64.7	±2.1	62.3–69.0	44.1	±2.2	40.5–48.4		8
(5)	64.0	±2.4		44.1	±1.5			77
(6)	66.0	±2.8	60.9–69.9	44.8	±0.7	43.6–45.9		7
(7)	60.4	±1.9	54.8–63.6	42.1	±1.1	39.6–44.1		20
(8)	60.8		55.6–66.5	42.3		38.9–45.3	59.0	100

INCUBATION: takes 39 days (range 36–43), with the period from first starring of the egg to hatching taking about 4 days (Myrberget 1962; Ashcroft 1976). May be irregular for the first few days, and occasional, brief periods of neglect continue throughout. At Lovunden, Norway, incubation shifts varied from 23 to 70 hrs (mean 32), with ♀♀ spending more time than ♂♂; most change-overs occurred at night (Myrberget 1962). Breeders develop two lateral brood patches, and older pre-breeders also develop brood patches.

HATCHING AND CHICK-REARING: the nestling period varies from 34 to more than 70 days, but most chicks leave within 50 days of hatching (mean durations 38–53 days, six studies, Harris 1984). Generally, slower growing chicks take longer to fledge than fast-growing chicks (Nettleship 1972; Harris 1980; Anker-Nilssen 1987). Food is brought to chicks in daylight, usually with a peak of deliveries in early morning. Chicks are fed one, or more usually several (up to 62), fish at each delivery. Deliveries vary from 0.2 to 29 g and the size of individual fish may be as much as 25 g, but is usually in the range 1–10 g. Rate of feeding normally peaks at 20–30 days old, falling to less than half of peak rate by the time the chick leaves, but contrary to some statements, the chicks are rarely deserted by their parents before fledging (Harris 1984). At Skomer, the peak rate was 10, and at Isle of May, 12 deliveries/day, with individual chicks receiving as many as 24 deliveries/day. Elsewhere, recorded feeding rates were generally lower; at Great I. from 2.4 meals/day on flat areas to 3.6 meals/day on sloping habitat, and at Lovunden 5.2 meals/day. The food delivered to chicks at Hornøy, Norway, where reproductive success was high and chicks reached normal growth rates, varied from 40 to 106 g/day, equivalent to 284–670 kJ/day energy. At colonies in the southern Lofoten Is., where food shortages have reduced reproductive success over many years, chicks received only 14–74 g daily, equivalent to 57–474 kJ energy. ♀♀ made more feeding

trips than ♂♂ (57% vs 43%) at Witless Bay, Newfoundland (Creelman and Storey 1991). At Skomer, weight increases among chicks tended to be synchronized, suggesting that feeding conditions varied considerably from day to day. In addition, weight increases were more synchronized among chicks in the same part of the colony than among those in different areas and arrivals of parents with food within a given area tended to be clumped in time (Ashcroft 1976). These observations suggested that local breeding groups tended to forage together and hence experienced similar feeding success. Synchrony of arrivals at the colony may help to swamp attacks by klepto-parasites. Birds departing to feed frequently did so from the rafts of birds socializing near the colony and these rafts appear to be formed of birds from the same part of the colony (Hudson 1979b). Cross-fostering experiments with chicks suggest that the rate of feeding changes on a fixed parental schedule, and is not affected by replacing the chick with one of older or younger age (Myrberget 1962; Hudson 1979b). This result suggests that the change in rate of feeding is a programmed pattern of behaviour adapted to the long-term needs of chicks and adults, rather than the result of changes in the availability of food over the course of the season. Moreover, even if fed unlimited food, chicks reduce their intake towards the end of the nestling period, presumably because, with growth almost complete, their energy needs are lower (Tschanz 1979). Although parents do not adjust feeding to the age of their chicks, they will deliver more in response to increased begging by the chick (Harris 1983b). Surprisingly, chick weight increases fluctuate from day to day even when fed a constant amount of food (Hudson 1983).

CHICK GROWTH AND DEVELOPMENT: chicks hatch at just under 10% of adult weight. Peak weight is reached at 31–43 days; earlier at colonies where chicks grow rapidly, after which there is a recession of up to 20% from the peak by the time the chick departs (Hudson 1979b). Wing feather pins appear at about 10 days, and contour feathers begin to lengthen a few days later; primaries burst their sheaths about 15 days and down begins to be shed from 21 days (Skokova 1990). The development of feathers and skeletal elements is affected by nutrition, with the growth of both wing and culmen length being slower in a year of reduced feeding rates (Barrett and Rikardsen 1992). Under good conditions, chicks reach a peak weight of 70–80% of adult weight after 30 days and decline thereafter to fledge at 60–70%. Both peak and fledging weights vary considerably among colonies and among years at the same colony. Generally, chicks at large colonies are lighter than those at small colonies. Chicks at Great I., Newfoundland were 56% and those at St Kilda 63–67% of adult weight at departure, both colonies having more than 100 000 pairs, whereas those at smaller colonies on Funk I., Newfoundland, Skomer I., and Lovunden and Hornøy, Norway fledged at 67–78% of adult weight. At Great I., chicks being fed by a single parent did not survive to fledge, whereas at Skomer, one parent succeeded in rearing a chick, albeit lighter than chicks with two parents (Ashcroft 1979), while two pairs reared artificial twins (Corkhill 1973). Chick behaviour in the burrow has been described by Boag and Alexander (1986); the only burrow-nesting auk for which this information is available. Chicks are active from soon after hatching and spend about 25% of their time in stretching, wing-flapping and performing various scraping and pecking exercises, including manipulating pebbles and roots with the bill. Older chicks wander into the tunnel, but retreat to the nest chamber at the sound of adults taking off; a possible sign that a predator is nearby. Chicks give low 'peep-peep' calls on the arrival of a parent, which they seem to anticipate by several seconds. After entering, the incoming adult replies with a deep growl. It may also give a series of soft 'clicks'. The chick seizes the bill of the parent, apparently to stimulate it to drop the fish it is carrying, as the chick makes no attempt to seize the fish while in the bill. They are eaten from the ground, smaller ones some-times tail-first. After feeding, the chick normally sleeps for some time. Chicks defecate in latrine areas in the burrow, but sometimes approach the burrow entrance in daylight, especially when poorly fed, and may be taken by gulls or corvids at such times. Starving chicks are prone to leave the burrow and wander on the surface, where they are very vulnerable to predation. Hence poor nutrition may also lead to increased predation.

KLEPTOPARASITISM: of adults bringing food to young is common and may be practised by large gulls, especially Herring Gulls, Jackdaws, and Parasitic Jaegers. Highest rates of piracy and losses were recorded at Great I., Newfoundland, where 30% of Puffins were attacked and 10% lost their loads to Herring Gulls, losses being greater on flat than on sloping areas (Nettleship 1972). Other studies at Great I. reported lower rates of piracy, especially in years when Puffin feeding rates were low, causing gulls to switch their attentions elsewhere (Pierotti 1983; Rice 1985). In Iceland, Parasitic Jaegers stole 4% of food loads, while at Skomer Herring Gulls took a maximum of 7% of loads of any one pair and Jackdaws took 1% (Corkhill 1973; Hudson 1979b).

CHICK DEPARTURE: occurs from 34 days. Mean age at Isle of May, Scotland, was 41 days (34–50, $n = 312$) and at St Kilda, 44 days (35–57, $n = 214$, Harris 1984). At departure, chick wing length is only 85% of adults', they usually still have some down attached, and they are capable of only very limited flight. Fledging age is inversely related to growth rate, so that slow growing chicks are older when they leave the colony. This leads to considerable inter-colony variation in mean fledging age (Ydenberg et al. 1995). Young birds generally depart after dark and reach the sea by flying, or scrambling and fluttering along the ground (Myrberget 1962). During departure they may be attracted to artificial lights on ships or buildings, leading to some mortality.

BREEDING SUCCESS: can be affected by burrow inspections during the incubation period, or sometimes by human activities in the vicinity of the colony (D. K. Cairns and W. A. Montevecchi, unpublished data), so exact measurements of reproductive success are hard to obtain. The proportion of eggs that hatch is generally about 75%, with most losses attributed to competition from other Puffins or shearwaters. Predation of eggs from shallow burrows by Jackdaws was observed at Skomer. At Skomer, Ashcroft (1978) estimated that chicks fledged from 73% of 'undisturbed' burrows where eggs were laid, or 64% of occupied burrows. Most losses occurred during incubation, many the result of disturbance by Manx Shearwaters competing for the same burrows. Most other estimates of reproductive success are between 60 and 90% (18 colony years, Harris 1984), with varying levels of disturbance. Reproductive success varies within colonies. At Great I., pairs breeding in burrows in densely settled, steeply sloping areas facing the sea had higher success (28–51%) than those in flatter, interior areas (10–24%), because of kleptoparasitism by Herring Gulls (Nettleship 1972). At St Kilda, burrow density was not related to slope, but pairs in sparse areas had lower success than those in dense areas, which also laid 2–5 days earlier (Harris 1980). However, densities in dense areas at St Kilda were lower than those in sparse areas at Great I. At Skomer, neither physical characteristics of the burrows and their surroundings, nor burrow density, had any effect on reproduction. At all colonies, early laying pairs tended to be the most successful. At colonies in central Norway, especially the very large one at Røst, many years of complete reproductive failure occurred from 1970 onwards (Tschanz 1979, Lid 1981). Years of poor reproductive success have also occurred in Newfoundland (Brown and Nettleship 1984) and, to a lesser extent at St Kilda (Harris 1980) and complete failure was observed at Hermaness and Foula, Shetland in 1987 and 1988 (Heubeck 1989). It may be significant that all the colonies involved were very large (>50 000 pairs).

MOULT: compared with murres and guillemots, primary moult appears to be relatively unsynchronized within the population. Feathers of the head are moulted soon after the end of breeding, along with the colourful plates on the bill, producing the grey winter cast. Primary feathers are shed simultaneously, beginning in Dec in Britain, or Jan further north. Duration of primary moult for individual birds is not known. Secondaries are also shed simultaneously, beginning after the primaries, and many rectrices are lost at the same time. Body moult begins once the primaries have started regrowing; underwing coverts are shed last (Harris and Yule 1977). Growth of the bill plates and body moult are completed by the time the primaries are fully grown. Juveniles may moult in the first summer, or in the following autumn. Immatures tend to moult their primaries later than breeders, so that their feathers are less worn when they appear at the colony in mid-season (Harris and Yule 1977).

Population dynamics

SURVIVAL: has been estimated from the rate of return of marked breeders as 96% at Isle of May, Scotland and 95% at Skomer, Wales (Ashcroft 1979; Harris 1983a). Mead (1974), using corrected ring recoveries, estimated overall survival of British Puffins as 94.5%. An estimate based on resightings of birds trapped, but not necessarily breeding, at Skomer was 89% (Hudson 1979b). Some breeding birds are killed at the colony by Great Black-backed Gulls (*Larus marinus*) (<2% of breeders, Harris 1980), Sea Eagles (*Haliaetus albicilla*), Peregrine Falcons, and Gyrfalcons (Myrberget 1959; Skokova 1967). The gulls took a higher proportion of non-breeders than breeders. The survival of young birds to age at first breeding was estimated at between 30 and 40%, with the higher figure taking into account those birds that emigrated to other colonies (Harris 1983a). Major causes of mortality away from the colony are oiling and drowning in fishing nets. The peak of adult recoveries in Britain coincides with the moult period (Feb–Apr) and suggests that this may be a period of stress for the Puffins. Recoveries of first years, which moult in summer, peak in Dec–Feb (Harris 1983a).

AGE AT FIRST BREEDING: a few birds breed at 3 years, but average is approximately 6 years (Harris 1984; Hudson 1985). Birds transported as chicks to colonies in the Gulf of Maine and reared to fledging in artificial burrows mostly returned at 2 years, and first bred at 4.2 years

(Eastern Egg Rock) and 6.1 years (Matinicus Rock, Kress and Nettleship 1988).

PAIR- AND NEST-SITE FIDELITY: At Skomer, 8% of breeders shifted burrows from one year to the next, the proportion being higher among those that failed in the first year, and many changes being caused by disturbance from the larger Manx Shearwaters. In addition, 8% of pairs where both members were alive the next year took new partners (Ashcroft 1979).

Horned Puffin *Fratercula corniculata*

Mormon corniculata (Naumann, 1821, *Isis* ix, pl. vii, figs 3–4, p 782—Kamschatka).

PLATE 7

Description

A stout, medium sized auk with a large head, massive, laterally compressed bill, and short neck. Confident, upright posture, standing on toes alone. Flight is direct, with rapid, shallow wing-beats; take off from the water after a short taxi. Often fly high above the sea (>30 m), unlike other auks, as if spotting surface features. Superficially very similar to the Atlantic Puffin, but bill is much larger (except for *F. a. naumanni*) and sharply bicoloured yellow and red.

ADULT SUMMER: back, upperwings, and tail, black, tinged with brown and distinctly greyish on crown and nape; underparts, white, but chin and throat black, dividing to join black collar on neck; head somewhat triangular from above, black, with white facial disks bisected by a narrow black line behind the eye. Face gives almost owl-like effect when seen head-on; bill triangular to onion-shaped in profile, yellow with red tip, with a variable number of vertical grooves (usually 3–6) on the upper mandible, the division between red and yellow coming at the innermost; bright orange rosette at rear of gape, and narrow, fleshy, red ring around eye, with small wattles of bare, black skin over eyes (up to 12 mm) that project ver-

tically to form the eponymous 'horns'; smaller black flaps also project below the eye; irides, grey-brown; legs, toes, and webs bright orange except back of tarsus sooty brown. As in other puffins, the inner toe is turned inwards and more sharply hooked than the others.

ADULT WINTER: as in summer, but bill ornaments (gape rosettes, and plate at base of upper mandible), absent. Bill is smaller, dull brownish horn at the base and reddish at the tip; eye-ring brown, legs pale flesh; facial disks are less well-defined, grading from greyish brown in front to silvery grey behind.

JUVENILE AND FIRST WINTER: similar to winter-plumage adult, but with shorter, all brown bill, and more sooty colour on the cheek patch; black eye-ring. Young birds in summer plumage show less developed nuptial bill ornamentation than breeders.

CHICK: down completely sooty brown, except for pale patch in centre of belly; bill black, legs dark brown.

VARIATION

No subspecies recognized. Considering the striking size variation found in the Atlantic

Puffin in relation to sea temperature, and the great geographical spread of the Horned Puffin, we might expect some differentiation. The lack of variation suggests that much of its range has been recently colonized after the Pleistocene.

MEASUREMENTS
♂♂ slightly larger than ♀♀ in wing, culmen length, and tarsus. See Table: (1) Russian Far East (Kharitonov 1990a); (2) St Lawrence I. (Sealy 1973c); (3) Cape Thompson, Alaska (Swartz 1966); (4) Alaska Peninsula, summer (J. Piatt, unpublished).

WEIGHTS
♂♂ tend to be slightly heavier than females and those collected in the Gulf of Alaska somewhat lighter than those from the Bering

Table: Horned Puffin measurements (mm)

♂♂	ref	mean	s.d.	range	n
wing	(1)	191.7		185–200	14
	(2)	190.2	6.9	183–200	6
	(3)	199.6	10.8	194–205	7
culmen	(1)	50.6		47.1–55.3	14
	(2)	51.6	2.8	47.3–55.9	8
	(3)	50.5	5.3	46.0–52.7	7
tarsus	(1)	30.9		29.0–33.2	14
	(2)	31.7	3.7	27.0–37.3	5
	(3)	32.4	6.9	27.5–35.6	7

♀♀	ref	mean	s.d.	range	n
wing	(1)	185.7		176–201	12
	(2)	187.4	3.5	182–193	8
	(3)	197.0		182–222	10
culmen	(1)	47.9		45.9–50.5	12
	(2)	50.8	2.1	47.3–53.4	9
	(3)	48.4	6.6	45.2–50.8	10
tarsus	(1)	29.2		26.9–31.8	12
	(2)	29.1	3.6	25.4–34.4	7
	(3)	31.4	4.7	29.0–34.1	10

UNSEXED	ref	mean	s.d.	range	n
wing	(4)	188.7	6.1		44
culmen	(4)	48.4	2.7		44
tarsus	(4)	30.7	1.3		38

Table: Horned Puffin weights (g)

♂♂	ref	mean	s.d.	range	n
	(1)	609	54	531–754	13
	(2)	648	78	610–669	7
	(3)	629	54	553–680	7

♀♀	ref	mean	s.d.	range	n
	(1)	589	54	499–691	9
	(2)	618	134	507–674	8
	(3)	581	48	500–636	11

UNSEXED	ref	mean	s.d.	range	n
	(4)	532	51		157

and Chukchi seas. Weight declines during breeding (Sealy 1973c). See Table: (1) St Lawrence I. (Sealy 1973c); (2) Cape Thompson, Alaska (Swartz 1966); (3) Nunivak I., Alaska (Sealy 1973c); (4) Alaska Peninsula, summer (J. Piatt, unpublished).

Range and status
Widespread in the Pacific; uncommon in boreal, but numerous in low Arctic waters, especially S of the Alaska Peninsula and throughout the Aleutian Is.

RANGE
BREEDING: on the coast of North America from British Columbia (rare and local summer visitor, breeding occasionally on Triangle I. and the Kerouard Is.), or possibly from Oregon (Pitman and Graybill 1985), through SE Alaska (Forrester I.), Kenai Peninsula, Barren Is., Kodiak I., Semidi and Shumagin Is., Alaska Peninsular, and in the Aleutian Is. W to the Rat Is. There are colonies in the Pribilof Is., St Mathew and Hall I., St Lawrence Is., and along the mainland coast of Seward Peninsula, Cape Thompson, and Cape Lisburne. A few pairs have attempted to breed at Barrow, on the N coast of Alaska (Divoky 1982). Also breeds on Wrangle I., and on N side of Chukchi Peninsula, as far W as Kolyuchin Bay and on

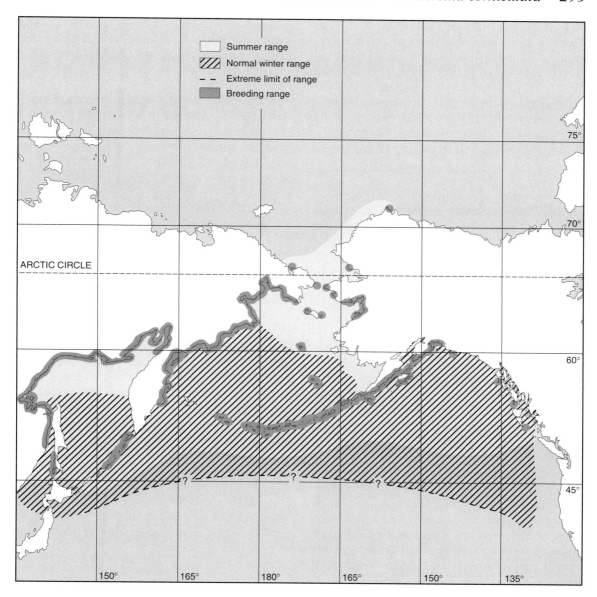

the entire Russian coast of the Bering Sea S to Kronotsky Bay, the Commander Is., the N and central Kuril Is., Penzhinskaya Gulf, the N coast of the Sea of Okhotsk, Talan I., northern Primorye, and parts of Sakhalin I. (Kharitonov 1990*a*). In some years 'hundreds' are seen off California in May and June, thought to be return migrants, but possibly summering non-breeders (Pitman and Graybill 1985).

WINTER: like other puffins, winters offshore, widespread in the N Pacific. Some birds remain within the breeding range, except where forced to do so by ice cover (Chukchi and N Bering Sea). Common in winter in the Kurils and off Sakhalin, Kamchatka, and the Commander Is. Uncommon off British Columbia and off Japan S to central Honshu, and rarely seen in winter off the W coast of

North America S to California and Baja California. However, commoner far offshore, being abundant between 185 and 260 km off San Francisco in Feb (Pitman and Graybill 1985). Occasional wrecks are recorded, including hundreds of birds washed ashore in Oregon in winter 1932–3 (Gilligan *et al.* 1993) and large numbers in the Queen Charlotte Is. in the 1940s (Sealy and Nelson 1973).

STATUS

Total population estimated at 1.2 million breeders, with the majority breeding on islands off the Alaska Peninsula (62%), in the Aleutian Is. (8%), the Bering Sea (8%), and the Sea of Okhotsk (16%, Byrd *et al.* 1993). The largest colony in Asia is at Talan I., Sea of Okhotsk (100 000–120 000 breeding birds). In Alaska, 17 colonies are thought to have 10 000 or more breeders. The largest are in the Semidi Is. which support 350 000 breeders, of which Suklik I. alone has 250 000; a further 160 000 breed in the Shumagin Is., and 140 000 on Amagat I. W of the Alaska Peninsula. Interestingly, much less numerous than Tufted Puffin in the area of the E Aleutians where that species is very abundant.

Field characters

A medium-sized, black-and-white auk, with a large, brightly coloured bill at all seasons. Unmistakable in breeding plumage, but distant birds in flight could be confused with Rhinoceros Auklet or murres. Compared with murres, this species has a much heavier bill that is visible even at long range and has rounder wings and a slower wing-beat that may be useful for speculating on the identity of specks on the horizon. Compared with Rhinoceros Auklets, Horned Puffins have much stronger contrast in their black-and-white plumage. The former have grey upperparts and always have a dirty grey upper breast and flanks, as well as a relatively narrower, more pointed bill.

Voice

Not very vocal, with a simple repertoire of low pitched, growling and groaning calls. A loud 'a-gaa-kah-kha-kha' may be given during aggressive displays. During head-flicking display a rhythmic 'op-op-op-op ...' is sometimes given (Kharitonov 1990*a*).

Habitat, food, and feeding behaviour
MARINE HABITAT
An offshore species throughout most of the year, but forages close to colonies while breeding (Sealy 1973*c*; Wehle 1982). In the Chirikov Basin of the northern Bering Sea, found most commonly in Bering Shelf water, highly stratified waters of intermediate temperature and salinity (Elphick and Hunt 1993).

DIET
Poorly known for adults, but good information available for chicks at a limited number of sites.

ADULTS: apparently rather varied, in summer including fish, squid, and inverebrates. Wehle (1982) collected birds returning to two breeding colonies in the Aleutians and found they contained mainly hard parts. At Buldir I., among breeders ($n = 41$), 27% contained fish remains, 88% squid, and 7% polychaete worms; subadults (presumed non-breeders, $n = 6$), 50% contained fish, 67% squid, and 33% polychaetes. At Ugaiushak, among breeders ($n = 12$) 42% contained fish, 8% polychaetes, and 8% crustacea; no squid were evident. These results could be substantially biased by differential digestion (Chapter 4). Winter diet not known.

CHICKS: almost entirely fish. In three seasons at the Semidi Is. and Sandman Reefs, Gulf of Alaska, chicks were fed 85% Pacific sandlance by weight, 6% greenling (*Hexagrammus* spp.), and 4% capelin ($n = 619$ samples, 4070 fish, Hatch and Sanger 1992). Prey other than fish constituted <1% by weight and included euphausiids and squid. The mean length of sandlance delivered was 70 mm (25–164, $n = 3746$); practically all fell within the 0 age-class. Other fish were within the same size range. At Buldir I. chick diet was predominantly Atka mackerel

(*Pleurogrammus monopterygius*), with some sand-lance, walleye pollock, and flatfish (ILJ).

Displays and breeding behaviour

From Kharitonov (1990a). Behaviour apparently similar to Atlantic Puffin. Attendance at the colony, and on the sea nearby is very variable throughout the breeding season, with 1–3 days of high numbers alternating with 1–2 days with very few present.

AGONISTIC BEHAVIOUR: pairs defend burrows and ♂♂ defend their mate, with a variety of threat displays: in 'forward threat' posture, the body is kept rather horizontal, with head low and neck stretched out. Escalated aggression includes chases, bill-grappling, and seizing feathers of head and nape.

SEXUAL BEHAVIOUR: seen on club aggregations, on flat boulders, especially in early morning and evening, and also adjacent to burrows. Head-flicking, both with bill closed and with it gaped at each flick, occurs on land and water; also billing display in which pair stand facing, with bills side by side, and waggle their heads, while opening and closing their bills; mutual bowing also occurs. Prior to mating, ♂ may perform swimming display on sea, in which fore part of body is raised from the water, neck extended upwards and bill held horizontal; this is followed by head-flicking and then immediately by mounting. Copulation almost entirely on water, but occasionally on land.

Breeding and life cycle

Rather poorly known compared with most other northern auks. Description taken mainly from Sealy (1973c), Baird and Gould (1983), and Kharitonov (1990a). Birds arrive at breeding colonies in Apr in the Kuril Is., in early May at Moneron I., and in late May, or early June at St Lawrence I. and Cape Thompson, Alaska.

BREEDING HABITAT AND NEST DISPERSION: on steep slopes or cliffs, sometimes among talus;

at Talan I. the minimum distance between sites is 1.5 m. Although some colonies are very large, many birds breed in small colonies of less than 1000 pairs, sometimes scattered more or less in isolation from other conspecifics. At Barrow, Alaska, where no suitable habitat was available, attempted to breed in artificial sites created for Black Guillemots (Divoky 1982).

NEST: usually in a rock crevice, or under boulders; occasionally in a burrow that is dug by the birds (Sealy 1973c). The nest is lined with grass and a few feathers, sometimes with fishing line or plastic debris.

EGG-LAYING: in the Gulf of Alaska peaks 14–26 June (5 studies, Baird and Gould 1983) and at St Lawrence I. between 20 and 27 June (Sealy 1973c).

EGGS: one, elliptical-ovate, dingy white with matt surface, somewhat granular, with faint brownish or lavender blotches, especially near the large end. Measurements 68.9×46.2 mm ($n = 5$, Sealy 1973c). Fresh mass 74 g (62–82, $n = 9$, Kharitonov 1990a); 75.3 g ($n = 71$, Johnsgaard 1987), about 12% of adult weight. Replacement layings occur 10–21 days after loss of first egg.

INCUBATION AND CHICK-REARING: incubation period 41.1 days (40–43, $n = 5$, Sealy 1973c); 41.2 ± 3.4 days (range 38–49, $n = 20$, Baird and Gould 1983). Change-overs occur in the evening (Manuwal 1984). The twin brood patches measure 55×27 mm (Sealy 1973). After hatching, brooding continues intermittently up to 6 days. Feeding diurnal, 3–6 times daily. Bill loads delivered consist of 1–16 fishes (mean 5.9), averaging 13.7 ± 1.0 g (9.6–25.4, Baird and Gould 1983).

CHICK GROWTH AND DEVELOPMENT: semi-precocial, achieving thermoregulation at 5–6 days. Hatching mass 58.6 g, maximum growth rate 12 g/day. Age at fledging 42.3 ± 2.9 days (37–46, Baird and Gould 1983). Fledglings depart at night, unassisted by parents; no evidence of post-fledging care.

BREEDING SUCCESS: for 18 colony-years in the Aleutians and Gulf of Alaska hatching success ranged from 56 to 93%; 11 colony-years were >80%. Fledging success in the same studies was 36–91% and reproductive success was 0.29–0.77 chicks/pair, with 10 colony-years >0.6 chicks/pair. Total reproductive failure occurred at Buldir I. in 1975 (Byrd *et al.* 1993).

MOULT: very little known, but adult post-breeding moult apparently occurs in late autumn or winter, as in Atlantic Puffin. Bill ornaments are dropped near the end of chick-rearing.

Population dynamics
Nothing is known of survival or age at first breeding. Some birds marked at Buldir I. in the 1970s were still present in the 1990s, suggesting that breeders regularly survive 20 years or more (ILJ).

Tufted Puffin *Fratercula cirrhata*

Alca cirrhata (Pallas, 1769, *Spicil Zool.*, I fasc. v, pl. i, v, figs 1–3—Bering Sea, Kamchatka).

PLATE 7

Description
A large, stocky auk with a very robust bill and entirely dark plumage apart from a white face and yellow facial ornaments. Face laterally compressed, forming a continuous wedge with the huge bill. A lumbering flier, with broad wings; not very manoeuvrable in the air; requires some slope to take off on land and in light winds must taxi a long way on the surface before taking off from water. On land, stalks with a laboured, rolling gait, its body leaning forward, or stands on the toes.

ADULT SUMMER: blackish-brown above, dark brown below. Face, including lores, cheeks, and area around eyes, as well as around base of bill, white; glossy, yellow, filamentous tufts up to 7 cm long curl backwards and downwards from above and behind eye, down sides of neck; underwing coverts brownish-grey, with white feathers along underside of leading edge; iris greyish-white, with bare eye-ring of red skin; distal two thirds of bill dull red; large olive-green to yellowish plate at base of bill, forming a saddle-like, bulbous ridge on top; fleshy, dull-red rosette at gape; up to four vertical grooves score the bill near the tip; legs and feet, red or orange-red. Second summer

birds similar to breeding adults, but with bill ornaments smaller and duller, usually with only two bill grooves, and facial plumes shorter and whitish, rather than golden; in first summer, similar to first winter, but breast and belly paler through feather wear.

ADULT WINTER: entirely blackish brown above and dark brown below, except for some paler flecks on belly and a grey-brown spot behind eye; plumes absent. Bill dull red at tip, shading to brown at base; the basal green plate is absent.

JUVENILE AND FIRST WINTER: similar to winter-plumage adult, but with throat and upper breast grey-brown and lower breast and belly white, sullied with brown; iris brownish-grey and bare skin around eye black; bill brown, triangular, and much shallower than the adult's.

CHICK: down dark brown, shading to greyish below, bill brown and feet brownish-black. Some chicks have white patches on the belly, or the entire belly may be off-white.

VARIATION
No subspecies recognized.

Tufted Puffin near its nesting burrow (ILJ photo).

MEASUREMENTS

Eastern Aleutians, summer, sexes combined (J. Piatt, unpublished) (mm): wing 203.5 ± 5.5 (n = 98); culmen 57.5 ± 4.1 (n = 78); tarsus 36.2 ± 2.8 (n = 86). North America (Johnsgaard 1987) (mm): ♂, wing 194.4 (187–206), culmen 57.1 (53–65, n = 11); ♀ wing 189.2 (179–196), culmen 57.1 (54.6–60.0, n = 9).

WEIGHTS

No sign of clinal variation is detectable, at least in Alaska. Weights of adults in summer (sexes combined, J. Piatt, unpublished): Alaska Peninsula 763 ± 70 g (n = 263); E Aleutians 776 ± 63 g (n = 263); W Aleutians 804 ± 66 g (n = 233). Triangle I., BC rearing chicks 746 ± 19 g (n = 27, Vermeer and Cullen 1979); Aleutians, ♂ 825 ± 46 g (758–913, n = 8), ♀ 778 ± 48 g (678–867, n = 13) (Shiomi and Ogi 1991).

Range and status

Found throughout boreal and low Arctic areas of the N Pacific, from Big Sur, California to Hokkaido, Japan, but abundant only from British Columbia to the Sea of Okhotsk. Small numbers penetrate the Chukchi Sea and as far as Wrangel I.; accidental in the Beaufort Sea.

RANGE

BREEDING: from California to Japan and as far N as Wrangel I. Abundant on islands off the Alaska Peninsula, in the Aleutians, and the Kuril Is. and the Sea of Okhotsk, with smaller numbers in the Bering Sea and only peripheral populations in Chukchi Sea as far N as Cape Thompson and Wrangel I. The centre of the species' distribution is in the E Aleutians, especially on islands in the major passes connecting the Bering Sea with the Pacific. Of the 58 islands in Alaska thought to support at least 10 000 breeders, nine are in the area of Akutan Pass (between Akutan and Unalaska Is.) and nine around Unimak Pass (between Unimak and Akutan Is.). In Asia, occurs as far S as Hokkaido.

WINTER: not well known, generally well offshore. Some birds remain close to breeding grounds except where forced to shift by ice conditions, but substantial numbers move further S. Occurs off California in winter, in the northern Sea of Japan and to N Honshu on the E coast of Japan (Brazil 1991). Only casual in winter off British Columbia (Campbell *et al.* 1990).

STATUS

The area of Akutan and Unimak passes supports 800 000–1 million breeders, including the largest colony, on Egg I. (163 000). The largest colonies elsewhere are Forrester I., SE Alaska (70 000 breeders), Castle Rock in the Shumagin Is. (80 000), Amagat I., off the Alaska Peninsula (100 000), The Triplets, off Kodiak I. (60 000), and E and W Amatuli Is. in Gulf of Alaska (93 000). The only sizable colony in British Columbia (50 000 breeders) is at Triangle I.; otherwise there are no colonies of

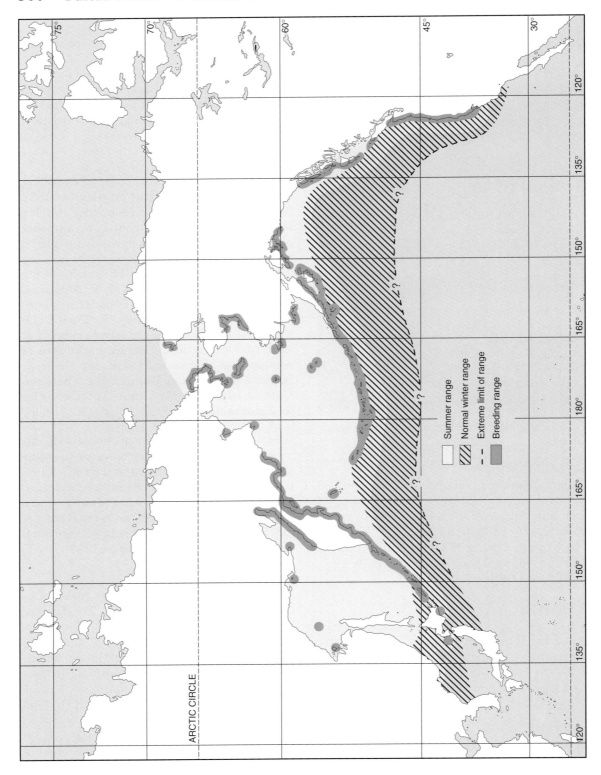

Summer range

Normal winter range

Extreme limit of range

Breeding range

ARCTIC CIRCLE

>10 000 pairs S of Alaska. Small numbers breed in Washington and Oregon, largest colony being 2300 pairs at Three-arches Rocks (Gilligan *et al.* 1993). Formerly bred in the California Channel Is., but not in this century (Small 1994); numbers have also declined at Farallon Is., the main station in California, from 'thousands' in the late nineteenth century to about 100 breeders in 1982. A few pairs breed on Wrangel I. and on the Russian Bering Sea coast. The largest colonies in Asia are on the Commander Is. (>20 000), and Talan I. (80 000). In Japan, moderate number bred on islands off Hokkaido up to about 1970, but numbers have much diminished since; the population may be in single figures (Brazil 1991). Main causes of decline probably drowning in gill-nets, introduced predators, and disturbance.

Field characters
The all-black body and wings, pale head, and huge, deep bill make this species unmistakable in summer. Possibly confused in winter or juvenile plumage with Rhinoceros Auklet, but Tufted Puffins are larger and darker and have a different head pattern.

Voice
Appears to have a limited range of calls, principally a low, grumbling sound heard on the breeding colony (Kharitonov 1990*b*).

Habitat, food, and feeding behaviour
MARINE HABITAT
Usually feeds offshore throughout the year, in winter, often well away from the continental shelf. Feeds further from colonies than Horned Puffin, where the two species nest together.

DIET
Chicks, for which most information is available, are fed almost entirely on fish, but stable isotope analysis suggests that adults feed themselves partly, perhaps primarily, on zooplankton (Hobson *et al.* 1994).

ADULTS: very variable. Net-drowned birds taken in the waters of the E Kamchatka current contained larval fish, squid, euphausiids, pteropods, and small numbers of amphipods and jellyfish ($n = 98$, Ogi 1980). Adults collected arriving at colonies in the Aleutians contained mostly hard parts, so data are probably biased; Wehle (1982) found that among breeding birds at Buldir I. ($n = 73$) 97% contained squid, but only 25% contained fish and 1% contained polychaete worms. Among subadults (presumed non-breeders, $n = 13$), 31% contained fish, 100% squid, and 23% polychaetes. At Ugaiushak I., 33% of breeders ($n = 20$) contained fish, 33% squid, and 33% crustacea, including crabs, shrimps, and euphausiids. Fish were found more commonly in birds collected during the chick-rearing period than earlier in the season.

CHICKS: almost entirely fish. Sampling of 13 colonies between the E Aleutians and Kodiak I., Gulf of Alaska over three seasons produced 32 species of fish and seven species of invertebrates, but three fish: Pacific sandlance (41% by weight), capelin (22%), and walleye pollock (19%) comprised the bulk of the diet (Hatch and Sanger 1992). All of these species made up >50% of the diet at least one colony in one year. The only other significant contributors were an unidentified salmonid contributing 76% (n samples = 2) of diet at Naked I., Prince William Sound, and squid, 24% (n samples = 12) at Noisy I., Shelikof Strait. Overall, invertebrates, including squid, octopus, and euphausiids, contributed <3% to the diet. Proportion of pollock in the diet was consistently higher in western part of area. Mean lengths of important species were: sandlance 72 mm (28–200, $n = 3530$), capelin 84 mm (26–147, $n = 445$), pollock 63 mm (33–112, $n = 1807$). Most sandlance and pollock taken were 0-age, but capelin were mainly 1-year and 2-year age-classes. At Buldir I., in W Aleutians, sandlance comprised 36% of fish delivered, squid 33%, yellow Irish lord (*Hemilepidotus jordani*) 21%, and Atka mackerel 6% ($n = 63$, Wehle 1983). At Triangle I., BC, 51% ($n = 57$) of deliveries were sandlance (60–110 mm long, in 1 year, 80–180 mm in another), 23% bluethroat argentine, 16% Pacific saury, 5% rockfish, and 4% squid (Vermeer 1979). In normal years at the Farallon Is. chicks receive mainly rockfish (Ainley *et al.* 1990*a*).

Displays and breeding behaviour

Not well described. This account mainly from Kharitonov (1990*b*). Attendance during pre-laying period alternates between 2–3 days with high numbers at the colony, followed by 2–3 days of absence. At Farallon Is., comes ashore from dawn to 15:00 hrs in Apr, but non-breeders remain ashore all day in June. In the pre-laying period large aggregations occur off-shore from the colony, engaging in intensive courtship with frequent copulations. On land, displays apparently similar to other puffins, with courtship involving posture displays such as skypointing and strutting and billing between pairs. Clubs form on rocks and other prominences within the colony. Pairs respond to intrusions close to their burrow by mutual billing. Higher intensity threat involves a head lowered posture with the bill gaped; this may be followed by bill grappling and kicking with the feet.

Breeding and life cycle

Arrives at breeding colonies Mar–May: at Farallon Is. first observations ranged from 12 March–6 April (13 years, Ainley and Boekelheide 1990); in Commander Is. in late Apr, in late May or early June on Chukchi Peninsula, St Lawrence I. and the Seward Peninsula, Alaska (Kessel 1989; Kharitonov 1990*a*).

BREEDING HABITAT AND NEST DISPERSION: generally breeds on islands with steep grassy slopes and good soil for burrowing, on turf-covered offshore stacks, on vegetated talus slopes, and (less commonly) on cliffs. Where foxes are present, nests are restricted to in-accessible crevices. At Triangle I., BC, most burrows are in *Deschampsia* grassland, with highest densities (1.2 burrows/m²) near the edge of seaward-facing cliffs; used mainly

areas sloping at 35–50° (Vermeer 1979). In Alaska coastal lagoons, breeds in low sand cliffs (<2 m), excavating burrows just below top of cliff (Gill and Sanger 1979). By con-trast, nests at the Barren Is., Alaska may be more than 600 m above sea level (Manuwal 1984). At the Farallon Is., may usurp nest-sites from Rhinoceros Auklets and Pigeon Guillemots (Ainley *et al.* 1990*a*). Density at Talan Island is 0.3–0.9 burrows/m² (Kharitonov 1990*a*). At colonies in the Gulf of Alaska, 44–70% of burrows were occupied in a given year (Baird and Gould 1983).

NEST: usually in earth burrows, in rock crevices, or in cavities below boulders or large talus. Burrows are dug with bill and feet and are used where an adequate depth of unfrozen soil is available (e.g. British Columbia, Gulf of Alaska, and Aleutian Is.) and may be up to 2 m long, with an entrance diameter of about 15 cm.

EGG-LAYING: Laying takes place from late Apr to late May (Ainley *et al.* 1990*a*). At Destruction I., Washington, laying occurs 6 May–6 June (mean 16 May, Burrell 1980) and at Triangle I., BC, and in Gulf of Alaska, in the first half of June (Vermeer *et al.* 1979; Baird and Gould 1983).

EGGS: one, usually unmarked with an off-white ground colour. Six out of 21 in the BMNH showed faint brown or bluish spots or scribbles. Measurements (various localities, BMNH): 71.5 ± 3.0 (65.0–77.7) $\times 48.6 \pm 1.2$ (45.9–50.9, $n = 21$) mm. At Triangle I., BC: 71.1 ± 2.9 (66.2–75.8) $\times 48.8 \pm 1.4$ (45.0–52.4, $n = 32$, ILJ and H. Knechtel, unpublished) mm.

INCUBATION: by both sexes; duration 45 days (41–54; Baird and Gould 1983).

HATCHING AND CHICK-REARING: at Triangle I., BC, chicks received an average of 3.5 feeds/day, with peak deliveries between 06:00 and 07:00 hrs and very few in the afternoon. Feeding also concentrated in the morning at colonies in Gulf of Alaska (Wehle 1983). Food loads delivered at Triangle I. averaged 14 ± 3 g ($n = 22$) and 22 ± 3 g ($n = 35$) in 2 years (Vermer 1979; Vermeer *et al.* 1979); at colonies in the Gulf of Alaska 14–20 g (Baird and Gould 1983). Kleptoparasitism by Glaucous-winged Gulls is frequent at some colonies.

Chick growth and development: chick growth very variable among colonies and from year to year, depending on local feeding conditions. Mean hatching weight 69.4 ± 10.3 g ($n = 30$; Baird and Gould 1983). Asymptotic weight reached at about 35 days. At Triangle I., few chicks fledged in 2 of 4 years and in the other 2 years fledging mass was 483 ± 24 g ($n = 18$) and 522 ± 19 g ($n = 23$, Vermeer and Cullen 1979); these were 90% of maximum chick mass. At Destruction I. Washington, maximum chick mass was 552 ± 59 g, fledging mass 497 ± 36 g and wing length at fledging 147 ± 4.3 mm ($n = 11$, Burrell 1980). In the Gulf of Alaska, mean fledging mass 274–609 g (8 colony-years, of which 7 >530 g; Baird and Gould 1983). Mean duration of nestling period in Gulf of Alaska 47 days (40–59; Baird and Gould 1983) and at Destruction I., Washington 50 days (43–59; Burrell 1980). Fledglings depart alone, at night, fluttering or flying to the sea, with no evidence of post-fledging parental care (Manuwal 1984).

Breeding success: in four seasons at Triangle I., the proportion of pairs fledging chicks was 45%, 1%, 1%, and 55% (Vermeer and Cullen 1979). In the Gulf of Alaska, hatching success at undisturbed sites ranged from 85 to 88%, fledging success from 90 to 91% and overall reproductive success 56–95% (6 colony-years). Success at sites visited regularly was much lower (Baird and Gould 1983).

Predation: taken by Bald Eagles and Snowy Owls. Very vulnerable to Arctic Foxes, which appeared to prefer them over other, more numerous, auks at St Paul I. (ILJ).

Moult: from Kharitonov (1990*a*). Juveniles undergo a post-juvenile (Pre-basic I) moult soon after going to sea and do not moult again until following autumn, when they moult completely to assume second winter plumage (Pre-basic II). This is followed by a spring body moult into a subadult breeding plumage (Pre-alternate I) that resembles adult plumage except that the plumes behind the eye are short and white, and the bill ornamentation is only partially developed. Adults undergo a complete post-breeding moult some time during the autumn and a pre-breeding moult, involving only the contour feathers and the development of the bill ornaments and yellow facial tufts, between Mar and May.

Population dynamics
No information: an excellent opportunity for research.

Bibliography

Adams, N. J. and Brown, C. R. (1990). Energetics of molt in penguins. In *Penguin biology* (ed. L. S. Davis and J. T. Darby), pp. 297–315. Academic Press, San Diego.

Ainley, D. G. (1976). The occurrence of seabirds in the coastal region of California. *Western Birds*, 7, 33–68.

Ainley, D. G. (1990). Farallon seabirds: patterns at the community level. In *Seabirds of the Farallon Islands*, (ed. D. G. Ainley and R. J. Boekelheide), pp. 349–80. Stanford University Press, Stanford, California.

Ainley, D. G. and Boekelheide, R. J. (1990). *Seabirds of the Farallon Islands*. Stanford University Press, Stanford, California.

Ainley, D. G. and Lewis, T. J. (1974). The history of the Farallon Islands marine bird populations, 1854–1972. Condor 76, 432–46.

Ainley, D. G., DeGange, A. R., Jones, L. L., and Beach, R. J. (1981). Mortality of seabirds in high seas salmon gill nets. *Fisheries Bulletin*, 79, 800–6.

Ainley, D. G., Morrell, S. H., and Boekelheide, R. J. (1990*a*). Rhinoceros Auklet and Tufted Puffin. In *Seabirds of the Farallon Islands*, (ed. D. G. Ainley and R. J. Boekelheide), pp. 339–48. Stanford University Press, Stanford, California.

Ainley, D. G., Strong, C. S., Penniman, T. M., and Boekelheide, R. J. (1990*b*). The feeding ecology of Farallon seabirds. In *Seabirds of the Farallon Islands* (ed. D. G. Ainley and R. J. Boekelheide), pp. 51–127. Stanford University Press, Stanford, California.

Ainley, D. G., Sydeman, W. J., Hatch, S. A., and Wilson, U. W. (1994). Seabird population trends along the west coast of North America: causes and the extent of regional concordance. *Studies in Avian Biology*, 15, 119–33.

Ainley, D. G., Sydeman, W. J., and Norton, J. (1996). Upper trophic level predators indicate interannual negative and positive anomalies in the California current food web. *Marine Ecology Progress Series*, 118, 69–79.

American Ornithologists' Union (1910). *Check-list of North American birds*. (3rd edn.). American Ornithologists' Union, New York.

American Ornithologists' Union (1983). *Check-list of North American birds*. (6th edn.). American Ornithologists' Union, New York.

Andersen-Harild, P. (1969). Nogle resultater of ringmaerkningen of Tejst (*Cepphus grylle*): Danmark. *Dansk Ornithologisk Forenings Tidsskrift*, 63, 105–10.

Andersson, M. (1994). *Sexual selection*. Princeton University Press, Princeton, New Jersey.

Anker-Nilssen, T. (1987). The breeding performance of the Puffin *Fratercula arctica* on Rost, northern Norway in 1979–85. *Fauna norvegica Series C, Cinclus*, 10, 21–38.

Anker-Nilssen, T. and Lorentsen, S-H. (1995). Size variation of Common Guillemots *Uria aalge* wintering in the northern Skagerrak. *Seabird*, 17, 64–73.

Anker-Nilssen, T. and Rostad, O. W. (1993). Census and monitoring of Puffins *Fratercula arctica* on Rost, N- Norway, 1979–1988. *Ornis Scandinavica*, 24, 1–9.

Armstrong, E. A. (1942). *Bird display and behaviour*. Lindsay Drummond, London.

Asbirk, S. (1979*a*). Some behaviour patterns of the Black Guillemot *Cepphus grylle*. *Dansk Ornithologisk Forenings Tidsskrift*, 73, 287–96.

Asbirk, S. (1979*b*). Breeding numbers and habitat selections of Danish Black Guillemots *Cepphus-*

grylle. Dansk Ornithologisk Forenings Tidsskrift, **72**, 161–78.

Asbirk, S. (1979c). A description of Danish Black Guillemots *Cepphus grylle* with remarks on the validity of the subspecies *atlantis*. *Dansk Ornithologisk Forenings Tidsskrift*, **73**, 207–14.

Asbirk, S. (1979d). The adaptive significance of the reproductive pattern in the Black Guillemot *Cepphus grylle*. *Videnskabelige Meddelelser Dansk Naturhistorisk Forening*, **141**, 29–80.

Ashcroft, R. E. (1976). Breeding biology and survival of puffins. Ph.D. thesis. University of Oxford.

Ashcroft, R. E. (1979). Survival rates and breeding biology of puffins on Skomer Island, Wales. *Ornis Scandinavica*, **10**, 100–10.

Asher, G. M. (1860). *Henry Hudson, the navigator*. The Hakluyt Society, London

Ashmole, N. P. (1963). The regulation of numbers of tropical oceanic birds. *Ibis*, **103b**, 458–73.

Ashmole, N. P. (1971). Seabird ecology and the marine environment. In *Avian biology* (ed. D. S. Farner and J. R. King), pp. 223–86. Academic Press, New York.

Audubon, J. J. (1827). *The birds of America*. Published by the Author (reproduced by Macmillan, New York, 1937).

Audubon, J. J. (1835). *Ornithological Biographies III*. Adam and Charles Black, Edinburgh.

Austin, O. L. (1948). The birds of Korea. *Bulletin of the Museum of Comparative Zoology, Harvard*, **101**, 1–301.

Austin, O. L. and Kuroda, N. (1953). The birds of Japan, their status and distribution. *Bulletin of the Museum of Comparative Zoology, Harvard*, **109**, 280–637.

Bailey, E. P. (1993). *Introduction of foxes to Alaskan islands—history, effects on avifauna and eradication*. U. S. Fish and Wildlife Service, Washington, DC.

Bailey, E. P. and Faust, N. H. (1980). Summer distribution and abundance of marine birds and mammals in the Sandman Reefs, Alaska. *Murrelet*, **61**, 6–19.

Bailey, E. P. and Kaiser, G. W. (1993). Impacts of introduced predators on nesting seabirds in the northeast Pacific. In *The status, ecology and conservation of marine birds of the North Pacific*, (ed. K. Vermeer, K. T. Briggs, K. H. Morgan,

and D. Siegel-Causey), pp. 218–26. Canadian Wildlife Service, Ottawa.

Baird, P. A. and Gould, P. J. (1983). The breeding biology and feeding ecology of marine birds in the Gulf of Alaska. *U.S. Dept. of Commerce, NOAA, OCSEAP final reports*, **45**, 121–505.

Balmford, A. and Read, A. F. (1991). Testing alternative models of sexual selection through female choice. *Trends in Ecology and Evolution*, **6**, 274–76.

Baltz, D. M. and Morejohn, G. V. (1977). Food habits and niche overlap of sea birds wintering on Monterey Bay, California. *Auk*, **94**, 526–43.

Bancroft, G. (1927). Notes on the breeding coastal and insular birds of central lower California. *Condor*, **29**, 188–95.

Bancroft, G. (1930). Eggs of Xantus' and Craveri's Murrelets. *Condor*, **32**, 247–54.

Barcena, F., Teixeira, A. M., and Bermejo, A. (1984). Breeding seabird populations in the Atlantic sector of the Iberian peninsula. In *Status and conservation of the world's seabirds*, (ed. J. P. Croxall, P. G. H. Evans, and R. W. Schreiber), pp. 335–45. International Council for Bird Preservation, Cambridge.

Barrett, R. T. (1984). Comparative notes on eggs, chick growth and fledging of the Razorbill *Alca torda* in North Norway. *Seabird*, **7**, 55–61.

Barrett, R. T. and Furness, R. W. (1990). The prey and diving depths of seabirds on Hornoy, North Norway after a decrease in the Barents Sea capelin stocks. *Ornis Scandinavica*, **21**, 179–86.

Barrett, R. T. and Rikardsen, F. (1992). Chick growth, fledging periods and adult mass loss of Atlantic Puffins *Fratercula arctica* during years of prolonged food stress. *Colonial Waterbirds*, **15**, 24–32.

Barrett, R. T. and Vader, W. (1984). The status and conservation of breeding seabirds in Norway. In *The status and conservation of the world's seabirds*, (ed. J. P. Croxall, P. G. H. Evans, and R. W. Schreiber), pp. 323–34. International Council for Bird Preservation, Cambridge.

Barrett, R. T., Fieler, R., Anker-Nilssen, T., and Rikardsen, F. (1985). Measurements and weight changes of Norwegian adult Puffins *Fratercula arctica* and Kittiwakes *Rissa tridactyla* during the breeeding season. *Ringing and Migration*, **6**, 102–12.

Barrett, R. T., Anker-Nilssen, T., Rikardsen, F., Valde, K., Rov, N., and Vader, W. (1987). The food, growth and fledging success of Norwegian Puffin chicks *Fratercula arctica* in 1980–83. *Ornis Scandinavica*, **18**, 73–83.

Bateson, P. P. G. (1961). Studies of less familiar birds 112: Little Auk. *British Birds*, **54**, 272–7.

Bédard, J. (1966). New records of Alcids from St Lawrence Island, Alaska. *Condor*, **68**, 503–6.

Bédard, J. (1969a). Adaptive radiation in Alcidae. *Ibis*, **111**, 189–98.

Bédard, J. (1969b). Histoire naturelle du Gode, *Alca torda*, L. dans le golfe Ste. Laurent, province de Quebec, Canada. *Etude du service Canadien de la Faune*, **7**, 1–79.

Bédard, J. (1969c). Feeding of the Least, Crested and Parakeet Auklets on St. Lawrence Island, Alaska. *Canadian Journal of Zoology*, **47**, 1025–50.

Bédard, J. (1969d). Nesting of the Least, Crested and Parakeet auklets on St. Lawrence Island, Alaska. *Condor*, **71**, 386–98.

Bédard, J. (1976). Coexistence, coevolution and convergent evolution in seabird communities: a comment. *Ecology*, **57**, 177–84.

Bédard, J. (1985). Evolution and characteristics of the Atlantic Alcidae. In *The Atlantic Alcidae*, (ed. D. N. Nettleship and T. R. Birkhead), pp. 1–51. Academic Press, London.

Bédard, J. and Sealy, S. G. (1984). Moults and feather generations in the Least, Crested and Parakeet auklets. *Journal of Zoology, London*, **202**, 461–88.

Beddard, F. G. (1898). *The structure and classification of birds*. Longmans, Green and Co., London.

Beebe, F. L. (1960). The marine peregrines of the northwest Pacific coast. *Condor*, **62**, 145–89.

Beja, P. R. (1989). A note on the diet of Razorbills *Alca torda* wintering off Portugal. *Seabird*, **12**, 11–3.

Belopol'skii, L. O. (1957). *Ecology of the sea colony birds of the Barents Sea* (trans. from Russian 1961), Israel Program for Scientific Translations, Jerusalem.

Bendire, C. (1895). Notes on the Ancient Murrelet (*Synthliboramphus antiquus*), by Chase Littlejohn. With annotations. *Auk* **12**, 270–8.

Bengtson, S. A. (1984). Breeding ecology and extinction of the Great Auk (*Pinguinus impennis*): anecdotal evidence and conjectures. *Auk*, **101**, 1–12.

Bent, A. C. (1919). Life histories of North American diving birds. *Smithsonian Institute Bulletin*, **107**, 1–245.

Bergman, G. (1971). [The Black Guillemot *Cepphus grylle*, in a peripheral area: food, breeding performance, diurnal rhythm and colonization]. *Commentationes Biologicae*: 42.

Bertram, D. F. (1988). The provisioning of nestlings by parent Rhinoceros Auklets (*Cerorhinca monocerata*). M.Sc. thesis. Simon Fraser University, Burnaby, BC.

Bertram, D. F. and Kaiser, G. W. (1993). Rhinoceros Auklet (*Cerorhinca monocerata*) nestling diet may gauge Pacific Sand Lance (*Ammodytes hexapterus*) recruitment. *Canadian Journal of Fisheries & Aquatic Sciences*, **50**, 1908–15.

Bertram, D. F., Kaiser, G. W., and Ydenberg, R. C. (1991). Patterns in the provisioning and growth of nestling Rhinoceros Auklets. *Auk*, **108**, 842–52.

Bird, C. G. and Bird, E. G. (1935). The birds of Jan Mayen Island. *Ibis*, **5**, 837–55.

Birkhead, T. R. (1976a). Effects of sea conditions on rates at which Guillemots feed chicks. *British Birds*, **69**, 490–2.

Birkhead, T. R. (1976b). Breeding biology and survival of Guillemots, *Uria aalge*. Ph. D. thesis. University of Oxford.

Birkhead, T. R. (1977a). Adaptive significance of the nesting period of guillemots *Uria aalge*. *Ibis*, **119**, 544–9.

Birkhead, T. R. (1977b). The effect of habitat and density on breeding success in the common guillemot (*Uria aalge*, Pontopp). *Journal of Animal Ecology*, **46**, 751–64.

Birkhead, T. R. (1978a). Attendance patterns of Guillemots *Uria aalge* at breeding colonies on Skomer Island. *Ibis*, **120**, 219–29.

Birkhead, T. R. (1978b). Behavioral adaptations to high density nesting in the common guillemot *Uria aalge*. *Animal Behaviour*, **26**, 321–31.

Birkhead, T. R. (1984). Distribution of the bridled form of the Common Guillemot *Uria aalge* in the North Atlantic. *Journal of Zoology, London*, **202**, 165–76.

Birkhead, T. R. (1985). Coloniality and social behaviour in the Atlantic Alcidae. In *The Atlantic Alcidae*, (ed. D. N. Nettleship and T. R. Birkhead), pp. 355–83. Academic Press, London.

Birkhead, T. R. (1993). *Great auk islands*. T. and A. D. Poyser, London.

Birkhead, T. R. and Gaston, A. J. (1988). The composition of Ancient Murrelet eggs. *Condor*, **90**, 965–6.

Birkhead, T. R. and Harris, M. P. (1985). Ecological adaptations for breeding in the Atlantic Alcidae. In *The Atlantic Alcidae*, (ed. D. N. Nettleship and T. R. Birkhead), pp. 205–32. Academic Press, London.

Birkhead, T. R. and Hudson, P. J. (1977). Population parameters for the common guillemot *Uria aalge*. *Ornis Scandinavica*, **8**, 145–54.

Birkhead, T. R. and Møller, A. P. (1992). *Sperm competition in birds*. Academic Press, London.

Birkhead, T. R. and Nettleship, D. N. (1981). Reproductive biology of Thick-billed Murres (*Uria lomvia*): an inter-colony comparison. *Auk*, **98**, 258–69.

Birkhead, T. R. and Nettleship, D. N. (1982a). The adaptive significance of egg size and laying date in Thick-billed Murres *Uria lomvia*. *Ecology*, **63**, 300–6.

Birkhead, T. R. and Nettleship, D. N. (1982b). Studies of alcids breeding at the Gannet Clusters, Labrador, 1981. Canadian Wildlife Service, Manuscript Report, Dartmouth, Nova Scotia.

Birkhead, T. R. and Nettleship, D. N. (1984a). Alloparental care in the Common Murre (*Uria aalge*). *Canadian Journal of Zoology*, **62**, 2121–4.

Birkhead, T. R. and Nettleship, D. N. (1984b). Egg size, composition and offspring quality in some Alcidae (Aves: Charadriiformes). *Journal of Zoology, London*, **202**, 177–94.

Birkhead, T. R. and Nettleship, D. N. (1985). Plumage variation in young Razorbills and Murres. *Journal of Field Ornithology*, **56**, 246–50.

Birkhead, T. R. and Nettleship, D. N. (1987a). Ecological relationships between Common Murres, *Uria aalge*, and Thick-billed Murres, *Uria lomvia*, at the Gannet Islands, Labrador. I. Morphometrics and timing of breeding. *Canadian Journal of Zoology*, **65**, 1621–9.

Birkhead, T. R. and Nettleship, D. N. (1987b). Ecological relationship between Common Murres, *Uria aalge*, and Thick-billed Murres, *Uria lomvia*, at the Gannet Islands, Labrador III feeding ecology of the young. *Canadian Journal of Zoology*, **65**, 1638–49.

Birkhead, T. R. and Nettleship, D. N. (1987c). Ecological relationships between Common Murres, *Uria aalge*, and Thick-billed Murres, *Uria lomvia*, at the Gannet Islands, Labrador II Breeding success and site characteristics. *Canadian Journal of Zoology*, **65**, 1630–7.

Birkhead, T. R. and Taylor, A. M. (1977). Moult of the Guillemot *Uria aalge*. *Ibis*, **119**, 80–5.

Birkhead, T. R., Biggins, J. D., and Nettleship, D. N. (1980). Non-random, intra-colony distribution of bridled guillemots *Uria aalge*. *Journal of Zoology, London*, **192**, 9–16.

Birkhead, T. R., Johnson, S. D., and Nettleship, D. N. (1985). Extra-pair matings and mate-guarding in the Common Murre *Uria aalge*. *Animal Behaviour*, **33**, 608–19.

Birkhead, T. R., Johnson, S. D., and Nettleship, D. N. (1986). Field observations of a possible hybrid Murre *Uria aalge* × *Uria lomvia*. *Canadian Field-Naturalist*, **100**, 115–7.

Birt, V. L. and Cairns, D. K. (1987). Kleptoparasitic interactions of Arctic Skuas *Stercorarius parasiticus* and Black Guillemots *Cepphus grylle* in northeastern Hudson Bay, Canada. *Ibis*, **129**, 190–6.

Birt-Friesen, V. L., Montevecchi, W. A., Cairns, D. K., and Macko, S. A. (1989). Activity-specific metabolic rates of free-living Northern Gannets and other seabirds. *Ecology*, **70**, 357–67.

Birt-Friesen, V. L., Montevecchi, W. A., Gaston, A. J., and Davidson, W. S. (1992). Genetic structure of Thick-billed Murre (*Uria lomvia*) populations examined using direct sequence analysis of amplified DNA. *Evolution*, **46**, 267–72.

Blake, B. F. (1983). A comparative study of the diet of auks killed during an oil incident in the Skagerrak in January 1981. *Journal of the Zoological Society, London*, **201**, 1–12.

Blake, B. F., Dixon, T. J., Jones, P. H., and Tasker, M. L. (1985). Seasonal changes in the feeding ecology of Guillemots (*Uria aalge*) off North and East Scotland. *Estuarine Coast and Shelf Sciences*, **20**, 559–68.

Boag, D. and Alexander, M. (1986). *The Atlantic Puffin*. Blandford Press, Poole, Dorset.

Bock, W. J. (1989). Principles of biological comparison. *Acta Morphologica Neerland-Scandinavica*, **27**, 17–32.

Bockstoce, J. R. (1986). *Whales, ice and men: the history of whaling in the western arctic*. University of Washington Press, Seattle.

Boekelheide, R. J., Ainley, D. G., Morrell, S. H., Huber, H. R., and Lewis, T. J. (1990). Common Murre. In *Seabirds of the Farallon Islands*, (ed. D. G. Ainley and R. J. Boekelheide), pp. 245–75. Stanford University Press, Stanford, California.

Bourne, W. R. P. (1963). A review of oceanic studies of the biology of seabirds. *Proceedings of the International Ornithological Congress*, **13**, 831–54.

Bourne, W. R. P. (1976). Seabirds and oil pollution. In *Marine pollution*, (ed. R. Johnston), pp. 403–502. Academic Press, London.

Bourne, W. R. P. (1993). The story of the Great Auk *Pinguinis impennis*. *Archives of Natural History*, **20**, 257–78.

Bradstreet, M. S. W. (1979). Thick-billed Murres and Black Guillemots in the Barrow Strait area, N. W. T., during spring: distribution and habitat use. *Canadian Journal of Zoology*, **57**, 1789–802.

Bradstreet, M. S. W. (1980). Thick-billed Murres and Black Guillemots in the Barrow Strait area, N. W. T., during spring: diets and food availability along ice edges. *Canadian Journal of Zoology*, **57**, 2120–40.

Bradstreet, M. S. W. (1982a). Occurrence, habitat use and behaviour of seabirds, marine mammals and arctic cod at the Pond Inlet ice edge. *Arctic* **35**, 28–40.

Bradstreet, M. S. W. (1982b). Pelagic feeding ecology of Dovekies *Alle alle* in Lancaster Sound and Western Baffin Bay. *Arctic*, **35**, 126–40.

Bradstreet, M. S. W. (1985). Feeding studies. In *Population estimation, productivity, and food habits of nesting seabirds at Cape Pierce and the Pribilof Islands, Bering Sea, Alaska*, (ed. S. R. Johnston). LGL Ecological Research Associates, Anchorage, Alaska.

Bradstreet, M. S. W. and Brown, R. G. B. (1985). Feeding ecology of the Atlantic Alcidae. In *The Atlantic Alcidae*, (ed. D. N. Nettleship and T. R. Birkhead), pp. 262–318. Academic Press, London.

Bradstreet, M. S. W. and Gaston, A. J. (1993). *The diet of Thick-billed Murres in the eastern Canadian arctic during the breeding season.* Canadian Wildlife Service Technical Report No. 180, Ottawa, Canada.

Brandt, J. F. (1837). Bulletin of the Imperial Academy of Sciences, St. Petersburg, **2**, 347.

Brandt, J. F. (1869). Erganzungen und Berichtigungen zur Naturgeschichte der Familie der Alciden. *Melanges Biologiques Bulletin of the Imperial Academy of Science, St. Petersburg*, **7**, 199–268.

Brazil, M. (1991). *Birds of Japan.* Smithsonian Institution Press, Washington, DC.

Breese, D., Tershy, B. R., and Craig, D. P. (1993). Craveri's Murrelet—confirmed nesting and fledging age at San Pedro-Martir Island, Gulf of California. *Colonial Waterbirds*, **16**, 92–4.

Brekke, B. and Gabrielsen, G. W. (1994). Assimilation efficiency of adult kittiwakes and Brunnich's Guillemots fed capelin and arctic cod. *Polar Biology*, **14**, 279–84.

Briggs, K. T. and Chu, E. W. (1987). Trophic relationships and food requirements of California seabirds: updating models of trophic impact. In *Seabirds: feeding ecology and role in marine ecosystems*, (ed. J. P. Croxall), pp. 279–301. Cambridge University Press, Cambridge.

Briggs, K. T., Tyler, W. B., Lewis, D. B., and Carlson, D. R. (1987). Seabird communities at sea off California: 1975–1983. *Studies in Avian Biology*, **11**, 1–74.

Brodkorb, P. (1960). Great Auk and Common Murre from a Florida midden. *Auk*, 77, 342–3.

Brodkorb, P. (1967). Catalogue of fossil birds. Part 3 (Ralliformes, Ichtyornithiformes, Charadriiformes). *Bulletin of the Florida State Museum, Biological Sciences*, **11**, 99–220.

Brooke, M. de L. (1990). *The Manx Shearwater.* T. and A. D. Poyser, London.

Brown, R. G. B. (1980). Seabirds as marine animals. In *Behavior of marine animals, 4 (marine birds)*, (ed. J. Burger, B. L. Olla, and H. E. Winn), pp. 1–39. Plenum Press, New York.

Brown, R. G. B. (1985). The Atlantic Alcidae at sea. In *The Atlantic Alcidae*, (ed. D. N. Nettleship and T. R. Birkhead), pp. 384–427. Academic Press, London.

Brown, R. G. B. (1986). *Revised atlas of eastern Canadian seabirds. 1. Shipboard surveys.* Canadian Wildlife Service, Ottawa.

Brown, R. G. B. and Nettleship, D. N. (1984). Capelin and seabirds in the Northwest Atlantic. In *Marine birds: their feeding ecology and commercial fisheries relationships*, (ed. D. N. Nettleship, G. A. Sanger, and P. F. Springer), pp. 184–94. Canadian Wildlife Service, Ottawa.

Bruemmer, F. (1972). Dovekies. *The Beaver*, **303**, 40–7.

Brun, E. (1971). Breeding distribution and population of cliff-breeding seabirds in Sor-Varanger, north Norway. *Astarte*, **4**, 53–60.

Brun, E. (1979). Present status and trends in populations of seabirds in Norway. In *Conservation of marine birds of northern North America*, (ed. J. C. Bartonek and D. N. Nettleship), pp. 289–301. U. S. Fish and Wildlife Service Report No. 11, Washington, DC.

Brunnich, M. T. (1764). *Ornithologia Borealis*. J. C. Kall.

Burger, A. E. (1991). Maximum diving depths and underwater foraging in alcids and penguins. In *Studies of high-latitude seabirds, 1. Behavioural, energetic and oceanographic aspects of seabird feeding ecology*, (ed. W. A. Montevecchi and A. J. Gaston), pp. 9–15. Canadian Wildlife Service, Ottawa.

Burger, A. E. and Piatt, J. F. (1990). Flexible time budgets in breeding Common Murres as buffers against variable prey abundance. *Studies in Avian Biology*, **14**, 71–83.

Burger, A. E. and Powell, D. (1990). Diving depths and diets of Cassin's Auklet at Reef Island, British Columbia. *Canadian Journal of Zoology*, **68**, 1572–7.

Burger, A. E. and Simpson, M. (1986). Diving depths of Atlantic Puffin and Common Murre. *Auk*, **103**, 828–30.

Burger, A. E. and Wilson, R. P. (1988). Capiliary tube depth gauges for diving animals: an assessment of their accuracy and methods for deployment. *Journal of Field Ornithology*, **59**, 345–54.

Burger, A. E., Wilson, R. P., Garnier, D., and Wilson, M. P. T. (1993). Diving depths, diet, and underwater foraging of Rhinoceros Auklets in British Columbia. *Canadian Journal of Zoology*, **71**, 2528–40.

Burkett, E. E. (1995). Marbled Murrelet food habits and prey ecology. In *Ecology and conservation of the marbled murrelet*, (ed. C. J. Ralph, G. L. Hunt, M. G. Raphael, and J. Piatt), pp. 223–46. U. S. Forest Service, Albany, NY.

Burness, G. P. and Montevecchi, W. A. (1992). Oceanographic-related variation in the bone sizes of extinct Great Auks. *Polar Biology*, **11**, 545–51.

Burrell, G. C. (1980). Some observations on nesting Tufted Puffins, Destruction Island, Washington. *Murrelet*, **61**, 92–4.

Byrd, G. V. and Day, R. H. (1986). The avifauna of Buldir Island, Aleutian Islands, Alaska. *Arctic*, **39**, 109–18.

Byrd, G. V. and Gibson, D. D. (1980). Distribution and population status of Whiskered Auklet in the Aleutian Islands. *Western Birds* **11**, 135–40.

Byrd, G. V., Day, R. H., and Knudtson, E. P. (1983). Patterns of colony attendance and censusing of auklets at Buldir Island, Alaska. *Condor*, **85**, 274–80.

Byrd, G. V., Murphy, E. C., Kaiser, G., Kondratyev, A. Y., and Shibaev, Y. V. (1993). Status and ecology of offshore fish-feeding alcids (murres and puffins) in the North Pacific. In *The status, ecology and conservation of marine birds of the North Pacific*, (ed. K. Vermeer, K. T. Briggs, K. H. Morgan, and D. Siegel-Causey), pp. 176–86. Canadian Wildlife Service, Ottawa.

Cairns, D. K. (1978). Some aspects of the biology of the Black Guillemot *Cepphus grylle* in the estuary and the Gulf of St. Lawrence. M.Sc. thesis. Université de Laval.

Cairns, D. K. (1980). Nesting density, habitat structure and human disturbance as factors in Black Guillemot reproduction. *Wilson Bulletin*, **92**, 352–61.

Cairns, D. K. (1981). Breeding, feeding and chick growth of the Black Guillemot (*Cepphus grylle*) in southern Quebec. *Canadian Field-Naturalist*, **95**, 312–8.

Cairns, D. K. (1986). Plumage colour in pursuit-diving seabirds: why do penguins wear tuxedos? *Bird Behaviour*, **6**, 58–65.

Cairns, D. K. (1987a). The ecology and energetics of chick provisioning by Black Guillemots. *Condor*, **89**, 627–35.

Cairns, D. K. (1987b). Diet and foraging ecology of Black Guillemots in northeastern Hudson Bay. *Canadian Journal of Zoology*, **65**, 1257–63.

Cairns, D. K. (1987c). Seabirds as indicators of marine food supplies. *Biological Oceanography*, **5**, 261–71.

Cairns, D. K. (1992a). Diving behaviour of Black Guillemots in northeastern Hudson Bay. *Colonial Waterbirds*, **15**, 245–8.

Cairns, D. K. (1992b). Population regulation of seabird colonies. *Current Ornithology*, 9, 37–62.

Cairns, D. K. (1992c). Bridging the gap between ornithology and fisheries science: use of seabird data in stock assessment models. *Condor*, 94, 811–24.

Cairns, D. K. and DeYoung, B. (1981). Back-crossing of a Common Murre (*Uria aalge*) and a Common Murre-Thick-billed Murre hybrid (*U. aalge × U. lomvia*). *Auk*, 98, 847.

Cairns, D. K. and Schneider, D. C. (1990). Hot spots in cold water: feeding habitat selection by Thick-billed Murres. *Studies in Avian Biology*, 14, 52–60.

Cairns, D. K., Bredin, K. A., and Montevecchi, W. A. (1987). Activity budgets and foraging ranges of breeding common murres. *Auk*, 104, 218–24.

Cairns, D. K., Montevecchi, W. A., Birt-Friesen, V. L., and Macko, S. A. (1990). Energy expenditures, activity budgets, and prey harvest of breeding Common Murres. *Studies in Avian Biology*, 14, 84–92.

Cairns, D. K., Chapdelaine, G., and Montevecchi, W. A. (1991). Prey exploitation by seabirds in the Gulf of St. Lawrence. *Canadian Special Publications in Fisheries and Aquatic Science*, 113, 277–91.

Calder, W. A. and King, J. R. (1974). Thermal and caloric relations of birds. In *Avian biology IV*, (ed. D. S. Farner, J. R. King, and K. C. Parkes), pp. 157–285. Academic Press, New York.

Calder, W. A. I. (1984). *Size, function and life history*. Harvard University Press, Cambridge, Massachusetts.

Campbell, R. W., Dawe, N. K., McTaggart-Cowan, I., Cooper, J. M., Kaiser, G. W., and McNall, M. C. E. (1990). *The birds of British Columbia*. Royal British Columbia Museum and the Canadian Wildlife Service, Victoria, BC.

Camphuysen, C. J. (1989). Biometrics of auks at Jan Mayen. *Seabird*, 12, 7–10.

Camphuysen, C. J. and Leopold, M. F. (1994). *Atlas of seabirds in the southern North Sea*. Institute for Forestry and Nature Research, Texel, Netherlands.

Camphuysen, C. J., Calvo, B., Durinck, J., Ensor, K., Follestad, A., Furness, R. W., Garthe, S., Leaper, G., Skov, H., Tasker, M. L., and Winter, C. J. N. (1995). *Consumption of discards by seabirds in the North Sea*. Netherlands Institute for Sea Research, Texel, Netherlands.

Cantelo, J. and Gregory, P. A. (1975). Feeding association between Common Tern and Razorbill. *British Birds*, 68, 296–7.

Carboneras, C. (1988). The auks in the western Mediterranean. *Ringing and Migration*, 9, 18–26.

Carter, H. (1984). At-sea biology of the Marbled Murrelet in Barkley Sound, British Columbia. M.Sc. thesis. University of Manitoba, Winnipeg, Manitoba.

Carter, H. R. (1986). Year-round use of coastal lakes by Marbled Murrelets. *Condor*, 88, 473–7.

Carter, H. R. and Sealy, S. G. (1984). Marbled Murrelet mortality due to gill-net fishing in Barkley Sound, British Columbia. In *Marine birds: their feeding ecology and commercial fisheries relationships*, (ed. D. N. Nettleship, G. A. Sanger, and P. F. Springer), pp. 212–20. Canadian Wildlife Service, Ottawa.

Carter, H. R. and Stein, J. L. (1995). Moults and plumages in the annual cycle of the Marbled Murrelet. In *Ecology and conservation of the marbled murrelet*, (ed. C. J. Ralph, G. L. Hunt, M. G. Raphael, and J. F. Piatt), pp. 99–109. U.S. Forest Service, Albany, CA.

Carter, H. R., Jaques, D. L., McChesney, G. J., Strong, C. S., Parker, M. W., and Takekawa, J. E. (1990). Breeding populations of seabirds on the northern and central California coasts in 1989 and 1990. (Abstract)

Cayford, J. (1981). The behaviour and dispersal of the Guillemot (*Uria aalge*) and Razorbill (*Alca torda*) on the sea at Lundy. Ms. report in EGI library, Oxford.

Chandler, R. M. (1990). Recent advances in the study of neogene fossil birds II. Fossil birds of the San Diego Formation, late Pliocene. Blancan, San Diego Co., California. *Ornithological Monographs*, 44, 75–161.

Chapdelaine, G. (1995). Fourteenth census of seabird populations in the sanctuaries of the North Shore of the Gulf of St. Lawrence. *Canadian Field-Naturalist*, 109, 220–6.

Chapdelaine, G. and Brousseau, P. (1991). Thirteenth census of the seabird populations in the sanctuaries of the North shore of the Gulf of St. Lawrence. *Canadian Field-Naturalist*, 105, 60–6.

Chapdelaine, G. and Brousseau, P. (1992). Distribution, abundance and changes of seabird populations of the Gaspe Peninsula, Quebec, 1979 to 1989. Canadian Field-Naturalist **106**, 427–34.

Chapdelaine, G. and Brousseau, P. (1996). Diet of Razorbill (*Alca torda*) chicks and breeding success in the St. Mary's Islands, Gulf of St. Lawrence, Quebec, Canada (1990–92). In *Studies of high latitude seabirds 4. trophic relationships and energetics of endotherms in cold ocean systems*, (ed. W. A. Montevecchi), pp. 27–36. Canadian Wildlife Service Occasional Paper No., Ottawa.

Chen, Z. -Q. (1988). Niche selection of breeding seabirds on Chenlushan Island in the Yellow Sea, China. *Colonial Waterbirds*, **11**, 306–7.

Chilton, G. and Sealy, S. G. (1987). Species roles in mixed species flocks of seabirds. *Journal of Field Ornithology*, **58**, 456–63.

Christy, M. (1894). On an early notice and figure of the Great Auk. *Zoologist*, **1894**, 142–5.

Clarke, W. E. (1912). The Little Auk visitation. *Scottish Naturalist*, **1912**, 77–81.

Clowater, J. S. and Burger, A. E. (1994). The diving behaviour of Pigeon Guillemots (*Cepphus columba*) off southern Vancouver Island. *Canadian Journal of Zoology*, **72**, 863–72.

Clutton-Brock, T. H. and Harvey, P. H. (1979). Comparison and adaptation. *Proceedings of the Royal Society, London, B*, **205**, 547–65.

Cody, M. L. (1971). Ecological aspects of reproduction. In *Avian biology*, (ed. D. S. Farner, and J. S. King), pp. 461–512. Academic Press, New York.

Cody, M. L. (1973). Coexistence, coevolution and convergent evolution in seabird communities. *Ecology*, **54**, 31–44.

Cohen, J. E., Pimm, S. L., Yodzis, P., and Saldana, J. (1993). Body sizes of animal predators and animal prey in food webs. *Journal of Animal Ecology*, **62**, 67–78.

Collar, N. J., Crosby, M. J., and Stattersfield, A. J. (1994). *Birds to watch 2: the world list of threatened birds*. Birdlife International, Cambridge.

Conder, P. J. (1950). On the courtship and social displays of three species of auk. *British Birds*, **43**, 65–9.

Connell, R. (1887). *St. Kilda and the St. Kildans*. Hamilton Adams and Co., London.

Corkhill, P. (1972). Measurements of puffins as criteria of sex and age. *Bird Study*, **19**, 193–201.

Corkhill, P. (1973). Food and feeding ecology of Puffins. *Bird Study*, **20**, 207–20.

Coues, E. (1868). A monograph of the Alcidae. *Proceedings of the Academy of Natural History, Philadelphia*, 1–81.

Coyle, K. O., Hunt, G. L., Jr., Decker, M. B., and Weingartner, T. J. (1992). Murre foraging, epibenthic sound scattering and tidal advection over a shoal near St. George Island, Bering Sea. *Marine Ecology Progress Series*, **83**, 1–14.

Cracraft, J. (1981). Towards a phylogenetic classification of the recent birds of the world. *Auk* **98**, 681–714.

Cramp, S. (1985). *The Birds of the Western Palaearctic*. Oxford University Press, Oxford.

Creelman, E., & Storey, A. E. (1991). Sex differences in reproductive behavior of Atlantic Puffins. *Condor*, **93**, 390–8.

Croll, D. A. (1990). Physical and biological determinants of the abundance, distribution, and diet of the Common Murre in Monterey Bay, California. *Studies in Avian Biology*, **14**, 139–48.

Croll, D. A. and Mclaren, E. (1993). Diving metabolism and thermoregulation in Common and Thick-billed Murres. *Journal of Comparative Physiology—B, Biochemical, Systemic, & Environmental Physiology*, **163**, 160–6.

Croll, D. A., Gaston, A. J., and Noble, D. G. (1991). Adaptive loss of mass in Thick-billed Murres. *Condor*, **93**, 496–502.

Croll, D. A., Gaston, A. J., Burger, A. E., and Konnoff, D. (1992). Foraging behavior and physiological adaptation for diving in Thick-billed Murres. *Ecology*, **73**, 344–56.

Croxall, J. P. (1991). *Seabird status and conservation: a supplement*. International Council for Bird Preservation, Cambridge.

Croxall, J. P. and Gaston, A. J. (1988). Patterns of reproduction in high-latitude northern- and southern-hemisphere seabirds. *Proceedings of the International Ornithological Congress*, **19**, 1176–94.

Croxall, J. P., Evans, P. G. H., and Schreiber, R. W. (1984). *Status and Conservation of the world's seabirds*. International Council for Bird Preservation, Cambridge.

Cullen, M. (1954). The diurnal rhythm of birds in the arctic summer. *Ibis*, **96**, 31–46.

Danchin, E. (1983). La posture de post-aterissage chez le Macareux Moine (*Fratercula arctica*). *Biology of Behaviour*, **8**, 3–10.

Darwin, C. (1859). *On the origin of species by means of natural selection*. John Murray, London.

Darwin, C. (1871). *The descent of man and selection in relation to sex*. John Murray, London.

Davis, M. B. and Guderley, H. (1987). Energy metabolism in the locomotor muscles of the Common Murre (*Uria aalge*) and the Atlantic Puffin (*Fractercula arctica*). *Auk*, **104**, 733–9.

Dawson, A. G. (1992). *Ice age earth: late quaternary geology and climate*. Routledge, London and New York.

Day, R. H. (1995). New information on Kittlitz's Murrelet nests. *Condor*, **97**, 271–3.

Day, R. H. (1996). Nesting phenology of Kittlitz's Murrelet. *Condor*, **98**, 433–7.

Day, R. H. and Byrd, G. V. (1989). Food habits of the Whiskered Auklet on Buldir Island, Alaska. *Condor*, **91**, 65–72.

Day, R. H., Oakley, K. L., and Barnard, D. R. (1983). Nest sites and eggs of Kittlitz's and Marbled Murrelets. *Condor*, **85**, 265–73.

Day, R. H., DeGange, A. R., Divoky, G. J., and Troy, D. M. (1988). Distribution and subspecies of the Dovekie in Alaska. *Condor*, **90**, 712–4.

de Forest, L. N. (1993). The effect of age, timing of breeding, and breeding site characteristics on the reproductive success of the Thick-billed Murre, *Uria lomvia*. M.Sc. thesis. University of Ottawa.

de Forest, L. N. and Gaston, A. J. (1996). The effect of age on timing of breeding and reproductive success in the Thick-billed Murre. *Ecology*, **77**, 1001–11.

de Wijs, W. J. R. (1978). De geographiese varietie van der zeekoet (*Uria aalge*, Pontoppidan) en de mogelijke relatie hiervan met de laat-Pleistocene geschiedenis van de Noor-delijke Atlantise Ocean. Ph.D. thesis. University of Amsterdam, Netherlands.

DeGange, A. R. and Nelson, J. W. (1982). Bald Eagle predation on nocturnal seabirds. *Journal of Field Ornithology*, **53**, 407–9.

DeGange, A. R., Possardt, E. E., and Frazer, D. A. (1977). Breeding biology of seabirds on the Forrester Island National Wildlife Refuge, 15 May–1 September 1976. U.S. Fish and Wildlife Service Ms. Report, Anchorage, Alaska.

Desrocher, A. E., Gill, M. J., Kaiser, G. W., Manley, I. A., and Cooke, F. (1994). *Ecology of Marbled Murrelets in Desolation Sound during the 1994 breeding season*. Canadian Wildlife Service, Delta, BC.

DeWeese, L. R. and Anderson, D. W. (1976). Distribution and breeding biology of Craveri's Murrelet. *Transactions of the San Diego Natural History Society*, **18**, 155–68.

Diamond, A. W. (1983). Feeding overlap in some tropical and temperate seabird communities. *Studies in Avian Biology*, **8**, 24–46.

Diamond, A. W., Gaston, A. J., and Brown, R. G. B. (1993). Studies of high latitude seabirds 3, A model of the energy demands of seabirds of eastern and arctic North America. *Canadian Wildlife Service Occasional Papers*, **77**, 1–39.

Dick, M. H. and Donaldson, M. (1978). Fishing vessel endangered by Crested Auklet landings. *Condor*, **80**, 235–6.

Dinetz, V. L. (1992). Nesting of the Spectacled Guillemot on the Chukotka Peninsula. In *Studies of marine birds in the USSR*. Institute of Biological Problems of the North, Magadan, Russia.

Divoky, G. J. (1982). The occurrence and behavior of non-breeding Horned Puffins at Black Guillemot colonies in northern Alaska. *Wilson Bulletin*, **94**, 356–8.

Divoky, G. J., Watson, G. E., and Bartonek, J. C. (1974). Breeding of the Black Guillemot in northern Alaska. *Condor*, **76**, 339–43.

Dolphin, W. F. and McSweeney, D. (1983). Incidental ingestion of Cassin's Auklets by Humpback Whales. *Auk*, **100**, 214–5.

Donaldson, G., Chapdelaine, G., and Andrew, J. D. (1995). Predation of Thick-billed Murres, *Uria lomvia*, at two breeding colonies by Polar Bears, *Ursus maritimus*, and Walruses, *Odobenus rosmarus*. *Canadian Field-Naturalist*, **109**, 112–4.

Donaldson, G. M., Gaston, A. J., Kampp, K., Chardine, J., Nettleship, D. N., and Elliot, R. D. (1997). Winter distributions of Thick-billed Murres from the eastern Canadian arctic and western Greenland in relation to age and time of year. *Canadian Wildlife Service Occasional Papers*, No. 96, 26 pp.

Drent, R. H. (1965a). Breeding biology of the Pigeon Guillemot. Ph.D. thesis. University of British Columbia.

Drent, R. H. (1965*b*). Breeding biology of the Pigeon Guillemot, *Cepphus columba*. *Ardea*, **53**, 99–160.

Drost, C. A. and Lewis, D. B. (1995). Xantus' Murrelet. In *The birds of North America No. 164* (ed. A. Poole and F. Gill), pp. 1–24. Academy of Natural Sciences of Philadelphia and the American Ornithologists' Union, Washington, DC.

Duchaussoy, M. H. (1897). *Le Grand Pengouin*. Musee d'histoire Naturelle d'Amiens, Amiens.

Duffy, D. C., Todd, F. S., and Siegfried, W. R. (1987). Submarine foraging behavior of Alcids in an artificial environment. *Zoo Biology*, **6**, 373–8.

Duncan, D. C. and Gaston, A. J. (1990). Movements of Ancient Murrelet broods away from a colony. *Studies in Avian Biology*, **14**, 109–13.

Elliot, R. D. (1991). The management of the Newfoundland turr hunt. In *Studies of high latitude seabirds. 2. Conservation biology of the thick-billed murre in the Northwest Atlantic*, (ed. A. J. Gaston, and R. D. Elliot), pp. 29–35. Candian Wildlife Service, Ottawa.

Elliot, R. D., Ryan, P. C., and Lidster, W. W. (1990). The winter diet of the Thick-billed Murre in coastal Newfoundlan waters. *Studies in Avian Biology*, **14**, 125–38.

Elliot, R. D., Collins, B. T., Hayakawa E. G., and Metras, L. (1991). The harvest of murres in Newfoundland from 1977–78 to 1987–88. In *Studies of high latitude seabirds. 2. Conservation biology of the thick-billed murre in the Northwest Atlantic*, (ed. A. J. Gaston and R. D. Elliot), pp. 36–44. Canadian Wildlife Service, Ottawa.

Elphick, C. S. and Hunt, G. L., Jr. (1993). Variations in the distributions of marine birds with water mass in the northern Bering Sea. *Condor*, **95**, 33–44.

Emms, S. K. and Morgan, K. H. (1989). The breeding biology and distribution of the Pigeon Guillemot *Cepphus columba* in the Strait of Georgia. In *Ecology and status of marine birds in the Strait of Georgia, British Columbia*, (ed. K. Vermeer and R. Butler), pp. 100–6. Canadian Wildlife Service, Ottawa.

Emms, S. K. and Verbeek, N. A. M. (1991). Brood size, food provisioning and chick growth in the Pigeon Guillemot *Cepphus columba*. *Condor*, **93**, 943–51.

Emslie, S. D., Sydeman, W. J., and Pyle, P. (1992). The importance of mate retention and experience on breeding success in Cassin's auklet (*Ptychoramphus aleuticus*). *Behavioral Ecology*, **3**, 189–95.

Eppley, Z. A. (1984). Development of thermoregulatory abilities in Xantus' Murrelet chicks *Synthliboramphus hypoleucus*. *Physiological Zoology*, **57**, 307–17.

Evans, P. G. H. (1981). Ecology and behaviour of the Little Auk *Alle alle* in West Greenland. *Ibis*, **123**, 1–18.

Evans, P. G. H. and Kampp, K. (1991). Recent changes in Thick-billed Murre populations in West Greenland. In *Studies of high latitude seabirds. 2. conservation biology of the thick-billed murre in the Northwest Atlantic* (ed. A. J. Gaston and R. D. Elliot), pp. 7–14. Canadian Wildlife Service, Ottawa.

Evans, P. G. H. and Nettleship, D. N. (1985) Conservation of the Atlantic Alcidae. In *The Atlantic Alcidae*, (ed. D. N. Nettleship and T. R. Birkhead), pp. 428–88. Academic Press, London.

Everett, W. T. (1992). Within and among-year nest site sharing by five seabird species at Islas Los Coronados, Baja California, Mexico. *Proceedings PSG Annual meeting*, 1992, (abstract).

Everett, W. T. and Anderson, D. W. (1991). Status and conservation of the breeding seabirds of offshore Pacific islands of Baja California and the Gulf of California. In *Seabird status and conservation: a supplement*, (ed. J. P. Croxall), pp. 115–40. International Council for Bird Preservation, Cambridge.

Ewins, P. J. (1985). Colony attendance and censusing of Black Guillemot *Cepphus grylle* in Shetland. *Bird Study*, **32**, 176–85.

Ewins, P. J. (1986). The ecology of Black Guillemots *Cepphus grylle* in Shetland. Ph.D. thesis. Oxford University.

Ewins, P. J. (1987). Opportunistic feeding of Black Guillemots *Cepphus grylle* at fishing vessels. *Seabird*, **10**, 58–9.

Ewins, P. J. (1988). The timing of moult in Black Guillemots *Cepphus grylle* in Shetland. *Ringing and Migration*, **9**, 5–10.

Ewins, P. J. (1992). Growth of Black Guillemot *Cepphus grylle* chicks in Shetland in 1983–84. *Seabird*, **14**, 3–14.

Ewins, P. J. (1993). Pigeon Guillemot. In *The birds of North America, No. 49*, (ed. A. Poole and F. Gill), pp. 1–24. Academy of Natural Sciences of Philadelphia and the American Ornithologists' Union, Washington, DC.

Ewins, P. J. and Kirk, D. A. (1985). Autumn and winter distribution and ringing recoveries of Shetland Tysties *Cepphus grylle*. Unpublished report to the Shetland Oil Terminal Group.

Ewins, P. J., Carter, H. R., and Shibaev, Y. (1993). The status, distribution and ecology of inshore fish-feeding alcids (*Cepphus* guillemots and *Brachyramphus* murrelets) in the North Pacific. In *The status ecology and conservation of marine birds of the North Pacific*, (ed. K. Vermeer, K. T. Briggs, K. H. Morgan, and D. Siegel-Causey), pp. 164–75. Canadian Wildlife Service, Ottawa.

Falk, K., Jensen, J. -K., and Kampp, K. (1992). Winter diet of Atlantic Puffins (*Fratercula arctica*) in the northeast Atlantic. *Colonial Waterbirds*, **15**, 230–5.

Falk, K. and Durinck, J. (1993). The winter diet of Thick-billed Murres, *Uria lomvia*, in western Greenland, 1988–1989. *Canadian Journal of Zoology*, **71**, 264–72.

Farrand, J., Jr. (1983). *The Audubon Society master guide to birding*. Alfred A. Knopf, New York.

Feilden, H. W. (1872). Birds of the Faeroe Islands. *Zoologist*, **7** (Ser. 2): 3277–94.

Ferdinand, L. (1969). Some observations on the behaviour of the Little Auk, *Plautus alle*, on the breeding ground, with special reference to voice production. *Dansk Ornithologisk Forenings Tidsskrift*, **63**, 19–45.

Fisher, J. (1952). *The fulmar*. Collins, London.

Fisher, J. and Lockley, R. M. (1954). *Seabirds*. Collins, London.

Fisher, R. A. (1930). *The genetical theory of natural selection*. Clarendon Press, Oxford.

Fitzhugh, W. W. and Crowell, A. (1988). *Crossroads of continents*. Smithsonian Institute, Washington, DC.

Fleming, J. H. (1912). The Ancient Murrelet (*Synthliboramphus antiquus*) in Ontario. *Auk*, **29**, 387–8.

Flint, V. E. and Golovkin, A. N. (1990). *Birds of the USSR: Auks (Alcidae)*. Nauka, Moscow.

Follestad, A. 1990. The pelagic distribution of the Little Auk *Alle alle* in relation to a frontal system off central Norway, March–April 1988. *Polar Research*, **8**, 23–8.

Forssgren, K. and Sjolander, S. (1978). Communal diving in the Guillemot (*Uria aalge*). *Astarte*, **11**, 55–60.

Freuchen, P. and Salomonsen, F. (1958). *The arctic year*. G. P. Putnam's Sons, New York.

Friedmann, H. (1935). The birds of Kodiak Island, Alaska. *Bulletin of the Chicago Academy of Sciences*, **5**, 13–54.

Friesen, V. L., Barrett, R. T., Montevecchi, W. A., and Davidson, W. S. (1993). Molecular identification of a backcross between a female Common Murre × Thick-billed Murre hybrid and a male Common Murre. *Canadian Journal of Zoology*, **71**, 1474–7.

Friesen, V. L., Baker, A. J., and Piatt, J. F. (1996*a*). Phylogenetic relationships within the Alcidae (Aves: Charadriiformes) inferred from total molecular evidence. *Molecular Biology and Evolution*, **13**, 359–67.

Friesen, V. L., Piatt, J. F., and Baker, A. J. (1996*b*). Evidence from allozymes and cytochrome b sequences for a new species of alcid, the long-billed murrelet (*Brachyramphus perdix*). *Condor*, **98**, 681–90.

Friesen, V. L., Montevecchi, W. A., Gaston, A. J., Barrett, R. T., and Davidson, W. S. (1996*c*). Molecular evidence for kin groups in the absence of large-scale genetic differentiation in a migratory bird. *Evolution*, **50**, 924–30.

Fujimaki, Y. (1986). Seabird colonies on Hokkaido Island. In *Ptitsy Dalnego Vostoka (Seabirds of the far east)*, (ed. N. M. Litvinenko), pp. 82–7. Far East Science Centre, USSR Academy of Sciences, Institute of Biology and Soil Sciences, Vladivostok.

Furness, R. W. (1978). Energy requirements of seabird communities: a bioenergetics model. *Journal of Animal Ecology*, **47**, 39–53.

Furness, R. W. (1983). *The birds of Foula*. Brathay Hall Trust, Cumbria.

Furness, R. W. (1987). *The skuas*. T. and A. D. Poyser, Calton.

Furness, R. W. (1991). The occurrence of burrow-nesting among birds and its influence on soil fertility and stability. In *The environmental impact* of burrowing animals and animal burrows, (ed. P. S. Meadows and A. Meadows), pp. 53–67. Zoological Society of London, Oxford.

Furness, R. W. and Barrett, R. T. (1985). The food requirement and ecological relationships of a seabird community in North Norway. *Ornis Scandinavica*, **16**, 305–13.

Furness, R. W. and Nettleship, D. N. (1991). Seabirds as monitors of changing marine environments. *Proceedings of the International Ornithological Congress*, **20**, 2239–40.

Furness, R. W. Thompson, D. R., and Harrison, N. (1994). Biometrics and seasonal changes in body composition of common guillemots, *Uria aalge*, from northwest Scotland. *Seabird*, **16**, 22–9.

Gabrielsen, G. W. (1994). Energy expenditure in Arctic seabirds. Ph.D. thesis. University of Tromso, Norway.

Gabrielsen, G. W., Taylor, J. R. E., Konarzewski, M., and Mehlum, F. (1991). Field and laboratory metabolism and thermoregulation in Dovekies (*Alle alle*). *Auk*, **108**, 71–8.

Gadow, H. (1892). On the classification of birds. *Proceedings of the Zoological Society of London*, **1892**, 229–56.

Gardarsson, A. (1982). Icelandic seabirds. In *Icelandic birds*, (ed. A. Petersen), pp. 15–60. Rit Landvernander 8, Reykjavik, Iceland.

Gardarsson, A. (1985). The huge bird-cliff, Latrabjarg, in Western Iceland. *Environmental Conservation*, **12**, 83–4.

Garrod, A. H. (1873). On certain muscles of the thigh of birds and on their value in classification. *Proceedings of the Zoological Society, London*, **1873**, 626–44.

Gaston, A. J. (1980). Populations, movements and wintering areas of Thick-billed Murres *Uria lomvia* in eastern Canada. Canadian Wildlife Service Progress Notes, 110, 1–10.

Gaston, A. J. (1982). Migration of juvenile Thick-billed Murres through Hudson Strait in 1980. *Canadian Field-Naturalist*, **96**, 30–4.

Gaston, A. J. (1984). How to distinguish first-year murres, *Uria* spp., from older birds in winter. *Canadian Field-Naturalist*, **98**, 52–5.

Gaston, A. J. (1985*a*). Development of the young in the Atlantic Alcidae, In *The Atlantic alcidae*,

(ed. D. N. Nettleship and T. R. Birkhead), pp. 319–54. Academic Press, London.

Gaston, A. J. (1985*b*). Energy invested in reproduction by Thick-billed Murres (*Uria lomvia*). *Auk*, **102**, 447–58.

Gaston, A. J. (1985*c*). The diet of Thick-billed Murre chicks in the eastern Canadian arctic. *Auk*, **102**, 727–34.

Gaston, A. J. (1988). The mystery of the murres: Thick-billed Murres, *Uria lomvia*, in the Great Lakes region, 1890–1986. *Canadian Field-Naturalist*, **102**, 705–11.

Gaston, A. J. (1990). Population parameters of the Ancient Murrelet. *Condor*, **92**, 998–1011.

Gaston, A. J. (1991). Seabirds of Hudson Bay, Hudson Strait and adjacent waters. In *Seabird status and conservation: a supplement*, (ed. J. P. Croxall), pp. 7–16. International Council for Bird Preservation, Cambridge.

Gaston, A. J. (1992*a*). *The ancient murrelet*. T. and A. D. Poyser, London

Gaston, A. J. (1992*b*). Annual survival of breeding Cassin's Auklets in the Queen Charlotte Islands, British Columbia. *Condor*, **94**, 1019–21.

Gaston, A. J. (1994). Status of the Ancient Murrelet, *Synthliboramphus antiquus*, in Canada and the effects of introduced predators. *Canadian Field-Naturalist*, **108**, 211–22.

Gaston, A. J. and Bradstreet, M. S. W. (1993). Intercolony differences in the summer diet of Thick-billed Murres in the eastern Canadian Arctic. *Canadian Journal of Zoology*, **71**, 1831–40.

Gaston, A. J. and Brown, R. G. B. (1991). Dynamics of seabird distributions in relation to variations in the availability of food on a landscape scale. *Proceedings of the International Ornithological Congress*, **20**, 2306–12.

Gaston, A. J. and Dechesne, S. B. C. (1996). Rhinoceros Auklet. In *Birds of North America No. 212*, (ed. A. Poole and F. Gill), pp. 1–20. Academy of Natural Science of Philadelphia and the American Ornithologists' Union, Washington, DC.

Gaston, A. J. and Donaldson, G. (1995). Peat deposits and Thick-billed Murre colonies in Hudson Strait and northern Hudson Bay: clues to post-glacial colonization of the area by seabirds. *Arctic*, **48**, 354–8.

Gaston, A. J. and Elliot, R. D. (1996). Predation by ravens *Corvus corax* on Brunnich's Guillemots *Uria lomvia* eggs and chicks and its possible impact on breeding site selection. *Ibis*, **138**, 742–8.

Gaston, A. J. and Jones, I. L. (1989). The relative importance of stress and programmed anorexia in determining mass loss by incubating Ancient Murrelets. *Auk*, **106**, 653–8.

Gaston, A. J. and Masselink, M. (1997). The impact of raccoons, *Procyon lotor*, on breeding seabirds at Englefield Bay, Haida Gwaii, Canada. *Bird Conservation International*, 7, 35–51.

Gaston, A. J. and McLaren, P. L. (1990). Winter observations of Black Guillemots in Hudson Bay and Davis Strait. *Studies in Avian Biology*, **14**, 67–70.

Gaston, A. J. and Nettleship, D. N. (1981). *The Thick-billed Murres of Prince Leopold Island*. Canadian Wildlife Service Monograph No. 6, Canadian Wildlife Service, Ottawa.

Gaston, A. J. and Noble, D. G. (1985). The diet of Thick-billed Murres (*Uria lomvia*) in west Hudson Strait and northeast Hudson Bay. *Canadian Journal of Zoology*, **63**, 1148–60.

Gaston, A. J. and Perin, S. (1993). Loss of mass in breeding Brunnich's Guillemots *Uria lomvia* is triggered by hatching. *Ibis*, **135**, 472–4.

Gaston, A. J. and Powell, D. W. (1989). Natural incubation, egg neglect, and hatchability in the Ancient Murrelet. *Auk*, **106**, 433–8.

Gaston, A. J., Chapdelaine, G., and Noble, D. G. (1983a). The growth of Thick-billed Murre chicks at colonies in Hudson Strait: inter- and intra-colony variation. *Canadian Journal of Zoology*, **61**, 2465–75.

Gaston, A. J., Goudie, R. I., Noble, D. G., and MacFarlane, A. (1983b). Observations on turr hunting in Newfoundland: age, body composition and diet of Thick-billed Murres (*Uria lomvia*) and proportions of other birds killed off Newfoundland in winter. *Canadian Wildlife Service Technical Report*, **141**, 1–7.

Gaston, A. J., Chapdelaine, G., and Noble, D. G. (1984). Phenotypic variation among Thick-billed Murres from colonies in Hudson Strait. *Arctic*, **37**, 284–7.

Gaston, A. J., Cairns, D.K., Elliot, R. D., and Noble, D. G. (1985). *A natural history of Digges Sound*. Canadian Wildlife Service, Ottawa.

Gaston, A. J., Jones, I. L., and Noble, D. G. (1988a). Monitoring Ancient Murrelet breeding populations. *Colonial Waterbirds*, **11**, 58–66.

Gaston, A. J., Jones, I. L., Noble, D. G., and Smith, S. A. (1988b). Orientation of ancient murrelet, *Synthliboramphus antiquus*, during their passage from the burrow to the sea. *Animal Behaviour*, **36**, 300–3.

Gaston, A. J., Carter, H. R., and Sealy, S. G. (1993a). Winter ecology and diet of Ancient Murrelets off Victoria, British Columbia. *Canadian Journal of Zoology*, **71**, 64–70.

Gaston, A. J., de Forest, L. N., Gilchrist, G., and Nettleship, D. N. (1993b). *Monitoring Thick-billed Murre populations at colonies in northern Hudson Bay, 1972–1992*. Canadian Wildlife Service, Ottawa.

Gaston, A. J., Deforest, L. N., and Noble, D. G. (1993c). Egg recognition and egg stealing in murres (*Uria* spp). *Animal Behaviour*, **45**, 301–6.

Gaston, A. J., Deforest, L. N., Donaldson, G., and Noble, D. G. (1994). Population parameters of Thick-billed Murres at Coats Island, Northwest Territories, Canada. *Condor*, **96**, 935–48.

Gaston, A. J., Eberl, C., Hipfner, M., and Lefevre, K. (1995). Adoption of chicks among Thick-billed Murres. *Auk*, **112**, 508–10.

Gill, R. E., Jr. and Sanger, G. A. (1979). Tufted Puffins nesting in estuarine habitat. *Auk*, **96**, 792–4.

Gilligan, J., Smith, M., Rogers, D., and Contreras, A. (1993). *Birds of Oregon*. Cinclus Publications, McMinnville, OR.

Gmelin, (1789). *Caroli a Linne Systema Naturae, Tom I*. Par 554–555. G. M. Beer, Leipzig.

Godfrey, W. E. (1986). *The birds of Canada*. National Museums of Canada, Ottawa.

Golovkin, A. N. (1990a). Whiskered Auklet. In *Birds of the USSR: Auks (Alcidae)* (ed. V. E. Flint and A. N. Golovkin), pp. 43–56. Nauka, Moscow.

Golovkin, A. N. (1990b). Thick-billed Murre. In *Birds of the USSR: Auks (Alcidae)* (ed. V. E. Flint and A. N. Golovkin), pp. 43–56. Nauka, Moscow.

Gore, M. E. J. and Pyong-Oh, W. (1971). *The birds of Korea*. Royal Asiatic Society, Korea.

Grant, P. R. and Nettleship, D. N. (1971). Nesting habitat selection by Puffins *Fratercula arctica* in Iceland. *Ornis Scandinavica*, **2**, 81–7.

Greenwood, J. J. D. (1964). The fledging of the Guillemot *Uria aalge* with notes on the Razorbill *Alca torda*. *Ibis*, **106**, 469–81.

Greenwood, J. J. D. (1972). The attendance of Guillemots and Razorbills at a Scottish colony. *Proceedings of the International Ornithological Congress*, **15**, 648.

Greenwood, J. G. (1987). Winter visits by Black Guillemots *Cepphus grylle* to an Irish breeding site. *Bird Study*, **34**, 135–6.

Greenwood, J. G. and L. Marshall. (1989). Breeding season attendance by Black Guillemots at Bangor, Co. Down. *Irish Birds*, **4**, 13–8.

Grover, J. J. and Olla, B. L. (1983). The role of the Rhinoceros Auklet (*Cerorhinca monocerata*) in mixed-species feeding assemblages of seabirds in the Strait of Juan de Fuca, Washington. *Auk* **100**, 979–82.

Guermeur, Y. and Monnat, J. (1980). *Histoire et geographie des oiseaux nicheurs de Bretagne*. Societé pour l'étude et la protection de la nature en Bretagne, Ar Vran.

Gurney, J. H. (1895). The recent abundance of the Little Auk in Norfolk. *Transactions of the Norfolk and Norwich Naturalists Society*, **6**, 67–70.

Gysels, H. and Rabaey, M. (1964). Taxonomic relationships of *Alca torda*, *Fratercula arctica* and *Uria aalge* as revealed by biochemical methods. *Ibis*, **106**, 536–40.

Halley, D. J. (1992). Behaviour, ecology and recruitment of immature Guillemots. Ph.D. thesis. University of St Andrews.

Halley, D. J. and Harris, M. P. (1993). Intercolony movements and behaviour of immature guillemots *Uria aalge*. *Ibis*, **135**, 264–70.

Halley, D. J. and Harris, M. P. (1994). Age-related changes in the agonistic behaviour of Common Guillemots *Uria aalge*. *Seabird*, **16**, 8–14.

Halley, D. J., Harrison, N., Webb, A., and Thompson, D. R. (1995). Seasonal and geographical variations in the diet of Common Guillemots *Uria aalge* off Western Scotland. *Seabird*, **17**, 12–20.

Hamer, T. E. and Nelson, S. K. (1995*a*). Nesting biology and behavior of the Marbled Murrelet. In *Ecology and conservation of the marbled murrelet*, (ed. C. J. Ralph, G. L. Hunt, M. G. Raphael, and J. F. Piatt), pp. 49–56. U. S. Forest Service, Albany, CA.

Hamer, T. E. and Nelson, S. K. (1995*b*). Characteristics of Marbled murrelet nest trees and nesting stands. In *Ecology and conservation of the marbled murrelet*, (ed. C. J. Ralph, G. L. Hunt, M. G. Raphael, and J. F. Piatt), pp. 69–82. U.S. Forest Service, Albany, CA.

Haney, J. C. (1991). Influence of pycnocline topography and water column structure on marine distributions of alcids (Aves: Alcidae) in Anadyr Strait, northern Bering Sea, Alaska. *Marine Biology*, **110**, 419–35.

Haney, J. C. and Schauer, A. E. S. (1994). Environmental variability facilitates coexistence within an alcid community at sea. *Marine Ecology Progress Series*, **103**, 221–37.

Haney, J. C. and Solow, A. R. (1992). Analysing quantitative relationships between seabirds and marine resource patches. *Current Ornithology*, **9**, 105–61.

Haney, J. C., Fristrup, K. M., and Lee, D. S. (1992). Geometry of visual recruitment by seabirds to ephemeral foraging flocks. *Ornis Scandinavica*, **23**, 49–62.

Hardy, F. P. (1888). Testimony of early voyagers on the Great Auk. *Auk*, **5**, 380–4.

Harfenist, A. (1991). Effects of growth rate and parental predation risk on fledging of Rhinoceros Auklets (*Cerorhinca monocerata*). M. Sc. thesis. Simon Fraser University, Burnaby, BC.

Harfenist, A. (1995). Effects of growth rate variation on fledging of Rhinoceros Auklets (*Cerorhinca monocerata*). *Auk*, **112**, 60–7.

Harfenist, A., and Ydenberg, R. C. (1996). Parental provisioning and predation risk in Rhinoceros Auklets (*Cerorhinca monocerata*) — effects on nestling growth and fledging. *Behavioral Ecology*, **6**, 82–6.

Harris, M. P. (1976). Lack of a 'desertion period' in the nestling life of the Puffin *Fratercula arctica*. *Ibis*, **118**, 115–7.

Harris, M. P. (1979). Measurements and weights of British Puffins. *Bird Study*, **26**, 179–86.

Harris, M. P. (1980). Breeding performance of puffins *Fratercula arctica* in relation to nest density, laying date and year. *Ibis*, **122**, 193–209.

Harris, M. P. (1981). Age determination and first breeding of British puffins. *British Birds*, **74**, 246–56.

Harris, M. P. (1983*a*). Biology and survival of the immature puffin *Fratercula arctica*. *Ibis* **125**, 56–73.

Harris, M. P. (1983*b*). Parent–young communication in the Puffin *Fratercula arctica*. *Ibis*, **125**, 109–14.

Harris, M. P. (1984). *The puffin*. T. and A. D. Poyser, London.

Harris, M. P. (1991). Population changes in British Common Murres and Atlantic Puffins, 1969–88. In *Studies of high-latitude seabirds. 2, Conservation biology of thick-billed murres in the Northwest Atlantic*, (ed. A. J. Gaston and R. D. Elliot), pp. 52–61. Canadian Wildlife Service, Ottawa.

Harris, M. P. and Hislop, J. R. G. (1978). The food of young Puffins *Fratercula arctica*. *Journal of Zoology, London*, **185**, 213–36.

Harris, M. P. and Murray, S. (1977). Puffins on St. Kilda. *British Birds*, **70**, 50–65.

Harris, M. P. and Wanless, S. (1985). Fish fed to young Guillemots *Uria aalge*, and used in displays on the Isle of May, Scotland. *Journal of Zoology, London*, **207**, 441–58.

Harris, M. P. and Wanless, S. (1986*a*). A comparison of the biology of the normal and bridled forms of the Guillemot *Uria aalge* at a single colony. *Journal of Zoology, London*, **210**, 121–30.

Harris, M. P. and Wanless, S. (1986*b*). The food of young Razorbills on the Isle of May and a comparison with that of young Guillemots and Puffins. *Ornis Scandinavica*, **17**, 41–6.

Harris, M. P. and Wanless, S. (1988*a*). Measurement and seasonal changes in weight of guillemots *Uria aalge* at a breeding colony. *Ringing and Migration*, **9**, 32–6.

Harris, M. P. and Wanless, S. (1988*b*). The breeding biology of Guillemots *Uria aalge* on the Isle of May over a six year period. *Ibis*, **130**, 172–92.

Harris, M. P. and Wanless, S. (1990*a*). Moult and autumn colony attendance of auks. *British Birds*, **83**, 55–66.

Harris, M. P. and Wanless, S. (1990*b*). Breeding status and sex of Common Murres (*Uria aalge*) at a colony in autumn. *Auk* **107**, 603–28.

Harris, M. P. and Yule, R. F. (1977). The moult of the puffin *Fratercula arctica*. *Ibis*, **119**, 535–41.

Harris, M. P., Webb, A., and Tasker, M. L. (1991). Growth of young Guillemots *Uria aalge* after leaving the colony. *Seabird*, **13**, 40–4.

Harris, M. P., Halley, D. J., and Wanless, S. (1992). The postfledging survival of young guillemots *Uria aalge* in relation to hatching date and growth. *Ibis*, **134**, 335–9.

Harrison, C. S., He Fen-Qi, Kyong Su Choe, and Shibaev, Y. V. (1992). The laws and treaties of North Pacific rim nations that protect seabirds on land and at sea. *Colonial Waterbirds*, **15**, 264–77.

Harrison, N. M. (1990). Gelatinous zooplankton in the diet of the Parakeet Auklet: comparisons with other Auklets. *Studies in Avian Biology*, **14**, 114–24.

Harrison, N. M. Hunt, G. L., and Cooney, R. T. (1990). Front affecting the distribution of seabirds in the northern Bering Sea. *Polar Research*, **8**, 29–31.

Hartley, C. H. and Fisher, J. (1936). The marine foods of birds in an inland fjord region in west Spitzbergen. *Journal of Animal Ecology*, **5**, 370–89.

Harvey, P. H. and Pagel, M. D. (1991). *The comparative method in evolutionary biology*. Oxford University Press, Oxford.

Hashimoto, H. (1993). Winter feeding of Thick-billed Murres in coastal waters off the Shikotan Peninsular, Hokkaido. *Journal of the Yamashina Institute for Ornithology*, **25**, 166–73.

Hatch, S. A. (1983). The fledging of Common and Thick-billed Murres on Middleton Island, Alaska. *Journal of Field Ornithology*, **54**, 266–74.

Hatch, S. A. (1984). Nestling diet and feeding rates of rhinoceros auklets in Alaska. In *Marine birds, their feeding ecology and commercial fisheries relationships*, (ed. D. N. Nettleship, G. A. Sanger, and P. F. Springer), pp. 106–15. Canadian Wildlife Service, Ottawa.

Hatch, S. A. (1993). Population trends of Alaskan seabirds. *Pacific Seabird Group Bulletin*, **20**, 3–12.

Hatch, S. A. (1994). Review of: Gaston, A. J. The Ancient Murrelet. *Auk*, **111**, 242–3.

Hatch, S. A. and Hatch, M. A. (1989). Attendance patterns of murres at breeding sites: implications for monitoring. *Journal of Wildlife Management*, **53**, 483–93.

Hatch, S. A. and Hatch, M. A. (1990). Components of breeding productivity in a marine bird community: key factors and concordance. *Canadian Journal of Zoology*, **68**, 1680–90.

Hatch, S. A. and Sanger, G. A. (1992). Puffins as samplers of juvenile pollock and other forage fish in the Gulf of Alaska. *Marine Ecology Progress Series*, **80**, 1–14.

Hatchwell, B. (1988*a*). Population biology and coloniality of Common Guillemots *Uria aalge*. Ph.D. thesis. University of Sheffield.

Hatchwell, B. J. (1988*b*). Intraspecific variation in extra-pair copulation and mate defence in common guillemots *Uria aalge*. *Behaviour*, **107**, 157–85.

Hatchwell, B. J. (1991). An experimental study of the effects of timing of breeding on the reproductive success of Common Guillemots *Uria aalge*. *Journal of Animal Ecology*, **60**, 721–36.

Hatchwell, B. J. and Birkhead, T. R. (1991). Population dynamics of Common Guillemots *Uria aalge* on Skomer Island, Wales. *Ornis Scandinavica*, **22**, 55–9.

Hatchwell, B. J. and Pellatt, J. (1990). Intraspecific variation in egg composition and yolk formation in the common guillemot (*Uria aalge*). *Journal of Zoology, London*, **220**, 279–86.

Haury, L. R., McGowan, J. A., and Weibe, P. H. (1978). Patterns and processes in the time-space scales of plankton distributions. In *Spatial pattern in plankton communities*, (ed. J. H. Steel), pp. 277–327. Plenum Press, New York.

Hayes, D. L. (1995). Recovering monitoring of Pigeon Guillemot populations in Prince William Sound, Alaska. *Pacific Seabirds*, 22, 32.

Hayward, J. L., Galusha, J. G., and Frais, G. (1993). Analysis of Great Horned Owl pellets with Rhinoceros Auklet remains. *Auk*, **110**, 133–5.

Heath, H. (1915). Birds observed on Forrester Island, Alaska, the summer of 1913. *Condor*, **17**, 20–41.

Hedgren, S. (1980). Reproductive success of Guillemots *Uria aalge* on the island of Stora Korlso. *Ornis Fennica*, **57**, 49–57.

Hedgren, S. (1981). Effects of fledging weight and time of fledging on survival of Guillemot *Uria aalge* chicks. *Ornis Scandinavica*, **12**, 51–4.

Heubeck, M. (1989). *Seabirds and sandeels: proceedings of a seminar held in Lerwick, Shetland, 15–16 October 1989*. Shetland Bird Club, Lerwick, Shetland.

Higuchi, Y. (1993*a*). Status of breeding grounds on Mikomotojima, Mimianajima, Kojima, the Danjo archipelago and the Nanatsujima Islands.

In *The Japanese murrelet: its status and conservation. Proceedings of the 1993 meeting of the Japan Ornithological Society*, (trans. by J. Fries) (ed. K. Ono). Japan Ornithological Society, Toho University, Japan.

Higuchi, Y. (1993*b*). The Japanese Murrelet: its status and conservation. In *The Japanese murrelet: its status and conservation. Proceedings of the 1993 meeting of the Japan Ornithological Society*, (trans. by J. Fries) (ed. K. Ono). Japan Ornithological Society, Toho University, Japan.

Hipfner, J. M. and Byrd, G. V. (1993). Breeding biology of the Parakeet Auklet compared to other crevice-nesting species at Buldir Island, Alaska. *Colonial Waterbirds*, **16**, 128–38.

Hirsch, K. V., Woodby, D. A., and Astheimer, L. B. (1981). Growth of a nestling Marbled Murrelet. *Condor*, **83**, 264–5.

Hislop, J. R. G. and Harris, M. P. (1985). Recent changes in the food of young Puffins *Fratercula arctica* on the Isle of May in relation to fish stocks. *Ibis*, **127**, 234–9.

Hobson, K. A. (1990). Stable isotope analysis of marbled Murrelets: evidence for freshwater feeding and determination of trophic level. *Condor*, **92**, 897–903.

Hobson, K. A. (1991). Stable isotopic determinations of the trophic relationships of seabirds: preliminary investigations of alcids from coastal British Columbia. In *Studies of high latitude seabirds, I. Behavioural, energetic and oceanographic aspects of seabird feeding ecology*, (ed. W. A. Montevecchi and A. J. Gaston), pp. 16–20. Canadian Wildlife Service, Ottawa.

Hobson, K. A. and Driver, J. C. (1989). Archaeological evidence for the use of the Straits of Georgia by marine birds. In *The ecology and status of marine and shoreline birds in the Straits of Georgia, BC*, (ed. K. Vermeer and R. W. Butler), pp. 168–73. Canadian Wildlife Service, Ottawa.

Hobson, K. A. and Montevecchi, W. A. (1991). Stable isotopic determinations of trophic relationships of Great Auks. *Oecologia*, **87**, 528–31.

Hobson, K. A. and Welch, H. E. (1992). Determination of trophic relationships within a high Arctic marine food web using $\delta^{13}C$ and $\delta^{15}N$ analysis. *Marine Ecology Progress Series*, **84**, 9–18.

Hobson, K. A., Piatt, J. F., and Pitocchelli, J. (1994). Using stable isotopes to determine seabird trophic relationships. *Journal of Animal Ecology*, **63**, 786–98.

Hoffman, W., Heinemann, D., and Wiens, J. A. (1981). The ecology of seabird feeding flocks in Alaska. *Auk*, **98**, 437–56.

Holgersen, H. (1951). Hvor kommer Alken fra? *Transactions of the Stavanger Museum*, **1951**, 1–12.

Hornung, M. and Harris, M. P. (1976). Soil water levels and delayed egg-laying of puffins. *British Birds*, **69**, 402–8.

Houston, A. I., Thompson, W. A., and Gaston, A. J. (1996). An analysis of the 'intermediate' nestling growth strategy in the Alcidae: a time and energy budget model of a parent Thick-billed Murre. *Functional Ecology*, 10, 432–9.

Howard, H. (1968). Tertiary birds from Laguna Hills, Orange County, California. *Los Angleles County Museum Contributions to Science*, **142**, 1–21.

Howard, H. (1971). Pliocene avian remains from Baja California. *Los Angeles County Museum Contributions to Science*, **217**, 1–17.

Howard, H. (1978). Late miocene marine birds from Orange County, California. *Los Angeles County Museum Contributions to Science*, **290**, 1–26.

Howard, H. (1982). Fossil birds from tertiary marine beds at Ocean Side, San Diego County, California, with descriptions of two new species of the genera *Uria* and *Cepphus* (Aves: Alcidae). *Los Angeles County Museum Contributions to Science*, **341**, 1–15.

Hudson, G. E., Hoff, K. M., Vandenberg, J., and Trivette, E. C. (1969). A numerical study of the wing and leg muscles of Lari and Alcae. *Ibis*, **111**, 459–524.

Hudson, P. J. (1979*a*). The parent–chick feeding relationship of the Puffin *Fratercula arctica*. *Journal of Animal Ecology*, **48**, 889–98.

Hudson, P. J. (1979*b*). Survival rates and behaviour of British Auks. D. Phil. thesis. University of Oxford.

Hudson, P. J. (1982). Nest-site characteristics and breeding success in the Razorbill *Alca torda*. *Ibis*, **124**, 355–8.

Hudson, P. J. (1983). The variation and synchronization of daily weight increments of puffin chicks *Fratercula arctica*. *Ibis*, **125**, 557–61.

Hudson, P. (1984). Plumage variation of Razorbill chicks. *British Birds*, **77**, 208–9.

Hudson, P. J. (1985). Population parameters for the Atlantic Alcidae. In *The Atlantic Alcidae*, (ed. D. N. Nettleship and T. R. Birkhead), pp. 233–261. Academic Press, London.

Hunt, G. L., Jr. (1990). The pelagic distribution of marine birds in a heterogeneous environment. *Polar Research*, **8**, 43–54.

Hunt, G. L., Jr. (1995). Oceanographic processes and marine productivity in waters offshore of Marbled Murrelet breeding habitat. In *Ecology and conservation of the marbled murrelet*, (ed. C. J. Ralph, G. L. Hunt, M. G. Raphael, and J. F. Piatt), pp. 219–22. U. S. Forest Service, Albany, CA.

Hunt, G. L., Jr. and Butler, J. L. (1980). Reproductive ecology of Western Gulls and Xantus' Murrelets with respect to food resources in the southern California bight. *CalCOFI Report*, **21**, 62–7.

Hunt, G. L., Jr. and Harrison, N. M. (1990). Foraging habitat and prey taken by Least Auklets at King Island, Alaska. *Marine Ecology Progress Series*, **65**, 141–50.

Hunt, G. L., Burgeson, B., and Sanger, G. A. (1981). Feeding ecology of seabirds of the eastern Bering Sea. In *The eastern Bering Sea shelf: oceanography and resources*, (ed. D. W. Hood and J. A. Calder), pp. 629–48. University of Washington Press, Seattle.

Hunt, G. L., Jr., Eppley, Z. A., and Schneider, D. C. (1986). Reproductive performance of seabirds: the importance of population and colony size. *Auk* **103**, 306–17.

Hunt, G. L., Jr., Harrison, N. M., Hamner, W. M., and Obst, B. S. (1988). Observations of a mixed species flock of birds foraging on euphausiids near St. Matthew island, Bering Sea. *Auk*, **105**, 345–9.

Hunt, G. L., Jr., Harrison, N. M., and Cooney, R. T. (1990). The influence of hydrographic structure and prey abundance on foraging of Least Auklets. *Studies in Avian Biology*, **14**, 7–22.

Hunt, G. L., Jr., Harrison, N. M., and Piatt, J. F. (1993). Foraging ecology as related to the distribution of planktivorous auklets in the Bering Sea. In *The status, ecology, and conservation of marine birds of the North Pacific*, (ed. K. Vermeer,

K. T. Briggs, K. H. Morgan, and D. Siegel-Causey), pp. 18–26. Canadian Wildlife Service, Ottawa.

Hunt, G. L., Jr., Russell, R. W., Coyle, K. O., and Weingartner, T. (1995). Foraging ecology of planktivorous auklets in an Aleutian islands pass. *Colonial Waterbird Society Bulletin*, **19**, 48.

Hunt, G. L., Jr., Bakken, V., and Mehlum, F. (1996). Marine birds in the marginal ice zone of the Barents Sea in late winter and spring. *Arctic*, **49**, 53–61.

Ingolfsson, A. (1961). The taxonomy of Black Guillemots (*Cepphus grylle* (L.)) from Iceland and the Foeroes. M.Sc. thesis. University of Aberdeen.

Ishizawa, T. (1933). Life history of *Synthliboramphus antiquus*. *Plants and Animals, Tokyo*, **1**, 279–80 (in Japanese).

Jehl, J. R. and Everett, W. T. (1985). History and status of the avifauna of Isla Guadalupe, Mexico. *Transactions of the San Diego Society for Natural History*, **20**, 313–36.

Jehl, J. R., Jr. and Bond, S. I. (1975). Morphological variation and species limits in murrelets of the genus *Endomychura*. *Transactions of the San Diego Society for Natural History*, **18**, 9–24.

Johnsgaard, P. A. (1987). *Diving birds of North America*. University of Nebraska Press, Lincoln, Nebraska.

Johnson, R. A. (1941). Nesting behavior of the Atlantic Murre. *Auk*, **58**, 153–63.

Johnson, S. R. and West, G. C. (1975). Growth and development of heat regulation in nestlings, and metabolism of adult Common and Thick-billed Murres. *Ornis Scandinavica*, **6**, 109–15.

Johnson, S. R., Renaud, W. E., Richardson, W. J., Davis, R. A., Coldsworth, C., and Hollingdale, P. D. (1976). Aerial surveys of birds in eastern Lancaster Sound, 1976. Unpublished report for Norlands Petroleum Ltd., L. G. L. Environmental Research Associates, Toronto.

Jones, I. L. (1985). Structure and function of vocalizations a related behaviour of the Ancient Murrelet (*Synthliboramphus antiquus*). M.Sc. thesis. University of Toronto.

Jones, I. L. (1989). Status signalling, sexual selection and the social signals of Least Auklets (*Aethia pusilla*). Ph.D. thesis. Queen's University, Kingston, Ontario.

Jones, I. L. (1990). Plumage variability functions for status signalling in least auklets. *Animal Behaviour*, **39**, 967–75.

Jones, I. L. (1992*a*). Colony attendance of Least Auklets at St. Paul Island, Alaska: implications for population monitoring. *Condor*, **94**, 93–100.

Jones, I. L. (1992*b*). Factors affecting survival of adult Least Auklets (*Aethia pusilla*) at St. Paul Island, Alaska. *Auk*, **109**, 576–84.

Jones, I. L. (1993*a*). Least Auklet (*Aethia pusilla*). In *The birds of North America, No. 69*, (ed. A. Poole and F. Gill), pp. 1–16. The Academy of Natural Sciences of Philadelphia and the American Ornithologists' Union, Washington, DC.

Jones, I. L. (1993*b*). Sexual differences in bill shape and external measurements of crested auklets *Aethia cristatella*. *Wilson Bulletin*, **105**, 525–9.

Jones, I. L. (1993*c*). Crested Auklet (*Aethia cristatella*). In *The birds of North America No. 70*, (ed. A. Poole and F. Gill), pp. 1–16. The Academy of Natural Sciences of Philadelphia and the American Ornithologists' Union, Washington, DC.

Jones, I. L. (1994). Mass changes of Least Auklets *Aethia pusilla* during the breeding season—evidence for programmed loss of mass. *Journal of Animal Ecology*, **63**, 71–8.

Jones, I. L. and Hunter, F. M. (1993). Mutual sexual selection in a monogamous seabird. *Nature*, **362**, 238–9.

Jones, I. L. and Montgomerie, R. (1991). Mating and remating of Least Auklets (*Aethia pusilla*) relative to ornamental traits. *Behavioural Ecology*, **2**, 249–57.

Jones, I. L. and Montgomerie, R. D. (1992). Least Auklet ornaments: do they function as quality indicators? *Behavioral Ecology and Sociobiology*, **30**, 43–52.

Jones, I. L., Falls, J. B., and Gaston, A. J. (1987*a*). Vocal recognition between parents and young of ancient murrelets, *Synthliboramphus antiquus* (Aves: Alcidae). *Animal Behaviour*, **35**, 1405–15.

Jones, I. L., Falls, J. B., and Gaston, A. J. (1987*b*). Colony departure of family groups of Ancient Murrelets. *Condor*, **89**, 940–3.

Jones, I. L., Falls, J. B., and Gaston, A. J. (1989). The vocal repertoire of the Ancient Murrelet. *Condor*, **91**, 699–710.

Jones, I. L., Gaston, A. J., and Falls, J. B. (1990). Factors affecting colony attendance by Ancient Murrelets. *Canadian Journal of Zoology*, **68**, 433–41.

Jouventin, P. and Mougin, J. L. (1981). Les strategies adaptatives des oiseaux de mer. *Terre Vie*, **35**, 217–72.

Kaftanovskii, Y. M. (1951). *Chistikovye ptitsy vostochnoi Atlantiki. Materiely k poznaniyu fauni y flori SSSR (Birds of the Murre group of the eastern Atlantic. Studies on the fauna and flora of the USSR)*. Moscow Society of Naturalists, New Series, Zoology, Moscow.

Kaiser, G. W. (1989). Nightly concentration of Bald Eagles at auklet colony. *Northwestern Naturalist*, **70**, 12–3.

Kaiser, G. W., Bertram, D., and Powell, D. (1984). A band recovery for the Rhinoceros Auklets. *Murrelet*, **65**, 57.

Kampp, K. (1991). Mortality of Thick-billed Murres in Greenland inferred from band recovery data. In *Studies of high latitude seabirds. 2. Conservation biology of the thick-billed murre in the Northwest Atlantic*, (ed. A. J. Gaston and R. D. Elliot), pp. 15–22. Canadian Wildlife Service, Ottawa.

Kampp, K., Nettleship, D. N., and Evans, P. G. H. (1995). Thick-billed Murres of Greenland: status and prospects. In *Seabirds on islands: threats, case studies, and action plans*, (ed. D. N. Nettleship, J. Burger, and M. Gochfeld), pp. 133–54. Birdlife International, Cambridge.

Karnovsky, N. J., Ainley, D. G., Nur, N., and Spear, L. B. (1995). Distribution, abundance and behaviour of Xantus' Murrelets off northern California and southern Oregon. *Colonial Waterbird Society Bulletin*, **19**, 50.

Kartaschew, N. N. (1960) *Die Alkenvogel des Nordatlantiks*. Ziemsen Verlag, Wittenberg Lutherstadt.

Kazama, T. (1971). Mass destruction of *Synthliboramphus antiquus* by oil pollution of Japan Sea. *Yamashina Chorui Kenkyuko Hokoku*, **6**, 389–98.

Kessel, B. (1989). *Birds of the Seward Peninsula, Alaska*. University of Alaska Press, Fairbanks, Alaska.

Kharitonov, S.P. (1990a). Horned Puffin. In *Birds of the USSR: Auks (Alcidae)*, (ed. V. E. Flint and A. N. Golovkin), pp. 164–73. Nauka, Moscow.

Kharitonov, S.P. (1990b). Tufted Puffin. In *Birds of the USSR: Auks (Alcidae)*, (ed. V. E. Flint and A. N. Golovkin), pp. 173–82. Nauka, Moscow.

Kidd, M. G. and Friesen, V. (1995). A guillemot intraspecific phylogeny: inferring population history from patterns of geographical variation in *Cepphus* mitochondrial control regions. *Colonial Waterbird Society Bulletin*, **19**, 50.

King, W. B. (1984). Incidental mortality of seabirds in gill nets in the North Pacific. In *Status and conservation of the world's seabirds*, (ed. J. P. Croxall, P. G. H. Evans, and R. W. Schreiber), pp. 709–16. International Council for Bird Preservation, Cambridge.

Kirkpatrick, M. (1982). Sexual selection and the evolution of female choice. *Evolution*, **36**, 1–12.

Kitaysky, A. S. (1994). Breeding biology, feeding ecology and growth energetics of the Spectacled Guillemot (*Cepphus carbo*). *Pacific Seabirds*, **21**, 43.

Kitaysky, A. S. (1995). Relationship between predictability of food and juvenile traits in planktivorous and piscivorous alcids. *Colonial Waterbird Society Bulletin*, **19**, 51.

Knudtson, E. P. and Byrd, G. V. (1982). Breeding biology of Crested, Least and Whiskered auklets at Buldir Island, Alaska. *Condor*, **84**, 197–202.

Konarzewski, M., Taylor, J. R. E., and Gabrielsen, G. W. (1993). Chick energy requirements and adult energy expenditures of Dovekies (*Alle alle*). *Auk*, **110**, 343–53.

Konyukhov, N. (1990a). Crested Auklet. In *Birds of the USSR: Auks (Alcidae)*, (eds. V. E. Flint and A. N. Golovkin), pp. 112–21. Nauka, Moscow.

Konyukhov, N. (1990b). Least Auklet. In *Birds of the USSR: Auks (Alcidae)*, (ed. V. E. Flint and A. N. Golovkin), pp. 125–31. Nauka, Moscow.

Konyukhov, N. (1990c). Parakeet Auklet. In *Birds of the USSR: Auks (Alcidae)*, (ed. V. E. Flint and A. N. Golovkin), pp. 131–9. Nauka, Moscow.

Konyukhov, N. B. and Kitaysky, A. S. (1995). The Asian race of the Marbled Murrelet. In *Ecology and conservation of the marbled murrelet*, (ed. C. J. Ralph, G. L. Hunt, M. G. Raphael, and J. F. Piatt), pp. 23–9. U.S. Forest Service, Albany, CA.

Kooyman, G. L., Cherel, Y., Le Maho, Y., Croxall, J. P., Thorsen, P. H., and Ridoux, V. (1992). Diving behaviour and energetics during foraging cycles in king penguins. *Ecological Monographs*, **62**, 143–63.

Kozlova, E. V. (1957). *The fauna of the USSR: Birds, Vol II (3) Charadriiformes, suborder Alcidae*. pp. 119–43. USSR Academy of Science, Moscow.

Krasrow, L. D. and Sanger, G. A. (1982). *Feeding ecology of birds in the nearshore waters of Kodiak Island*. National Oceans and Atmospheric Administration, Anchorage, Alaska.

Kress, S. W. and Nettleship, D. N. (1988). Re-establishment of Atlantic Puffins (*Fratercula arctica*) at a former breeding site in the Gulf of Maine. *Journal of Field Ornithology*, **59**, 161–70.

Kuletz, K. (1983). Mechanisms and consequences of foraging behaviour in a population of breeding Pigeon Guillemots. M.Sc. thesis. University of California.

Kuroda, N. (1954). On some osteological and anatomical characters of Japanese Alcidae (Aves). *Japanese Journal of Zoology*, **11**, 311–27.

Kuroda, N. (1963). A survey of the seabirds of Teuri Island, Hokkaido, with notes on land birds. *Miscellaneous reports of the Yamashina Institute*, **3**, 363–83.

Kuroda, N. (1967). Morpho-anatomical analysis of parallel evolution between Diving-Petrel and Ancient Auk, with comparative osteological data of other species. *Miscellaneous Reports of the Yamashina Institute*, **5**, 111–37.

Lack, D. (1966). *Population studies of birds*. Clarendon Press, Oxford.

Lack, D. (1968). *Ecological adaptations for breeding in birds*. Methuen and Co., London.

Lande, R. (1981). Models of speciation by sexual selection on polygenic traits. *Proceedings of the National Academy of Sciences, USA*, **78**, 3721–5.

Lenhausen, W. A. (1980). Nesting habitat relationships of four species of alcid at Fish Island, Alaska, M.Sc. thesis. University of Alaska.

Leschner, L. L. (1976). The breeding biology of the Rhinoceros Auklet on Destruction Island. M.Sc. thesis. University of Washington.

Lid, G. (1981). Reproduction of the Puffin on Rost in the Lofoten Islands in 1964–1980. *Fauna norvegica Series C, Cinclus*, **4**, 30–9.

Linnaeus, C. (1758). *Systema Naturae, 10th edn*. Stockholm.

Litvinenko, N. M. and Shibaev, Y. V. (1987). The Ancient Murrelet *Synthliboramphus antiquus* (Gm.): reproductive biology and raising of young. In *Distribution and biology of seabirds of the Far East*, (ed. N. M. Litvinenko), pp. 72–84. Far Eastern Science Centre, Vladivostok.

Livezey, B. C. (1988). Morphometrics of flightlessness in the Alcidae. *Auk*, **105**, 681–98.

Lloyd, C. S. (1971). Observations on the incubation and fledging period of the razorbill *Alca torda*. *Skokholm Bird Observatory Report*, **1971**, 35–9.

Lloyd, C. S. (1976). The breeding biology and survival of the Razorbill *Alca torda* L. Ph.D. thesis. University of Oxford.

Lloyd, C. S. (1977). The ability of the Razorbill *Alca torda* to raise an additional chick to fledging. *Ornis Scandinavica*, **8**, 155–9.

Lloyd, C. S. (1979). Factors affecting breeding of Razorbills *Alca torda* on Skokholm. *Ibis*, **121**, 165–76.

Lloyd, C. S. and Perrins, C. M. (1977). Survival and age at first breeding in the razorbill (*Alca torda*). *Bird Banding* **48**, 239–52.

Lloyd, C. S., Tasker, M. L., and Partridge, K. (1991). *The status of seabirds in Britain and Ireland*. T. and A. D. Poyser, London.

Lockley, R. M. (1953). *Puffins*. J. M. Dent and Son, London.

Lovenskiold, H. L. (1954). Studies on the avifauna of Spitzbergen. *Norsk Polarinstitutt Skrifter*, **103**, 1–131.

Lovvorn, J. R. and Jones, D. R. (1991). Effects of body size, body fat, and change in pressure with depth on buoyancy and costs of diving in ducks (*Aythya* spp.). *Canadian Journal of Zoology*, **69**, 2879–87.

Lucas, F. A. (1890). The expedition to Funk Island, with observations upon the history and anatomy of the Great Auk. *Report of the National Museum*, **1867–88**, 493–529.

Lydersen, C., Gjertz, I., and Weslawski, J. M. (1985). Aspects of vertebrate feeding in the marine ecosystem in Hornsund, Svalbard. *Oslo*, **21**, 1–57.

Lyngs, P. (1994). The effects of disturbance on growth rate and survival of young Razorbills *Alca torda*. *Seabird*, **16**, 46–9.

Mahon, T. E., Kaiser, G. W., and Burger, A. E. (1992). The role of Marbled Murrelets in mixed-species feeding flocks in British Columbia. *Wilson Bulletin*, **104**, 738–43.

Mahoney, S. P. (1979). Breeding biology and behaviour of the Common Murre *Uria aalge* (Pont.) on Gull Island, Newfoundland. M.Sc. thesis. Memorial University of newfoundland, St John's, Newfoundland.

Makatsch, K. (1974). *Die eier der Vogel europas, I.* Neumann Verlag, Radebuel, Germany.

Malaurie, J. (1982). *The last kings of Thule.* University of Chicago Press, Chicago.

Manuwal, D. A. (1974*a*). Effects of territoriality on breeding in a population of Cassin's Auklet. *Ecology*, 55, 1399–406.

Manuwal, D. A. (1974*b*). The natural history of Cassin's Auklet (*Ptychoramphus aleuticus*). *Condor*, 76, 421–31.

Manuwal, D. A. (1978). Criteria for aging Cassin's Auklets. *Bird Banding*, 49, 157–61.

Manuwal, D. A. (1979). Reproductive commitment and success of Cassin's Auklet. *Condor*, 81, 111–21.

Manuwal, D. A. (1984). Alcids—Dovekie, Murres, Guillemots, Murrelets, Auklets, and Puffins. In *Seabirds of eastern North Pacific and Arctic waters*, (ed. D. Haley), pp. 168–87. Pacific Search Press, Seattle, WA.

Manuwal, D. and Manuwal, N. (1979). Habitat specific behaviour of the Parakeet Auklet. *Western Birds*, 10, 189–200.

Manuwal, D. A. and Thoresen, A. C. (1993). Cassin's Auklet. In *The birds of North America*, (ed. A. Poole and F. Gill), pp. 1–20. The Academy of Natural Sciences of Philadelphia and the American Ornithologists' Union, Washington, DC.

Martin, M. (1698). *A late voyage to St. Kilda, the remotest of the Hebrides, or western isles of Scotland.* Gent, London.

Mathews, D. R. (1983). Feeding ecology of the Common Murre, *Uria aalge*, off the Oregon coast. M.Sc. thesis. University of Oregon.

Mayr, E. and Amadon, D. (1951). A classification of recent birds. *American Museum Novitates*, 1496, 1–42.

McLaren, P. L. (1982). Spring migration and habitat use by seabirds in eastern Lancaster Sound and Western Baffin Bay. *Arctic*, 35, 88–111.

Mead, C. J. (1974). The results of ringing auks in Britain and Ireland. *Bird Study*, 21 45–86.

Mehlum, F. and Giertz, I. (1984). Feeding ecology of seabirds in the Svalbard area—a preliminary report. *Oslo*, 16, 3–41.

Meldgaard, M. (1988). The Great Auk *Pinguinus impennis* (L.) in Greenland. *Historical Biology*, 1, 145–78.

Meyer de Schauensee, R. (1984). *The Birds of China.* Oxford University Press, Oxford.

Minami, H., Shiomi, K., and Ogi, H. (1991). Morphological characteristics and functions of the Spectacled Guillemot *Cepphus carbo*. *Bulletin of the Faculty of Fisheries, Hokkaido University*, 42, 160–81.

Minami, H., Aotsuka, M., Terasawa, T., Maruyama, N., and Ogi, H. (1995). Breeding ecology of the Spectacled Guillemot (*Cepphus carbo*) on Teuri Island. *Journal of the Institute for Ornithology*, 27, 30–40.

Mizutani, H., Kabaya, Y., and Wada, E. (1991). Linear correlation between latitude and soil ^{15}N enrichment at seabird rookeries. *Naturwissenschaften*, 78, 34–6.

Mlikovsky, J. and Kovar, J. (1987). A new auk species (Aves, Alcidae) from the Upper Oligocene of Austria. *Annalen des Naturhistorischen Museums in Wien. Serie A, Mineralogie und Petrographie, Geologie und Paleontologie, Anthropologie und Praehistorie*, 88, 131–48.

Moe, R. A. and Day, R. H. (1977). Populations and ecology of seabirds of the Koniuji Group, Shumagin Islands, Alaska. U. S. Fish and Wildlife Service Unpublished Report, Anchorage, Alaska.

Moen, S. M. (1991). Morphological and genetic variation among breeding colonies of the Atlantic Puffin (*Fratercula arctica*). *Auk* 108, 755–63.

Monaghan, P., Uttley, J. D., Burns, M. D., Tuaine, C., and Blackwood, I. (1989). The relationship between food supply, reproductive effort and breeding success in Arctic Terns. *Journal of Animal Ecology*, 58, 261–74.

Monaghan, P., Walton, P., Wanless, S., Uttley, J. D., and Burns, M. D. (1994). Effects of prey abundance on the foraging behaviour, diving efficiency and time allocation of breeding Guillemots *Uria aalge*. *Ibis*, 136, 214–22.

Montevecchi, W. A. (1993*a*). Birds as indicators of change in marine prey stocks. In *Birds as monitors of environmental change*, (ed. R. W. Furness and

J. J. D. Greenwood), pp. 217–66. Chapman and Hall, New York.

Montevecchi, W. A. (1993*b*). Seabird indication of squid stock conditions. *Journal of Cephalopod Biology*, **2**, 57–63.

Montevecchi, W. A. and Kirk, D. (1996). Great Auk (*Pinguinus impennis*). In *Birds of North America*, (ed. A. Poole and F. Gill), pp. 1–20. Academy of Natural Sciences of Philadelphia and American Ornithologists' Union, Washington, DC.

Montevecchi, W. A. and Tuck, L. M. (1987). *Newfoundland birds: exploitation, study, conservation*. Nuttall Ornithological Club, Cambridge, Massachusetts.

Morbey, Y. E. (1995). Fledging variability and the application of fledging models to the behaviour of Cassin's Auklets (*Ptychoramphus aleuticus*) at Triangle Island, British Columbia. M.Sc. thesis, Simon Fraser University, Burnaby, BC.

Morgan, K. H. (1989). Marine birds of Saanich Inlet, a Vancouver island fjord entering the Straits of Georgia. In *The ecology and status of marine and shoreline birds in the Straits of Georgia*, (ed. K. Vermeer, R. W. Butler, and K. H. Morgan), pp. 158–65. Canadian Wildlife Service, Ottawa.

Morgan, K. H., Vermeer, K., and McKelvey, R. W. (1991). *Atlas of pelagic birds of Western Canada*. Canadian Wildlife Service, Ottawa

Moum, T., Johansen, S., Erikstad, K. E., and Piatt, J. F. (1994). Phylogeny and evolution of the auks (subfamily Alcinae) based on mitochondrial DNA sequences. *Proceedings of the National Academy of Sciences, USA*, **91**, 7912–6.

Moyer, J. T. (1957). The birds of Miyake Jima, Japan. *Auk*, **74**, 215–25.

Murphy, E. C., Roseneau, D. G., and Bente, P. M. (1984). An inland nest record for the Kittlitz's Murrelet. *Condor*, **86**, 218.

Murphy, R. C. (1936). *Oceanic birds of South America*. American Museum of Natural History, New York.

Murphy, E. C., Springer, A. M., and Roseneau, D. G. (1986). Population status of Common Guillemots *Uria aalge* at a colony in Western Alaska: results and simulations. *Ibis*, **128**, 348–63.

Murray, J. M. (1968). The Newfoundland journal of Aaron Thomas. London:

Murray, K. G., Winnett-Murray, K., and Hunt, G. L., Jr. (1980). Egg neglect in Xantus' Murrelet. *Proceedings of the Colonial Waterbirds Group*, **3**, 186–95.

Murray, K. G., Winnett-Murray, K., Eppley, Z. A., Hunt, G. L., Jr., and Schwartz, D. B. (1983). Breeding biology of the Xantus' Murrelet. *Condor*, **85**, 12–21.

Myrberget, S. (1959). Variation in the numbers of puffins present at the colony. *Sterna*, **3**, 239–46.

Myrberget, S. (1962). Contribution to the breeding biology of the Puffin *Fratercula arctica* (L.), eggs, incubation and young. *Meddelelser fra Statens Viltundersokelser 2nd series*, **2**, 1–51.

Nakamura, Y. (1993). Results of the national survey on the distribution of the Japanese Murrelet. In *The Japanese murrelet: its status and conservation. Proceedings of the 1993 meeting of the Japan Ornithological Society* (trans. J. Fries), (ed. K. Ono). Japan Ornithological Society, Toho University, Japan.

Naslund, N. L., Piatt, J. F., and van Pelt, T. (1994). Breeding behaviour and nest site fidelity of Kittlitz's Murrelet. *Pacific Seabirds*, **21**, 46.

Naslund, N. L. and O'Donnell, B. P. (1995). Daily patterns of Marbled Murrelet activity at inland sites. In *Ecology and conservation of the marbled murrelet*, (ed. C. J. Ralph, G. L. Hunt, M. G. Raphael, and J. F. Piatt), pp. 129–34. U.S. Forest Service, Albany, CA.

Nazarov, Y. N. and Labzyuk, V. I. (1972). On the biology of the Spectacled Guillemot in Southern Ussuriland. *Nauch dokl vyssheyshk Biolia nauki*, **15**, 32–5.

Nechaev, V. A. (1986). New information about seabirds on Sakhalin Island. In *Morskie Ptitsy Dalnego Vostoka (Seabirds of the Far East)*, (ed. N. M. Litvinenko), pp. 71–81. Far East Science Centre, USSR Academy of Science, Institute of Biology and Soil Science, Vladivostok, USSR.

Nelson, D. A. (1981). Sexual differences in measurements of Cassin's Auklet. *Journal of Field Ornithology*, **52**, 233–4.

Nelson, D. A. (1982). The communication behaviour of the Pigeon Guillemot *Cepphus columba*. Ph.D. thesis. University of Michigan, Ann Arbor.

Nelson, D. A. (1984). Communication of intentions in agonistic contexts by the Pigeon Guillemot, *Cepphus columba*. *Behaviour*, **88**, 145–89.

Nelson, D. A. (1985). The systematic and semantic organization of Pigeon Guillemot (*Cepphus columba*) vocal behaviour. *Zeitschrift für Tierpsychologie*, **67**, 97–130.

Nelson, D. A. (1991). Demography of the Pigeon Guillemot on Southeast Farallon Island, California. *Condor*, **93**, 765–8.

Nelson, E. W. (1883). *Birds of Bering Sea and the Arctic Ocean In Arctic cruise of the revenue steamer Corwin 1881: notes and observations.* Government Printing Office, Washington, DC.

Nelson, E. W. (1887). *Report upon natural history collections made in Alaska between the years 1877 and 1881.* Government Printing Office, Washington, DC.

Nelson, S. K. (1996). Marbled Murrelet. In *Birds of North America*, (ed. A. Poole, P. Stettenheim, and F. Gill), pp. 1–24. Academy of Natural Sciences of Philadephia and the American Ornithologists' Union, Washington, DC.

Nelson, S. K. and Hamer, T. E. (1995). Nesting biology and behaviour of the Marbled Murrelet. In *Ecology and conservation of the marbled murrelet.* (ed. C. J. Ralph, G. L. Hunt, M. G. Raphael, and J. F. Piatt), pp. 57–67. U.S. Forest Service, Albany, CA.

Nettleship, D. N. (1972). Breeding success of the Common Puffin (*Fratercula arctica* L.) on different habitats at Great Island, Newfoundland. *Ecological Monographs*, **42**, 239–68.

Nettleship, D. N. (1974). Seabird colonies and distribution around Devon Island and vicinity. *Arctic*, **27**, 95–103.

Nettleship, D. N. and Birkhead, T. R. (1985). *The Atlantic Alcidae.* Academic Press, London.

Nettleship, D. N. and Evans, P. G. H. (1985). Distribution and status of the Atlantic Alcidae. In *The Atlantic Alcidae*, (ed. D. N. Nettleship and T. R. Birkhead), pp. 53–154. Academic Press, London.

Nettleship, D. N. and Gaston, A. J. (1978). Patterns of pelagic distribution of seabirds in western Lancaster Sound and Barrow Strait, NWT. *Canadian Wildlife Service Occasional Papers*, **39**, 1–40.

Newton, A. (1861). Abstract of Mr J. Wooley's researches in Iceland respecting the Garefowl or Great Auk (*Alca impennis*, Linn.). *Ibis* **3**, 374–99.

Newton, A. (1865). The Garefowl and its historians. *Natural History Revue*, **1865**, 467–90.

Nice, M. M. (1962). Development of behavior in precocial birds. *Transactions of the Linnaean Society of New York*, **8**, 1–211.

Nol, E. and Gaskin, D. E. (1987). Distribution and movements of Black Guillemots (*Cepphus grylle*) in coastal waters of the southwestern Bay of Fundy, Canada. *Canadian Journal of Zoology*, **65**, 2682–9.

Norderhaug, M. (1980). Breeding biology of the Little Auk (*Plautus alle*) in Svalbard. *Norsk Polarinstitutt Skrifter*, **173**, 1–45.

Norrevang, A. (1986). Traditions of seabird fowling in the Faeroes: an ecological basis for sustained fowling. *Ornis Scandinavica*, **17**, 275–81.

Oakley, K. L. (1981). Determinants of population size of Pigeon guillemots *Cepphus columba* at Naked Island, Prince William Sound, Alaska. M.Sc. thesis, University of Alaska, Fairbanks, Alaska.

Oberholzer, A. (1975). Angeborene Orientierungsvorgange im Fressablnf bei Trottellummen (*Uria aalge aalge* Pont.). *Zeitschrift für Tierpsychologie*, **39**, 150–72.

Ochikubo, L., Carter, H. R., and Fries, J. N. (1995). Status of Japanese Murrelet colonies in the Izu Islands, Japan in 1994. *Pacific Seabirds*, **22**, 39.

O'Connor, R. J. (1984). *The growth and development of birds.* John Wiley and Sons, Chichester.

O'Donald, P. (1980). *Genetic models of sexual selection.* Cambridge University Press, Cambridge.

Ogi, H. (1980). The pelagic feeding ecology of Thick-billed Murres in the North Pacific, March–June. *Bulletin of the Faculty of Fisheries, Hokkaido University*, **31**, 50–72.

Ogi, H. and Shiomi, K. (1991). Diet of murres caught incidentally in winter in Northern Japan. *Auk*, **108**, 184–5.

Ogi, H., Tanaka, H., and Tsujita, T. (1985). The distribution and feeding ecology of murres in the Northwestern Bering Sea. *Journal of the Yamashina Institute for Ornithology*, **17**, 44–56.

Ohlendorf, H. M., Bartonek, J. C., Divoky, G. J., Klaas, E. E., and Krynitsky, A. J. (1982). Organochloride residues in eggs of Alaskan seabirds. *U. S. Fish & Wildlife Service Special Scientific Report—Wildlife*, **245**, 1–41.

Olson, S. (1985). The fossil record of birds. In *Avian Biology, vol. 8*, (ed. D. Farner, J. King, and K. Parkes), pp. 79–238. Academic Press, New York.

Olson, S. and Hasegawa, Y. (1979). Fossil counterparts of giant penguins from the North Pacific. *Science*, **206**, 688–9.

Olson, S. L., Swift, C. C., and Mokhiber, C. (1979). An attempt to determine the prey of the Great Auk (*Pinguinus impennis*). *Auk*, **96**, 790–2.

Ono, K. (1993). *The Japanese murrelet: its status and conservation* (trans. from Japanese by J. Fries). Japan Ornithological Society, Toho University, Japan.

Ono, K. (1996). Japanese Murrelet. In *Status and conservation of rare alcids in Japan*, (ed. K. Ono), pp. 117–24. Japan Alcid Society, Tokyo.

Ono, K. and Nakamura, Y. (1993). The biology of nesting Japanese Murrelets on Biroto Island (Kadokawa-cho, Miyazaki Pref.). In *The Japanese murrelet: its status and conservation* (trans. J. Fries), (ed. K. Ono). Japan Ornithological Society, Toho University, Japan.

Ono, K., Fries, J. N., and Nakamura, Y. (1995). Crow predation on Japanese Murrelets on Biro Island, Japan. *Pacific Seabirds*, **22**, 39.

Paine, R. T., Wootton, J. T., and Boersma, P. D. (1990). Direct and indirect effects of Peregrine Falcon predation on seabird abundance. *Auk*, **107**, 1–9.

Pallas, P. (1811). *Zoographia Rosso-asiatica*, The Imperial Academy of Sciences, St Petersburg.

Paludan, K. (1947). *Alken. Dens ynglebiologi og dens forekomst i Denmerk (A complete biology of the Razorbill in the Baltic)* (in Danish). Ejnar Munksgaard, Copenhagen.

Parkin, T. (1894). The Great Auk or Garefowl (*Alca impennis*, Linn.). J. E. Budd, St Leonards-on-sea.

Parslow, J. L. F. and Jefferies, D. J. (1973). Relationship between organochlorine residues in livers and whole bodies of Guillemots. *Environmental Pollution*, **5**, 87–111.

Pennant, T. (1785). *Arctic zoology, II, class 2, Birds*. Henry Hughs, London.

Pennycuik, C. J. (1956). Observations on a colony of Brunnich's Guillemots, *Uria lomvia*, in Spitzbergen. *Ibis*, **98**, 80–99.

Pennycuik, C. J. (1987). Flight of auks (Alcidae) and other northern seabirds compared with southern Procellariiformes: ornithodolite observations. *Journal of Experimental Biology*, **128**, 335–48.

Percy, J. A. and Fife, F. J. (1985). Energy distribution in an arctic coastal macrozooplankton community. *Arctic*, **38**, 39–42.

Perry, D. A. (1995). Status of forest habitat of the Marbled Murrelet. In *Ecology and conservation of the marbled murrelet*, (ed. C. J. Ralph, G. L. Hunt, M. G. Raphael, and J. F. Piatt), pp. 381–4. U.S. Forest Service, Albany, CA.

Perry, R. (1940). *Lundy, isle of puffins*. L. Drummond, London.

Petersen, A. (1976). Size variables in Puffins *Fratercula arctica* from Iceland, and bill features as criteria of age. *Ornis Scandinavica*, **7**, 185–92.

Petersen, A. (1981). Breeding biology and feeding ecology of Black Guillemots. D.Phill. thesis. University of Oxford.

Petrie, M., Halliday, T., and Sanders, C. (1991). Peahens prefer peacocks with elaborate trains. *Animal Behaviour*, **41**, 323–31.

Piatt, J. F. (1990). The aggregative response of Common Murres and Atlantic Puffins to schools of capelin. *Studies in Avian Biology*, **14**, 35–51.

Piatt, J. F. and Ford, R. G. (1993). Distribution and abundance of Marbled Murrelets in Alaska. *Condor*, **95**, 662–9.

Piatt, J. F. and Gould, P. J. (1994). Postbreeding dispersal and drift-net mortality of endangered Japanese Murrelets. *Auk*, **111**, 953–61.

Piatt, J. F. and Lensink, C. J. (1989). Exxon Valdez bird toll. *Nature*, **342**, 865–6.

Piatt, J. F. and McLagan, R. L. (1987). Common Murre (*Uria aalge*) attendance patterns at Cape St. Mary's, Newfoundland. *Canadian Journal of Zoology*, **65**, 1530–4.

Piatt, J. F. and Naslund, N. L. (1995). Abundance, distribution, and population status of Marbled Murrelets in Alaska. In *Ecology and conservation of the Marbled Murrelet*, (ed. C. J. Ralph, G. L. Hunt, M. G. Raphael, and J. F. Piatt), pp. 285–94. U.S. Forest Service, Albany, CA.

Piatt, J. F. and Nettleship, D. N. (1985). Diving depths of four alcids. *Auk*, **102**, 293–7.

Piatt, J. F., Lensink, C. J., Butler, W., Kendziorek, M., and Nysewander, D. R. (1990*a*). Immediate impact of the Exxon Valdez oil spill on marine birds. *Auk*, **107**, 387–97.

Piatt, J. F., Roberts, B. D., and Hatch, S. A. (1990*b*). Colony attendance and population monitoring of Least and Crested Auklets on St. Lawrence Island, Alaska. *Condor*, **92**, 97–106.

Piatt, J. F., Roberts, B. D., Lidster, W. W., Wells, J. L., and Hatch, S. A. (1990*c*). Effects of human disturbance on breeding Least and Crested Auklets at St. Lawrence Island, Alaska. *Auk*, **107**, 342–50.

Piatt, J. F., Carter, H. R., and Nettleship, D. N. (1991*a*). Effects of oil pollution on marine bird populations. In *The effects of oil on wildlife*, (ed. J. White, L. Frink, T. M. Williams, and R. W. Davis), pp. Sheridan Press, Hanover, Pennsylvania.

Piatt, J. F., Wells, J. L., MacCharles, A., and Fadely, B. S. (1991*b*). The distribution of seabirds and fish in relation to ocean currents in the southeastern Chukchi Sea. In *Studies of high latitude seabirds, 1. Behavioural, energetic and oceanographic aspects of seabird feeding ecology*, (ed. W. A. Montevecchi and A. J. Gaston), pp. 21–31. Canadian Wildlife Service, Ottawa.

Piatt, J. F., Pinchuk, A., Kitaysky, A., Springer, A. M., and Hatch, S. A. (1992). *Foraging distribution and feeding ecology of seabirds at the Diomede islands*, Bering Sea, OCS Study MMS92–0041. Unpublished report to Minerals Management Service, Anchorage, Alaska.

Pierotti, R. (1983). Gull–Puffin interactions on Great Island, Newfoundland. *Biological Conservation*, **26**, 1–14.

Pierotti, R. (1987). Isolating mechanisms in seabirds. *Evolution*, **41**, 559–70.

Pitman, R. L. and Graybill, M. R. (1985). Horned Puffin sightings in the eastern Pacific. *Western Birds*, **16**, 99–102.

Pitman, R. L., Ballance, L. T., and Reilly, S. (1995). Distribution, movements and population status of Craveri's Murrelet: implications for ecology and conservation. *Pacific Seabirds*, **22**, 41.

Pitocchelli, J., Piatt, J., and Cronin, M. A. (1995). Morphological and genetic divergence among Alaskan populations of *Brachyramphus* murrelets. *Wilson Bulletin*, **107**, 235–50.

Plumb, W. J. (1965). Observations on the breeding biology of the Razorbill. *British Birds*, **58**, 449–56.

Pomiankowski, A. N. (1988). The evolution of female mate preference for male genetic quality. *Oxford Surveys in Evolutionary Biology*, **5**, 136–84.

Porter, J. M. and Sealy, S. G. (1981). Dynamics of seabird multi species feeding flocks: chronology of flocking in Barkley Sound, British Columbia, in 1979. *Colonial Waterbirds*, **4**, 104–13.

Porter, J. M. and Sealy, S. G. (1982). Dynamics of seabird multispecies feeding flocks: age-related feeding behaviour. *Behaviour*, **81**, 91–109.

Powers, K. D. (1983). Pelagic distribution of marine birds off the northeastern US. *National Oceanic and Atmosphere Administration, Technical Memorandum NMFS- F/NEC*, **27**, 1–201.

Prach, R. W. and Smith, A. R. (1992). Breeding distribution and numbers of Black Guillemots in Jones Sound, N. W. Territories. *Arctic*, **45**, 111–4.

Preston, F. W. (1962). The canonical distribution of commoness and rarity. *Ecology*, **43**, 185–215.

Preston, W. (1968). Breeding ecology and social behaviour of the Black Guillemot. Ph.D. thesis. University of Michigan, Ann Arbor.

Prince, P. A. and Harris, M. P. (1988). Food and feeding ecology of breeding Atlantic alcids and penguins. *Proceedings of the International Ornithological Congress*, **19**, 1195–204.

Rahn, H. and Ar, A. (1974). The avian egg: incubation time and water loss. *Condor*, **76**, 147–52.

Rahn, H., Paganelli, C. V., and Ar, A. (1975). Relation of avian egg weight to body weight. *Auk*, **92**, 750–65.

Ralph, C. J., Hunt, G. L., Raphael, M. G., and Piatt, J. F. (1995). Ecology and conservation of the Marbled Murrelet in North America: an overview. In *Ecology and conservation of the marbled murrelet*, (ed. C. J. Ralph, G. L. Hunt, M. G. Raphael, and J. F. Piatt), pp. 3–22. U.S. Forest Service, Albany, CA.

Ratcliffe, D. A. (1980). The *peregrine*. T. and A. D. Poyser, Calton.

Raymont, J. E. G. (1980). *Plankton and productivity in the oceans, 2nd edn*. Pergamon Press, Oxford.

Rayner, J. M. V. (1985). Vorticity and propulsion mechanics in swimming and flying vertebrates. *Sunderforschungsbereich, University of Stuttgart and Tubingen*, **230**, 89–118.

Reinsch, H. H. (1976). Zur Ökologie der Seevogel. *Beitrage zur Vogelkund*, **22**, 236–58.

Renaud, W. E. and Bradstreet, M. S. W. (1980). Late winter distribution of Black Guillemots in northern Baffin Bay and the Canadian high arctic. *Canadian Field-Naturalist*, **94**, 421–5.

Renaud, W. E., McLaren, P. L., and Johnson, S. R. (1982). The Dovekie, *Alle alle*, as a spring migrant in eastern Lancaster Sound and western Baffin Bay. *Arctic*, **35**, 118–25.

Rice, J. C. (1985). Interactions of variation in food supply and kleptoparasitism levels on the reproductive success of Common Puffins (*Fratercula arctica*). *Canadian Journal of Zoology*, **63**, 2743–7.

Richardson, F. (1961). Breeding biology of the Rhinoceros Auklet on Protection Island, Washington. *Condor*, **63**, 456–73.

Ricklefs, R. E. (1979). Adaptation constraint and compromise in avian post-natal development. *Biological Reviews*, **54**, 269–90.

Ricklefs, R. E. and Schew, W. A. (1994). Foraging stochasticity and lipid accumulation by nestling petrels. *Functional Ecology*, **8**, 159–70.

Ricklefs, R. E., White, S.C., and Cullen, J. (1980). Energetics of post-natal growth in Leach's Storm-Petrels. *Auk*, **97**, 566–75.

Ricklefs, R. E., Place, A. R., and Andersen, D. J. (1987). An experimental investigation of the influence of diet quality on growth rate in Leach's Storm-Petrel. *American Naturalist*, **130**, 300–5.

Ricklefs, R. E., Day, C. H., Huntington, C. E., and Williams, J. B. (1985). Variability in feeding rate and meal size of Leach's Storm-Petrel at Kent Island, New Brunswick. *Journal of Animal Ecology*, **54**, 883–98.

Riedman, C. (1990). *The pinnipeds: seals, sea lions and walruses*. University of California Press, Berkeley, CA.

Risebrough, R. W., Menzel, D. B., Martin, D. J., and Olcott, H. S. (1995). DDT residues in Pacific Seabirds: a persistent insecticide in marine food chains. *Nature*, **220**, 1098–102.

Robards, M. D., Piatt, J. R., and Wohl, K. D. (1995). Increasing frequency of plastic particles ingested by seabirds in the subarctic North Pacific. *Marine Pollution Bulletin*, **30**, 151–7.

Roberson, D. (1980). *Rare birds of the west coast of North America*. Woodcock Publishing, Pacific Grove, CA.

Roby, D. D. and Brink, K. L. (1986a). Breeding biology of Least Auklets on the Pribilof Islands, Alaska. *Condor*, **88**, 336–46.

Roby, D. D. and Brink, K. L. (1986b). Decline of breeding Least Auklets on St. George Island, Alaska. *Journal of Field Ornithology*, **57**, 57–9.

Roby, D. D. and Ricklefs, R. E. (1986). Energy expenditure in adult Least Auklets and Diving Petrels during the chick-rearing period. *Physiological Zoology*, **59**, 661–78.

Roby, D. D., Brink, K. L., and Nettleship, D. N. (1981). Measurements, chick meals and breeding distribution of Dovekies (*Alle alle*) in Northwest Greenland. *Arctic*, **34**, 241–8.

Rodway, M. S. (1990). *Attendance patterns, hatching chronology and breeding population of Common Murres on Triangle Island, British Columbia following the Nestucca oil spill*. Canadian Wildlife Service Technical Report Series No. 87, Pacific and Yukon Region, Delta, BC.

Rodway, M. S. (1991). Status and conservation of breeding seabirds in British Columbia, In *Supplement to the status and conservation of the world's seabirds*, (ed. J. P. Croxall), pp. 43–102. International Council for Bird Preservation, Cambridge.

Rodway, M. S. (1994). Intra-colony variation in breeding success of Atlantic Puffins: an application of habitat selection theory. M.Sc. thesis, Memorial University, St John's, Newfoundland.

Rodway, M. S., Lemon, M., and Kaiser, G. W. (1988). *Canadian Wildlife Service seabird inventory report No. 1: East coast of Moresby Island*. Canadian Wildlife Service *Technical Report Series No 50*. Pacific and Yukon Region, Delta.

Rodway, M. S., Lemon, M. F., and Summers, K. R. (1990a). *Scott Islands: census results from 1982– 1989 with reference to the Nestucca oil spill. British Columbia seabird colony inventory, report No. 4*. Canadian Wildlife Service, Pacific and Yukon Region, Delta, BC.

Rodway, M. S., Lemon, M. F., Savard, J., and McKelvey, R. (1990b). *Nestucca oil spill: impact assessment on avian populations and habitat*. Canadian Wildlife Service, *Technical Report Series No. 68*. Pacific and Yukon Region, Delta, BC.

Rodway, M. S., Carter, H. R., Sealy, S. G., and Campbell, R. W. (1992a). Status of the Marbled Murrelet in British Columbia. In *Status and conservation of the marbled murrelet in North America*, (ed. H. R. Carter and M. L. Morrison), pp. 17–41. Western Foundation of Vertebrate Zoology, San Diego.

Rodway, M. S., Lemon, M. J. F., and Summers, K. R. (1992b). Seabird breeding populations in the Scott islands on the west coast of Vancouver Island. In *The ecology, status and conservation of marine and shoreline birds on the west coast of Vancouver Island*, (ed. K. Vermeer, R. W. Butler, and K. H. Morgan), pp. 52–9. Canadian Wildlife Service, Ottawa.

Russell, R. W., Hunt, G. L., Jr., Coyle, K. O., and Cooney, R. T. (1992). Foraging in a fractal environment: spatial patterns in a marine predator–prey system. *Landscape Ecology*, 7, 195–209.

Ryan, M. J. (1990). *Sexual selection, sensory systems, and sensory exploitation*. Oxford University Press, Oxford.

Safina, C. and Burger, J. (1988). Ecological dynamics among prey fish, blue fish and foraging Common Terns in an Atlantic coastal system. In *Seabirds and other marine vertebrates, competition, predation and other interactions*, (ed. J. Burger), pp. 95–173. Columbia University Press, New York.

Salomonsen, F. (1941). The Black-winged Guillemot (*Uria grylle mut. motzfeldi* Benicken). *Meddelelser om Grønland*, 131, 1–20.

Salomonsen, F. (1944). The Atlantic Alcidae—the seasonal and geographical variation of the Auks inhabiting the Atlantic Ocean and the adjacent waters. *Goteborgs Kungliga Vetenskaps- och Vitterhetts-Samhalles Handlingar, Serie B*, 5, 4–138.

Salomonsen, F. (1945). Gejrfuglen. *Dyr i Natur og Museum*, 1944–45, 99–110.

Salomonsen, F. (1950). *Grønlands Fugle*. Munksgaard, Copenhagen.

Salomonsen, F. (1967). *Fuglene pa Grønland*. Rhodos, Copenhagen

Salomonsen, F. (1979). Ornithological and ecological studies in Southwest Greenland (59 46′–62 27′ N latitude). *Meddelelser om Grønland*, 204, 1–214.

Salvadori, T. (1865). Descrizione di altre nuove specie di uccelli esistenti nel Museo di Torino. *Memoria del Accademia Scientias, Torino, (II)*, 65, 1–49.

Sanger, G. A. (1972). Preliminary standing stock and biomass estimate of seabirds in the subarctic Pacific region. In *Biological oceanography of the northern North Pacific ocean*, (ed. A. Y. Takenouti), pp. 589–611. Idemitsu Shoton, Tokyo.

Sanger, G. A. (1986). Diets and food web relationships of seabirds in the Gulf of Alaska and adjacent marine regions, In *Environmental assessment of the Alaskan continental shelf. Final reports of the principal investigators*, 45, pp. 631–771. National Oceans and Atmosphere Administration, Boulder, CO.

Sanger, G. A. (1987a). Trophic levels and trophic relationships of seabirds in the Gulf of Alaska, In *Seabirds: feeding ecology and role in marine ecosystems*, (ed. J. P. Croxall), pp 229–57. Cambridge University Press, Cambridge.

Sanger, G. A. (1987b). Winter diets of Common Murres and Marbled Murrelets in Katchemak Bay, Alaska. *Condor*, 89, 426–30.

Schneider, D. C., Hunt, G. L., Jr., and Powers, K. D. (1987). Energy flux to pelagic seabirds: a comparison of Bristol Bay (Bering Sea) and George's Bank (Northwest Atlantic). In *Seabirds: feeding ecology and role in marine ecosystems*, (ed. J. P. Croxall), pp. 259–72. Cambridge University Press, Cambridge.

Schneider, D. C., Harrison, N. M., and Hunt, G. L., Jr. (1990a). Seabird diet at a front near the Pribilof Islands, Alaska. *Studies in Avian Biology*, 14, 61–6.

Schneider, D. C., Pierotti, R., and Threlfall, W. (1990b). Alcid patchiness and flight direction near a colony in East Newfoundland. *Studies in Avian Biology*, 14, 23–35.

Sclater, P. L. (1880). Remarks on the present state of the systema avium. *Ibis*, 22, 340–50.

Scott, J. M. (1973). Resource allocation in four syntopic species of marine diving birds. Ph.D. thesis. Oregon State University.

Scott, J. M. (1990). Offshore distribution patterns, feeding habits and adult–chick interactions of the Common Murre in Oregon. *Studies in Avian Biology*, 14, 103–8.

Scott, J. M., Hoffman, W., Ainley, D. G., and Zeillemaker, C. F. (1974). Range expansion and activity patterns in Rhinoceros Auklets. *Western Birds*, 5, 13–20.

Sealy, S. G. (1968). A comparative study of breeding ecology and timing in plankton-feeding alcids (*Cyclorhynchus* and *Aethia* spp.) on St. Lawrence

Island, Alaska. M.Sc. thesis. University of British Columbia, Vancouver.

Sealy, S. G. (1973*a*). Interspecific feeding assemblages of marine birds off British Columbia. *Auk*, **90**, 796–802.

Sealy, S. G. (1973*b*). The adaptive significance of post-hatching developmental patterns and growth rates in the Alcidae. *Ornis Scandinavica*, **4**, 113–21.

Sealy, S. G. (1973*c*). Breeding biology of the Horned Puffin on St. Lawrence Island, Bering Sea, with Zoogeographical Notes on the North Pacific Puffins. *Pacific Science*, **27**, 99–119.

Sealy, S. G. (1975*a*). Aspects of the breeding biology of the Marbled Murrelet in British Columbia. *Bird Banding*, **46**, 141–54.

Sealy, S. G. (1975*b*). Feeding ecology of the Ancient and Marbled Murrelets near Langara Island, British Columbia. *Canadian Journal of Zoology*, **53**, 418–33.

Sealy, S. G. (1975*c*). Influence of snow on egg-laying in auklets. *Auk*, **92**, 528–38.

Sealy, S. G. (1976). Biology of nesting Ancient Murrelets. *Condor*, **78**, 294–306.

Sealy, S. G. (1977). Wing moult of the Kittlitz's Murrelet. *Wilson Bulletin*, **89**, 467–9.

Sealy, S. G. (1982). Voles as a source of egg and nestling loss among nesting auklets. *Murrelet*, **63**, 9–14.

Sealy, S. G. (1984). Interruptions extend incubation by Ancient Murrelets, Crested Auklets and Least Auklets. *Murrelet*, **65**, 53–6.

Sealy, S. G. and Bédard, J. (1973). Breeding biology of the Parakeet Auklet (*Cyclorhynchus psittacula*) on St. Lawrence Island, Alaska. *Astarte*, **6**, 59–68.

Sealy, S. G. and Nelson, R. W. (1973). The occurrences and status of the horned puffin in British Columbia. *Syesis*, **6**, 51–5.

Sealy, S. G., Carter, H. R., and Alison, D. (1982). Occurrences of the asiatic Marbled Murrelet (*Brachyrhamphus marmoratus perdix* (Pallas) in North America. *Auk*, **99**, 778–80.

Searing, C. F. (1977). Some aspects of the ecology of cliff-nesting seabirds at Kongkok Bay, St. Lawrence island, Alaska, during 1976. In *Environmental assessment of the Alaskan continental shelf, annual reports.*, pp. 263–412. National Oceans and Atmosphere Administration, Boulder, CO.

Sharpe, F. (1995). Return of the killer bubbles: interactions between alcids and fish schools. *Pacific Seabirds*, **22**, 43.

Shea, R. E. and Ricklefs, R. E. (1985). An experimental test of the idea that food supply limits growth rate in tropical pelagic seabirds. *American Naturalist*, **125**, 116–22.

Shea, R. E. and Ricklefs, R. E. (1996). Temporal variation in growth performance in six species of tropical, pelagic seabirds. *Journal of Animal Ecology*, **65**, 29–42.

Shibaev, Y. V. (1987). Census of bird colonies and survey of bird species in Peter the Great Bay. In *Distribution and biology of seabirds of the Far East*, (ed. N. M. Litvinenko), pp. 43–59. Far East Science Centre, USSR Academy of Science, Institute of Biology and Soil Sciences, Vladivostok.

Shibaev, Y. V. (1990*a*). Marbled Murrelet. In *Birds of the USSR: Auks (Alcidae)*, (ed. V. E. Flint and A. N. Golovkin), pp. 82–8. Nauka, Moscow.

Shibaev, Y. V. (1990*b*). Spectacled Guillemot. In *Birds of the USSR: Auks (Alcidae)*, (ed. V. E. Flint and A. N. Golovkin), pp. 74–82. Nauka, Moscow.

Shibaev, Y. V. (1990*c*). Kittlitz's Murrelet. In *Birds of the USSR: Auks (Alcidae)*, (ed. V. E. Flint and A. N. Golovkin), pp. 88–92. Nauka, Moscow.

Shibaev, Y. V. (1990*d*). Ancient Murrelet. In *Birds of the USSR: Auks (Alcidae)*, (ed. V. E. Flint and A. N. Golovkin), pp. 92–104. Nauka, Moscow.

Shibaev, Y. V. (1990*e*). Rhinoceros Auklet. In *Birds of the USSR: Auks (Alcidae)*, (ed. V. E. Flint and A. N. Golovkin), pp. 139–48. Nauka, Moscow.

Shiomi, K., and Ogi, H. (1991). Sexual morphological differences based on functional aspects of skeletal and muscular characteristics in breeding Tufted Puffins. *Journal of Yamashina Institute for Ornithology*, **23**, 85–106.

Shuntov, V. P. (1986). Seabirds of the sea of Okhotsk. In *Seabirds of the Far East*, (ed. N. M. Litvinenko), pp. 6–20. Far East Science Centre, USSR Academy of Science, Institute of Biology and Soil Science, Vladivostok.

Sibley, C. G. and Ahlquist, J. E. (1990). *Phylogeny and classification of birds*. Yale University Press, New Haven and London

Simons, T. R. (1980). Discovery of a ground-nesting Marbled Murrelet. *Condor*, **82**, 1–9.

Simpson, G. G. (1976). *Penguins, past and present, here and there.* Yale University Press, New Haven, Connecticut.

Skokova, N. N. (1967). On the factors which determine the state of the population of puffins during the nesting period. *Transactions of the kandalakshshii State Reserve*, **5**, 155–77.

Skokova, N. N. (1990). Atlantic Puffin. In *Birds of the USSR: Auks (Alcidae)*, (ed. V. E. Flint and A. N. Golovkin), pp. 148–64. Nauka, Moscow.

Slater, P. J. B. (1976). Tidal rhythm in a seabird. *Nature*, **264**, 636–8.

Small, A. (1994). *California birds: their status and distribution.* Ibis Publishing Co., Vista, CA.

Smith, T. G. and Hammill, M. O. (1980). Distribution and food habits of the birds along the southeastern Baffin Island coast. *Technical Reports of Fisheries and Aquatic Science*, **1573**, 1–27.

Sowls, A. L., Hatch, S. A., and Lensink, C. J. (1978). *Catalog of Alaskan seabird colonies.* U. S. Fish and Wildlife Service, Anchorage, Alaska

Speich, S. and Manuwal, D. A. (1974). Gular pouch development and population structure of Cassin's Auklet. *Auk*, **91**, 291–300.

Speich, S. M. and Wahl, T. R. (1989). *Catalogue of Washington seabird colonies.* U. S. Fish and Wildlife Service, Biological Report **88**(6), 1–510.

Spring, L. (1971). A comparison of functional and morphological adaptations in the Common Murre (*Uria aalge*) and the Thick-billed Murre (*Uria lomvia*). *Condor*, **73**, 1–27.

Springer, A. M. (1991). Seabird distribution as related to food webs and the environment: examples from the North Pacific Ocean. In *Studies of high latitude seabirds, 1. Behavioural, energetic and oceanographic aspects of seabird feeding ecology*, (ed. W. A. Montevecchi and A. J. Gaston), pp. 39–48. Canadian Wildlife Service, Ottawa.

Springer, A. M., Roseneau, D. G., Murphy, E. C., and Springer, M. I. (1984). Environmental controls of marine food webs: food habits of seabirds in the eastern Chukchi Sea. *Canadian Journal of Fisheries and Aquatic Science*, **41**, 1202–15.

Springer, A. M., Murphy, E. C., Roseneau, D. G., McRoy, C. P., and Cooper, B. A. (1987). The paradox of pelagic food webs in the northern Bering Sea – I. Seabird food habits. *Continental Shelf Research*, **7**, 895–911.

Springer, A. M., Kondratyev, A. Y., Ogi, H., Shibaev, Y. V., and van Vliet, G. B. (1993). Status, ecology and conservation of *Synthliboramphus* murrelets and auklets. In *Status, ecology and conservation of marine birds of the North Pacific*, (ed. K. Vermeer, K. T. Briggs, K. H. Morgan, and D. Siegel-Causey), pp. 187–201. Canadian Wildlife Service, Ottawa.

Stacey, P. J., Baird, R. W., and Hubbard-Morton, R. B. (1990). Transient killer whale (*Orcinus orca*) harassment and 'surplus killing' of marine birds in British Columbia. *Pacific Seabird Group Bulletin*, **17**, 38.

Stallcup, R. W. (1976). Pelagic birds of Monterey Bay, California. *Western Birds*, **7**, 113–36.

Stejneger, L. H. (1885). Results of ornithological exploration of the Commander Islands and Kamtschatka. *Bulletin of the U.S. National Museum*, **29**, 1–382.

Stejneger, L. H. (1887). Contributions to the natural history of the Commander Islands. *Proceedings of the U.S. National Museum*, **1887**, 117–45.

Stempniewicz, L. (1980). Breeding biology of the Little Auk *Plautus alle* in the Hornsund region of Spitzbergen. *Acta Ornithologia*, **18**, 141–65.

Stempniewicz, L. (1990). Biomass of Dovekie excretia in the vicinity of a breeding colony. *Colonial Waterbirds*, **13**, 62–6.

Stempniewicz, L. (1993). Polar Bear *Ursus maritimus* feeding in a seabird colony in Fraz Joseph Land. *Polar Research*, **12**, 33–6.

Stephenson, R., Hedrick, M. S., and Jones, D. R. (1992). Cardiovascular responses to diving and involuntary submergence in the Rhinoceros Auklet (*Cerorhinca monocerata* Pallas). *Canadian Journal of Zoology*, **70**, 2303–10.

Stettenheim, P. R. (1959). Adaptations for underwater swimming in the Common Murre (*Uria aalge*). Ph.D. thesis. University of Michigan.

Steventon, D. J. (1979). Razorbill survival and population estimates. *Ringing and Migration*, **2**, 105–12.

Steventon, D. J. (1982). Shiants Razorbills: movements, first year survival and age of first return. *Seabird Report*, **6**, 105–9.

Stewart, D. T. (1993). Sexual dimorphism in Thick-billed Murres, *Uria lomvia*. *Canadian Journal of Zoology*, **71**, 346–51.

Storer, R. W. (1945). The systematic position of the murrelet genus *Endomychura*. *Condor*, **47**, 154–60.

Storer, R. W. (1952). A comparison of variation, behaviour and evolution in the sea bird genera *Uria* and *Cepphus*. *University of California Publication in Zoology*, **52**, 121–222.

Storer, R. W. (1960). Evolution in the diving birds. *Proceedings of the International Ornithological Congress*, **12**, 694–707.

Storer, R. W. (1971). Classification of birds. In *Avian Biology vol. 1*, (ed. D. S. Farner and J. R. King), pp. 1–18. Academic Press, New York.

Stotskaya, Y. E. (1990a). Cassin's Auklet. In *Birds of the USSR: Auks (Alcidae)*, (ed. V. E. Flint and A. N. Golovkin), pp. 107–12. Nauka, Moscow.

Stotskaya, Y. E. (1990b). Pigeon Guillemot. In *Birds of the USSR: Auks (Alcidae)*, (ed. V. E. Flint and A. N. Golovkin), pp. 66–74. Nauka, Moscow.

Strauch, J. G., Jr. (1978). The phylogeny of the Charadriiformes (Aves): a new estimate using the method of character compatibility analysis. *Transactions of the Zoological Society of London*, **34**, 263–345.

Strauch, J. G., Jr (1985). The phylogeny of the Alcidae. *Auk*, **102**, 520–39.

Stresemann, E. and Stresemann, V. (1966). Die Mauser der Vögel. *Journal Für Ornithologie*, **107**, 1–447.

Summers, K. R. and Drent, R. H. (1979). Breeding biology and twinning experiments of Rhinoceros Auklets on Cleland Island British Columbia. *Murrelet*, **60**, 16–22.

Sutton, G. M. (1932). Birds of Southampton Island. *Memoirs of the Carnegie Museum*, **12**, 1–275.

Swann, R. L., Harris, M. P., and Aiton, D. G. (1991). The diet of some young seabirds at Canna 1981–1990. *Seabird*, **13**, 54–8.

Swartz, L. G. (1966). Sea-cliff birds, In *Environment of the Cape Thompson Region, Alaska*, (ed. N. J. Wilimovsky and J. N. Wolfe), pp. 611–78. U. S. Atomic Energy Commission, Oak Ridge, Tennessee.

Swartz, L. G. (1967). Distribution and movement of birds in the Bering and Chukchi Seas. *Pacific Science*, **21**, 332–47.

Swennen, C. (1977). *Laboratory research of seabirds*. Netherlands Institute for Sea Research, Texel, Netherlands.

Swennen, C. and Duiven, P. (1977). Size of food objects of three fish eating seabird species: *Uria aalge*, *Alca torda*, and *Fratercula arctica* (Aves: Alcidae). *Netherlands Journal of Sea Research*, **11**, 92–8.

Sydeman, W. J. (1993). Survivorship of Common Murres on Southeast Farallon Island, California. *Ornis Scandinavica*, **24**, 1–7.

Takekawa, J. E., Carter, H. R., and Harvey, T. E. (1990). Decline of the Common Murre in central California, 1980–1986. *Studies in Avian Biology*, **14**, 149–63.

Tasker, M. L., Webb, A., Hall, A. J., Pienkowski, M. W., and Langslow, D. R. (1988). *Seabirds in the North Sea*. Nature Conservancy Commission, Peterborough.

Taylor, J. R. E. and Konarzewski, M. (1989). On the importance of fat reserves for Little Auk (*Alle alle*) chicks. *Oecologia*, **81**, 551–8.

Taylor, J., & Konarzewski, L. (1992). Budget of elements in Little Auk (*Alle alle*) chicks. *Functional Ecology*, **6**, 137–44.

Taylor, K. (1984). Puffin behaviour. In *The puffin*, (ed. M. P. Harris), pp. 96–105. T. and A. D. Poyser, Calton.

Taylor, K. (1993). *Puffins*. Whittet Books, London.

Taylor, K. and Reid, J. B. (1981). Earlier colony attendance by Guillemots and Razorbills. *Scottish Birds*, **11**, 173–80.

Teixeira, A. M. (1986). Razorbill *Alca torda* losses in Portugese nets. *Seabird*, **9**, 11–4.

Tershy, B. R., Van Gelder, E., and Breese, D. (1993). Relative abundance and seasonal distribution of seabirds in the Canal de Ballenas, Gulf of California. *Condor*, **95**, 458–64.

Thompson, S. P., McDermond, D. K., Wilson, U. W., and Montgomery, K. (1985). Rhinoceros Auklet burrow count on Protection Island. *Murrelet*, **66**, 62–5.

Thoresen, A. C. (1964). The breeding behaviour of the Cassin Auklet. *Condor*, **66**, 456–76.

Thoresen, A. C. (1983). Diurnal activity and social displays of Rhinoceros Auklets on Teuri Island, Japan. *Condor*, **85**, 373–5.

Thoresen, A. C. and Booth, E. S. (1958). Breeding activities of the Pigeon Guillemot, *Cepphus columba columba* (Pallas). *Walla Walla College Publications of Department of Biological Sciences and the Biological Station*, **23**, 1–37.

Tomkinson, P. M. L. and Tomkinson, J. W. (1966). Eggs of the Great Auk. *Bulletin of the British Museum (Natural History) History Series*, **3**, 95–128.

Trivers, R. L. (1972). Parental investment and sexual selection. In *Sexual selection and the descent of man*, (ed. B. Campbell), pp. 136–79. Aldine, Chicago, Illinois.

Troy, D. and Bradstreet, M. S. W. (1991). Marine bird abundance and habitat use. In *Marine birds and mammals of the Unimak Pass area: abundance, habitat use and vulnerability*, (ed. C. Truitt and K. Kertell), pp. 1–70. Minerals Management Sevice, Anchorage, Alaska.

Tschanz, B. (1959). Zur Brutbiologie der Trottellumme (*Uria aalge aalge* Pont). *Behaviour*, **14**, 1–108.

Tschanz, B. (1968). Trottellummen (*Uria aalge aalge* Pont.). *Zeitschrift für Tierpsychologie*, **4**, 1–103.

Tschanz, B. (1979). Zur entwicklung von Papageitaucherkuken *Fratercula arctica* in Frielandund Labor bei unzulanglichem und ausreichendem Futerangebot. *Fauna Norvegica Series C, Cinclus*, **2**, 70–94.

Tuck, L. M. (1961). *The murres*. Canadian Wildlife Service Monograph No 1, Canadian Wildlife Service, Ottawa.

Tuck, L. M. and Squires, H. J. (1955). Food and feeding habits of Brunnich's Murre (*Uria lomvia lomvia*) on Akpatok Island. *Journal of the Fisheries Research Board, Canada*, **12**, 781–92.

Udvardy, M. D. F. (1963). Zoogeographical study of the Pacific Alcidae. *Proceedings of the Pacific Science Congress*, **10**, 85–111.

Unitt, P. (1984). The birds of San Diego County. *Memories of the San Diego Society for Natural History*, **13**, 1–276.

U. S. Fish and Wildlife Service. (1993). *Catalogue of Alaskan seabird colonies—computer archives*. U. S. Fish and Wildlife Service, Anchorage, Alaska.

Uspenski, S. M. (1956). *The bird bazaars of Novaya Zemlya*. Canadian Wildlife Service Translations of Russian Game Reports, No. 4, Ottawa.

Uttley, J. D., Walton, P., Wanless, S., and Burns, M. D. (1994). Effects of prey abundance on the foraging behaviour, diving efficiency and time allocation of breeding Guillemots *Uria aalge. Ibis*, **136**, 214–22.

Vader, W., Barrett, R. T., Erikstad, K. E., and Strann, K.-B. (1990). Differential response of Common and Thick-billed murres to a crash in the capelin stock of the southern Barents Sea. *Studies in Avian Biology*, **14**, 175–80.

Valiela, I. (1995). *Marine ecological processes*. Springer, New York.

van Vliet, G. (1994). Status, ecology and conservation of Kittlitz's Murrelet. *Pacific Seabirds*, **21**, 50–1.

van Vliet, G. and McAllister, M. (1994). Kittlitz's Murrelet: the species most impacted by direct mortality from the Exxon Valdez oil spill. *Pacific Seabirds*, **21**, 5–6.

Varoujean, D. H., Sanders, S. D., Graybill, M. R., and Spear, L. (1979). Aspects of Common Murre breeding biology. *Pacific Seabird Group Bulletin*, **6**, 28.

Vaughan, R. (1992). *In search of arctic birds*. T. and A. D. Poyser, London.

Vaurie, C. (1965). *The birds of the Palaearctic fauna. Non-Passeriformes*. H. F. & G. Witherby, London.

Verheyen, R. (1958). Contribution a la systematique des Alciformes. *Bulletin Institute Royal des Sciences Naturelles Belgique*, **34**, 1–15.

Verheyen, R. (1961). A new classification for the non-passerine birds of the world. *Bulletin Institute Royal des Science Naturelle Belgiques*, **37**, 1–36.

Vermeer, K. (1979). Nesting requirements, food and breeding distribution of Rhinoceros Auklets, *Cerorhinca monocerata* and Tufted Puffins, *Lunda cirrhata. Ardea*, **67**, 101–10.

Vermeer, K. (1980). The importance of timing and type of prey to reproductive success of Rhinoceros Auklets, *Cerorhinca monocerata. Ibis*, **122**, 343–50.

Vermeer, K. (1984). The diet and food consumption of nestling Cassin's Auklets during summer, and a comparison with other plankton-feeding alcids. *Murrelet*, **65**, 65–77.

Vermeer, K. (1992). The diet of birds as a tool for monitoring the biological environment. In *The ecology, status and conservation of marine and shoreline birds of the west coast of Vancouver Island*, (ed. K. Vermeer, R. W. Butler, and K. H. Morgan), pp. 41–50. Canadian Wildlife Service, Ottawa.

Vermeer, K. and Cullen, L. (1979). Growth of Rhinoceros Auklets and Tufted Puffins, Triangle island, British Columbia. *Ardea*, **67**, 22–7.

Vermeer, K. and Lemon, M. (1986). Nesting habits and habitats of Ancient Murrelets and Cassin's Auklets in the Queen Charlotte Islands, British Columbia. *Murrelet*, **67**, 33–44.

Vermeer, K. and Westrheim, S. J. (1984). Fish changes in the diet of nestling Rhinoceros Auklets and their implications. In *Marine birds: their feeding ecology and commercial fisheries relationships*, (ed. D. N. Nettleship, G. A. Sanger, and P. F. Springer), pp. 96–105. Canadian Wildlife Service, Ottawa.

Vermeer, K., Cullen, L., and Porter, M. (1979). A provisional explanation of the reproductive failure of Tufted Puffins *Lunda cirrhata* on Triangle Island, British Columbia. *Ibis*, **121**, 348–54.

Vermeer, K., Fulton, J. D., and Sealy, S. G. (1985). Differential use of zooplankton prey by Ancient Murrelets and Cassin's Auklets in the Queen Charlotte Islands. *Journal of Plankton Research*, 7, 443–59.

Vermeer, K., Sealy, S. G., and Sanger, G. A. (1987). Feeding ecology of Alcidae in the eastern North Pacific Ocean. In *Seabirds: feeding biology and role in marine ecosystems*, (ed. J. Croxall), pp. 189–227. Cambridge University Press, Cambridge.

Vermeer, K., Morgan, K. H., and Smith, G. E. J. (1992). Habitat analysis and co-occurrence of seabirds on the west coast of Vancouver Island. In *The ecology, status and conservation of marine and shoreline birds of the west coast of Vancouver Island*, (ed. K. Vermeer, R. W. Butler, and K. H. Morgan), pp. 78–85. Canadian Wildlife Service, Ottawa.

Vermeer, K., Morgan, K. H., and Smith, G. E. J. (1993*a*). Colony attendance of Pigeon Guillemots as related to tide height and time of day. *Colonial Waterbirds*, 16, 1–8.

Vermeer, K., Morgan, K. H., and Smith, G. E. J. (1993*b*). Nesting biology and predation of Pigeon Guillemots in the Queen Charlotte Islands, British Columbia. *Colonial Waterbirds*, **16**, 119–27.

Vermeer, K., Briggs, K. T., Morgan, K. H., and Siegel-Causey, D. (1993*c*). *The status, ecology and conservation of marine birds of the North Pacific*. Canadian Wildlife Service, Ottawa.

Verspoor, E., Birkhead, T. R., and Nettleship, D. N. (1987). Incubation and brooding shift duration in the Common Murre, *Uria aalge*. *Canadian Journal of Zoology*, **65**, 247–52.

Vezina, A. F. (1985). Empirical relationships between predation and prey-size among terrestrial vertebrate predators. *Oecologia*, **67**, 555–65.

Violani, C. (1975). L'Alca Impenne (*Alca impennis* L.) nelle collezioni Italiane. *Natura Milan*, **66**, 18–24.

Violani, C. and Boano, G. (1990). L'Uria di Craveri *Synthliboramphus craveri*. *Riv. Piem. St. Nat.* **11**, 155–62.

Voronov, G. A. (1982). *Acclimatization of mammals on Sakhalin and Kurile Islands*. Nauka, Moscow.

Vyatkin, P. S. (1986). Nesting colonies of colonial birds in the Kamchatka region. In *Morskie Ptitsy Dalnego Vostoka (Seabirds of the Far East)*, (ed. N. M. Litvinenko), pp. 20–36. Far East Science Centre, USSR Academy of Science, Institute of Biology and Soil Science, Vladivostok, USSR.

Wagner, R. H. (1991*a*). Evidence that female razorbills control extra-pair copulations. *Behaviour*, **118**, 157–69.

Wagner, R. H. (1991*b*). Pairbond formation in the Razorbill. *Wilson Bulletin*, **103**, 682–5.

Wagner, R. H. (1991*c*). The use of extra-pair copulations for mate appraisal by razorbills, *Alca torda*. *Behavioral Ecology*, **2**, 198–203.

Wagner, R. H. (1992*a*). The pursuit of extra-pair copulations by monogamous female razorbills: how do females benefit? *Behavioural Ecology and Sociobiology*, **29**, 455–64.

Wagner, R. H. (1992*b*). Behavioural and breeding habitat related aspects of sperm competition in Razorbills. *Behaviour*, **123**, 1–26.

Wagner, R. H. (1992*c*). Extra-pair copulations in a lek: the secondary mating system of monogamous razorbills. *Behavioral Ecology and Sociobiology*, **31**, 63–71.

Wallace, G. E., Collier, B., and Sydeman, W. J. (1992). Interspecific nest-site competition among cavity nesting alcids on SE Farallon Island, California. *Colonial Waterbirds*, **15**, 241–4.

Walsh, P. M., Brindley, E., and Heubeck, M. (1995). *Seabird numbers and breeding success in Britain and Ireland, 1994*. Joint Nature Conservation Committee, Peterborough.

Walter, H. (1979). *Eleanora's falcon: adaptations to prey and habitat in a social raptor*. University of Chicago Press, Chicago.

Wanless, S. and Harris, M. P. (1986). Time spent at the colony by male and female Guillemots *Uria aalge* and Razorbills *Alca torda*. *Bird Study*, **33**, 168–76.

Wanless, S., Morris, J. A., and Harris, M. P. (1988). Diving behaviour of guillemot *Uria aalge*, puffin *Fratercula arctica* and razorbill *Alca torda*, as shown by radio-telemetry. *Journal of Zoology, London*, **216**, 73–81.

Wanless, S., Harris, M. P., and Morris, J. A. (1990). A comparison of feeding areas used by individual Common Murres (*Uria aalge*), Razorbills (*Alca torda*) and an Atlantic Puffin (*Fratercula arctica*) during the breeding season. *Colonial Waterbirds*, **13**, 16–24.

Ward, P. and Zahavi, A. (1973). The importance of certain assemblages of birds as 'information centres' for food-finding. *Ibis*, **115**, 517–24.

Warham, J. (1990). *The petrels: their ecology and breeding systems*. Academic Press, San Diego, CA.

Warheit, K. I. (1992). A review of fossil seabirds from the Tertiary of the North Pacific: plate tectonics, paleoceanography, and faunal change. *Paleobiology*, **18**, 401–24.

Warheit, K. I. and Lindberg, D. R. (1988). Interaction between seabirds and marine mammals through time: interference competition at breeding sites. In *Seabirds and other marine vertebrates: competition, predation and other interactions*, (ed. J. Burger), pp. 292–328. Columbia University Press, New York.

Watada, M., Kakizawa, R., Kuroda, N., and Utida, S. (1987). Genetic differentiation and phylogenetic relationships of an avian family, Alcidae (Auks). *Journal of the Yamashina Institute of Ornithology*. **19**, 79–88.

Watanuki, Y. (1987). Breeding biology and foods of Rhinoceros Auklets on Teuri Island, Japan. *Proceedings of the NIPR Symposium on Polar Biology*, **1**, 175–83.

Watanuki, Y., Aotsuka, M., and Terasawa, T. (1986). Status of seabirds breeding on Teuri Island. *Tori*, **34**, 146–50.

Watanuki, Y. (1990). Daily activity pattern of Rhinoceros Auklets and kleptoparasitism by Black-tailed Gulls. *Ornis Scandinavica*, **21**, 28–36.

Watanuki, Y., Kondo, N., and Nakagawa, H. (1988). Status of seabirds breeding in Hokkaido. *Japanese Journal of Ornithology*, **37**, 17–32.

Wehle, D. H. S. (1980). The breeding biology of the puffins: Tufted Puffin (*Lunda cirrhata*), Horned Puffin (*Fratercula corniculata*), Common Puffin (*F. arctica*) and Rhinoceros Auklet (*Cerorhinca monocerata*). Ph.D thesis. University of Alaska, Fairbanks.

Wehle, D. H. S. (1982). Food of adult and subadult Tufted and Horned Puffins. *Murrelet*, **63**, 51–8.

Wehle, D. H. S. (1983). The food, feeding and development of young Tufted and Horned Puffins in Alaska. *Condor*, **85**, 427–42.

Welch, H. E., Bergmann, M. A., Siferd, T. D., Martin, K. A., Curtis, M. F., Crawford, R. E., Conover, R. J., and Hop, H. (1992). Energy flow through the marine ecosystem of the Lancaster Sound Region, Arctic Canada. *Arctic*, **45**, 343–57.

Welch, H. E., Crawford, R. E., and Hop, H. (1993). Occurrence of Arctic Cod (*Boreogadus saida*) schools and their vulnerability to predation in the Canadian high arctic. *Arctic* **46**, 331–9.

Welham, C. V. J. and Bertram, D. F. (1993). The relationship between previous meal size and begging vocalizations of nestling Rhinoceros Auklets, *Cerorhinca monocerata*. *Animal Behaviour*, **45**, 827–9.

White, P. (1995). *The Farallon Islands: sentinels of the Golden Gate*. Scottwall Assoc., San Francisco.

Whittam, T. S. and Siegel-Causey, D. (1981). Species interactions and community structure in Alaskan seabird colonies. *Ecology*, **62**, 1515–24.

Wiens, J. A. (1989). *The ecology of bird communities*. Cambridge University Press, Cambridge.

Wiens, J. A. and Scott, J. M. (1975). Model estimation of energy flow in Oregon coastal bird populations. *Condor*, **77**, 439–52.

Williams, A. J. (1972). The social behaviour of guillemots. D.Phil. thesis. University of Oxford.

Williams, A. J. (1975). Guillemot fledging and predation on Bear Island. *Ornis Scandinavica*, **6**, 117–24.

Williams, J. C. and Byrd, G. V. (1994). Whiskered Auklet. In *Birds of North America, No. 76*, (ed. A. Poole, P. Stettenheim, and F. Gill), pp. 1–12. Academy of Natural Sciences of Philadelphia, and the American Ornithologists' Union, Washington, DC.

Williams, T. D. (1995). *The penguins*. Oxford University Press, Oxford.

Wilson, R. P. (1991). The behaviour of diving birds. *Proceedings of the International Ornithological Congress*, **20**, 1853–67.

Wilson, R. P., Grant, W. S., and Duffy, D. C. (1986). Recording devices on free-ranging marine animals: does measurement affect foraging performance. *Ecology*, **67**, 1091–3.

Wilson, R. P., Ryan, P. G., James, A., and Wilson, M. -P. (1987). Conspicuous coloration may enhance prey capture in some piscivores. *Animal Behaviour*, **35**, 1558–60.

Wilson, R. P., Burger, A. E., Wilson, B. L. H., Wilson, M. T., and Noldeke, C. (1989). An inexpensive depth gauge for marine animals. *Marine Biology*, **103**, 275–83.

Wilson, R. P., Wilson, M.-P., and Noldeke, E. C. (1992). Pre-dive leaps in diving birds: why do kickers sometimes jump? *Marine Ornithology*, **20**, 7–16.

Wilson, U. W. (1977). A study of the biology of the Rhinoceros Auklet on Protection Island, Washington. M.Sc. thesis. University of Washington, Seattle.

Wilson, U. W. (1986). Artificial Rhinoceros Auklet burrows: a useful tool for management and research. *Journal of Field Ornithology*, **57**, 295–9.

Wilson, U. W. (1995). Rhinoceros Auklet burrow use, breeding success, and chick growth: gull-free vs. gull-occupied habitat. *Journal of Field Ornithology*, **64**, 256–61.

Wilson, U. W. and Manuwal, D. A. (1986). Breeding biology of the Rhinoceros Auklet in Washington. *Condor*, **88**, 143–55.

Winnett, K. A., Murray, K. G., and Wingfield, J. C. (1979). Southern race of Xantus' Murrelet breeding on Santa Barbara Island, California. *Western Birds*, **10**, 81–2.

Witherby, H. F., Jourdain, F. C. R., Ticehurst, N. F., and Tucker, B. W. (1941). *Handbook of British birds*. H. F. and G. Witherby, London.

Wittenberger, J. F. and Hunt, G. L., Jr. (1985). The adaptive significance of coloniality in birds. In *Avian Biology, vol. 8*, (ed. D. S. Farner, J. R. King, and K. C. Parkes), pp. 2–78. Academic Press, Orlando, Florida.

Woodby, D. A. (1984). The April distribution of murres and prey patches in the southeastern Bering Sea. *Limnological Oceanography*, **29**, 181–8.

Yarrell, W. (1884). *A history of British birds*. John van Voorst, London.

Ydenberg, R. C. (1989). Growth–mortality trade-offs and the evolution of juvenile life-histories in the Alcidae. *Ecology*, **70**, 1494–506.

Ydenberg, Y., Clark, C. W., and Harfenist, A. (1995). Intraspecific fledging mass variation in the Alcidae, with special reference to the seasonal fledging mass decline. *American Naturalist*, **145**, 412–33.

Zahavi, A. (1975). Mate selection — a selection for a handicap. *Journal of Theoretical Biology*, **53**, 205–14.

Zubakin, V. A. (1990). Some aspects of the nesting biology and social behaviour of the Crested Auklet (*Aethia cristatella*). In *Study of colonial seabirds of the USSR*, pp. 9–13. Academy of Sciences of the USSR, Magadan.

Zubakin, V. A. and Zubakina, E. V. (1993). Observations of colour-banded alcidae at Talan Island, Okhotsk Sea. *Beringian Seabird Bulletin*, **1**, 43–4.

Zubakin, V. A. and Zubakina, E. V. (1994). Some results of a marked population study of Crested Auklets, Parakeet Auklets and Tufted Puffins at Talan Island (Tanskaya Bay, Sea of Okhotsk). *Bering Bulletin*, **1**, 43–4.

Zweifelhofer, D. C. and Forsell, D. J. (1989). *Marine birds and mammals wintering in selected bays of Kodiak Island, Alaska: a five year study*. US Fish and Wildlife Sevice Report, Anchorage, Alaska.

Zweifelhofer, D. C. and Forsell, D. J. (1995). Population trends of seabirds wintering in Kodiak Alaska—a fifteen year study. *Pacific Seabirds*, **22**, 47.

Index

Note: Page numbers in *italics* refer to figures and tables. Plate numbers are prefixed 'Pl.'